Y0-BVO-584

Motion: Representation and Perception

Motion: Representation and Perception

Proceedings of the ACM SIGGRAPH/SIGART Interdisciplinary Workshop on Motion: Representation and Perception, held in Toronto, Ontario, Canada, 1983.

Norman I. Badler

Department of Computer and Information Science
University of Pennsylvania
Philadelphia, Pennsylvania, U.S.A.

John K. Tsotsos

Fellow, Canadian Institute for Advanced Research
Department of Computer Science
University of Toronto
Toronto, Ontario, Canada

North-Holland

New York • Amsterdam • London

Elsevier Science Publishing Co., Inc.
52 Vanderbilt Avenue, New York, New York 10017

Sole distributors outside the United States and Canada:
Elsevier Science Publishers B.V.
P.O. Box 211, 1000 AE Amsterdam, the Netherlands

Library of Congress Cataloging-in-Publication Data

ACM SIGGRAPH/SIGART Interdisciplinary Workshop (1983 :
Toronto, Ont.)
Motion: representation and perception.

Includes index.
1. Motion perception (Vision)—Congresses. 2. Computer Graphics—Congresses. 3. Animation (Cinematography) I. Badler, Norman I. II. Tsotsos, John K.
III. SIGGRAPH. IV. SIGART. V. Title.
BF245.A25 1983a 006.3′7 86-8491
ISBN 0-444-01079-3

Current printing (last digit)
10 9 8 7 6 5 4 3 2 1

Manufactured in the United States of America

CONTENTS

Foreword

In 1979, the first gathering of researchers interested in the problems of motion analysis was held in Philadelphia, at the University of Pennsylvania. Workshop on Computer Analysis of Time-Varying Imagery was the excellent idea and hard work of Norman I. Badler and J.K. Aggarwal. In their words, the meeting was conceived as "a cozy gathering of at most two dozen people"; however, it mushroomed into a much larger event. The meeting provided the first forum for the presentations on motion research and led to the inclusion of sessions on motion in the current large conferences. The papers presented revealed that there was much more to motion research and many more researchers active in the field than previously believed. Overall, the meeting was an extremely successful event.

Since then, the area of motion research has rapidly expanded. We are realizing that the field includes not only computer scientists and electrical engineers, but researchers in artificial intelligence, psychophysics of motion perception, computer graphics, and neurophysiology. As there was no existing forum for the interaction of such a diverse group of researchers, this Workshop was designed to bring together such people in an attempt to foster communication between disciplines and to make each discipline aware of the relevant studies of motion analysis being performed by other disciplines.

The generous support and sponsorship of the Association for Computing Machinery (ACM), particularly the Special Interest Groups in Computer Graphics (SIGGRAPH) and Special Interest Groups in Artificial Intelligence (SIGART), enabled us to present such an interdisciplinary forum.

I wish to extend my sincere gratitude to the Program Chairman, Norman I. Badler, for his hard work and advice. In addition, the Program Committee and Local Arrangements Committee deserve praise for their efforts in ensuring that we present the highest quality program possible and the most pleasant and smoothly functioning environment possible.

The goals of the Workshop were to provide a broad view of motion research and in this way to attempt to provide direction for future motion research. The burden is now on you, the participants, for only by your active participation in discussion and by actively pursuing contacts in disciplines other than your own, will such a direction emerge. I hope that the interdisciplinary aspects of motion research will excite each of you as much as it has excited the organizers of this Workshop.

John K. Tsotsos
General Chairman
Toronto, Canada

From the Program Chair

This Proceedings contains those papers that were accepted for presentation at the ACM SIGGRAPH/SIGART Interdisciplinary Workshop on Motion. These papers were selected from among the forty-eight submitted to the Program Committee. We feel that the contents of this Proceedings form a collection of papers of considerable value and arranged to have the abstracts published as a special issue of the SIGGRAPH Newsletter.

Among the activities of the Workshop, this Proceedings contains the Keynote Address that was delivered by Hans Wallach of Swarthmore College and the Workshop Summaries; these items have not appeared in any other publication.

Plenty of time was allocated throughout the Workshop for discussion of individual papers and each session, thus encouraging the free exchange of questions, comments, and criticism.

Due to the surprising volume of papers submitted, the members of the Program Committee and additional reviewers were pressed into service to read and comment on the papers within a short period of time. I wish to personally express my appreciation and thanks for the remarkable and ungrudging effort they all put forth and the careful readings and judgments they provided. The members of the Program Committee and other reviewers are acknowledged below.

We all hope you enjoy the Workshop and these Proceedings.

Norman I. Badler
Program Chair

<u>Program Committee</u>

James Allen
J.K. Aggarwal
Paul Kolers
Bernd Neuman
John Tsotsos
Shimon Ullman

<u>Reviewers</u>

Aaron Bobick
Tom Calvert
Alain Fournier
Eric Grimson
Don Hoffman
John Hollarbach
K. Ikeuchi
Jim Korein
M.J. Magee
John Mylopoulos
Ken Nielson
Keith Nishihara
Ray Perrault
Steve Platt
John Ruban
Dimitri Terzopoulos
Joshua Tsal
Jeremy Wolfe
Steve Zucker

KEYNOTE ADDRESS

How human perception deals with motion

Hans Wallach
Department of Psychology
Swarthmore College
Swarthmore, PA

We know of at least three conditions of stimulation that mediate the visual perception of motion in the environment. Two of them represent the change of the egocentric visual direction of a moving object; one is the displacement of the retinal image of a moving object that takes place when the eyes rest on a stationary point and the other the ocular pursuit that occurs when the eyes track the moving object. For obvious reasons, these two are called the subject-relative stimuli. The third stimulus condition makes use of the fact that a moving object changes its position relative to stationary objects that surround it. It consists of a change in the configuration of the retinal projection of the region in which the moving object is located.

Configurational change causes motion perception quite independently of the other two stimulus conditions and, since it functions according to rules of its own, that independence has peculiar consequences.

I. It mediates only relative displacement; what is really moving or stationary is not transmitted by this stimulus condition. This ambiguity is, however, not represented in perception. When the relative displacement between an object and its surround is given, the configurational change that results will tend to cause motion of the object and immobility of the surround. This goes under the name of Duncker's rule (Duncker, 1929) and may result in the well-known misperception called induced motion, which occurs when a stationary object is given in a surround that is in

Notion: Representation and Perception
N.I. Badler and J.K. Tsotsos, Editors
Published by Elsevier Science Publishing Co., Inc.

translatory motion: The object will be seen to move in the direction opposite to the motion of the surround.

Under ordinary circumstances, the result of configurational change agrees with the result of one of the subject-relative stimuli that is then also present, but when configurational change leads to induced motion, it is in conflict with the subject-relative stimuli, for they represent the objective conditions veridically. Under such circumstances the relative potency of configurational change becomes manifest. When a stationary object is seen in a moving surround, induced motion of the object occurs in the majority of instances, although subject-relative stimulation represents the object correctly as stationary. When an object in a moving surround has a motion of its own that is in a direction different from the motion direction of the surround, induction always takes place. Here, configurational change represents a relative displacement that is the resultant of the real motion of the object and its relative motion in the direction opposite to the motion of the surround. The perceived motion that results differs from the real motion in direction: Induced motion consists in a motion component that is added to a real motion and changes its direction.

II. Perceived speed of motion that results from configurational change is proportional to the rate of configurational change. This can cause a misperception of speed known as J. F. Brown's transposition phenomenon (Brown, 1931). When motion speed is given by configurational change and is proportional to the rate of configurational change, there may be a conflict with ocular pursuit, since ocular pursuit results in perceived speed being proportional to the rate of change of the visual direction of the moving object. This happens in Brown's transposition experiments where subjects are asked to match the speed of objects that move in identically shaped apertures of different sizes. In such experiments perceived speeds turn out to result to a high degree from the rate of configurational change, that is, they vary with aperture size. The same objective speeds will, for instance, be perceived as different when the motions occur in apertures of different size. In such experiments, the information provided by ocular pursuit affects speed only by 10 to 20 percent. (1)

Recently, Wallach, O'Leary and McMahon (1982) used an induced motion display to compare the relative effectiveness of the three conditions of stimulation for motion. The moving surround consisted of a pattern of tall vertical lines that moved left and right in reciprocating simple harmonic motion. The surrounded object was a dot that moved up and down at the same rates as the line pattern and in phase with it. Because the surrounding pattern consisted of vertical lines, the vertical motion of the dot encountered no landmarks and was not given by configurational change. Also, because the moving pattern consisted of lines only, the relative displacement between the dot and the lines did not cuase motion with a specific direction. (2) It defined only motion with a horizontal component. Therefore the relative motion between the dot and the line pattern could cause any motion of the dot that had a horizontal component opposite to the line motion. This horizontal component of the dot's motion altered its vertical motion and caused its perceived motion to be oblique. The conflict that is always present in induced motion consisted in configurational change causing this oblique motion path of the dot on the one hand, and subject-relative stimulation representing the objective vertical motion of the dot on the other. Therefore, the slope angle of the perceived oblique motion measured the outcome of this conflict and permitted a comparison of the relative effectiveness of the three stimulus conditions. The dot's real motion was either given by ocular pursuit, or, when the eyes fixated a stationary mark, as image displacement and was in either case in conflict with configurational change.

When the vertical motion of the dot was tracked and given by ocular pursuit, the horizontal component of the oblique motion path was equal in amount to the horizontal motion of the line pattern. That means that configurational change was fully effective and completely prevailed over the subject-relative stimulation. When, on the other hand, the eyes fixated a stationary mark so that the vertical motion of the dot was given by image displacement, the perceived motion path of the dot was much steeper. The horizontal component of the path amounted only to half the extent of the motion of the line pattern. When they were in conflict with each other, configurational change and image displacement seemed to be about equally effective.

Robert Becklen and I recently obtained corresponding results with a display where the two motions were no longer in phase with each other but were combined with a 90° phase shift. An oval motion path of the dot resulted, whose shape represented the relative effectiveness of the conflicting stimuli. The height of the oval represented subject-relative stimulation and its width configurational change. When the vertical motion of the dot was given by ocular pursuit, the oval path showed complete dominance of configurational change. Estimated width over height was equal to the ratio of the extents of the horizontal and vertical motions. On the other hand, when the vertical motion of the dot was given as image displacement, the ratio of width to height was one half of what it had been in the other combination. Again, image displacement and configurational change turned out to be about equally effective.

That these displays yield well-shaped motion paths is in itself a remarkable fact, since the horizontal and vertical dimension of these paths are given by different stimulation. Ocular pursuit and configurational change require different processes in the nervous system before they yield commensurate components of a unified path. There does not seem to be an analogue to our display in ordinary experience, instances where the results of different processes must combine to provide a single motion experience. In terms of the underlying functions, perceiving such a combined motion was a novel event for our subjects. The question is to what degree it was novel. When, as is usual under ordinary conditions, a motion is simultaneously given by ocular pursuit and by configurational change, with the two conditions of stimulation in agreement, it is not necessary that the two processes combine. Some people may habitually use one stimulus condition and some the other. On the other hand, the two processes may operate in concert nevertheless.

On the oscilloscope, one can put together a vertical and a horizontal motion component and obtain an oblique line or an oval motion path. They are the simplest Lissajous figures. But when our display caused these motion paths to be seen, they existed only in perception. By changing the frequency relation between the motion of the dot and the motion of the line pattern, O'Leary and I were able to produce the psychological counterparts of the more complex Lissajour figures. We wanted to find out whether the accuracy with which such psychological Lissajous figures are perceived reflects the complexity of process combination. We had subjects reproduce more complex Lissajous figures, those with phase relations 2 to 1 and 3 to 2, both open and closed ones. The subjects made these reproductions by putting down a narrow hose made of cloth and filled with fine shot. After the subjects had shaped the hose so that it resembled the motion path they had seen, the hose was photographed. We compared the reproductions of these psychological Lissajous figures with reproductions subjects made after observing real Lissajous paths, that were given either as image displacements or by ocular pursuit. We also obtained reproductions of perceived Lissajous paths of a stationary dot that underwent induced motion in both dimensions. There were some interesting findings: Reproductions were not better when the Lissajous paths were given as image displacement rather than by ocular pursuit. They were slightly better when the perceived paths consisted altogether of induced motion, that is, when a stationary dot was fixated that was seen on a pattern of vertical and horizontal lines, that as a whole underwent the Lissajous motions. However, the finding which is important in the present discussion concerned the psychological Lissajous paths. There was no striking loss of accuracy of the reproductions that would have been expected if combining of different perceptual processes that takes place when these paths are perceived had never occurred before. It seemed unlikely that the combination of the two processes was altogether a novel event. This suggests that, in ordinary motion perception, the different processes that result simultaneously from configurational change and from ocular pursuit do not remain independent of each other.

Visual vector analysis

The recognition that three different conditions of stimulation operate in motion perception helps in understanding a puzzling type of motion event, Gunnar Johansson's visual vector analysis. It is in this volume discussed by Pomerantz and Toth, who use as an example a display of Johansson, the motion of two dots one of which moves vertically and the other horizontally to form an L-shaped configuration. The dots appear to move on the same oblique path toward each other while they simultaneously move as a group in the direction perpendicular to that path. These two simultaneous perceived motions of each dot are kinematic components of the objectively given vertical and horizontal motions. Pomerantz and Toth call the apparent motions of the dots toward each other "relative" and the other perceived motion "common." Obviously, the two perceived motions are given by different stimulus conditions, the relative motions by configurational change and the common motion by one of the subject-relative stimulus conditions. Rather than being the result of an analysis, the two perceived motions are manifestations of different stimulus conditions operating simultaneously. The objective vertical and horizontal motions of the dots are given by subject-relative stimulation. But configurational change is apt to play a dominant role, as we have seen, and results in the perception of the relative dot motions, while the vertical and horizontal dot paths are not seen. This corresponds to what happens in the experiments by Wallach, O'Leary and McMahon (see above) where the real vertical motion of the dot is not seen when configurational change operates. The common motion of the two dots that is evoked by subject-relative stimulation is perceived independently. (3) The different motion processes that are caused by different stimuli, remain here independent of each other because they pertain to different entities, the one caused by configurational change to single dots and the other to the group of dots.

Learning in motion perception

The three stimulus conditions of motion perception are redundant. Any one of them would suffice for correct motion perception. Then, why are there three? I believe that, like other stimuli in space perception, two are learned.

It seems likely that configurational change and ocular pursuit are acquired and that image displacement is the innate stimulus. That the first two are learned is suggested by the finding that their relation to perceived motion can be easily altered by perceptual adaptation. In the case of configurational change, such an alteration took the form of cue-conflict adaptation. When two stimuli evoke the same perceptual property they usually both operate veridically. Altering one of them artificially, for instance, with spectacles, produces conflicting stimulation, and the outcome of the resulting perceptual processes will disagree. At that point one or both of these processes may gradually change in such a way that the disagreement is diminished (Wallach & Huntington, 1973). In the case of configurational change, no artificial intervention is necessary. As we have seen, the condition that yields induced motion in itself presents a cue conflict, and prolonged exposure to a case of induced motion may therefore result in an adaptation where the effectiveness of configurational change becomes diminished. Wallach, Bacon, and Schulman (1978) succeeded with such experiments. We measured the extent of the horizontal component of the oblique path of the dot in the line pattern display before and after a ten minute exposure to induced motion and found that afterwards the induced motion component was diminished by 15 percent. Later Bacon got similar results using tests in which estimates of the extent of the induced motion of a stationary dot were obtained before and after such an exposure.

A very different procedure was used to obtain an alteration of the perceived motion evoked by ocular pursuit. It was based on a hypothesis about the manner in which ocular pursuit becomes a stimulus for perceived motion (Wallach, 1976, p. 98). When an object in our environment starts to move, the eyes remain still for a brief time before they start to track the moving object to keep its image near the fovea. Thus, pursuit is typically preceded by image displacement. In the case of straight motion, image displacement and subsequent pursuit have the same direction. There is no reason why the same sequence should not have taken place before pursuit had become a stimulus for perceived motion; tracking of a moving object may well occur independently of motion perception. My hypothesis assumes that this sequence is the occasion where the connection between ocular pursuit and perceived motion becomes established. Pursuit always follows image displacement that causes perceived motion, and pursuit eventually comes to evoke the perceived motion that is caused by the preceding image displacement. Bacon and Wallach (1982a) attempted to obtain support for this notion with an adaptation experiment.

The acquisition of the capacity by which ocular pursuit evokes motion perception can probably not be undone. But it seemed possible that the direction of the perceived motion that results from ocular pursuit could be changed. Suppose reality were such that motion were always horizontal but changed to oblique at the moment when the eyes take up pursuit. Horizontal image displacement would then always be followed by oblique pursuit. If the hypothesis is correct that ocular pursuit eventually comes to evoke the sort of perceived motion that is produced by the preceding image displacement, the oblique pursuit should cause perceived horizontal motion when it becomes a stimulus for motion.

The adaptation experiment asked whether such learning would occur in adults, that is, even after the normal relation between ocular pursuit and perceived motion had been established, and would manifest itself in partial adaptation. During the exposure period the subject saw a dot 75 times move horizontally and change direction by 45° upwards as soon as his eyes started to move. Adaptation should cause the second phase of this motion sequence to be gradually perceived as more horizontal. Adaptation was measured by obtaining estimates of the apparent slope of the second motion phase in tests where the direction change was smaller, only 25° or 10°. These tests were given before and after the 75 adaptation exposures. There was a mean change toward horizontal of 12° at the 25° test and a mean 9° change at the 10° test. Since the slope angles were overrated, these changes amounted only to 38 and 57 percent respectively. There was a control experiment in which the image displacement had the same slope of 45° as the pursuit phase. Here the tests registered no perceived slope change, a result that was significantly different from the previous one at the .001 level.

This experiment may be the model for the learning process by which ocular pursuit becomes a stimulus condition that causes perceived motion. A sequence of events that happens to occur with great frequency provide the basis for this learning. Other instances of acquisition of the capacity to function as stimulation probably operate in a different fashion. Configurational change becoming a stimulus for motion is an example. When an object moves under ordinary circumstances, that is, when its stationary surround is visible, its motion may be given as image displacement as well as by configurational change. Assume that configurational change has not yet become a stimulus for motion. Image displacement will cause perceived motion of the object and configurational change will occur simultaneously.

A connection will be established between configurational change and perceived motion. I am not suggesting that this happens by continguity accounting. Contiguity accounting would be unnecessary, if covariance played a role in establishing new connections in perception: Image displacement, perceived motion, and configurational change start and stop simultaneously, and that may bring them together.

If configurational change becomes a stimulus for motion because, under ordinary condition, it occurs together and covaries with image displacement, the origin of Duncker's rule is evident. The rule which makes configurational change, which represents relative displacement, unambiguous, says: When configurational change represents a relative displacement between an object and its surround, the object will be perceived to move and the surround will tend to be stationary. That is what is perceived under the ordinary conditions that are the occasions where configurational change becomes a stimulus for motion: a moving object and a stationary surround. Configurational change and Duncker's rule are acquired together.

If covariance had the capacity to bring about connections, it would make some other learning in space perception plausible. It would explain how such diverse things as eye movements and perceived distance become related when convergence of the eyes becomes a cue for distance. Very rapid adaptation that changes the relation between oculomotor adjustments and perceived distance makes it likely that they are learned cues for distance (Wallach & Frey, 1972). When we approach an object or vice versa, other cues for distance as well as the perceived distances they produce covary with the oculomotor adjustments.

Further duplicating stimulation

If covariance is the cause of conditions becoming stimuli for perceptual properties with which they vary, there should exist more stimulus conditions that duplicate known stimulation, just as ocular pursuit and configurational change duplicate image displacement in motion perception. I shall briefly describe two instances that fit this postulate.

If covariance can cause a condition that varies with a perceptual property to become a stimulus for it, it may not be necessary that occasions at which such covariance makes itself felt are very frequent. The following instance serves as an example. When Wallach and O'Leary (1982) became aware that the slope angle with which an object on the ground in front of an observer is viewed varies with distance, they tested whether it functions as a cue for distance and found that it does. But they also found that it is a weak cue; it readily yielded to a set effect. A scarcity of learning occasions apparently makes this cue weak but does not prevent its acquisition.

The most important instance of duplicating stimulation occurs in stereoscopic vision. Because the eyes view a tridimensional object or a depth interval between objects from different directions, depth is represented by small differences in the configuration with which solid objects or a tridimensional scene is given on the two retinae. The nervous system extracts these differences in terms of retinal disparities: Environmental points that are nearer or farther from the eyes than the plane of the point on which the eyes converge have their images fall on non-corresponding locations on the retinae of the eyes, and these retinal disparities cause these points to be perceived as nearer or farther than the points in the plane of convergence. That is the well-known way in which the innate mechanism of stereoscopic vision operates. But along with the retinal disparities occur the differences in the configurations of the retinal projections in the two eyes that give rise to the disparities, and they are, of course, covariant with the disparities. Configurational differences in the two eyes should therefore be expected to function as stimulation conditions that duplicate retinal disparities. It has gradually become clear that they do.

First it was found that more conspicuous configurational differences can produce greater perceived depth than less conspicuous ones, when the disparities caused by the configurational differences were of equal amount (Wallach, 1972, pp. 57-67). Then we were able to show that one kind of configurational difference was more effective than another kind when disparities were the same: When vertical alignment differences were present, the delay with which depth in stereoscopic charts was perceived was shortened, and vertical alignment differences made stereoscopic depth perception more effective when it was in conflict with other conditions of stimulation that determine perceived depth (Wallach & Bacon, 1976a). Finally, we did an experiment where an added configurational difference that did not increase disparity did increase perceived depth (Bacon & Wallach, 1982b).

Motion stimulation produced by movements of the observer

When one is moving one causes a variety of relative motions between the environment and the eyes, which are represented by stimulation that under other circumstances would lead to motion perception. But when that stimulation is caused by one's own movements, it does not, in most cases, bring about perceived motion of the environment. Turning or nodding of the head is an example; it causes relative motion of the environment, which is given by subject-relative stimulation but is not perceived. A very accurate compensation process matches up visual stimulation with kinesthetic stimulation that represents the head movements and only unmatched portions of visual stimulation are perceived. To obtain measurements of the accuracy of this compensation process we produced these unmatched portions artificially by adding real motion to the relative motion that the visual environment undergoes during head turning. The added real motion could be in the same direction as the relative motion and increase it, or it could be in the opposite direction and diminish it. What the subject saw before him could be made to move in variable amounts dependent on the same head movements that caused the relative motion. Very slight head-movement-dependent real motions were correctly perceived; these unmatched portions of the relative environmental motion were seen when they amounted to as little as 3 percent of the head rotation. The range of real motion that can increase or diminish the stationary environment's relative motion without being perceived measures the accuracy of the compensation process. It is called immobility range. Much is known about this compensation process that deals with the effects of head rotations.(4)

Another compensation process that prevents perception of relative motion caused by our own movements may be of greater interest to the reader, the one connected with one's translatory movements. Solid objects we pass when we move forward are successively seen from different directions, that is, they are given in partial rotation. While the rotation by way of the changing shape of the object's projection provides information for the perception of the object's tridimensional form, one is not aware of the rotation itself. (5) This is due to a compensating process that takes the change in the observer's vantage point into account. (6) The accuracy of this compensation was measured by Wallach, Stanton, and Becker (1974). They had subjects walk past solid objects that could be made to rotate dependent on the subjects changing vantage point, either in the direction of the relative rotation caused by the subject's movement, so that it added to it, or in the opposite direction so that the real rotation diminished the relative rotation. Large real rotations during the subject's movements remained undetected, but here, too, unmatched motions of the test object were perceived, namely, when they, on the average, exceeded the relative rotation by about 40 percent.

This compensating process becomes conspicuous when it operates where it is not needed. When one passes by a painting that renders perspective depth realistically, the scene in the painting appears to turn, clockwise when the painting is on one's right. If the scene were real, it would rotate counterclockwise relative to the observer's eyes and the compensation process would prevent one from seeing this relative rotation. But the painted scene does not rotate, and that is equivalent to an unmatched portion of the relative rotation that diminishes it by 100 percent. That is well outside of the measured immobility range that showed that added real rotation in excess of about 40 percent of the relative rotation would be perceived.

There are more such compensation processes. (7) They clean up perception so that only informative visual inputs reach the stage of awareness and are preserved in memory. Some are likely to be acquired; two of them have been altered by adapting subjects to novel, artifically produced relations between head movements and environmental motions. (8)

Notes

1. Implicit in speed perception based on ocular pursuit is that the distance of the moving object from the observer is taken into account. That such a process can actually take place was shown by Rock, Hill and Fineman (1968). Speed perception based on configurational change, on the other hand, is not affected by observation distance (Wallach, 1939).

   J. F. Brown's experiments (Brown, 1931) have been misrepresented in several secondary sources. Citing O.W. Smith and L. Sherlock (1957) they attribute the transposition phenomenon to the impression of rhythmic speed that may develop as rows of evenly spaced dots enter the apertures. This explanation overlooks that Brown (1931) reported on page 212 that all experiments were done with only one dot visible at a time when most earlier experiments were repeated at Yale. He did this for the expressed purpose of excluding an impression of rhythmic speed. Virtually the same results were obtained with single dots that had been earlier obtained with rows of evenly spaced dots.

2. For another consequence of this property of straight lines see E. H. Adelson and I. A. Movshon in this volume.

3. This explanation of vector analysis was first presented in my tutorial "Eye movement and motion perception" at the NATO Conference "Symposium on the Study of Motion Perception; Recent Developments and Applications," 1980. Thi paper was published in Wertheim, et al., Tutorials on Motion Perception (1983). Vector analysis is discussed on pages 8 and 9.

4. For a fuller explanation see the introduction to Wallach and Kravitz (1968). A report on the most recent investigation of this compensation process can be found in Wallach and Bacon (1977).

5. Tridimensional form perception that results from rotation is discussed in this volume by Myron L. Braunstein.
6. Failure of this compensation process during passive locomotion can be observed when one looks from the window of a fast moving train.
7. Mack (1970) and Whipple and Wallach (1978) investigated compensation for the effects of eye movements and Wallach and Bacon (1976b) for the effect of head tilting.
8. See Wallach and Bacon (1977) which also provides references of earlier work related to head turning, and Wallach and Bacon (1967b).

References

Bacon, J. & Wallach, H. Adaptation in motion perception: Alteration of motion evoked by ocular pursuit. _Perception & Psychophysics_, 1982, _31_, 251-255.(a)

Bacon, J. & Wallach, H. Configurational differences in stereovision. _Perception_, 1982, _11_, No. 5. (b)

Brown, J. F. The visual perception of velocity. _Psychologische Forschung_, 1931, _14_, 199-232.

Duncker, K. Uber Inducierte Bewegung. _Psychologische Forschung_, 1929, _12_, 180-259.

Mack, A. An investigation of the relationship between eye and retinal image movement in the perception of movement. _Perception & Psychophysics_, 1970, _8_, 291-297.

Rock, H., Hill, A. L. & Fineman, M. Speed constancy as a function of size constancy. _Perception & Psychophysics_, 1968, _4_, 37-40.

Smith, O. W. & Sherlock, L. A new explanation of the velocity-transposition phenomenon. _American Journal of Psychology_, 1957, _70_, 102-105.

Wallach, H. On constancy of visual speed. _Psychological Review_, 1939, _46_, 541-552.

Wallach, H. _On perception_. New York, N.Y.: Quadrangle/The New York Times Book Co., 1976.

Wallach, H. & Bacon, J. Two forms of retinal disparity. _Perception & Psychophysics_, 1976, _19_, 375-382. (a)

Wallach, H., & Bacon, J. The constancy of the orientation of the visual field. _Perception & Psychophysics_, 1976, _19_, 492-498. (b)

Wallach, H. & Bacon, J. Two kinds of adaptation in the constancy of visual direction and their different effects on the perception of shape and visual direction. Perception & Psychophysics, 1977, 21, 227-242.

Wallach, H., Bacon, J. & Schulman, P. Adaptation in motion perception: Alteration of induced motion. Perception & Psychophysics, 1978, 24, 509-514.

Wallach, H. & Frey, K. J. Adaptation in distance perception based on oculomotor cues. Perception & Psychophysics, 1972, 11, 77-83.

Wallach, H., Frey, K. J. & Bode, K. A. The nature of adaptation in distance perception based on oculomotor cues. Perception & Psychophysics, 1972, 11, 110-116.

Wallach, H. & Huntington, D. Counteradaptation after exposure to displaced visual direction. Perception & Psychophysics, 1973, 13, 519-524.

Wallach, H. & Kravitz, H. H. Adaptation in the constancy of visual direction tested by measuring the constancy of auditory direction. Perception & Psychophysics, 1968, 4, 299-303.

Wallach, H. & O'Leary, A. Slope of regard as a distance cue. Perception & Psychophysics, 1982, 31, 145-148.

Wallach, H., O'Leary, A. & McMahon, M. L. Three stimuli for visual motion compared. Perception & Psychophysics, 1982, 32, 1-6.

Wallach, H., Stanton, L., & Becker, D. The compensation for movement-produced changes in object orientation. Perception & Psychophysics, 1974, 15, 339-343.

Wertheim, A. H., Wagenaar, W. A. & Leibowitz, H. W. Tutorials on motion perceptions. New York and London: Plenum Press.

Whipple, W. R. & Wallach, H. Direction-specific motion thresholds for abnormal image shifts during saccadic eye movement. Perception & Psychophysics, 1978, 24, 349-355.

THE SCOPE OF RESEARCH ON MOTION: SENSATIONS, PERCEPTION, REPRESENTATION AND GENERATION

John K. Tsotsos
Dept. of Computer Science
University of Toronto
Toronto, Ontario, Canada

Abstract

This paper will attempt to briefly outline the broad scope of research endeavours that involve visual temporal change. It is acknowledged that several classes of researchers with differing backgrounds such as psychophysics and neurophysiology, as well as computer vision, artificial intelligence and computer graphics, are active in the area of motion research, and that their interaction may be of benefit. In general terms, these endeavours are those of quantification of intensity changes in image sequences, representation of visual temporal information, temporal reasoning and event description, motion generation, animation, control structures for coordinating the above into a unified computer system whether it be a vision system or an animation system, and studies of biological visual systems. Motion research seems to encompass several broad classes of efforts: sensing motion, perception, representation, interpretation and description of motion; and, motion generation.

1.0 Introduction and Motivation

In the domain of time-varying image analysis by computer, the vast majority of researchers are concerned primarily with investigations of optical flow or correspondence. Elsewhere in this volume, Bernd Neumann provides an introduction to the area of optical flow research, and points out that the goal of such research is the computation of 3D motion parameters and depth values for point or edge-like features in imagery. The question arises: Then what? There are two aspects to the answer to this question. The first is application oriented. Given some specific application, how does a computer system use such information in order to solve the problem at hand? The second involves deeper issues. What representations are most appropriate for such data? How does grouping of flow patterns take place? How does motion computation interact with form, colour and depth computation? (See the paper by Paul Kolers in this volume for a brief description of some of these interactions.) What kind of control structure can manipulate these representations in order to interpret the changes over time in terms of everyday phenomena? What is the relationship of such representations and control structures to human motion perception? How does graphics animation research impact research on motion interpretation? This paper will not answer these questions, but will attempt to provide motivation and background for research on these topics.

During this workshop, one particular question was asked several times: Why should computer scientists be interested in the work done by the psychologists and neurophysiologists studying motion perception? Why should we try to relate our computational models to models of biological visual systems? After all, aren't we in the business of building machines that solve specific problems? These are valid questions and the answers are not entirely clear. One thing is definitely clear however. A thorough understanding of vision is not is not currently within our grasp. Moreover, if a computer scientist were to attempt to outline the complexity of the general vision problem without benefit of the centuries of thought that has already gone into it, he would be very hard pressed. A scholarly approach to a scientific endeavour does not ignore past efforts, both successes and failures, even when those efforts may not be from within one's own discipline. The main reason, in my mind, for studying both past work and work in differing disciplines is therefore to better enable us to grapple with the complexity of the problem. Many different kinds of processing, as well as interactions between processing steps, seemingly exist in biological vision, and have been studied previously by non-computer scientists. Psychologists and neurophysiologists concerned with vision seek to understand biological visual systems in their totality. In simple terms, the studies of psychologists reveal functional characteristics of various elements of the visual system, while those of neurophysiologists suggest possible structures of processing elements.

Motion: Representation and Perception
N.I. Badler and J.K. Tsotsos, Editors
Published by Elsevier Science Publishing Co., Inc.

In one sense, it is true that computer scientists are indeed in the vision machine building business - in the same way that aircraft engineers are in the airplane building business. Man had attempted to fly for centuries by flapping his arms, emulating the birds - yet that was not to be the final solution to manned flight. The basic characteristics of airfoil design and function were abstracted from biological flight machines, but that is all. In the same way, it is not clear to what degree the final computer vision solution will have similarities with biological vision systems. The basic characteristics and function will of course, have to be the same. Even if hypotheses of processing elements from biological vision systems are the motivation for several current research activities, rather than claiming that computational vision models are actually models of biological vision systems, it may be more correct in the long run, to simply claim that their design was biologically motivated. After all, a biologically correct machine vision system would suffer from all the illusions that our own visual systems suffer from - is this desirable?

The biological motivation will be evident in the remainder of this paper. A number of different aspects of visual motion perception will be related in order to demonstrate the scope of motion research, and to outline some of the characteristics that computational models of motion understanding must possess. Moreover, these characteristics will not be presented in a species-specific manner, although most will be taken from cat, monkey or human experimentation. They should be considered simply as relevant facts and hypotheses about visual information processing.

II.0 Motion Understanding - Basic Components

I will divide this discussion into three basic parts covering the main aspects of motion computation: the sensing of motion; the perception, interpretation and description of motion; and the generation of motion. Each will be elaborated in turn. More detailed overviews of computational motion interpretation research can be found in [Scacchi 79], [Ullman 81], [Martin & Aggarwal 78], and the most thorough review of applications in [Nagel 81].

II.1 Sensing Motion

A number of interesting mechanisms for early processing of visual change have appeared in the neurophysiology literature, among them [Dowling & Werblin 71], [Orban 75], and a series of papers by Regan and colleagues whose results are summarized in [Regan et al. 79]. Dowling and Werblin present a model of chained contrast change detectors, implementing directional selectivity by using inhibition in the null direction between detectors. The output of the change detectors feeds another detector whose output is a measure of contrast intensity change in the given direction. Orban presents a hypothesized hierarchy of motion computation, whose units of processing exhibit different combinations of selectivity and sensitivity to degree of contrast change, direction, velocity, and displacement. Units that are selective for a certain parameter have high output when the sensed value of the parameter falls within a given narrow range, and their output falls rapidly outside that range. Sensitivity to a given parameter means that the unit's output increases as the sensed value of the parameter increases, for example, the higher the velocity the higher the response. The hierarchy begins with directionally selective, velocity sensitive contrast change detectors. These also exhibit spatial blur since they temporally integrate their inputs. Temporal integration seems to present throughout this system, and is a feature conspicuously missing from most computational models. Next change detectors are found that are selective for both velocity and direction. The third layer of detectors add displacement to the selectivity list. It seems that in biological vision systems, a great deal of processing must be done before displacement is even considered, let alone being the key determinant in correspondence [Ullman 79]. The model incorporates both units that are typically associated with form computation as well as motion units, and includes the fact that such computations proceed in parallel. Compare the model of directionally selective motion cells that is presented in [Marr & Ullman 79]. Their model proposes the measurement of the time derivative of their Laplacian operator output at the zero-crossings. This measurement determines the local motion direction to within 180 degrees. The true velocity is then measured by combining these local measurements in a later stage of processing. In [Nagel & Enkelmann 82], second order greyvalue corner points are used as correspondence tokens in the determination of a displacement vector field, with the claim that second order tokens are more reliable matching tokens than those based on image intensities since their characteristics remain constant across occlusion or shadow edges.

The series of papers by Regan and his colleagues point out two other very different mechanisms for motion computation, namely those of stereoscopic motion in depth detectors, and changing size (or "looming") detectors. Both seem to require input from more primitive detectors such as those described in the previous paragraph. Stereoscopic motion detectors act largely independently of size change, yet both are hypothesized to feed a single "sensation of motion in depth" detector. There seem to be two separate batteries of size change detectors, one for increase in size and the other for decrease. Each detector, of either type, is driven by two, perhaps overlapping, regions of retina whose most sensitive points have some specific separation. The detector responds to changes of intensity contrast moving in "antiphase" within those two regions, that is, away from or towards each other. The determination of whether the two intensity contrasts belong to opposite sides of the same object requires further processing, and one may hypothesize that interactions with other motion and form units may be necessary, as well as more cognitive processing. The stereo motion units are binocularly driven and are selective for direction of motion in depth, the computation involving the ratio of apparent velocities from each eye. Static disparity and absolute velocity does not seem to play a role.

Psychophysics also has much to offer with respect to the functional characteristics of the motion system. One group of papers, summarized in [Ross & Burr 83], deals with the characteristics of motion-form system interactions. Interestingly enough, their investigations support a view of a motion system that not only computes motion information, but also form information although at the expense of some detail. The tuning of the motion system's processing units not only has a temporal frequency component but also a spatial frequency one, with units tuned to progressively higher temporal frequencies also tuned to progressively lower spatial frequencies. The effect is that low contrast edges which in static imagery are not discernable are easily detected when those edges are in motion. There are many other examples of interesting motion psychophysics in the first two sets of papers of this volume.

Current computational schemes for early motion processing are mainly concerned with optical flow or three-dimensional correspondence computations. Introductions to these areas appear in this volume by Bernd Neumann and Jake Aggarwal, respectively. It is clear however, that no computational model currently deals with the entire range of motion information described above. More importantly, current models in general, do not incorporate time as an integral component thus not addressing the temporal integration nor the sampling issues. Some considerations on sampling issues for a temporal cooperative process for motion classification are presented in [Tsotsos 81b].

II.2 Perception, Interpretation and Description of Motion

Concurrently with the motion sensing that was briefly described in the previous section, processing for colour, form, and depth information takes place. The different visual components in the perceptual system do not function independently of each other, but rather have important interactions that play a role in the determination of their function. I will take the position that a major aspect of perception is the interaction and integration of these different visual modalities. Some of the many interactions will be briefly outlined below.

An important factor in motion interpretation by humans, is that in the face of ambiguity, the visual system prefers to achieve as high a degree of object constancy or rigidity as possible [Johansson 75]. The rigidity assumption was key in the structure from motion work described in [Ullman 79]. Two examples of such ambiguity are: uniform change in size vs. motion in depth, and change in size in one dimension vs. rotation in depth. Further work on rotations and on the perception of rotating objects can be found in [Shepard & Cooper 82].

The motion sensing units described earlier simply cannot handle such interpretations. Moreover, it was noted earlier that even at the first level of change information detection, a certain amount of spatial blurring occurs, thus implying that in order to compute accurate motion-position information, there must be some interaction between the change and form abstraction mechanisms. In addition, interaction between form and motion information is necessitated because of the motion aperture problem [Marr & Ullman 79]. If the motion of an oriented element is detected by a unit that is small compared to the size of the moving element, the only information that can be obtained is the component of motion perpendicular to the local orientation of the element. This means that for higher order interpretations, there is a need for a motion combination stage, perhaps guided by short range motion correspondence information [Ullman 81].

This motion combination stage must combine local temporal information; but also there must be further combination of motion with depth, colour and form information. Some aspects of motion combination for change and depth were mentioned earlier with the work of Regan and colleagues. The experiments described by Kolers and his colleagues reveal some other interactions. Their main result is that the coding of form, depth, colour and motion information cannot be completed in all cases independently of each other. These four systems constantly monitor each others results and compute a consistent interpretation in parallel.

The interactions between translation, form changes and rigid rotation in depth were investigated in [Kolers & Pomerantz 71]. They discovered that if the visual system is given enough time, a rigid interpretation is the preferred interpretation. In contrast, if the amount of time allotted to interpretation is decreased below a certain point, the interpretation will involve a form change, thus implying that interpretations of rigid rotations are more demanding computationally. If the transformation involves both a size change and a rotation, there is no difference in processing efficiency, implying that computations are performed in parallel. In apparent motion with collision the preferred interpretation is motion in depth to avoid collision. This implies that the spatial characteristics of the traversed path are being monitored and have an effect on the interpretation. In [Kolers & von Grunau 76] results were obtained providing evidence for the interactions between form and motion and form and colour systems.

From a computational point of view, the interactions between form and motion fields are explored in the work of [Hoffman 80], [Hildreth & Ullman 82], and others, while those between depth and optical flow for rigid body motion derivation in [Ballard & Kimball 81].

Beyond these combinations of stimuli, whose mechanisms are not well understood, are computations involving object segmentation, identification and naming, motion classification and description. The past literature on psychology and neurophysiology particularly, gives little guidance here. The majority of past computational attempts at these aspects rely on complex representations of the semantics of domain knowledge and interactions between so-called "high" and "low" levels of processing [Hanson & Riseman 78], [Badler 75], [Okada 80], [Marburger et al. 81], [O'Rourke & Badler 80], [Brooks et al 79], [Tsotsos et al 80]. A discussion of representation and control schemes for vision systems is beyond the scope of this paper. See [Tsotsos 82] for more on issues relating representation and control to vision systems. It is clear that none of these vision systems employ the sophistication of motion sensation that was described earlier nor the richness of the required motion combination.

II.3 Generation of Motion

There are at least three main aspects of motion generation by computer: graphical animation, robotics, and generation of eye motions or foci of attention for image scanning or motion pursuit. The first is introduced in the paper by Norman Badler later in this volume. The second, although there is a paper in this volume on robotics, is beyond the scope of this presentation. The third, I will try to relate to the generation and use of expectations that guide visual information processing.

The use of expectations or predictions as aids in the processing of visual information is not a new idea. In this discussion, the purpose of expectations will be to direct the attention of the system to particular sets of concepts and events, regardless of whether they be "high level" or "low level". Two well-known models that support their use are those of Mackworth [Mackworth 78] and Kanade [Kanade 80]. In each of those papers, the production and use of expectations was a vital link in the cycle of visual processing, and involved a transduction from the "model space" into the "image space" of hypothesized image contents. Expectations were used in the SEER system of [Freuder 77] to guide region growing and identification of specific portions of a hammer. A thorough understanding of human body motions and a model of the allowed joint configurations enabled the design of a constraint propagation network that integrated current motions and known body positions with hypothesized ones, producing expected locations in 3D for given body joints [O'Rourke & 75 Badler 80].

Hay [Hay 66], furthering work presented in [Gibson 57], related object motions in 3D to their resulting image projections. This includes image-specific changes such as translation, stretch, shear, foreshortening and magnification related to object-specific changes such as motion in depth, rotations and translations. These results provide a starting point for the transduction between spaces mentioned earlier, that is, translating between image specific and object specific characteristics - the projection between the two domains. The generation of expectations from a knowledge base of motion concepts in a manner relating work described by Hay to motion understanding, was explored in [Down 83]. He showed how the generalization/specialization hierarchy of the knowledge base can be effectively used to recover from incorrect expectations by moving up the hierarchy to more general hypotheses, thus relaxing constraints.

Evidence from psychophysical experimentation for both the use of generalization as a concept organizational tool and for the use of expectations and their link to the generalization relationship comes from the following. The experiments described in [Cooper & Shepard 82] show the strong positive effect of a priori expectations on time for interpretation, while those of [Bugelski & Alampay 61] and [Palmer 75] show the effects of generalization of expectation classes. Cooper and Shepard reported that in the identification of letters presented at varying orientations, the time taken to identify the letter varied with the amount of rotation (to a maximum value at 180 degrees), implying that mental rotation and matching was being performed by the visual system. If identity and orientation were given previous to the stimulus, the response time was flat across all orientations, as long as sufficient time was allowed before the stimulus was presented for expectation formation.

Bugelski and Alampay showed that if a subject is conditioned to expect a given category (or generalization) of stimulus, then the identification time of the stimulus is reduced. They presented stimuli all belonging to the same class of concept (animals), and when non-animal stimuli were presented, the response time increased. This was further examined by Palmer who also noted the impairment of identification if the context is mis-leading. It should be pointed out here that the mechanisms that produce such behaviour are not understood.

The generation and use of expectations for vision have not been extensively studied partially because of the start-up expense in constructing a large unified vision system. However, in graphical animation, the problems of transduction between model space and image space have received some attention. The impact of such work on computational models of visual interpretation is unclear.

Summary

I have presented a short overview of some important aspects of motion research, and have briefly introduced some of the many hypotheses of biological visual system function and structure. There is no current computational vision system that encompasses all of the capabilities that biological visual systems seem to have for motion analysis, particularly those capabilities that require interactions among several kinds of visual information. Moreover, it is unlikely that a complete model will appear soon. The main conclusion we draw is that interaction among the sciences has been shown to be very profitable and is the key motivation for this volume of papers.

References

Badler, N.I., "Temporal Scene Analysis: Conceptual Descriptions of Object Movements", TR-80, Dept. of Computer Science, University of Toronto, 1975.

Ballard, D., Kimball, O., "Rigid Body Motion and Depth from Optical Flow", TR 70, Dept. of Computer Science, University of Rochester, 1981.

Brooks, R., Cereiner, R., Binford, T., "The ACRONYM Model-Based Vision System", Proceedings International Joint Conference on Artificial Intelligence, Tokyo, 1979.

Bugelski, B., Alampay, D., "The Role of Frequency in Developing Perceptual Sets", *Canadian Journal of Psychology 15*, 1961.

Cooper, L., Shepard, R., "Chronometric Studies of The Rotation of Mental Images", in *Mental Images and Their Transformations*, Bradford-MIT Press, 1982.

Dowling, J.E., Werblin, F.S., "Synaptic Organization of the Vertebrate Retina", *Vision Research Supp. 3*, 1971.

Down, B., "Using Feedback in Motion Understanding", M.Sc. thesis, Dept. of Computer Science, University of Toronto, 1983.

Freuder, E., "A Computer System for Visual Recognition Using Active Knowledge", Proceedings International Joint Conference on Artificial Intelligence, Cambridge, 1977.

Gibson, J.J., "Optical Motions and Transformations as Stimuli for Visual Perception", *Psychological Review 64*, No. 5, 1957.

Hanson, A., Riseman, E., "VISIONS: A Computer System for Interpreting Scenes", in *Computer Vision Systems*, Academic Press, 1978.

Hay, J., "Optical Motions and Space Perception: An Extension of Gibson's Analysis", *Psych. Review 73*, 1966.

Hildreth, E., Ullman, S., "The Measurement of Visual Motion", MIT AI LAB Memo 699, 1982.

Hoffman, D., "Inferring Shape from Motion Fields", MIT AI LAB memo 592, 1980.

Johansson, G., "Visual Motion Perception", *Scientific American 232-6*, 1975.

Kanade, T., "Region Segmentation: Signal vs Semantics", *Computer Graphics and Image Processing 13*, 1980.

Kolers, P., Pomerantz, J., "Figural Change in Apparent Motion", *Journal of Experimental Psychology 87*, 1971.

Kolers, P., von Grunau, M., "Shape and Colour in Apparent Motion", *Vision Research 16*, 1976.

Mackworth, A.K., "Vision Research Strategy: Black Magic, Metaphors, Mechanisms, Miniworlds, and Maps", in *Computer Vision Systems*, Academic Press, 1978.

Marburger, H., Neumann, B., Novak, H.-J., "Natural Language Dialogue about Moving Objects in an Automatically Analyzed Traffic Scene", Proc. IJCAI-81, Vancouver, 1981.

Marr, D., Ullman, S., "Directional Selectivity and its use in Early Visual Processing", MIT AI LAB, memo 524, 1979.

Martin, W., Aggarwal, J.K., "Dynamic Scene Analysis: A Survey", *Computer Graphics and Image Processing 7*, 1978.

Nagel, H.-H., "Image Sequence Analysis: What Can we Learn from Applications?", in *Image Sequence Analysis*, Springer-Verlag, 1981.

Nagel, H.-H., Enkelmann, W., "Investigation of Second Order Greyvalue Variations to Estimate Corner Point Displacements", Proc. Pattern Recognition, Munich, 1982.

Okada, N., "Conceptual Taxonomy of Japanese Verbs for Understanding Natural Language and Picture Patterns", Proc. COLING-80, 1980.

O'Rourke, J., Badler, N.I., "Model-Based Image Analysis of Human Motion Using Constraint Propagation", *IEEE Pattern Analysis and Machine Intelligence 2*, No. 6, 1980.

Orban, G.A., "Visual Cortical Mechanisms of Movement Perception", Ph. D. Thesis, Dept. of Brain and Behaviour Research, Katholieke Universiteit Leuven, Belgium, 1975.

Palmer, S., "The Effects of Contextual Scenes on the Identification of Objects", *Memory and Cognition 3*, 1975.

Regan, D., Beverley, K., Cynader, M., "The Visual Perception of Motion in Depth", *Scientific American 241-7*, 1979.

Ross, J., Burr, D., "The Psychophysics of Motion", Proc. Workshop on Vision, Brain and Cooperative Computation, University of Massachusetts at Amherst, May, 1983.

Scacchi, W., "Visual Motion Perception by Intelligent Systems", Proc. IEEE Pattern Recognition and Image Processing, Chicago, 1979.

Shepard, R., Cooper, L., (eds.) *Mental Images and Their Transformations*, Bradford-MIT Press, 1982.

Tsotsos, J.K., "Temporal Event Recognition: An Application to Left Ventricular Performance Assessment", Proc. IJCAI-81, Vancouver, 1981a.

Tsotsos, J.K., Mylopoulos, J., Covvey, H.D., Zucker, S.W., "A Framework for Visual Motion Understanding", *IEEE Pattern Analysis and Machine Intelligence 2*, No. 6, 1980.

Tsotsos, J.K., "On Classifying Time-Varying Events", Proc. Pattern Recognition and Image Processing,

Dallas, 1981b
Tsotsos, J.K., "Knowledge of the Visual Process: Content, Form and Use", Proc. Pattern Recognition, Munich, 1982.
Ullman, S., "Analysis of Visual Motion by Biological and Computer Systems", *IEEE Computer 14*, 1981.
Ullman, S., *The Interpretation of Visual Motion*, MIT Press, 1979.

Psychophysics

THE FOX AND THE FOREST: toward a Type I/Type II constraint for early optical flow

Steven W. Zucker
Computer Vision and Robotics Laboratory
Department of Electrical Engineering
McGill University
Montreal, Quebec, Canada

1. Introduction

Imagine a fox chasing a rabbit through the forest. The fox's eyes will be focused on the rabbit, and most of his body will remain in view. The fox must be sensitive to small changes in the path of the rabbit, perhaps as they are defined by outlines of major body parts, or the rabbit could effect evasive manoeuvers. But, at the same time, the fox need not be very sensitive to the structure of the forest; it is only gross changes, such as the emergence of a large obstacle or ravine, to which he must react. Indeed, given the structure of the forest as a dense collection of grasses and trees going into and out of occlusion relationships and lighting changes, it may not even be possible for the fox to accurately infer its detailed structure. Unlike the rabbit, the forest is more like a waterfall flowing past him. These two events, the rabbit and the forest, define two conceptually different optical flow problems for the fox engaged in a chase. Our goal in this paper is to explore and clarify this distinction.

Our principle thesis -- that there are (at least) two conceptually different early optical flow computations -- is founded on the belief that the earlier stages of visual information processing reflect essential aspects of the physical structure of the visual environment. Of course, the specific aspects will depend on the precise nature of the sub-problem with

Motion: Representation and Perception
N.I. Badler and J.K. Tsotsos, Editors
Published by Elsevier Science Publishing Co., Inc.

which the visual system is engaged. This general statement will become more concrete as we develop it into an analogy connecting certain problems in early orientation selection with problems in optical flow. In particular, the above distinction for the fox between the rabbit and the forest is directly analagous to a distinction in orientation selection between what we have called Type I and Type II patterns. Briefly, it is the distinction between locating contours as they bound distinct objects, as contrasted with the contour arising from an individual hair on one's head. This distinction will be developed next, in terms of both pattern and (computational) process, so that we can extend it more clearly into the domain of optical flow.

2. Type I and Type II Patterns in Orientation Selection

Optical flow results in paths through space and time, but for now we shall restrict ourselves to purely spatial problems. Once we establish the framework in space, it will be readily extendible to include the temporal dimension. We shall concentrate our efforts on dot patterns, because they carry the essential geometric structure within richer, more natural images, and because they are invariant to certain other aspects of even earlier visual processing such as receptor physiology. Indeed, dot patterns provide a representation not unlike what is available from neurons with only slightly overlapping receptive-fields: an array of point responses [Zucker, Stevens, and Sander, in press]. They are simpler in the sense that specific color and intensity variations are all taken to be nil.

Orientation selection can be defined as the task of inferring an orientation from a collection of spatially-distributed entities. To show that it is an essential task for early vision, suppose that the entities correspond to retinal locations at which there are strong intensity changes (or, as they are sometimes called, edges). Such intensity changes typically lie along smooth contours, and these contours are significant because they indicate explicit structure in the physical environment. They can arise from object boundaries, surface creases, or lighting changes; it is extremely unlikely that they would arise purely by chance. Clearly it is advantageous to be able to recover these contours as accurately as possible from the isolated samples. If we assume that the contour passes through each of the sample points, then, in order to actually recover the contour, further specification of the direction in which it is going as it passes through each point is required. Otherwise, it could wind indifferently into any of the other points. Stated differently, as the contour passes through a point, it must be oriented toward another point. Specifying this point, or this orientation, is the process of orientation selection. Mathematically, the goal is to specify the tangent to the contour at every point. Conceptually, it indicates which point is to be taken as the neighbor of a given point along the contour.

To start the discussion of orientation selection, suppose that we are given an arrangement of dots representative of a physically meaningful contour such as an object's outline. The grouping of these dots into a contour was first studied by Wertheimer [1923], who was able to show that smooth curves, with quite a high degree of curvature, are readily perceived, as are small, but abrupt, changes in orientation; see Fig. 1a. In fact, these orientation changes then provide the anchors on which further processes of grouping can be based (cf. the horizontal line of orientation discontinuities in Fig. 1a.) Clearly it is advantageous to be sensitive to small changes or deviations within these patterns, and the evidence suggests that we are able to align dots to the accuracy of hyperacuity [Beck and Schwartz, 1978]. Experiments in our laboratory have shown that humans are able to discriminate sections of sinusoidally smooth curves from sharp V-shapes as long as the radius of curvature is at least 1.5 - 2.0 minutes of visual angle (for dots subtending 1.3 minutes of angle) [Zucker and Lamoureux, 1983]. This is a prototypical Type I pattern.

A second class of physical events in space is given by highly complex, but still oriented patterns such as hair or forests (from the point of view of the trees). These patterns are dominated by a rich web of interacting events -- the individual hairs are woven into and out of occlusion relationships; the highlights jump from hair to hair, and the sheer number of hairs is overwhelming. There is virtually no hope of recovering the detailed geometry of each hair. But, fortunately, there is also no need to. All that is perceived is a sense of the "flow" of the patterns. This is a prototypical Type II situation.

Techniques for constructing Type II dot patterns must preserve this random complexity, and one approach can be based on random dot Moire patterns (RDMP's) [Glass, 1969; 1980]. Such patterns are constructed by taking a pattern of random dots, making a copy of it, transforming the copy, and then superimposing the transformed copy directly on the original. For example, if the transformation is a linear displacement at a certain angle, then the result is a pattern that seems to flow at that angle. Fig. 1b shows exactly this sort of pattern, in which the angular displacements are identical to the ones in Fig. 1a. In actuality, this pattern was constructed from Fig. 1a, by simply moving the dot pairs perpendicularly to their axis of orientation. Thus the interdot distances for the nearest dots, as well as their general distribution, have been preserved. But note that the sharp orientation-discontinuity contour is no longer apparent; rather, it appears as if the pattern is smoothly changing orientation from the top to the bottom. For much higher orientation changes the discontinuity is, however, apparent; see Fig. 1c,d.

We are now in a position to summarize the differences between Type I and Type II patterns. Type I patterns are spatially very precise, and we are psychophysically very sensitive to small changes in them. Type II patterns are much less precise; changes in position of single entities are extremely hard to detect, and the threshold for discriminating smooth sinusoidal curves from abrupt changes is a full order of magnitude lower than for Type I patterns [Zucker and Lamouraux, 1983]. Such sensitivities are exactly what one would expect from the complexity of the configurations from which they naturally arise. Many additional examples of these differences are in Zucker [1982].

Figure 1a. An illustration of a Type I pattern composed of dots. Note that the dots all lie along well-defined contours (in fact, straight lines), and that the orientation of the lines changes abruptly along the center. Such dotted contours could be presented in a 2-D arrangement, such as this one, or they could stand in isolation.

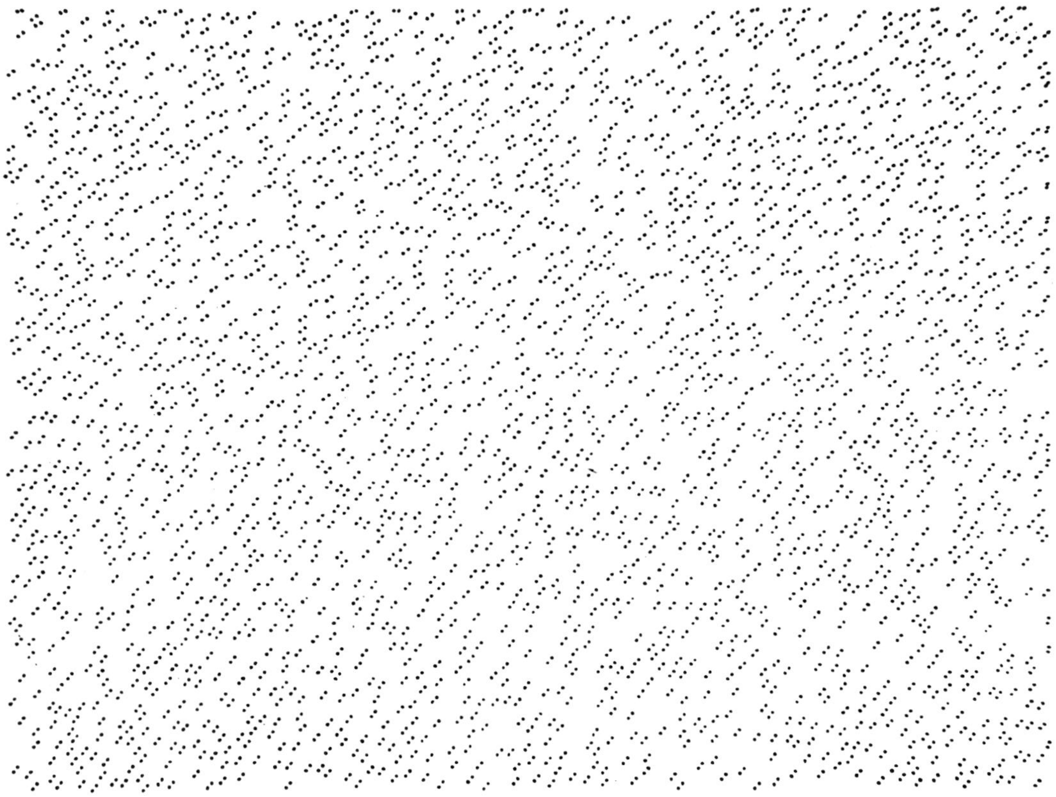

Figure 1b. An illustration of a Type II pattern composed of dots. It can be viewed either as a RDMP (see text), or as a version of Fig. 1a in which the dot pairs have been displaced randomly in a direction perpendicular to their orientation. The pairs remain in the proper orientation, however, so that the local orientation information is identical to that in Fig. 1a, as are the global statistics. Note, however, that the apparent line of orientation discontinuity has disappeared.

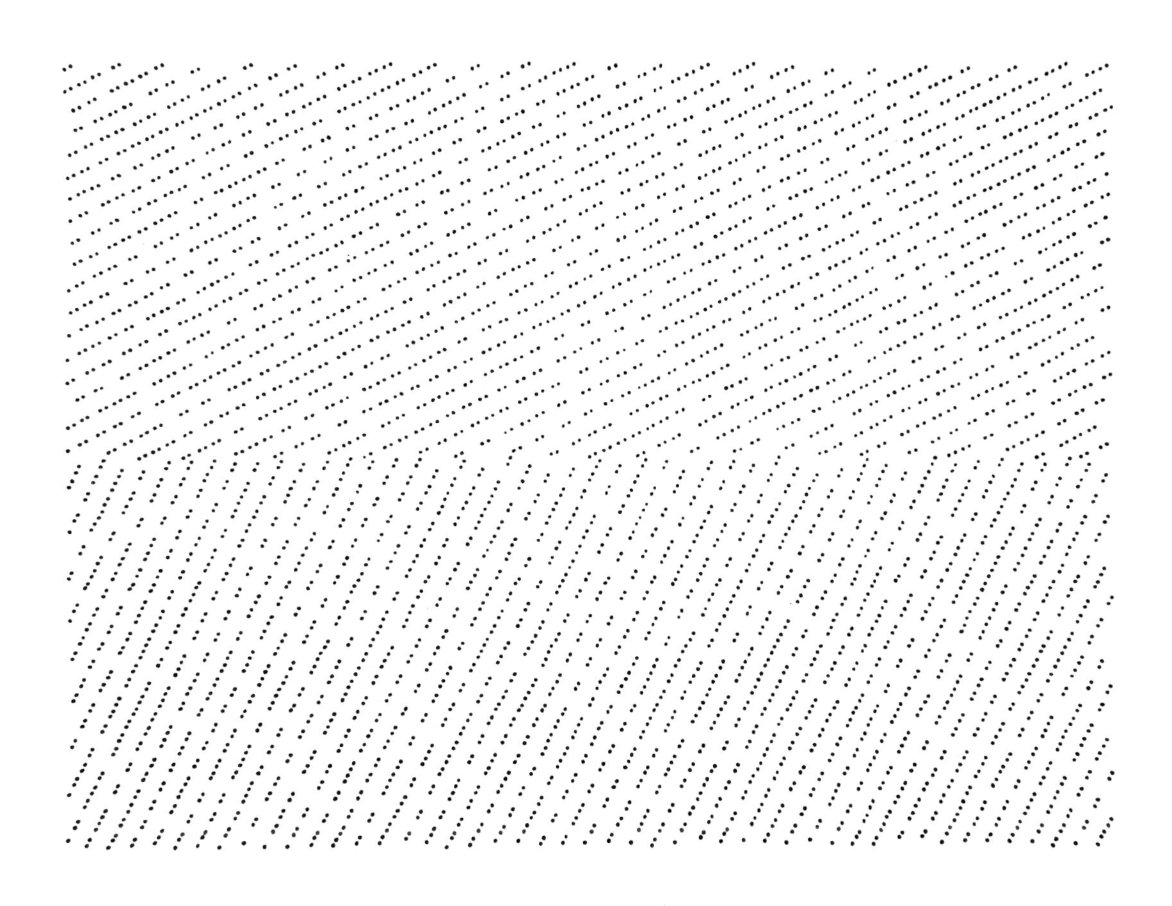

Figure 1c. A Type I pattern identical to Fig. 1a, but for which the angular change is much greater.

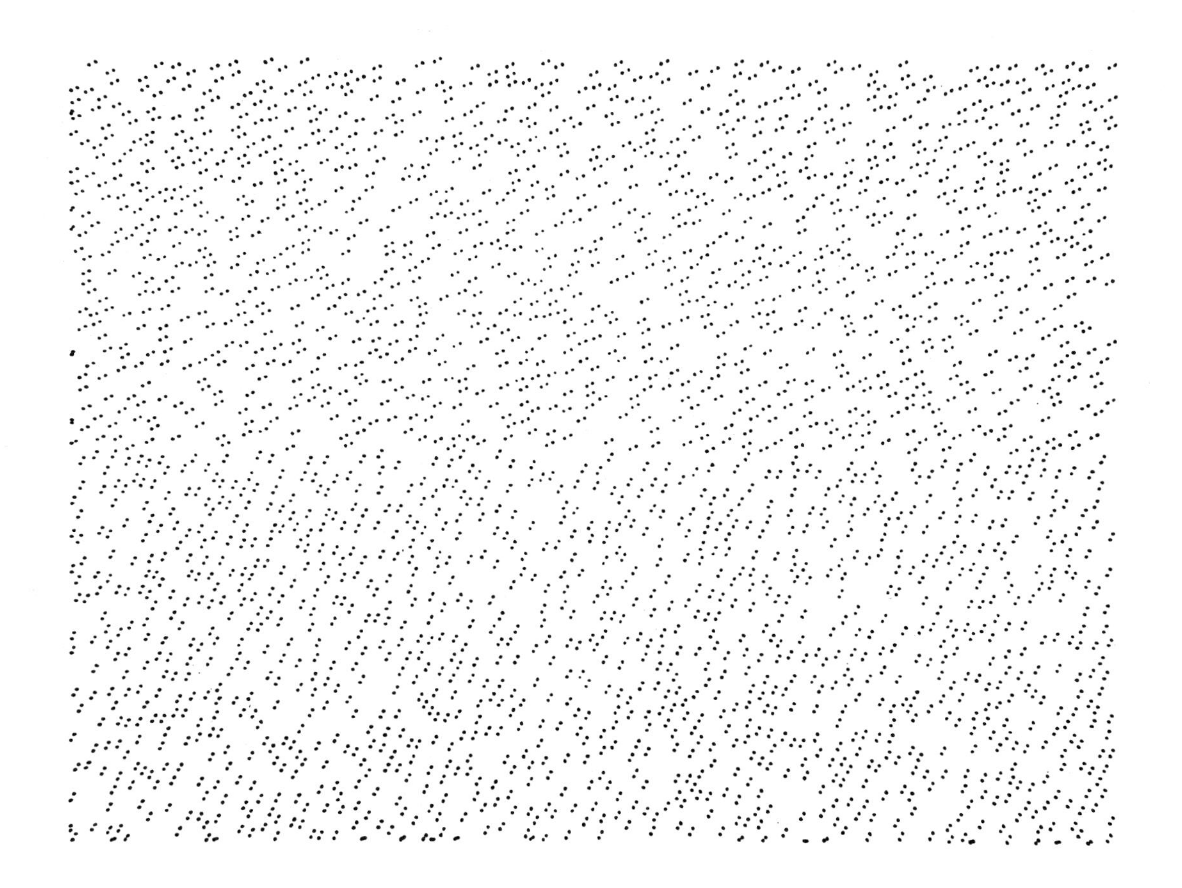

Figure 1d. A Type II pattern identical to Fig. 1b, but for which the angular change is the same as in Fig. 1c. Note that now the line of orientation discontinuity is apparent, illustrating that our sensitivity for Type II patterns is much less than for Type I.

3. A Model for Early Orientation Selection

The difference between Type I and Type II patterns can be characterized computationally within a model for early orientation selection. The model begins with measurements over the dot patterns, and is given its power by an optimal strategy for interpreting these measurements. Its goal is to produce a vector field of tangents, i.e., a collection of oriented entities distributed over "retinal" coordinates. The model is described in full in Zucker [1982]; it is only sketched here to provide a basis for its optical flow analogue.

Orientation selection begins with the initial measurements. The first requirement is that these measurements be asymmetrical, so that they can signal orientation, and we shall model them as convolutions with (a numerical approximation to) second-directional derivatives; see Fig. 2. These operators were formed as the difference of two (2-D) Gaussians, and were chosen both because of their simplicity and because of their neurophysiological significance; they approximate the receptive fields of the first orientationally selective cells in the visual cortex [Hubel and Wiesel, 1977]. The strategy is to take a collection of these operators, distributed in orientation (say, every 40 degrees) and size, and then to convolve (a representation of) the dot pattern with each of these operators. The result will be a large collection of measurements, some

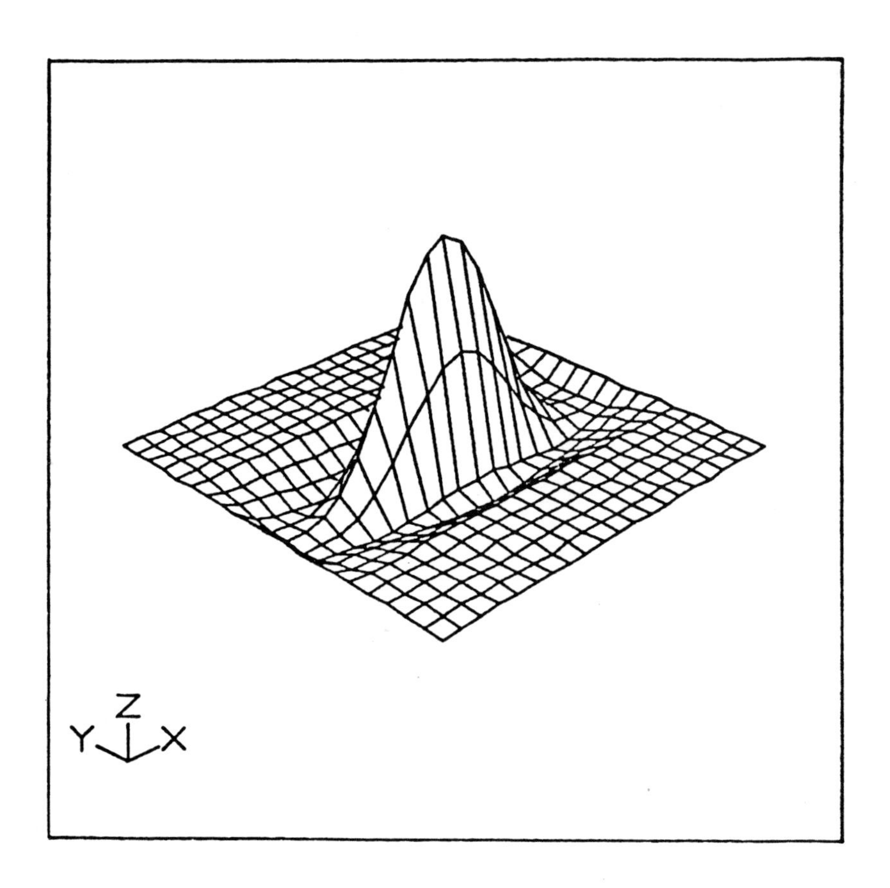

Figure 2. The orientationally sensitive operators, shown in relief, used in the first stage of the model for orientation selection. They can be viewed either as differences of Gaussians, with a 2:1 aspect ratio, or a numerical approximations to second directional derivatives.

strong and some weak, which have extracted and encoded information about the local spatial structure of these patterns. As we shall now show, these local measurements contain sufficient information for inferring the relevant orientation information.

To motivate the inferential process, consider the response of one of the operators to a straight line of dots (Fig. 3a). When the major axis of the operator is aligned with the line, the response (or measurement) will be maximal. It will then drop off monotonically as the operator is rotated over the line, and will be minimal when they are orthogonal. Thus it would seem that a strategy of selecting the strongest response at every point would give the correct orientation at that point. While this is true for sparse, straight lines, it no longer holds when the lines are curved or dense. The interpretation of the responses is more complex, since more of the values will be intermediate, and a more elaborate strategy is required.

The key to this more elaborate strategy lies in the expected response profiles of the operators. Just as they could be determined for straight lines, they can be determined for curved ones as well; see Fig. 3b. The stragegy then becomes one of matching these expected response profiles against the profiles observed from an unknown pattern. The solution is then the vector field which, if it were actually present, would yield a set of expected measurements that are as close to the observed ones as possible. To develop a symbolic representation of this

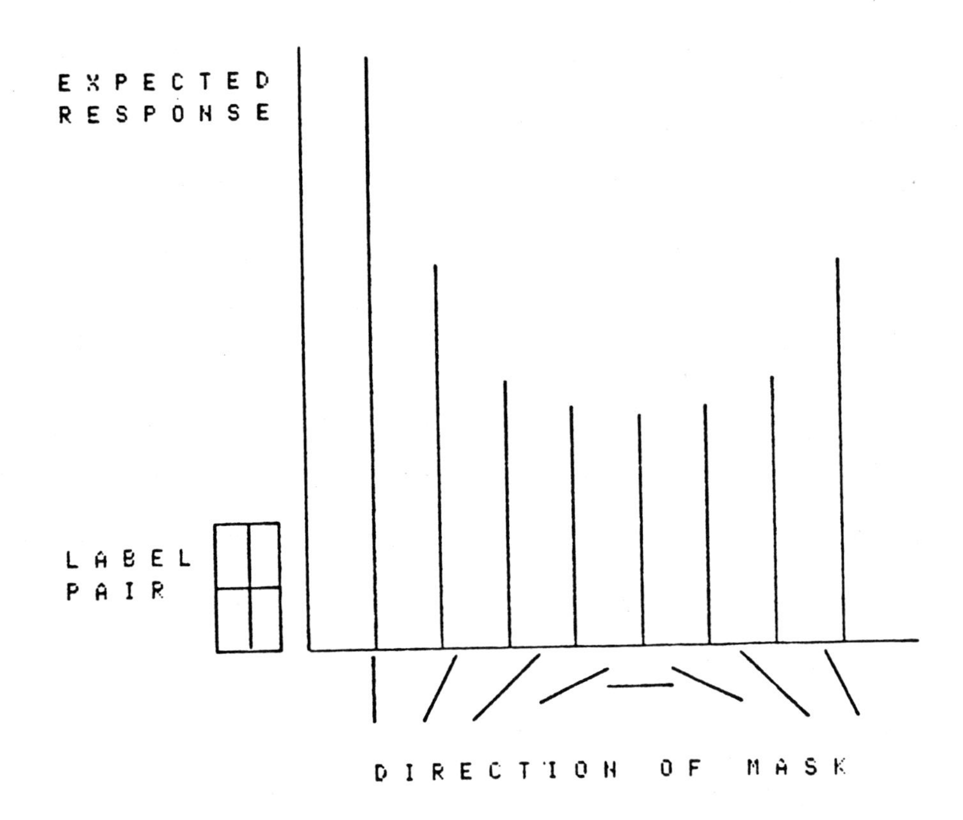

Figure 3a. The expected response of the operator to a straight contour, as a function of the orientation of the operator. Note that the response is maximal when the two are aligned, and falls off monotonically until the operator is perpendicular to the contour.

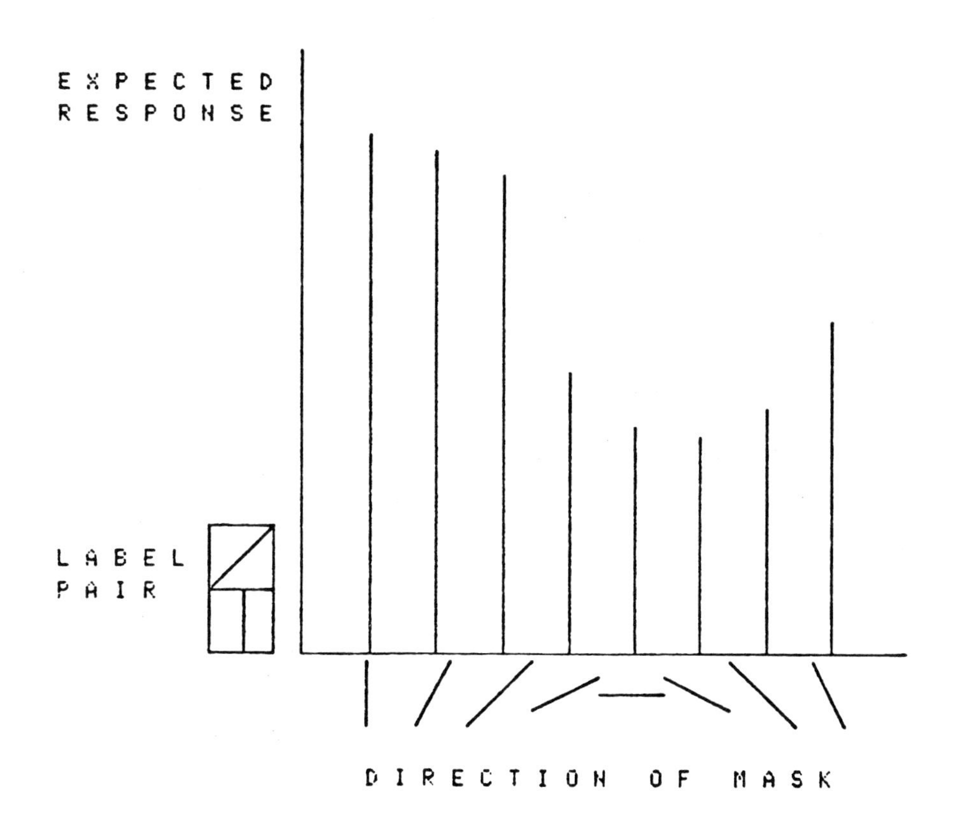

Figure 3b. The expected response of the operator to a curved contour. The contour is illustrated within the two unit boxes on the left, and it changes orientation by 45 degrees within the support of the operator. Note the difference in shape of the response profile from the one in Fig. 3a; it is this difference that provides the key to matching the expected profiles with those obtained from new patterns.

minimization problem, suppose that both orientation and spatial position are discretely quantized, and consider operators whose spatial support covers exactly two unit primitives. In other words, actual curves under the operator will be approximated by two straight segments. Now, if we let Rexp(theta, s; or1, or2; x) denote the expected response of an operator at orientation theta and size s to a curve of orientation or1 and or2 centered at location x, then we desire as solution to the problem:

Find an orientation OR(x), at every point x = (X,Y), such that

$$||Rexp(theta, s; or1, or2; x) - Robs(theta, s; ; x)|| \quad (1)$$

is minimized simultaneously at all positions x.

This minimization problem can be solved cooperatively using a relaxation network [Hummel and Zucker, 1983], and amounts to an excitatory and inhibitory network running among the operator responses. The fact that the expected responses Rexp are a function of the tangent orientation (or1) at the postion x and the neighboring orientation (or2) one discrete location away is what guarantees the contextual continuity of the solution. It can be shown to work sufficiently over Type I patterns; see Fig. 4.

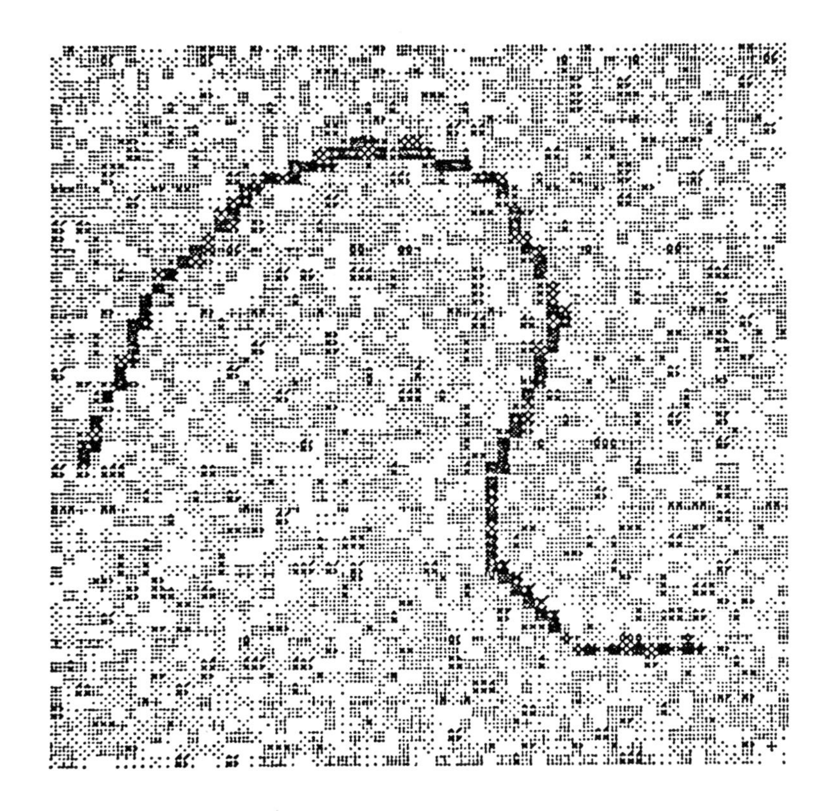

Figure 4a. A contour embedded in random noise.

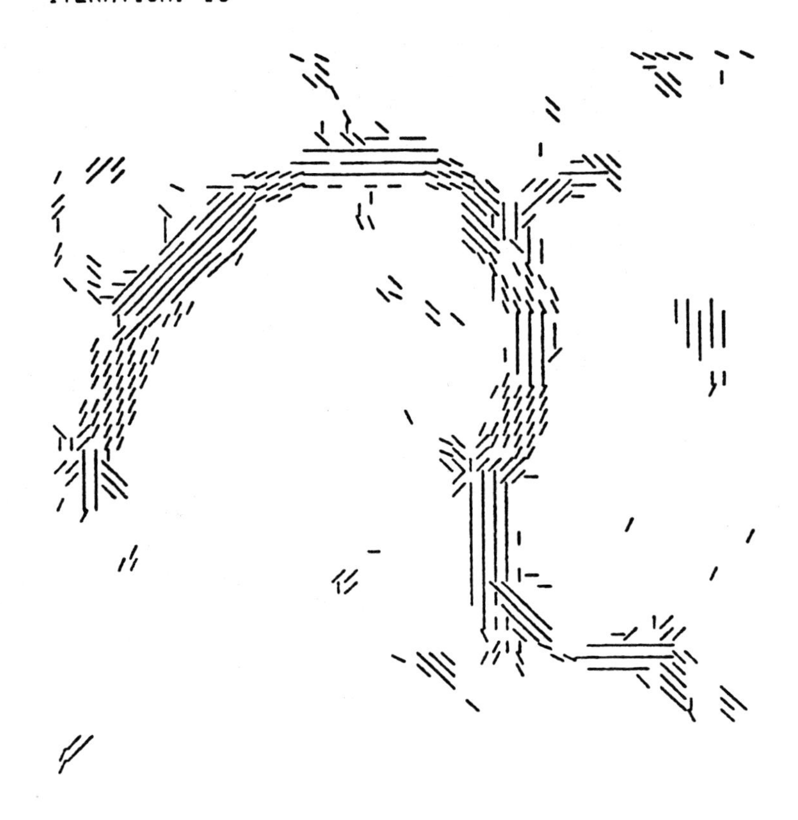

Figure 4b. The result of a Type I process that solves the minimization problem (1) on the image in Fig. 4a. The small amount of thickening in the vector field results from the coarse quantization of orientation into only 8 directions.

But the minimization problem (1) will not work for Type II patterns. The result is a sparse vector field of mainly artifactual orientations [Zucker 1982;1983]. The problem is that the random occlusion and spatial structure of Type II patterns is not built into (1); on the contrary, the responses must be matched at exact positions as is implicit within Type I patterns. To build in the spatial structure of Type II patterns, it is sufficient to relax the spatial specificity. Instead of matching expected responses at a point with observed reponses precisely at that point, the match can take place over a small neighborhood N containing that point:

Find an orientation OR(x), at every point x, such that

$$||Rexp(theta, s; or1, or2; x) - Robs(theta, s; ; x \text{ in } N)|| \qquad (2)$$

is minimized over all x.

This relaxed version of the problem can also be solved by a cooperative network; see Fig. 5.

In summary, then, the model for orientation selection amounts to the optimal interpretation of the responses of asymmetric (i.e., elongated) operators by a process that is (i) spatially specific for Type I patterns and (ii) relatively spatially unspecific for Type II patterns. This spatial specificity was explicitly characterized within a matching problem: expected responses were matched (i) point to point for

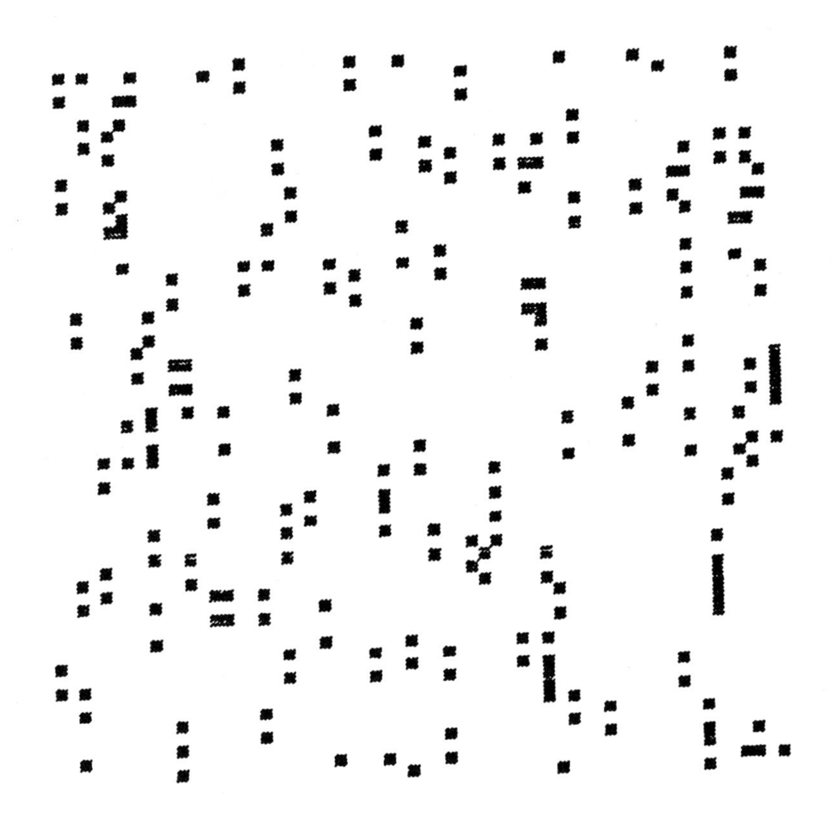

Figure 5a. A small section of an RDMP in which the displacement is vertical.

Figure 5b. The result of a Type II process on the RDMP in Fig. 5a. Note that the vector field is dense and vertical (or nearly so) almost everywhere.

Type I patterns, and (ii) point to region N for Type II patterns. This difference in the matching process relects the random structure of RDMP's perfectly; recall the random positions of each of the original dots, but the perfectly-correlated displaced dots. We now show that an analogous distinction holds for early optical flow.

4. Type I and Type II Optical Flow

Consider, again, the image of the fox and the rabbit sketched at the beginning of the paper. The fox needed an accurate, Type I view of the rabbit, but a much less detailed "flow" of the forest past him was sufficient. Such flows are archtypical Type II patterns. They derive from situations in which the physical and the photometric (lighting) structure change extremely rapidly. In particular, since motion involves temporal changes at some level of description, starting from the raw image intensities and going up to some rather abstract tokens, an essential part of the process of computing motion will involve matching entities across time from these different levels. In particular, correlation [Rosenfeld and Kak, 1982] is a standard technique for matching intensity tokens, and more elaborate correspondence schemes have been proposed for more abstract tokens [Ullman and Hildreth, 1983]. While the detailed structure of plausible matching processes still remain to be specified for primate perception, our basic thesis is that, analagously to orientation selection, they will partition into two general types for computing early optical flow. (A subsequent paper will suggest one such matching problem in detail.)

The essential difference between Type I and Type II optical flows can be nicely illustrated by waterfalls. While the complexity of the forrest flow could be traced to a large number of densely arranged rigid -- but out-of-focus -- objects interfering with one another, waterfalls are dominated by the highly deformable structure of water. Very few local intensity patterns or image tokens persist long enough to provide the spacially specific basis necessary for Type I computations. When they do, however, they stand out in sharp contrast to the general flow of water, often attracting one's attention directly to them. The more general case, however, is for slight protrusions/intrusions to develop in the flow of water, which cast rapidly changing highlights that persist (while changing) for only a short time. Each of these highlights can give an indication of flow while it lasts, but then the indications must come from other areas when it disappears. Note that this random shifting of flow "tokens" is directly analagous to the random spatial shifting of the dot pairs in Fig. 1b; each of the local flows (or local orientations) is in the correct direction. The pairs of entities, in optical flow, are the tokens presented at the present and immediately past time instants; their (image or retinal) velocities are given by changes in their spatial positions during this time interval.

A simple Gedanken experiment can illustrate the essential random component in Type II flows such as waterfalls. Imagine taking a succession of images, closely spaced in time, of the waterfall. If the movement effects due to the flow of water could be determined, and somehow "subtracted out", then all that would remain is a time-varying view of the changing microstructure of the highlights. In this view they would simply come and go, or flicker on and off. If they were on sufficiently long, then they would define a Type I sub-pattern.

It follows, from the study of orientation selection, that Type I processes should have very high sensitivity, while Type II should be much lower. There is some published evidence in support of the extremely high sensitivity of Type I processes in motion, such as the following experiment of Lappin, Doner, and Kottas [1980]. Imagine a large number of dots distributed on the surface of an invisible sphere. Project these dots onto a flat "screen" to make an image, which we shall refer to as frame 1, and then rotate the sphere slightly and re-project the dots to make frame 2. If frame 1 and frame 2 are now presented in alteration, with appropriate exposure (e.g., 200 msec.) and interval (0 isi) timing, the impression will be one of a palindromically rotating sphere. Such effects are very powerful and well-known [Wallach and O'Connell, 1953]. The relevant variation of this paradigm studied by Lappin et al. involves a slight random perturbation of frame 2. They were able to show, in particular, that when the correlation between the altered frame 2 and the perfect one differed by only 0.062, the crisp percept of a rotating sphere was destroyed. A slight perturbation of the data was sufficient, in other words, to drastically change the percept, exactly the kind of sensitivity required for Type I processing.

On the other hand, the sensitivity for Type II flows should be much less, and, again, there is supporting psychophysical evidence. Consider, as above, two alternating frames of dots, but this time in which the dots in one frame are simply displaced (say, along the x-axis) with respect to the other. This gives rise to the percept of a region moving back and forth, and a number of investigators have studied the conditions for "coherence" in such patterns [Braddick, 1974; Lappin and Bell, 1976; Morgan and Ward, 1980]. They have found, as should be expected, that coherent motion -- of, say, a central square patch -- can be seen over a wide range of temporal and spatial displacements. Moreover, this motion persists even when a substantial amount of random noise, or snow, is added to corrupt the display. In particular, van Doorn and Koenderink [1982] noted that sharp local changes in the velocity field disappear almost immediately with the addition of noise, while an overall impression of flow persists through even more noise. This observation is, in a rough sense, analagous to the one in Fig. 1, where the line of orientation discontinuity disappears for small changes in orientation, but not for large ones. The noise introduced by van Doorn and Koenderink does not correspond exactly to the natural physical situations responsible for Type II patterns, and experiments in our laboratory are currently in progress to quantify the difference between Type I and Type II motion sensitivities.

There is also a more theoretical argument in support of the Type I/II distinction in optical flow. Close examination of mathematical models for computing the motion parameters of rigid bodies reveals that, in general, there are so many degrees of freedom that either drastically limiting constraints must be introduced, such as that of a rigid body undergoing planar motion perpendicular to the viewer's line of sight, or else quantities such as the second derivatives of the flow field must be computed [Longuet-Higgins and Pradzny, 1980; Waxman, 1983]. Such high-order derivatives are extremely difficult to compute with reliability when any noise or numerical inaccuracy is present -- the normal situation -- which implies that the visual system must be highly selective in attempting it. Such detailed, highly accurate fields should therefore correspond to (likely) Type I situations, and there may be additional connections with mechanisms such as pursuit eye-tracking and phenomena such as figure/ground segmentation.

5. Discussion

In an earlier study of motion, Braddick [1974; 1980] hypothesized that motion consists of a short-range and a long-range process. This is a rather different distinction than the one for which we are arguing in this paper. His distinction is defined in terms of the maximal extent (in space and time) over which motion processes can operate; ours is defined over structural features of the patterns from which motion is perceived, or from the structure of the natural environment from which these patterns arise. All of the structure in these patterns is (in Braddick's terms) short-range, and we take the Type I/II distinction to be a further refinement of the kinds of processing that take place within the short-range class of processes.

In another study, Marr and Ullman [1981] conceptually separated processes of segregation from processes of integration. Again, the Type I/II distinction should be viewed within the context of processes of segregation. We prefer to emphasize the inferential side of these computations, however, since both orientation selection and optical flow require the inference of oriented or directed entities (or both) from local data.

6. Conclusions

This paper began with the metaprinciple that the structure of processing in the visual system reflects, in a non-trivial manner, essential structure in the visual environment. Such structure is, of course, problem dependent, and we described two basic problems in early visual information processing -- orientation selection and optical flow -- that stand in a fundamental analogue to one another. Both are problems involving the inference of abstract quantities from image data, in the first case over space and, in the second, over space and time. At least in abstract terms, both have similar geometric structure.

The strength of the analogy derives from a detailed consideration of the physical situations underlying both early orientation selection and early optical flow. In each case the physical structure partitions into two types; (I) those for which detailed information is available and useful, such as the bounding contour of a moving object (say, a rabbit being chased by a fox); and (II) those for which such information is neither available nor very useful, such as the structure of the forrest flowing back past the fox. Since physical structure projects to image (retinal) structure, these configurations defined what we refered to as Type I and Type II patterns.

The argument that Type I and Type II patterns require different processing was supported primarily by orientation selection. Both psychophysical evidence and a computational model were put forth to express the differences. Stated briefly, there are many formulations of orientation selection in terms of matching problems, and we used one based on the match between the expected responses of operators (to known patterns of dots) with those responses obtained from observed (i.e., un-interpreted) dot patterns. The aim is to find that pattern which, if it were actually present, would give a response as close as possible to those observed from the uninterpreted pattern. This expected pattern is then taken as the interpretation. Within this framework, then, the difference between Type I and Type II processing can be precisely specified as a difference in the spatial region over which matching is accomplished: for Type I situations, matching is point-to-point, while, for Type II situations, it is point-to-region. Such abstract processing differences would lead, through any number of implementations, to the psychophysical differences already noted.

The case for optical flow is not worked out to this degree, but there is both physical and psychophysical evidence for the distinction. Experiments are currently in progress in our laboratory to detect and to quantify it precisely, and preliminary results favour the Type I/II distinction in optical flow strongly. Whether such analogies hold for more abstract processing than those considered in this paper still remains to be determined, although, as processing becomes more abstract, process may not reflect physical structure so homomorphically.

ACKNOWLEDGEMENTS

This research was supported by the Natural Sciences and Engineering Research Council. I thank Norah Link and Sheldon Davis for comments.

REFERENCES

Beck, J., and Schwartz, T., Vernier acuity with test dot objects, Vision research, 1978, 19, 313-319.

Braddick, O., A short range process in apparent motion, Vision Research, 1974, 14, 519-527.

Braddick, O., Low level and high level processes in apparent motion Phil. Trans. Roy. Soc. Lond., B, 1980, 290, 137-151.

Glass, L., Moire effect from random dots, Nature, 1969, 243, 578-580.

Glass, L., Physiological mechanisms for the perception of random dot Moire patterns, in H. Haken (ed.), Proc. of Synergetics in Physics and Biology, Springer, Berlin, 1980.

Hubel, D., and Wiesel, T., Receptive fields and functional architecture of monkey striate cortex, J. Physiol. (Lond.), 1968, 195, 215-243.

Hummel, R.A., and Zucker, S.W., On the foundations of relaxation labeling processes, IEEE Trans. Pattern Analysis and Machine Intelligence, 1983, PAMI-5, 267-287.

Lappin, J., and Bell, H., The detection of coherence in moving random-dot displays, Vision Res., 1976, 16, 161 - 167.

Lappin, J., Donner, J., Kottas, B., Minimal conditions for the visual detection of structure and motion in three dimensions, Science, 1980, 209, 717-719.

Longuet-Higgins, H. C., and Pradzny, K., The interpretation of a moving retinal image, Proc. Roy. Soc. Lond., B, 1980, 208, 385-397.

Marr, D., and Ullman, S., Directional selectivity and its use in early visual processing, Proc. Roy. Soc. Lond., B, 1981, 211, 151-180.

Morgan, M., and Ward, R., Conditions for motion flow in dynamic visual noise, Vision Res., 1980, 20, 431-435.

Rosenfeld, A., and Kak, A., Digital Image Processing, New York, Academic Press, 1982.

Ullman, S., and Hildreth, E., The measurement of visual motion, in Physical and Biological Processing of Images, O. Braddick and A. Sleigh (eds.), New York, Springer, 1983.

van Doorn, A., and Koenderink, J., Visibility of movement gradients, Biol. Cybern., 1982, 44, 167-175.

Wallach, H., and O'Connell, D., The kinetic depth effect, Journal Experimental Psychology, 1953, 45, 205-217.

Waxman, A., Kinematics of image flows, Proc. Image Understanding Workshop, Arlington, Va, June, 1983.

Wertheimer, M., Laws of organization in perceptual forms, Psych. Forsch., 1923, 4, 301-350; trans. in Ellis, W., A Source Book of Gestalt Psychology, Routledge and Kegan Paul, London, 1938, 71-88.

Zucker, S.W., Cooperative grouping and early orientation selection, in Physical and Biological Processing of Images, O. Braddick and A. Sleigh (eds.), New York, Springer, 1983.

Zucker, S.W., Early orientation selection and grouping: Evidence for Type I and Type II processes, T. R. 82-6, Computer Vision and Robotics Lab., McGill University, August, 1982b.

Zucker, S., and Lamauroux, P., Optimal curvature sensitivity in random dot Moire patterns, in preparation, 1983.

Zucker, S.W., Stevens, K., and Sander, P., The relation between brightness and proximity in dot patterns, A. I. Memo 670, Artificial Intelligence Laboratory, Massachusetts Institute of Technology, May, 1982; Perception and Psychophysics, in press.

MOTION PERCEPTION: SECOND THOUGHTS ON THE CORRESPONDENCE PROBLEM

by George Mather and Stuart Anstis

York University, Downsview, Ont.

If two pictures are presented rapidly one after another in superimposition, like two successive frames of a movie, then apparent motion can be seen between them, provided that the pictures are sufficiently alike. How similar must they be? The visual system must decide which features in the two pictures correspond to each other, so that they can be matched up and linked into apparent motion (Anstis 1970, 1978). Ullman (1979) has dubbed this visual task the 'correspondence problem'. This paper discusses the correspondence problem, and suggests that features in two pictures are linked up on the basis of: 1. Proximity: Other things being equal, a point in one picture tends to be seen as moving toward the nearest (suitable) point in the next picture. 2. Similarity: Other things being equal, a feature in one picture tends to be seen as moving to the most similar feature in the next picture. We shall discuss what 'similarity' can mean, and what happens when these criteria of similarity and proximity are placed in experimental conflict.

SHAPE INFORMATION WITHIN PICTURES AND MOTION INFORMATION BETWEEN PICTURES.

We shall first review some of the literature on the correspondence problem (sections 1-3) and then describe some demonstrations (section 4), as follows:

Motion: Representation and Perception
N.I. Badler and J.K. Tsotsos, Editors
Published by Elsevier Science Publishing Co., Inc.

1. Extraction of information from a stationary picture: What information is available to the visual system in any single interval of time which can form the basis for perceiving motion over many successive intervals of time? In other words, what are the inputs to the motion-processing system?

2. Extraction of correspondences between successive pictures: Given the information available in a static picture, which information is actually used by the visual system to extract apparent motion? Which of the available inputs does the visual system use, how are these inputs manipulated, and what information is output as a result?

3. Braddick (1974) suggested that correspondences can be established between pictures by means of two different processes. The short-range process operates on short spatial jumps between luminance edges at short time intervals, whilst the long-range process operates on large spatial jumps between edges defined by luminance or texture at longer time intervals. Evidence for this theory will be discussed.

4. Correspondence determined by proximity vs. similarity. This discussion will be based on a metaphor.

1. SEEING SHAPES WITHIN A PICTURE.

The earliest stages in the extraction of features from stationary images can be grouped into three levels. At the earliest level, in the eye and lateral geniculate nucleus, gray-level intensity changes in the image are coded by receptive fields which are small in extent (less than 9 min arc in diameter: Wilson and Bergen 1979) and have concentric excitatory and inhibitory sub-fields. These receptive fields are essentially differential operators, extracting information about local changes in image intensity (Marr 1982).

Converging evidence from several sources suggests that at the next level, namely in the visual cortex, the visual system has a small set of elementary tokens which it can use to describe the features present in the image. Marr (1976, 1982) called them 'spatial primitives' and Julesz (1981) has called them 'textons'. These tokens are congruent with the visual elements isolated psychophysically, as reviewed by Braddick (1980), and with the preferred stimuli of the neural detectors found by neurophysiologists (Hubel 1980: Zeki 1977). They include: oriented edges and bars, cyclopean depth, colour, brightness, line terminators, crossed lines, etc. These can be used by both the short-range and the long-range process (Braddick, 1974).

At the third level lie extended zones defined by groups of primitive features, for example, the shapes made visible in random dot stereograms by cyclopean depth, or the shapes made visible in Julesz' texton demonstrations, or shapes defined by equiluminous colour. These can be used only by the long-range, not the short-range process.

Higher levels of processing recover information about the solid structure and identity of the objects being viewed (Marr 1982).

2. SEEING MOTION BETWEEN SUCCESSIVE PICTURES

Braddick (1974) suggested that there are two processes responsible for seeing motion between pictures, and his theory has become widely accepted. One process called the 'short-range' (SR) process is assumed to operate over short intervals in time and space with interstimulus intervals (ISIs) below 100 ms and displacements smaller than 15 min arc. The other process, called the 'long-range' (LR) process, is supposed to catch movements beyond the range of the SR system and can operate over longer intervals in space and time (ISIs of 300 ms or longer, and displacements up to many degrees of arc.) The SR process might be implemented in hard-wired motion sensing cells, whereas the LR process might be implemented in higher level, more central processes.

The two processes are thought to differ in their correspondence requirements, i.e. the stimulus features in each interval of time which they use to detect the movement. The SR process uses inputs from elementary features (spatial primitives or textons), and requires strict, elementary correspondences across successive pictures, e.g. luminance edges of the same orientation and luminance polarity. SR motion may actually constitute a texton or primitive itself. Mechanisms which might code SR motion have been proposed by Reichardt (1961), Barlow and Levick (1964) and Marr and Ullman (1981).

The LR process, on the other hand, can operate both at and beyond the texton level, on the shapes defined by texton zones regardless of the particular defining features in each picture. The correspondence requirements are supposed to be very loose and flexible, perhaps because false positives are preferable to false negatives, i.e. misses.

3. EVIDENCE FOR THE TWO-PROCESS THEORY.

a. Braddick (1974) examined responses to random-dot kinematograms and found that motion detection based on local pairing of dots in successive pictures was limited spatially to 1/4 deg. arc, and temporally to ISIs well below 100 ms. Dichoptic presentation of the stimuli, or bright fields presented during the ISIs, were also found to impair motion detection (see also Braddick 1973). Baker and Braddick (1982a, b) confirmed that the spatial limit is determined by absolute, not relative displacements, and that displacement in terms of retinal angle, not in terms of multiples of picture elements (pixels), was the critical parameter. Classical studies of apparent movement using isolated stimulus elements find no such severe restrictions (Neuhaus 1930). So Braddick proposed that the process operating in the kinematogram experiments was different from that operating in the classical experiments on apparent motion.

b. Banks and Kane (1972) measured motion aftereffects produced by viewing circles undergoing contraction in small, discrete steps. They found aftereffects only for the smallest displacements (about 10 min arc or smaller.) This result has been interpreted in terms of adaptable, SR detectors operating over small displacements, and non-adaptable LR mechanisms operating over larger displacements. However, Pantle (personal communication) has pointed out that Banks and Kane's stimuli consisted of widely spaced thin lines, which contain many Fourier harmonics (high spatial frequency components). He repeated their observations using sinusoidal gratings, and found that any jump size up to half a spatial period could give rise to a motion aftereffect, even for gratings of low spatial frequency whose half-period was much greater than 10 min arc. Pantle's finding knocks out one of the major pieces of evidence for Braddick's theory.

c. Ramachandran, Madhusudhan and Rao (1973) used stimuli in which a shape was defined by a texture boundary and had a different location in two successively viewed pictures (Figure 1b). Crucially, none of the individual texture elements used in one picture to define the shape were also present

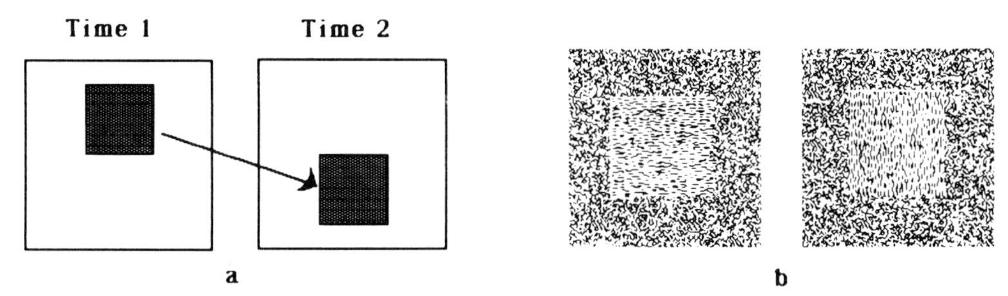

Figure 1. a, the square differs in luminance from its surround. For short jumps (<10-15 min arc) the short-range motion system senses the displacement of the luminance edges (Braddick, 1974).

b, the square differs from the surround only in texture, not in mean luminance. Only the long-range motion process can use such texture edges to sense motion (Ramachandran, Madhusudhan and Rao, 1973).

in the other picture -- they were completely uncorrelated. The apparent motion was mediated by a system, identified with the LR system, which can handle shape corespondences regardless of how these shapes are defined. Similar phenomena have been reported by Pantle (1973).

d. A small change in timing can radically change the apparent motion seen in a display. Pantle and Picciano (1976), Braddick and Adlard (1978) and Pantle and Petersik (1980) report experiments on a display containing isolated stimulus elements. It consisted of three elements (dots or lines) in a row separated by approximately 1 deg. arc and shown alternately in two positions. The positions of the two rightmost elements in one frame coincided with the position of the two leftmost elements in the other frame (Figure 2). Two percepts could be elicited by this stimulus. With short (20-40 ms) ISIs the outer element appeared to jump around or over the two apparently stationary inner elements ('element motion'). But with long ISIs (100-120 ms) the whole group of three elements was seen to jump back and forth ('group motion'). Braddick and Adlard (1978) interpreted these results as follows: The SR system here gives rise to the element motion

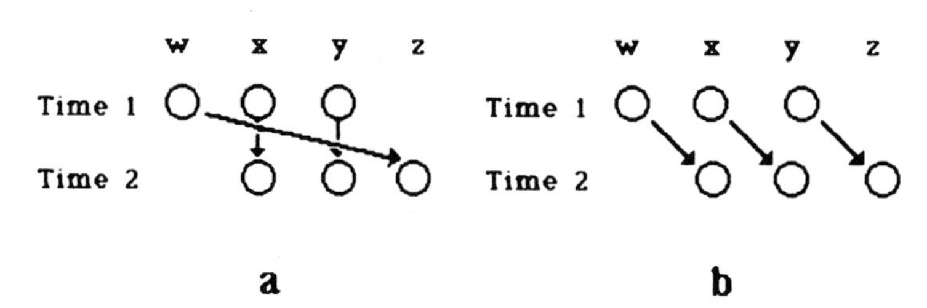

Figure 2: Ambiguous apparent motion. Three dots w, x, y at time 1 were replaced by three dots x', y', z at time 2. Rows of dots were superimposed, not one above the other. a, at inter stimulus intervals of 20-40 ms the two inner dots appeared to keep still while the outer dot jumped back and forth ('element motion'). b, at inter stimulus intervals of 100-120 msec the trio of dots jumped back and forth as a group ('group motion') (Pantle and Picciano, 1976).

percept by signalling 'no motion' of the inner elements, and the LR system causes the jumps of the outer element. The LR system alone gives rise to the group motion percept, by registering the three elements as a group and signalling their motion as one unit across 1 deg arc. Several experiments have taken stimuli which normally produce good element motion (ISIs below 20 ms), presumed to be mediated by the SR system, and perturbed them in ways expected to be damaging to the response of SR detectors, e.g. by using dichoptic presentation of the alternating frames (Braddick and Adlard 1978), or by separating them with a uniform bright field (Braddick and Adlard 1978), or by introducing tilt perturbations in the elements across frames (Pantle and Petersik 1980). All such perturbations greatly increase the percentage of group motion percepts reported, usually to well over 50%. These findings, along with the explanation for element motion, force the conclusion that the LR system can, in fact, respond over the same range of stimulus parameters as the SR system, and more, but the SR system dominates the percept if the stimulus conditions allow it to operate.

e. The SR process always cares about the polarity of luminance edges but the LR system does not. An experiment in our laboratory (Anstis and Mather, 1985) employed the following stimulus: Two adjacent dots, which were aligned vertically, alternated in successive frames with another pair of dots which were aligned horizontally (Figure 3). The four dots were positioned so that they formed the corners of an imaginary diamond. One dot in each frame was black and the other was white, on a mid-gray background. If the ISI was short and the displacement of the dots was small (4 min arc) motion was invariably seen only between dots of the same luminance polarity (white to white and black to black), as shown by the solid arrows in Figure 3a. As the displacement was increased to 2 deg arc, at the same ISI, the percentage of "same polarity" responses declined to approximately 65%. If a uniform gray field was inserted as an ISI for 117 ms, the "same polarity"

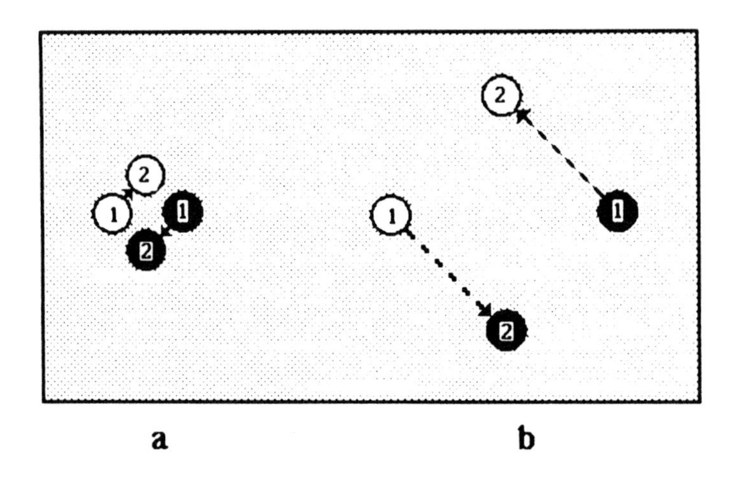

Figure 3: The two dots labelled '1' were presented at time 1, followed at time 2 by the two dots labelled '2'. Each pair comprised a white dot and a black dot. a, when the dots were close together and at short ISIs, white jumped to white and black to black (solid arrows). b, when the dots were further apart and at long ISIs, white jumped to black and black to white (dotted arrows) on about one trial in every three. Conclusion: in solving the correspondence problem, the short-range process takes luminance and contrast into account, the long range process largely does not.

responses occurred on only 75% of trials, even for the smallest displacement, and again fell to approximately 60% for the largest displacement. We interpret these results as follows: Over short temporal and spatial intervals, apparent motion is mediated by the SR system, which is strongly dependent on correspondences between luminance polarity and so produces a high percentage of "same polarity" responses. Over long temporal and spatial intervals the LR system operated and since it is not so crucially dependent on luminance polarity, there is a tendency for a large proportion of "different polarity" responses to occur. Notice that a large number of LR-type responses CAN occur for short ISIs provided that the spatial interval is long, or for short spatial intervals provided that the ISI is long. In other words, the same stimulus parameters covered by the

SR system are covered by the LR system. But if both systems can detect a particular stimulus, the SR response appears to take priority over the LR response. Both our four-dot experiment, and the three-dot experiments reviewed in section 3d above, point to this conclusion.

All this evidence suggests that the SR and LR systems are not specialised to cover different portions of the spatiotemporal spectrum in a complementary fashion, but overlap considerably. Indeed, natural stimuli almost always move smoothly and continuously rather than jumping in large discrete steps, since there are no stroboscopes in nature. Why, then, do we need two systems at all? Perhaps they serve different functions: the SR system may keep us informed about motion itself, while the LR system may be more concerned with the recovery of three-dimensional structure (Ullman 1979) and the preservation of object identity (Marr 1982).

4. LOVE AND CORRESPONDENCE: DEMONSTRATIONS OF PROXIMITY VS. SIMILARITY.

To illustrate the roles of proximity and similarity in AM, we shall use the metaphor that picture elements or dots in successive pictures which are linked into apparent motion are like people who are linked when they fall in love and get married. Let us pretend that each dot in the first picture is a male person and each dot in the second picture is a female person. How do they pair off? What rules determine who marries whom? We can think of the visual system as a matchmaking computer, pairing off competing candidates according to their compatibility.

The sociologist Bossard looked at the addresses on 5000 marriage licences in Philadelphia in the 1930's and found that 7 % of couples who got married lived within 1 city block of each other, and 34 % lived within 4 blocks. Clearly proximity is an important factor in finding a marriage partner. But it is also true that 'like likes like': you tend to marry someone whose background, tastes and education are similar to your own. Some visual demonstrations point out the relative importance of proximity and similarity in apparent motion:

a. A single dot at time t1 is followed by a single, displaced dot at time t2. Result: one perceives a single jumping dot in apparent motion. The two dots are paired off visually because there is no competition or rivalry (Figure 4a). A single man and woman on a desert island have no choice and no rivals to distract them.

b. A dot appears at t1, followed by a dot shifted upward at time t2, followed by a dot shifted to the right of the first dot at time t2. Result: horizontal motion to the right is no longer seen. Instead the dot appears to jump first upward, then obliquely down to the right (Figure 4b). The horizontal link between dots 1 and 3 is broken because it is pre-empted by the links 1-2 and 2-3. Moral: in love, timing is important. You must be not only in the right place but also at the right time.

c. But the correspondence just described (Figure 4b) is itself dependent on getting the timing right. Suppose the dots are now flashed in the sequence 1-2-3-2-1-2-3-2.... (Figure 4c). If the alternation is too rapid the oblique motion links 1-2 and 2-3 break down and again one sees horizontal motion between 1 and 3. Note that the dot at position 2 is flashed twice as often (at times t2 and t4) as the dots at positions 1 and 3. At slow rates (SOA > 160 msec) motion is still seen back and forth along the V-shaped trajectories which are marked with solid arrows in Figure 4b, c. At rapid rates, however, these motion paths disappear (dotted lines in Figure 4d) and one spot appears to jump back and forth horizontally between positions 1 and 3, while the other dot (shaded grey in Figure 4d) appears to flicker in place. Even at slow rates, percept Figure 4c is seen at first, but prolonged inspection of the apparent motion produces adaptation

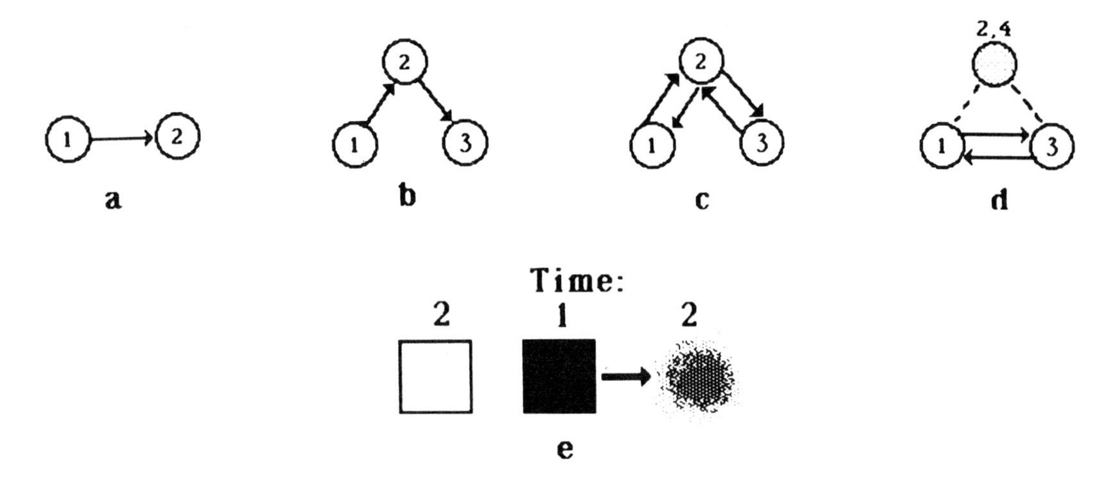

Figure 4: Correspondence in apparent motion depends on timing. The number inside each spot refers to the time at which it was presented.

a, a spot jumps from left to right.

b, same stimulus as a but now an extra spot is interpolated and the previous horizontal motion disappears and is replaced by a V-shaped motion. Hiding the middle dot restores the horizontal motion.

c, same as b except that the spot jumped repetitively through the positions 1-2-3-2-1-2-3-2..... At slow rates motion was still seen back and forth along the V-shaped trajectory marked with solid arrows, much as in b.

d, at rapid rates these motion paths disappeared (dotted lines) and one spot appeared to jump back and forth horizontally between positions 1 and 3, while the other dot (shaded grey) appeared to flicker in place.

e, Low spatial frequencies can determine correspondence (Ramachandran, Ginsburg and Anstis, 1983). When a solid square at time t1 was followed by an outline square on the left and a round blob on the right, both at time t2, motion was seen toward the blob, which shares low frequency content with the solid square.

which affects rapid alternations more than slow ones (Anstis, Giaschi and Cogan, 1985), so the percept changes to Figure 4d. This shows that solutions to the correspondence problem are affected by stimulus timing and also by the state of adaptation of the motion system.

d. Correspondences in an ambiguous situation can be affected by what has happened just before and just after. When two dots were flashed up at the northwest and southeast corners of an imaginary square, and replaced by a pair of dots at the northeast and southwest corners, one is equally likely to see horizontal or vertical apparent motion (Figure 5). But when this display is embedded in a sequence of dots moving horizontally along two rows, the horizontal motion predominates (Ramachandran and Anstis, 1983). Motion is seen to continue along the direction of the embedding sequence. We called this effect "visual inertia". Moral: inertia can keep people linked who might otherwise have separated. Ask any married couple.

e. When correspondences compete the outcome can also depend on the relative strengths of the correspondences as defined by their luminance contrasts. When a black and a white spot exchange places, so that the black spot suddenly becomes white and the white spot became black, one can ask a subject whether he sees a black spot jumping up or a white spot jumping down (Figure 6). We have found (Anstis and Mather, 1985) that the percept depended on the luminance of the background; on a light

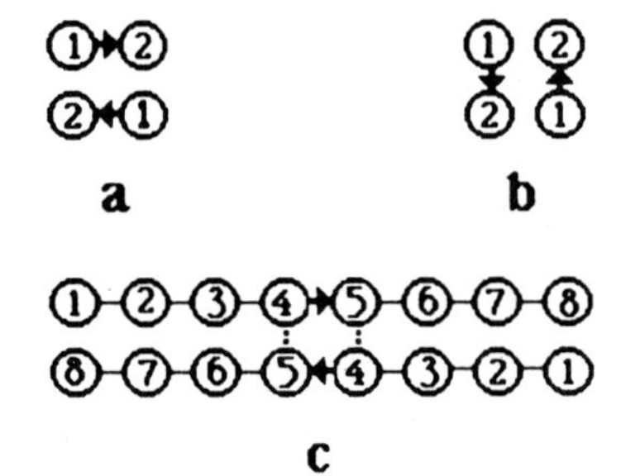

Figure 5. Visual momentum. a, b: The alternating dot pairs provide an ambiguous stimulus which can be seen as moving either horizontally (a) or vertically (b).

c, when these same dots are embedded in two rows of dots jumping horizontally, to the right in the top row and the left in the bottom row, horizontal apparent motion is nearly always seen (Ramachandran and Anstis, 1983).

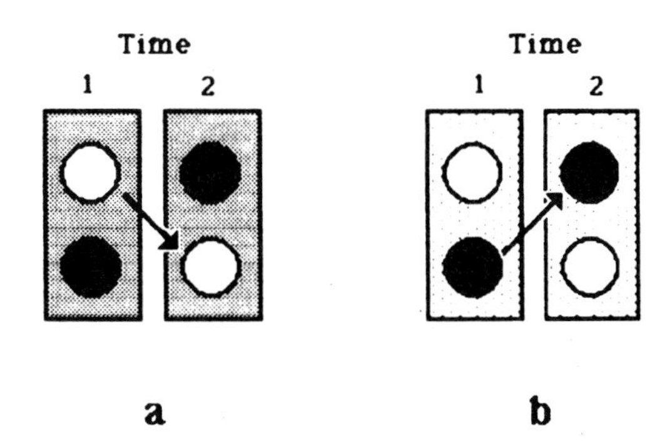

Figure 6. When a black and white spot exchanged places the perceived motion depended upon the surround luminance (Anstis and Mather, 1985).
a, on a dark background the white spot appeared to move, since it had a higher contrast relative to the surround than the black spot did.
b, on a light background the black spot appeared to move.

surround it was the black spot that appeared to move, and on a dark surround it was the white spot. The spot which differed most from the background, and thus had the higher contrast, seemed to move.

f. Proximity can beat similarity when they are placed in conflict. A row of people standing 1 meter apart were photographed. Each person then moved 90 cm to the right, and the row was re-photographed. Result: when the two pictures were presented in succession, the people appeared to jump 10 cm to the left, and change identities, rather than appearing to jump 90 cm to the right and retain their identities, as they had actually done.

g. The motion system takes account of Fourier components, and sometimes similarity can be defined in terms of Fourier components. A solid square at time t1 was replaced at time t2 by two objects placed symmetrically to left and right. On the left was an outline square of the same size and shape, but hollow. On the right was a round, filled in blob, without a sharp square contour (indeed, without any sharp contour) round it (Figure 4e). Result: motion was seen from the solid square to the solid blob (Ramachandran, Ginsburg and Anstis, 1983). This suggests that motion is carried only by certain properties of the stimuli, namely the low spatial frequencies, which the solid square shares with the blob, and not by the high spatial frequencies, which are shared by the solid and the outline square. Moral: Love is not blind. Individuals prefer partners with features that attract them, even when alternative partners with less attractive features are equally available.

ACKNOWLEDGEMENTS.

This work was supported by Grant # A 0260 to SMA from the Natural Science and Engineering Research Council of Canada (NSERC).

REFERENCES

Anstis S. M. (1970) Phi movement as a subtraction process. Vision Res. 10, 1411-1430.

Anstis S. M. (1978) Apparent movement. In Handbook of Sensory Physiology (ed. R. Held, H. W. Leibowitz and H-L. Teuber) pp. 655-673. Springer-Verlag, New York.

Anstis S. M., Giaschi, D. and Cogan, A. (1985) Adaptation to apparent motion. Vision Res, 25, 1051-1062.

Anstis, S. M. and Mather, G. (1985) Effects of luminance and contrast on direction of ambiguous apparent motion. Perception, 14, 167-180.

Baker C. and Braddick O. L. (1982a) Does segregation of differently moving areas depend on relative or absolute displacement? Vision Res. 22, 851-856.

Baker C. and Braddick O. L. (1982b) The basis of area and dot number effects in random dot motion perception. Vision Res. 22, 1253-1260.

Banks W. P. and Kane D. A. (1972) Discontinuity of seen motion reduces the visual motion aftereffect. Perception & Psychophysics 12, 69-72.

Barlow H. B. and Levick W. R. (1964) Retinal ganglion cells responding selectively to direction and speed of image motion in the rabbit. J. Physiol. (Lond). 173, 377-407.

Braddick O. J. (1974) A short-range process in apparent motion. Vision Res. 14, 519-527.

Braddick O. J. (1980) Low-level and high-level processes in apparent motion. Phil. Trans. Roy. Soc. Lond. B 290, 137-151.

Braddick O. J. and Adlard A. J. (1978) Apparent motion and the motion detector. In Visual psychophysics: its physiological basis (ed. J. Armington, J. Krauskopf and B. R. Wooten), pp. 417-426. Academic Press, New York.

Hubel D. H.(1979) The Visual Brain. Scientific American, 241, 44-53.

Lappin J. S. and Bell H. H. (1976) The detection of coherence in moving random-dot patterns. Vision Res. 16, 161-168.

Marr D. (1976) Early processing of visual information. Phil. Trans. R. Soc. Lond. B 275, 483-524.

Marr D. (1982) Vision. W. H. Freeman, San Francisco.

Marr D. and Ullman S. (1981) Directional selectivity and its use in early visual processing. Proc. R. Soc. Lond. B 211, 151-180.

Neuhaus W. (1930) Experimentelle Untersuchung der Scheinbewegung. Archiv fur die gesamte Psychologie, 75, 315-458.

Pantle A. (1973) Stroboscopic movement based upon global information in successively presented visual patterns. J. opt. Soc. Am. 63, 1280 A.

Pantle A. and Petersik J. T. (1980) Effects of spatial parameters on the perceptual organization of a bistable motion display. Perception & Psychophysics 27, 4, 307-312.

Pantle A. and Picciano L. (1976) A multistable movement display: evidence for two separate motion systems in human vision. Science 193, 500-502.

Ramachandran, V. S. and Anstis, S. M. (1983) Extrapolation of motion path in human visual perception. Vision Res 23, 83-85.

Ramachandran, V. S., Ginsburg, A. P. and Anstis, S. M. (1983) Low spatial frequencies dominate apparent motion. Perception, 12.

Ramachandran V. S., Madhusudhan, V. R. and Vidyasagar, T. S. (1973) Apparent movement with subjective contours. Vision Res. 13, 1399-1401.

Reichardt, W. (1961) Autocorrrelation, a principle for the evaluation of sensory information by the nervous system. In: Rosenblith, W. (Ed): Sensory Communication, pp. 303-317. MIT Press, Wiley, New York.

Ullman S. (1979) The interpretation of visual motion. MIT Press, Cambridge, Mass.

Wilson H. R. and Bergen J. R. (1979) A four mechanism model for spatial vision. Vision Res. 19, 19-32.

Zeki S. M. (1977) Colour coding in the superior temporal sulcus of rhesus monkey visual cortex. Proc. R. Soc. Lond. B 197, 195-223.

The Representation and Perception of Geometric Structure in Moving Visual Patterns [1]

Joseph S. Lappin

Vanderbilt University

Theoretical Background: The visual measurement of space and time.

A principal function of vision is to measure the environment. Virtually all visual tasks, involving either visual-motor coordination or recognition of objects, demand metric information -- about three-dimensional (3D) spatial distances between points within an object or between separate objects, about the motion of one object relative to another, or about the position and motion of the observer in relation to the environment. In guiding a car through rushing city traffic, in returning a fast-moving tennis ball, in watching the changing form of ocean waves rolling over a beach, or in performing any of a million everyday activities so routine we scarcely think about them, vision provides remarkably detailed geometric information about the structure, location, and motion of environmental objects and events.

The precision and speed with which such geometric information can be visually acquired are suggested by the skills of athletes and other rapidly moving animals in coordinating motor activities with the space-time positions and trajectories of moving objects. Comparable achievements are suggested by the object recognition skills of many animals in discriminating and identifying subtly different geometric shapes in widely varying environmental contexts from continually changing viewing perspectives. Such geometric representations of environmental structure are fundamental in the control of rapidly moving animals and machines, but the specific mechanisms by which they can be acquired

[1]Supported by NSF Grant BNS 81-12473

Motion: Representation and Perception
N.I. Badler and J.K. Tsotsos, Editors
Published by Elsevier Science Publishing Co., Inc.

remain poorly understood.

Although vision is seldom considered as a measurement process, this seems a proper and useful conception: As in other more familiar cases of physical measurement, some relational structure of empirical objects must be mapped into a corresponding representation in another relational system. In physical measurement the representation is in the relational system of real numbers, but in visual measurement the representation is only tacitly revealed by the behavioral capabilities for form discrimination and visual-motor control. (Such behavioral capabilities can be precisely and quantitatively evaluated by psychophysical tests, however.) In both cases the formal problem in constructing a theory of the measurement process involves a specification of (a) the conditions under which qualitative observations of empirical relations can properly be given a quantitative representation, (b) the permissible transformations of the empirical objects and events under which their relational structure is preserved, and (c) the permissible transformations of the representational structure (or behavioral measures) under which the empirical relational structure is preserved.

The psychophysical indicators of perceived geometric relations employed in the present study are measures of the acuity for detecting the displacement of a spot of light from a position specified in relation to the geometric structure of some visual pattern. Thus, the acuity for such perturbations of position provides an evaluation of the precision of the visual perception of the geometric structure. More specifically, the acuity for 3D distances among points moving as a rigid structure in 3D space is evaluated in stimulus displays of motion parallax. A basic experimental question is, what is the precision with which motion parallax provides for the perception of 3D spatial relations? The answer to this question bears on the origins of perceived geometric structure in visual stimulus patterns.

Before turning to specific experimental methods and findings, it is helpful to consider two alternative types of procedures for measuring and representing geometric structure. These two different types of measurement correspond to two alternative definitions of the geometric information initially available to the visual system. Depending on one's assumptions about the initial representation of geometric structure in optical patterns, one is led to very different conceptions of the perceptual or computational processes necessary for determining the geometric structure of environmental objects and events.

The most familiar type of measurement procedure is _extensive_ measurement, in which a numerical value corresponding to some extensive property of an observed object is determined by comparison of the object with an _extrinsic_ reference system consisting of a concatenated sequence of standard units. The length of a board, for example, might be measured by aligning a tape measure in parallel with it and counting the number of units (e.g., centimeters) that lie between the two ends of the board.

A corresponding extensive measurement procedure is usually adopted in defining the spatial and temporal structure of optical data patterns for biological and machine vision systems. The spatial representation of data for a machine vision system is usually specified by reference to a two-dimensional (2D) Cartesian coordinate system corresponding to the 2D mosaic of photosensitive elements. The position of any given point of light is defined by the particular photosensitive element it excites, with the position of the photosensor taken as defined a priori with great precision. Similarly, the spatial structure of stimulus patterns for biological vision systems is usually specified as a 2D Cartesian coordinate system that is presumed to be determined by the spatial arrangement of photoreceptors on the retinal surface. In both cases the spatial representation of optical patterns is 2D and given a priori by reference to a coordinate framework that is extrinsic to the optical patterns being measured. Perception of the 3D structure and motion of these patterns then necessarily requires interpretive processes for inferring the 3D structure and motion corresponding to the changing optical patterns defined in this framework of 2D coordinates.

An alternative measurement procedure is what might be called <u>intensive</u> or <u>intrinsic</u>, where structural relations are specified by intrinsic structural constraints or symmetries in the optical pattern associated with some particular form. A circle, for example, can be specified as a set of points in the plane that are equidistant from a given point (in contrast to the functional equation in Cartesian coordinates, $x^2 + y^2 = r^2$). A well-known intrinsic measure determined by the structure of a circle is the so-called "dimensionless constant", π, which is defined by a ratio of two extensive measures and has a value that is independent of the unit of measure. Physical laws provide a variety of such "dimensionally invariant" intrinsic measures, and indeed it is frequently conjectured (e.g., Luce, 1978) that all laws of nature satisfy the property of being dimensionally invariant. In any case, the geometric structure of natural objects and events is often associated with the symmetries produced by constraints on the space-time patterns of positions of the components of an object undergoing transformation. The spherical surface of a lens, for example, is produced by rubbing an initially rough piece of glass with an irregular sequence of translations and rotations; the smooth spherical surface evolves as a structure symmetrical under the group of translations and rotations.

The potential for applying intrinsic measurement procedures to the representation of visual patterns arises in part from geometric constraints on the distribution of space-time positions in optical patterns undergoing perspective transformations. That is, the transformations may be used to measure the structures that remain invariant under the transformations. Psychophysical work on the "kinetic depth effect" and motion parallax has made increasingly clear in recent years that small amounts of motion in geometrically simple displays of only a few points of light are sufficient for human observers to obtain apparently accurate perceptions of 3D structure and motion. These results are often conceptualized as involving the operation of an interpretive process that translates the sequence of 2D coordinate positions of a given point at successive intervals of time into a 3D representation (e.g., Ullman, 1979). This conception, however, implicitly assumes that the position of each point in the optical pattern is precisely defined in 2D space and in time. Alternatively, the primary visual task might be regarded as the measurement of space-time distances between points just so that the overall pattern has a coherent geometric structure and motion. The coherence or symmetry of the pattern over time may often tightly constrain the possible distance measures that can yield an internally consistent representation of the geometric structure. The primary property detected by the visual system may be the symmetry of the pattern and the measures of space and time may be derived from it.

The distinction between these extrinsic and intrinsic standards of measurement is analogous to the distinction between Newtonians and relativistic approaches to measurement in physics. The classical Newtonian conception was that the framework of 3D space and time was given a priori and that lawful relations and symmetries were to emerge from the regularities of extensive measures within that framework. The relativistic approach adopted by Einstein, however, begins with the symmetry as fundamental and derives the measures of space and time to conform with these constraints. By rough analogy with the logic of special relativity, visual measures of distance in space-time may be derived from their symmetry under perspective transformation.

Consider, for example, the simple case of a rigid object rotating in 3D space and seen from two slightly different perspectives at two consecutive moments in time. Let Δ designate the 3D spatial distance between any two points on the object and let $\underline{\Delta}'$ designate the transformation of this quantity as produced by viewing from another perspective at another moment in time. By definition of rigid motion in space and of Euclidean distance, the two quantities are equivalent, even though they may be measured from different coordinate systems: $\Delta = \underline{\Delta}'$. The retinal distance between two points, however, does not generally remain invariant under motion in 3D space. Suppose that $\underline{r} = (\Delta x^2 + \Delta y^2)^{\frac{1}{2}}$ is the projected retinal distance between two points, and that $\underline{r}'$ is the result of a perspective transformation. Then $\underline{\Delta}^2 = \underline{r}^2 + (\underline{\Delta z})^2$ and $\underline{r}^2 + (\underline{\Delta z})^2 = \underline{r}'^2 + (\underline{\Delta z}')^2$. If the visual system can detect the automorphism of an optical pattern undergoing perspective transformation, then this symmetry constraint on the distance relations provides an opportunity for measuring the 3D structure of the pattern. Indeed, experimental evidence indicates that vision is exquisitely sensitive to such automorphisms (e.g., Lappin, Doner, & Kottas, 1981; Ullman, 1979).

Extending the analogy with special relativity, several pieces of experimental evidence suggest that the visually detected symmetries include not only rigid structures but also flexible and moving structures that are undergoing perspective transformation. That is, the motion of a component substructure may be seen as invariant under the perspective transformations induced by the motion of a larger structure or of the observer (e.g., Johansson, 1973). Thus, if $m(t) = (r^2 + z^2)^{\frac{1}{2}}$ is the 3D distance moved by some element in *t* units of time, then we have the intrinsic relation

$$r^2 + z^2 - (m\{t\})^2 = 0$$

that must be satisfied by any perspective of the event. Accordingly, the symmetry constraint expressed by the functional equation

$$r^2 + z^2 - (m\{t\})^2 = r'^2 + z'^2 = (m\{t'\})^2$$

permits the determination of space-time distance measures that will satisfy the constraint imposed by the symmetry of the pattern over time. Several studies lend support to this type of formulation by demonstrating the predicted perceptual tradeoff between space and time (Burr & Ross, 1979; Burr, 1979; Lappin, Bell, Harm, & Kottas, 1975; McKee, 1981).

Experimental Strategy: Evaluate acuity for 3D distance under perspective transformation

Experimental questions raised by the preceding theoretical background involve the acuity with which 3D distances can be discriminated on the basis of visual information from perspective transformations and the relative acuity for 2D and 3D positions. Are the subjective impressions of 3D structure in moving patterns mainly just subjective, or can observers accurately detect differences in distance as defined by perspective transformations? Is the acuity for 3D position as good as for 2D position? If acuity for 3D position were actually better than for 2D position, then the perceived structure would seem to derive from intrinsic measures of 3D structure as constrained by the perspective transformations. If the acuity for 2D position were better than for 3D, then the primary visual information would seem to derive from the extensively defined positions on the 2D retinal surface. Equivalence of the two acuity tasks might be consistent with either theoretical conception of measurement.

Of course, the degree to which such experiments should be regarded as crucial tests of thest two alternative conceptions of visual measurement depends upon assumptions about the way in which behavioral discriminations are presumed to reflect sensory data on interpretations of sensory data. A first relatively weak assumption is that the resolution of the behavioral discriminations cannot exceed the resolution provided by the initial sensory data. Information cannot be increased by subsequent perceptual processing. A second and much stronger assumption would be that the resolution exhibited by various behavioral discriminations is monotonically related to the resolution available in the initially encoded sensory data. That is, the loss of information during perceptual processing might be assumed as occurring at the same rate for two or more geometric relationships. If for example, the acuity for 3D distances were greater than that for 2D distances, one might conclude that the initial sensory data involves the intrinsic measurement of 3D relationships rather than the extensive measurement of 2D positions, and correspondingly, that the perception of 3D structure does not derive from the interpretation of sensory data. This is a much stronger and less easily testable assumption, however.

In any case, evaluations of the acuity for geometric relationships and of the factors that influence this acuity should be informative about the representation and perceptual processing of geometric information in moving visual patterns.

The stimulus patterns employed in one series of experiments relevant to these issues are illustrated in Figure 1. Three collinear points of light on a computer-controlled CRT display were moved as if projected from a rigidly rotating rod in 3D space. A schematic illustration of the arrangement is given in sections A and B of Figure 1 and a time-lapse photograph of one of the displays is given in section C.

The observer's task was to detect (yes/no) the displacement of the middle point from the exact center of the distance between the two outer points. On half the trials the middle point was centered between the two outer points and on the other trials it was displaced a very small distance toward one of the outer points, maintaining the collinearity of the three points. As illustrated in Figure 1A, perspective projection yielded patterns in which a point centered in 3D space was not centered in the 2D projection. In most of the experiments the task was to detect displacements from the 3D center, but one experiment compared this 3D distance acuity with the acuity for 2D distance, where the middle point was positioned at or near the center of the 2D distance between the outer points as they were rotated in 3D space. Some experiments also compared this distance acuity with the vernier acuity for collinearity of the three points -- by comparing the detectability for displacements from the center that maintained the collinearity of the points with that for displacements in the perpendicular direction that altered the collinearity.

Principal result: Accurate discriminations of 3D distances.

The principal experimental finding was that human observers are extremely accurate in perceiving 3D distance. This main result is illustrated in Figure 2A, where a linear relation between detection accuracy and distance of displacement is shown to be unchanged by the average degree of tilt of the pattern

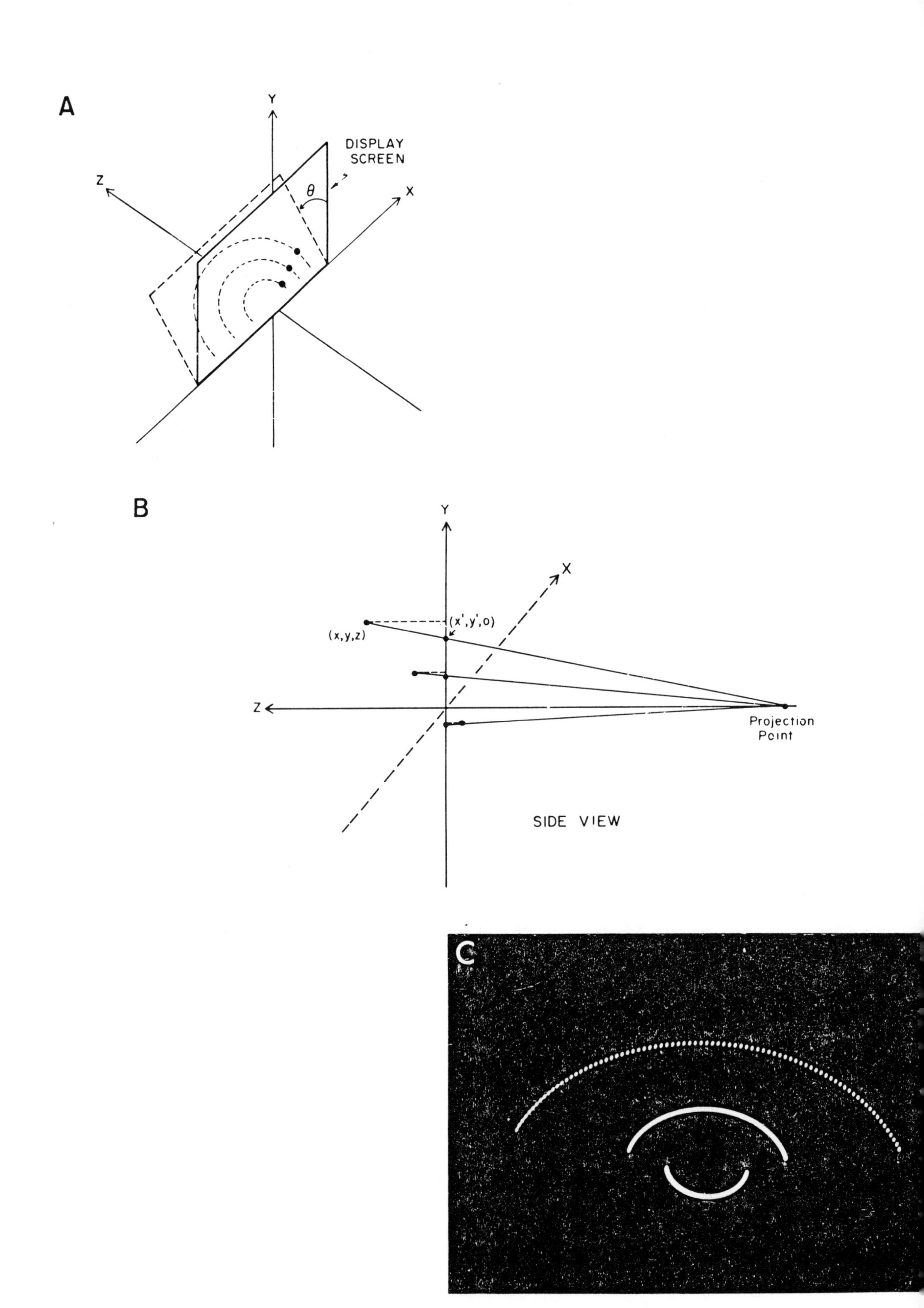
A
Y
DISPLAY
SCREEN
Z
X
θ
B
Y
X
(x',y',o)
(x,y,z)
Z
Projection
Point
SIDE VIEW
C

Figure 1. Stimulus patterns. (A) Schematic illustration of CRT displays of three collinear dots rotating as if in a plane tilted in 3D space. (B) Schematic side view of the perspective projection, drawn to scale for the first experiment. The distance of the perspective projection point from the center of rotation was 2.5 times the distance between the two end-points. (C) Time-lapse photograph of the one CRT display, showing the complete paths of three equally spaced dots in the 45° tilt condition. In the first experiment the rotation was through an angle of 160°. The visual subtense of the distance between the two end-points was 1.02° in the 0° tilt condition of the first experiment; with 30°, 45°, and 60° of tilt the projected distance between the end-points in the perspective displays was reduced to a minimum of .89°, .70°, and .48°, respectively, at the middle of the path. The displays were viewed monocularly in the dark from a distance of 94cm. The temporal duration of each rotation in the first experiment was .94 sec., and each display consisted of four successive rotations. In subsequent experiments, performance was unimpaired with the use of a single rotation through an angle of 120° over a temporal duration of .47 sec. Observers responded to each display as centered or displaced, and auditory feedback was given for each correct response.

in 3D space away from the 2D projective plane, unchanged by variability of the tilt, and unchanged by the perspectivity of the projection. Displacements as small or smaller than the center-to-center spacing between foveal photoreceptors (about 30 sec. arc) were detectable. Indeed, as indicated by results shown in Figure 2B and C, the appropriate measure of the displacement

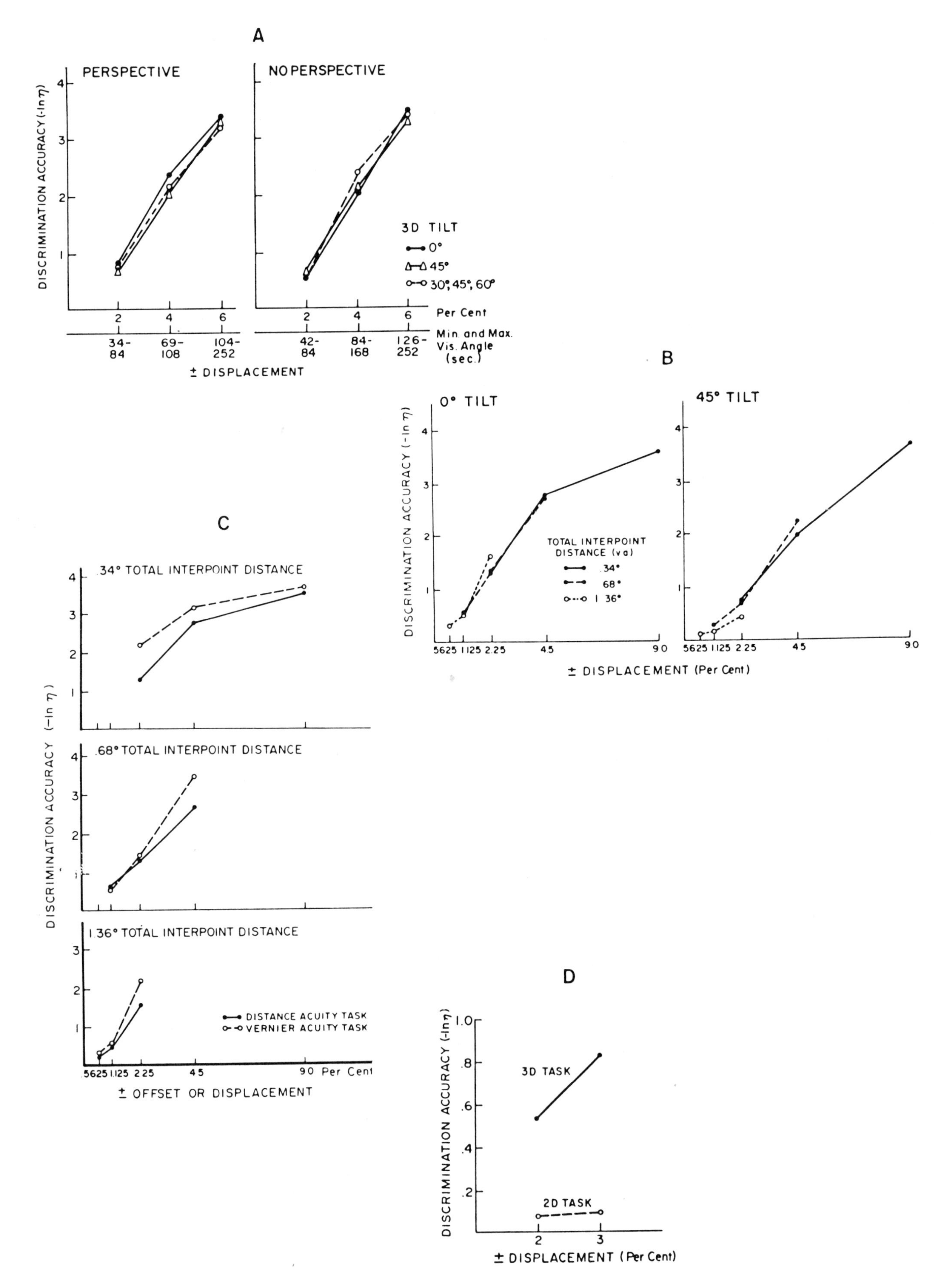
A
PERSPECTIVE
NO PERSPECTIVE
DISCRIMINATION ACCURACY (-ln η)
3D TILT
0°
45°
30°, 45°, 60°
Per Cent
Min. and Max. Vis. Angle (sec.)
± DISPLACEMENT
B
0° TILT
45° TILT
TOTAL INTERPOINT DISTANCE (va)
.34°
.68°
1.36°
± DISPLACEMENT (Per Cent)
C
.34° TOTAL INTERPOINT DISTANCE
.68° TOTAL INTERPOINT DISTANCE
1.36° TOTAL INTERPOINT DISTANCE
DISTANCE ACUITY TASK
VERNIER ACUITY TASK
± OFFSET OR DISPLACEMENT
D
3D TASK
2D TASK
± DISPLACEMENT (Per Cent)

Figure 2. Results of four experiments. (A) Results of the first and principal experiment, showing average detectability of a small displacement of the mid-point as a function of the percentage of the 3D distance between end-points. The range of absolute visual angles for a given percentage displacement is also given on the horizontal axis; the specific value within the range depended on the perspective, the tilt, and the position on the path of rotation. (B) Average detection accuracy as a function of the percentage displacement between end-points separated by three different amounts of visual subtense in the 0° tilt condition. In the 45° tilt condition, the distance between the end-points in the smallest display ranged from 0.2 to 0.25 deg. visual arc, with proportional expansion for the two larger displays. (C) Comparison of average detectabilities for vernier offsets from collinearity and for displacements that preserved collinearity and changed only distances between points. The three graphs are for three different amounts of separation between the end-points. The rotating patterns were presented only in the 0° tilt condition. (D) Comparison of average detectabilities for 3D and 2D displacements. The 2D displacement was a constant percentage of the 2D distance between the projected positions of the end-points, which were moving as if in 3D space and projected with perspective onto the 2D display screen. As in the previous experiments, the 3D displacement was a constant percentage of the 3D distance between the 3D positions of the end-points.

was the percentage of the distance between the two outer points, so that displacements of even smaller absolute magnitude could have been made detectable simply by reducing the distance between the two outer points.

The performance of this distance-acuity task was also found to be very similar to the vernier acuity for detecting displacements from collinearity of the three points, as shown in Figure 2C.

Finally, the detectability of displacements of the same magnitude from the 2D center were found to be not significantly better than chance.

The general result is that perspective transformations of even very simple patterns provide sufficient information for very precise visual measures of 3D spatial structure. The geometric information for these accurate visual discriminations seems to derive from the symmetry constraints on the space-time patterns undergoing perspective transformation.

THE PERCEPTION OF COHERENT MOTION IN TWO-DIMENSIONAL PATTERNS

Edward H. Adelson
RCA David Sarnoff Research Center

and

J. Anthony Movshon
New York University

Introduction

When one looks at a two-dimensional scene of moving objects, one can usually assign a velocity to each point in that scene with little effort. This suggests that some early visual processes are able to generate a two-dimensional velocity map using fast parallel computations. But it is not obvious how this should be done, and we are currently trying to understand how the human visual system does it.

If moving patterns were merely one-dimensional, the visual system's task would be easier. One dimensional motion can be signaled by cells such as those described by Barlow and Hill in rabbit (1963); these units compare responses of two adjacent regions on the retina at two successive moments of time. Combinations of such cells can signal the one-dimensional velocity of a moving edge, bar, or other contour.

But the retinal image is two-dimensional, and the problem becomes more complex than it first seems. Figure 1 illustrates a well-known visual phenomenon called the "barberpole effect," which is based on the inherent ambiguity of the motion of extended contours.

In fig. 1(a), the grating seems to move to the right when viewed through the horizontal window, even though the "true" motion of the physical grating is diagonally down and to the right. When a vertical window is used, as in fig. 1(b), the same physical motion leads to a percept of a downward motion.

The fact that one physical motion can be seen in different ways is to be expected because the observed motion of a stimulus like a grating does not, in and of itself, define the physical motion that produced it. The point is made clear in figure 2, where three different physical motions give rise to an identical pattern of stimulation when viewed through the window. In this case, all three patterns will be seen as moving down and to the right.

The ambiguity is not, of course, limited to the motion of gratings. Any extended pattern, such as an edge or a straight line, will offer the same problem. Consider the two moving diamonds shown in fig. 3. In fig. 3(a), the diamond moves down, yet a cell that "looks" at a local patch on the lower right edge will signal a contour moving down and to the right. In fig. 3(b), the diamond moves to the right, and the same cell will again signal contour motion down and to the right. So it is impossible to tell which way the diamond is moving by merely looking at one of its edges.

Marr and Ullman (1981) call this the "aperture problem," and have discussed how one can combine information from cells that signal only direction in order to constrain the underlying motion to lie within a limited range of directions. Fennema and Thompson (1979) showed that by taking advantage of local information about both the direction and speed of moving contours one could obtain a full solution for the object's motion using a Hough transform; we take a similar approach in the work we will discuss. (Note that this solution is only correct for the case of pure translation of a two dimensional pattern. For more complex transformations, such as rotations or deformations, more complex strategies are required (Horn and Schunck, 1981; Hildreth and Ullman, 1982)). An early discussion of the problem can be found in Wohlgemuth's (1911) classic work on the motion aftereffect; Hans Wallach (1935, 1976) did some very interesting experiments on its perceptual implications. It

Motion: Representation and Perception
N.I. Badler and J.K. Tsotsos, Editors
Published by Elsevier Science Publishing Co., Inc.

continues to interest both physiologists and psychologists today (Henry et al, 1974; Burt and Sperling, 1981; Adelson and Movshon, 1982).

There is both physiological and psychophysical evidence that the first stages of visual processing analyze the retinal image into a patchwork of localized one-dimensional components, which may variously be conceived as representing bars, edges, local Fourier components, Gabor functions, or what have you. In any event, such an analysis brings the aperture problem in with it from the very start. The visual system must go from the local motion of one-dimensional components to the percept of a single coherently moving pattern. In our discussion we will use the term "component motion" to refer to the motion of extended one-dimensional patterns such as lines, edges, and gratings; and we will use the term "pattern motion" to refer to the unambiguous two-dimensional motion of more complex patterns such as textures and objects

Resolving Ambiguity

While it is true that a single component motion is ambiguous, the ambiguity is limited to a single degree of freedom. Consider once again the motion of the lower right edge of a rightward-moving diamond; the edge is magnified in fig. 4. The set of object motions consistent with the observed edge motion are indicated by the arrows fanning out from the edge. This family of velocities may be depicted as a line in "velocity space," where any velocity is represented as a vector from the origin whose length is proportional to speed and whose angle corresponds to direction of motion. The family of motions consistent with the edge maps to a straight line in velocity space.

Figure 5 shows how a pair of such constraints can be used to determine the true motion of the diamond. Each of the edges in fig. 5(a) is associated with its own family of motions in velocity space, as shown by the lines in fig. 5(b). The lines intersect in a single point, and that point represents the object's motion.

The same analysis may be applied to combinations of graitngs, as shown in fig. 6. Two gratings move behind a circular window. One moves up and to the right, the other moves down and to the right. When they are added together, the resulting plaid moves rightward, as an apparently rigid pattern, in accord with the requirements of the velocity space construction. The appearance of this pattern is interesting, because one does not see either of the component grating motions in the combined pattern. Rather, the plaid seems to move rightward as a single coherent surface.

At first glance, it may appear that we have described a very complex method for doing vector addition. But the velocity space solution is generally different than one would get from a vector sum, a point that is exemplified in figure 7. Here we combine two gratings, each of which moves down and to the right, each with its own speed and direction. On a vector sum model, the combined pattern should move down and to the right as well, while on a velocity space model it should move up and to the right. And the stimulus does look like it is moving up and to the right, in accord with the velocity space requirements.

Crossed gratings do not always cohere into a single moving pattern. Figure 8 shows a case in which they cannot do so because there are three gratings which generate mutually incompatible constraints. The three gratings here set up three constraint lines; any two gratings can cohere, but then the third one cannot be included in the same pattern motion. When one views this stimulus, one sees a multistable display, in which any two of the gratings can be seen as a coherent pair moving in one direction, while the third grating (the odd man out) floats off by itself in the opposite direction. There are three such percepts, each corresponding to a particular intersection in velocity space. One can select out a particular percept by tracking the desired pattern motion with one's eyes; the tracking strongly biases the perception toward the tracked coherent pattern rather than the other two possibilities.

Determinants of Coherence

Even with just two gratings, coherence does not always occur. In some circumstances a pair of crossed gratings, each moving in a different direction, will be seen as just that -- a pair of crossed gratings, each moving in a different direction, each sliding across the other as if the other wasn't there. In such cases the visual system has elected not to combine the two component motions into a single pattern motion. By studying the conditions under which coherence does and does not occur we may learn something about the mechanisms that underly the perception of pattern motion.

One of the most striking determinants of coherence is the spatial frequency of the crossed gratings (Adelson and Movshon, 1982). In many of our experiments we use sine-wave gratings, i.e., gratings whose luminance profiles are sinusoidal (such stimuli have become popular in vision research because many early visual mechanisms are spatial frequency tuned, so that by using a sine-wave stimulus one can preferentially stimulate a relatively small subset of the mechanisms under study). We have found that two gratings will have a strong tendency to cohere if they are of the same spatial frequency, but have a rather weak tendency to cohere if their spatial frequencies differ by more than an octave. Thus, for example, if a 3 cycle/deg grating of one orientation is summed with a 9 cycle/deg grating of a different orientation, chances are that they will be seen as sliding over each other rather than moving as a single coherent plaid.

The contrasts of the two gratings are also important in determining coherence. If the first grating is of high contrast while the second one is of low contrast, then coherence may break down. Only when the contrast of the second grating is increased will the coherent percept return.

The contrast dependence can be used to derive a tuning curve for the frequency dependence, in the following way. Start with one moving grating of fixed spatial frequency and contrast. Add to it a second moving grating (of a different orientation), of a different spatial frequency. If the second grating's spatial frequency is substantially different from that of the first grating, then its contrast will have to be quite high in order for the two gratings to cohere. But if the second grating has a spatial frequency that is similar or identical to that of the first, then it will cohere with the first even when its contrast is quite low. By measuring the minimum contrast at which coherence occurs, one can trace out a tuning curve for this effect.

Figure 9 shows the results of two such experiments on one subject. The closed circles show the first experiment, in which the standard grating was 2.2 cycle/deg, moving at 3 deg/sec, and had a contrast of 0.3. The second grating had an orientation that differed from that of the first grating by 135 degrees; the second grating's spatial frequency was varied. The filled circles indicate the minimum contrast the second grating needed in order to cohere with the first, at various spatial frequencies. The contrast was noticeably elevated when the two frequencies differed by an octave, and became quite high when they differed by two octaves. The tendency to cohere was greatest (and the needed contrast was lowest) when the second grating's spatial frequency was matched to that of the first grating.

The open circles show the results of a similar experiment in which the standard grating had a spatial frequency of 1.2 cycle/deg. The results are much the same, and once again the tuning curve is centered on the spatial frequency of the standard grating. So these experiments indicate that the visual mechanisms underlying coherent motion perception are spatial frequency tuned, like many other aspects of early visual processing.

Global Versus Local Analyses

The velocity space construction is quite useful in understanding and predicting pattern motion phenomena, but this utility does not demonstrate that the human visual system is actually performing the rather global computations suggested in the velocity space diagrams. There are alternative approaches that will lead to the unambiguous perception of moving patterns, such as approaches based on "landmarks," or localized features in the moving patterns. In the case of the moving diamond, a corner could serve as a landmark, and by following its position over time one could correctly determine the motion of the diamond. Similarly, in the case of crossed sine-wave gratings, the peaks and troughs corresponding to the intersections of the light and dark grating bars could serve as landmarks by which the coherent motion direction could be inferred. The motion derived by these approaches is, of course, identical to that derived from velocity space, since in either case there exists a single pattern motion that is consistent with all the visual information in the display.

But there is an interesting display that we call the "split herringbone," for which a landmark model makes a different prediction than does a more global model based in velocity space. The display is shown in fig. 10(a); it consists of alternating columns of line segments that tilt left or right on the odd and even columns. The odd (right-tilting) columns move down, while the even (left-tilting) columns move up.

The most obvious landmarks in this display are the endpoints of the line segments. If they determine the motion percept, then one should see the split herringbone for what it is: a set of interleaved columns moving continuously up or down.

If, on the other hand, a more global process is at work, then it is possible that one will see an illusion of rightward motion, as illustrated in fig. 10(b). The odd columns produce one constraint line in velocity space, and the even ones produce another. The intersection corresponds to pure rightward motion. So this global approach predicts an illusion of motion to the right.

When one actually sets up the display, one finds that either of the two percepts is possible. The model based on local landmarks works when the display is of high contrast, is sharp, and is centrally fixated; in these conditions one sees the "correct" percept of vertically moving columns. On the other hand, when the display is of low contrast, or is blurred with a diffusion screen, or is viewed peripherally, then one can see the illusion of rightward motion predicted by the global model based in velocity space.

When the illusion does occur it is quite striking. The herringbone pattern seems to be moving continuously rightward, and yet it never gets anywhere, since the vertical "creases" where the columns abutt are fixed in position.

These observations suggest that both kinds of models -- the local landmark-based models, and the more global models based in velocity space--can be useful in understanding the way we perceive moving patterns.

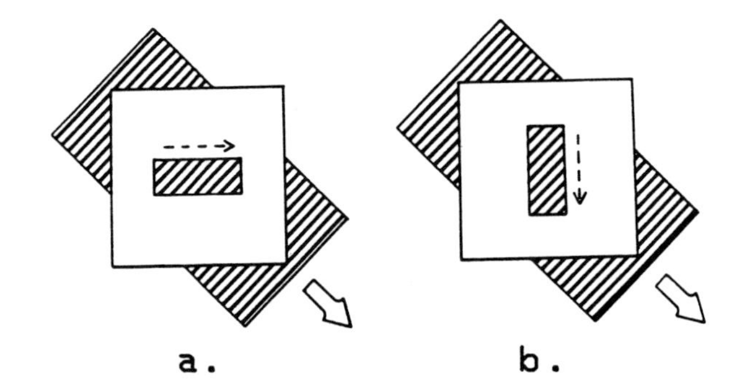

Fig. 1: The barberpole effect. The same physical motion of the grating behind the window can be seen as corresponding to different motions within the window.

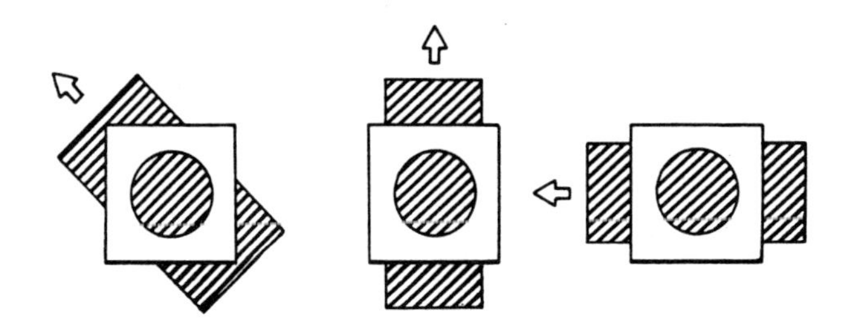

Fig. 2: The ambiguity of motion for extended contours. All three physical motions give rise to identical motion as viewed through the window.

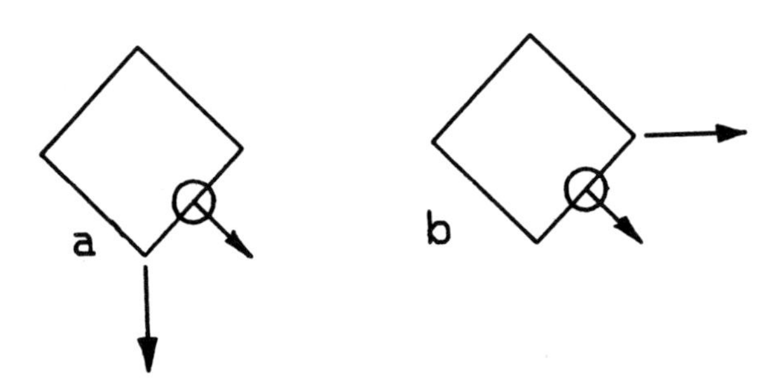

Fig. 3: A cell looking at a local edge of a diamond cannot determine which way the diamond is moving.

References

Adelson, E.H., and Movshon, J.A., Nature, 300, 523-525 (1982).

Barlow, H.B., and Hill, R.M., Science, 139, 412 (1963).

Burt, P., and Sperling, G., Psychological Review, 88, 171-195 (1981).

Henry, G.H., Bishop, P.O., and Dreher, B., Vision Research, 14, 767-777 (1974).

Hildreth, E.C., and Ullman,S. MIT AI Memo 699, December 1982.

Horn, B.K.P., and Schunck, B., Artificial Intelligence, 17, 185-203, (1981).

Fennema, C.L., and Thompson, W.B., Computer Graphics and Image Processing, 9, 301-315 (1979).

Marr, D., and Ullman, S., Proceedings of the Royal Society, London, B. 211, 151-180 (1981).

Wallach, H., Psychol. Forsch., 20, 325-380 (1935).

Wallach, H., On Perception, 201-216, Quadrangle, New York, 1976.

Wohlgemuth, A., British Journal of Psychology, monograph supplement 1 (1911).

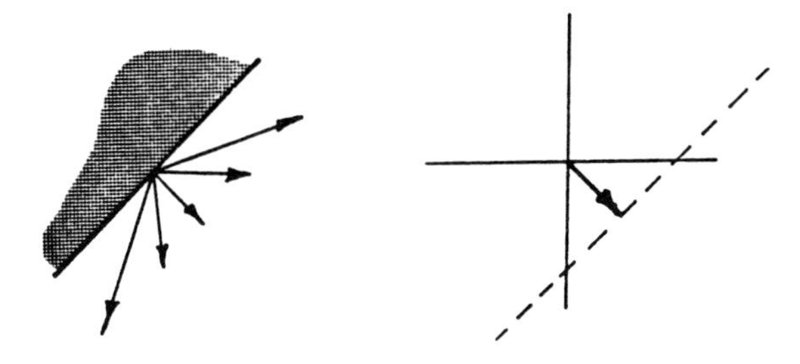

Fig. 4: Left: The set of possible object motions that could give rise to an observed motion of the edge. Right: Mapping the family of possible motions into velocity space.

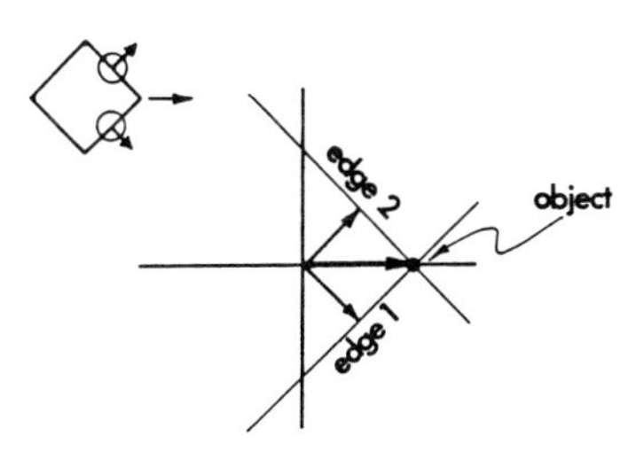

.Fig. 5: An exact solution for the diamond's motion can be determined by the intersection of two constraint lines from two edges.

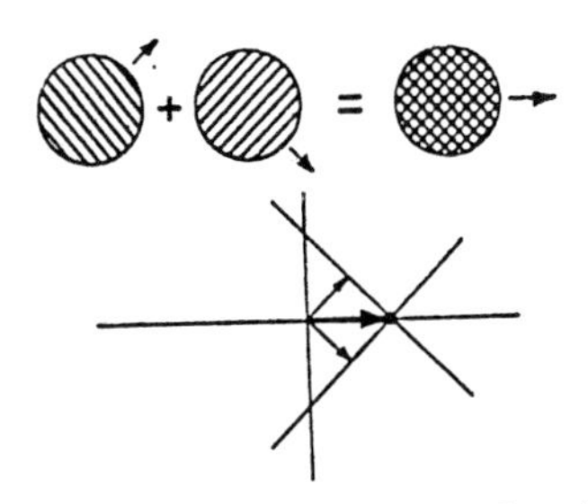

Fig. 6: The velocity space solution applied to the case of two moving gratings which are summed to form a moving plaid. The plaid moves to the right.

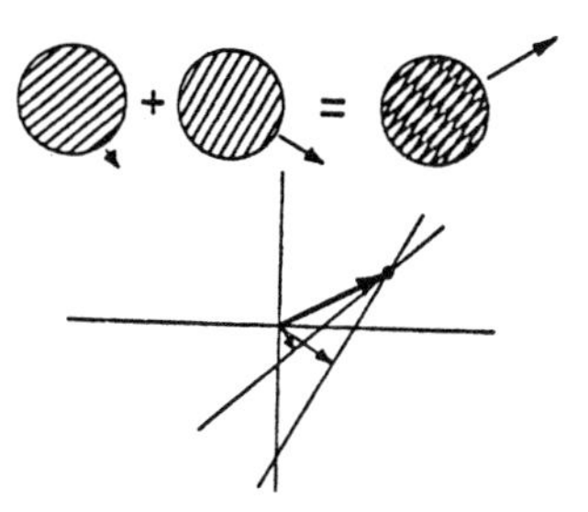

Fig. 7: An example where the velocity space solution is quite different from the solution based on a vector sum. Two gratings that move down and to the right give a pattern motion that is up and to the right.

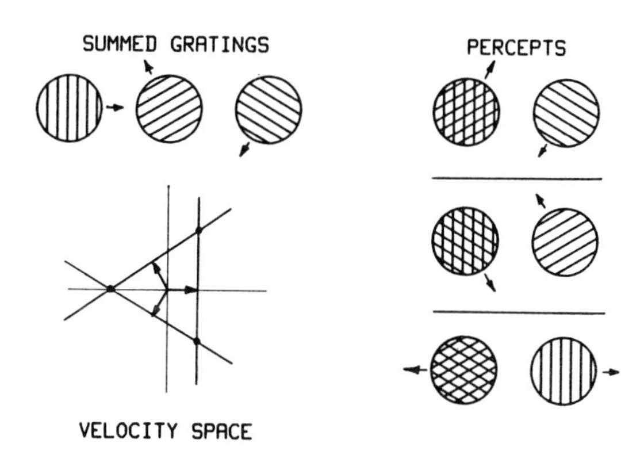

Fig. 8: A tristable display, composed of the sum of three moving gratings. There are three velocity space solutions, each of which is consistent with only two of the three grating motions. There are three possible percepts corresponding to these solutions.

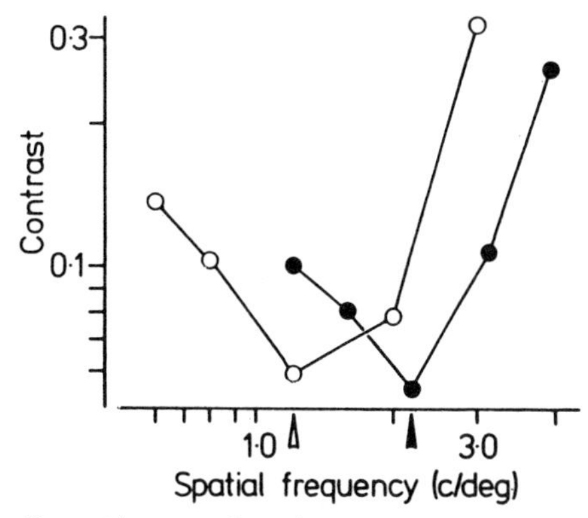

Fig. 9: The mechanisms responsible for coherent pattern perception are tuned for spatial frequency. Shown here are two tuning curves, in which contrast for coherence is measured as a function of the relative spatial frequencies of the two gratings. See text for details.

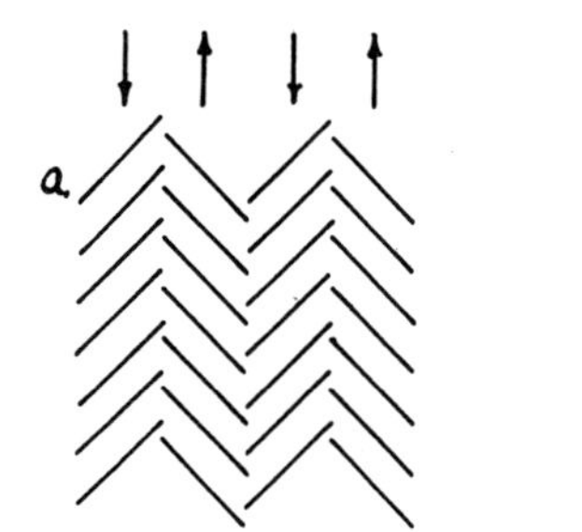

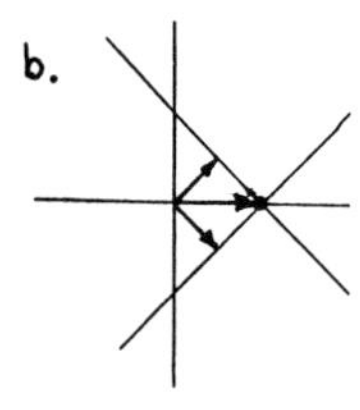

Fig. 10: The split herringbone illusion. Alternating columns of tilting line segments move up and down. Under certain conditions one has the illusion that the herringbone pattern is moving continuously to the right. See text.

REAL AND APPARENT MOTION: ONE MECHANISM OR TWO?

Marc Green, Psychology Department, University of Toronto, Mississauga, Ontario

Michael von Grunau, Psychology Department, Queen's University, Kingston, Ontario

ABSTRACT

Two direction selective adaptation experiments were conducted to investigate whether real and apparent motion are processed by a single visual mechanism. Previous studies with real motion have shown that adaptation to a grating drifting in one direction has an effect on perceived motion of subsequently viewed test gratings (the velocity aftereffect) and also selectively raises contrast threshold (direction-specific threshold elevation). We conducted analogous experiments in which observers adapted to real motion but were tested with apparent motion. In the "velocity aftereffect" study, we found that adaptation to real motion had a profound effect on the strength of apparent motion, suggesting a single mechanism. However, it was found in a second experiment that although adaptation to a moving grating produced a direction selective effect on perception of apparent motion, there was no direction selective threshold elevation for detection of the test stimuli. We conclude that although a single mechanism is responsible for mediate conscious motion perception, detection of objects in real and apparent motion is performed by two separate mechanisms.

INTRODUCTION

Motion may be perceived when the image of an object moves smoothly and continuously across the retina. This "real motion" situation is often contrasted with the case in which two stationary objects are sequentially presented at different locations in the visual field. If the interval between presentations is optimal, a compelling sensation of "apparent motion" is seen. Although the stimulus conditions which produce the two classes of motion differ, they can appear phenomenally identical (Wertheimer, 1912; Gibson, 1954). This has led many to wonder whether real and apparent motion percepts are mediated by the same visual mechanisms.

Supporters of the unitary mechanisms viewpoint note the phenomenal similarity as well as the fact that individual neurons may respond similarly to real and apparent motion (e.g., Barlow and Levick, 1965). Moreover, recent psychophysical evidence (Clatworthy and Frisby, 1973; Barbur, 1981) also suggests a common mechanism for real and apparent motion. However, the interpretation of these studies is not as clear as was suggested by their authors (see discussion below). On the other hand a two-process model is suggested by the demonstration (Smith, 1948) that apparent motion can be seen even when stimuli are alternately presented to the monocular crescent of the two eyes. Under these conditions, it is difficult to attribute the motion percept to neurons with conventional receptive fields. Other possible differences between real and apparent motion have also been proposed (Rock, 1975; Kolers, 1963).

One attempt to resolve the controversy of whether real and apparent motion are processed by a unitary mechanism has been the proposal of "long range" and short range" visual mechanisms (Braddick, 1974). According to the model, the short range system codes real motion as well as apparent motion when the displacement size is small. The long range system mediates motion perception when the displacement size is large. This means that there are in fact two motion detection systems, but they cannot be strictly distinguished on the basis of real vs. apparent motion but rather the size of displacement.

In the present experiment, we investigated the relationship between real (processed by the short range system) and long range apparent motion by means of the direction-selective adaptation paradigm. Two versions of the method have been frequently employed to study the properties of mechanisms underlying perception of real motion (see Sekuler, Pantle, and Levinson, 1978 for a review). In the first version, the aim is to determine the effect of adaptation to a drifting grating (or other moving field) on the perceived motion of a subsequently viewed suprathreshold test pattern. Although the test pattern is typically stationary (motion aftereffects), it is also possible to use moving test patterns stimuli (velocity aftereffects). Studies of the velocity aftereffect have shown that 1) when the adaptation pattern drifts faster than a test pattern moving in the same direction, the apparent velocity of the test pattern

Notion: Representation and Perception
N.I. Badler and J.K. Tsotsos, Editors
Published by Elsevier Science Publishing Co., Inc.

is reduced (Carlson, 1962; Scott, Jordan, and Powell, 1963; Clymer, 1973; Thompson, 1981), 2) when the adaptation velocity is slower than the test velocity, the apparent speed of the test pattern is increased (Scott et al., 1963; Clymer, 1973; Thompson, 1981) and 3) if the adaptation drifts in the opposite direction, then apparent velocity is increased. In the first experiment, we tested the effect of adaptation to a drifting grating on the apparent motion produced by sequential presentation of 1 c/deg grating patches at different retinal locations. Displacement of the patches was 1.5^o, a distance large enough to selectively stimulate the long-range system. Our logic was that if real and long range apparent motion were processed by a unitary mechanism, then adaptation to the drifting grating should have a direction selective effect on the strength of apparent motion. That is, prolonged viewing of a grating which was drifting to the left should selectively impair apparent motion when the grating patches are presented right to left. On the basis of the velocity aftereffect studies, it might also be expected that leftward adaptation would enhance opposite direction (left to right) apparent motion.

The second paradigm is the direction-selective threshold elevation (DSA). In this method, the goal is to determine the effect of motion adaptation on detection threshold for a test grating drifting in the same or opposite direction. It has been shown (Sekuler and Ganz, 1963), that adaptation to a real motion in one direction causes a greater loss of sensitivity for targets moving in the same dirction than for targets moving in the same direction than for targets moving in the opposite direction. In a second experiment, we measured contrast sensitivity for detecting the grating patches when their stroboscopic sequence was in the same or opposite direction as the adaptation real motion.

METHOD

Observers

One of the authors, MvG, and LC served as subjects. At the time of the experiment, only MvG was fully aware of the purpose of the study.

Apparatus

Stimuli were displayed on the face of a Tektronix 602 X-Y display scope. A raster (mean luminance 5.0 cd/m^2) was produced on the display by the standard television method. An opaque mask limited the viewing area of the screen to a rectangular area 6.5^o wide and 2.5^o high. This area, as shown in Figure 1, was divided by thin (5') white lines into three sections, a 1.5^o wide center area flanked by two 2.5^o wide regions. Long range apparent motion was created by presenting a 1 c/deg grating patch for 75 msec to the left (or right) section of the screen, waiting an appropriate stimulus onset asynchrony (SOA) and then switching on an identical grating to the right (or left) section. Under optimal conditions, this procedure produced a convincing percept of a single grating which moved from one side of the display across the middle section to the other side. Switching the gratings on and off produced no change in the mean luminance of the display.

Procedure

Experiment 1: Observers began each session with 5 minutes of adaptation to dim illumination followed by two minutes of viewing a fixation mark located 1° below the midline of the display. This fixation procedure was employed since it has been reported (Kolers and von Grunau, 1972) that apparent motion is stronger when it does not appear to pass through the fovea. After viewing the adaptation stimulus for 30 seconds, observers were tested in a series of trials as shown in Figure 1. An eight second adaptation period was followed by a 500 msec interstimulus interval (ISI) during which the raster was unmodulated. Next, the stroboscopic sequence was presented with the grating patch presented in the left (or right) section followed by an SOA and then the second patch. (If the SOA were less than 75 msec, then the gratings were present in both sections for some period.) After another ISI, the observer readapted for 8 seconds. Adaptation gratings were set 0.8 log unit above their own detection thresholds and the test patches contained gratings 0.5 log unit above threshold. After adaptation, the visibility of the test grating was reduced to a level close to threshold. We chose to use these barely visible test stimuli in order to facilitate comparison between the results that we obtained in the velocity aftereffect and threshold elevation experiments.

Each session consisted of several "runs". On each run, the adaptation stimulus was held constant and the observer indicated whether motion had been seen following each stroboscopic sequence. The criterion for a "yes" response was that the observers must perceive a single object moving smoothly and continuously across the screen. During a run, each of 10 SOA's was used 10 times, with half being left to right motion and the other half programmed as right to left. Both SOA and direction were varied randomly from trial to trial. Since the test stimuli were so low in apparent contrast, observers occasionally (5%) failed to see one or both grating patches. These trials were eliminated from the data and reprogrammed at the end of the run. Final data were obtained after 30 judgments had been obtained for each condition of adaptation and SOA.

Experiment 2:

Contrast thresholds were measured for detection of the grating patches by means of a two alternative forced choice procedure. A trial sequence consisted of the following events: 1) 8 second adaptation period, 2) 500 msec ISI, 3) test interval 1, 4) a 500 msec ISI, 5) 4 seconds of readaptation, 6) 500 msec ISI, 7) test interval 2, 8) 500 msec ISI. This sequence was immediately followed by the adaptation for the next trial. In one of the test intervals, the grating patches were presented with the SOA set at the optimal period (150 msec) for producing apparent motion. In the other test interval only the unmodulated raster was present for the same length of time. The observer's task was to indicate whether the grating patches were seen in the first or second interval. Thresholds were obtained by changing test contrast up or down contingent on the observer's accuracy. Using the procedures previously described (Green, 1982) contrast was increased in a 0.1 log unit step following three correct responses in a row and decreased 0.1 log unit whenever an error was made. By tracking threshold in this manner, a detection level equal to the 79.6% correct point on a psychometric function was obtained (Wetherill and Levitt, 1965).

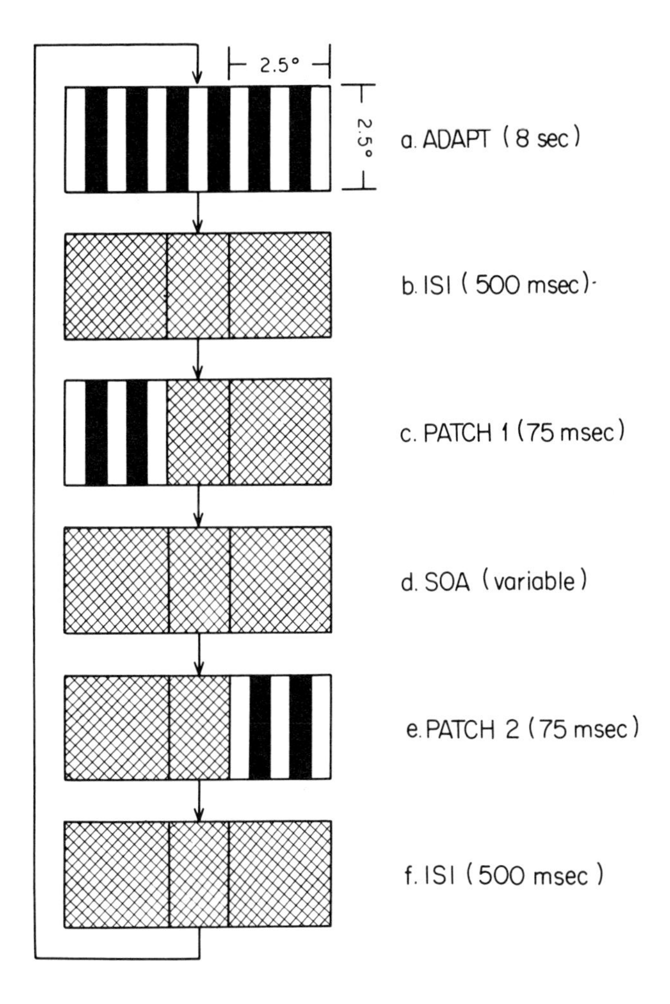

Figure 1. Schematic representation of the sequence of events on each trial. The adaptation grating drifted leftward, rightward, flickered or was stationary. The stroboscopic sequence could be left to right (as shown) or right to left.

RESULTS

Exeperiment 1

Figure 2 shows the percentage of apparent motion reports obtained when observers viewed same-direction motion, opposite-direction, or a uniform field between stroboscopic presentations. In the unadapted condition, strength of apparent motion exhibited the typical inverted-U function. At the shortest SOA's, the two test gratings appeared to be present simultaneously while the longest SOA produced the perception of succession. Functions for both observers showed a broad peak at middle intervals with a high percentage of seen motion. Optimal intervals for observer LC ranged from 75 to 150 msec while observer MvG reported motion on most trials with SOA's between 75 and 175 msec.

Adaptation to a grating drifting in the opposite direction of the stroboscopic sequence produced an enhancement of apparent motion. For LC, frequency of motion reports was higher at all SOA's. Although size of the facilitation was greatest at the long and short intervals, this may have been due to a ceiling effect: frequency of seen motion was already very high at the medium SOA's. Observer MvG also showed an enhancement of seen motion at the very long and short intervals but no systematic effect at the moderate SOA's. Magnitude of the effect is enormous at 45 and 60 msec intervals, where motion reports jump from near 0% without adaptation to 80% with the opposite-direction motion.

The effects of adapting to same-direction motion are more complex. Both observers show an enhancement of apparent motion at short intervals, and a reduction at long SOA's. This effect is particularly large for MvG. At the shortest interval seen motion increases from 3% to 53% while motion is seldom seen when the SOA is longer than 100 msec. Although this result seemed puzzling at first, it is quite consistent with results from velocity aftereffect studies (see discussion.)

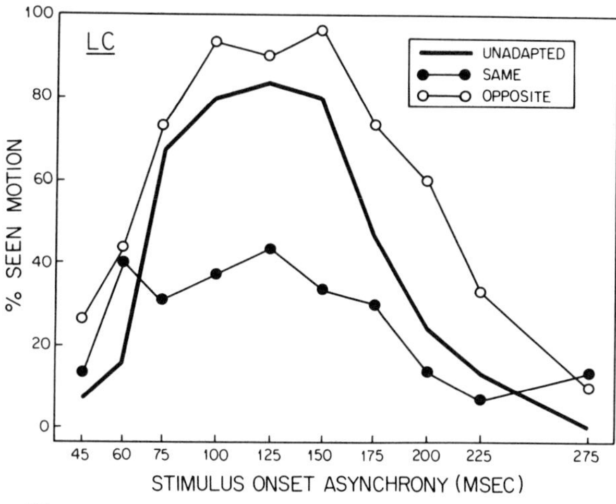

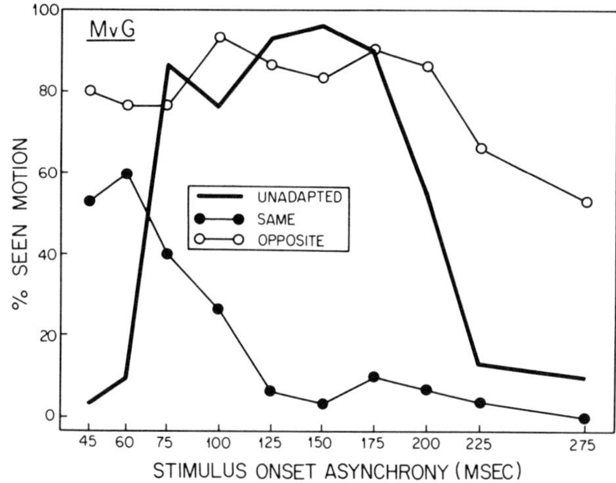

Figure 2. Percent of motion reports as a function of SOA condition. The bold line shows data for unadapted conditions while open circles show opposite direction conditions and closed circles indicate same direction adaptation.

Experiment 1b

Since drifting gratings contain spatial contrast, it was possible that part of the effects found in the previous experiment might be due to pattern adaptation *per se*. In order to assess the possibility, we conducted an additional experiment in which observers viewed a stationary adaptation grating. Further, observers were tested with a counterphase flickering adaptation grating in order to determine the effect of temporal change without a directional component.

Figure 3 shows the results that we obtained with these new adaptation conditions. Counterphase flicker produced results very similar to those found with same-direction adaptation. Observer LC exhibited an increase in motion reports at all SOA's. As before, MvG showed an enhancement at long and short SOA's and perhaps a small decrement at middle intervals. Adaptation to a stationary grating improved apparent motion for both observers at long SOA's. At short intervals LC showed no change while MvG reported motion slightly less often. However, no conditions produced the marked decrease that was found with same-direction motion.

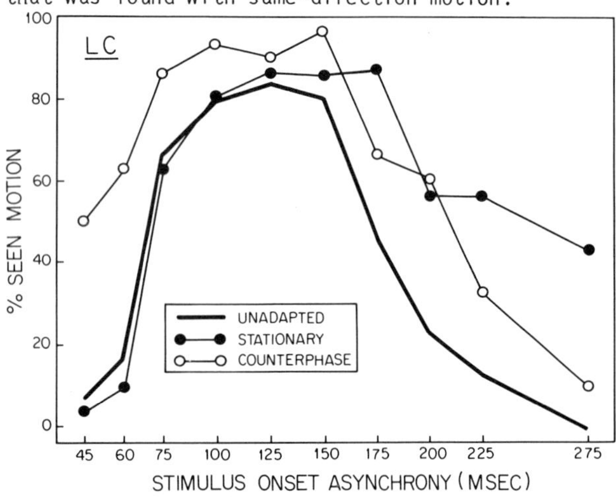

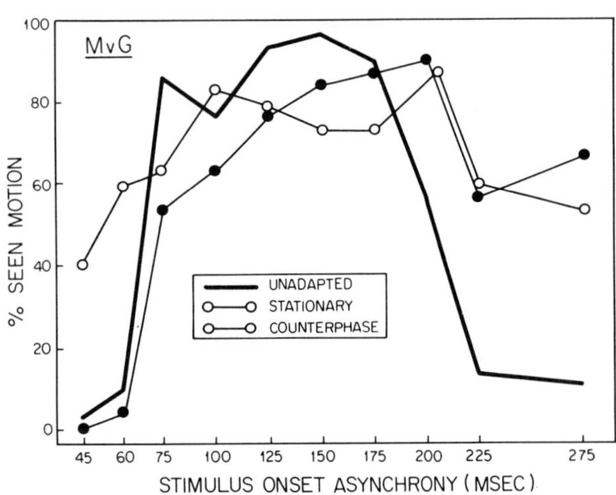

Figure 3. Percent of motion reports as a function of SOA condition. The bold line indicates unadapted conditions while open symbol shows counterphase results while closed symbols indicate data obtained with stationary grating adaptation.

Experiment 2

The effect of the real motion adaptation on detection of the test grating patches is shown in Table 1. These numbers represent the log of the ratio of contrast sensitivities obtained before and after adaptation. The adapting gratings produced a 0.3-0.4 log unit loss of sensitivity for the test patches. Unlike the case of real motion (Sekuler and Ganz, 1963), however, the magnitude of the threshold elevation did not depend on direction of the test gratings. Therefore, no direction specific threshold elevation occurs between real and apparent motion.

THRESHOLD ELEVATION (LOG UNITS)

		Same Direction	Opposite Direction
OBSERVER	LC	0.38 (±.04)	0.400 (±.05)
	MvG	0.325(±.05)	0.350 (±.05)

Table 1

DISCUSSION

Velocity Aftereffects

The results of our experiment in velocity aftereffects appear to support the view that real and long range apparent motion are processed by a common neural mechanism. We found that following adaptation to real motion, apparent motion in the opposite direction was enhanced. Moreover, when adaptation was in the same direction, apparent motion was facilitated at short SOA's and impaired at middle and long SOA's. If one accepts the view that apparent motion with a short SOA is equivalent to high "velocity" and apparent motion with longer SOA has a lower velocity, then our results are similar to those found when observers were adapted and tested with real motion (Carlson, 1962; Scott and Jordan, 1963; Clymer, 1973; Thompson, 1981).

Our results are also consistent with those of Clatworthy and Frisby (1973) who adapted observers to real motion, a grid which oscillated back and forth, and test with apparent motion. Since adaptation to the oscillating grid reduced strength of apparent motion, they concluded that there is a common mechanism for real and apparent motion. However, their study was not entirely convincing because the oscillating grid contained stationary endpoints, i.e., it stopped momentarily at the end of each excursion to change directions. It has been shown (Kaufman et al., 1971) that stationary endpoints confound the experiment becuase the apparent motion mechanism may be stimulated. Therefore, the adaptation stimulus used by Clatworthy and Frisby actually contains elements of real and apparent motion. We feel that the use of linearly drifting gratings in the present experiment precluded the possibility of stationary endpoints. Moreover, we were able to obtain facilitory as well as inhibitory effects analogous to those obtained with real motion alone.

The results of the velocity aftereffect study are also consistent with data from a previous study (Green, 1983) in which summation between real and apparent motion was measured. Observers viewed a display of grating patches, similar to those used here. However, the patches contained drift (real motion) in the same or opposite direction as the apparent motion. When real and apparent motion were in the same direction, summation was found as the frequency of apparent motion reports increased. However, when opposite-direction drift was added, apparent motion was completely eliminated. This experiment, along with the one reported here, indicates that perception of apparent motion is greatly influenced by real motion. On the other hand, we know of no evidence demonstrating that apparent motion can influence perception of real motion. In fact, we attempted to adapt observers to apparent motion and then test their perception of real motion. However, this proved impossible because prolonged viewing of stroboscopically presented stimuli results in a rapid breakdown of the apparent motion.

We further adapted observers to stationary and counterphase flickering gratings. Both of these adaptation stimuli produced an enhancement of apparent motion at long SOA's. The flicker improved apparent motion at short intervals while the stationary grating adaptation had little or no effect. Therefore, we cannot rule out the possibility that some of the enhancement produced by the opposite-direction or same-direction adaptation may have been due to the effects of nondirection (pattern or flicker) adaptation. (This also remains a possibility with the experiments in which observers adapted and tested with real motion.) However, our results seem to contradict the findings of Clatworthy and Frisby (1973) who failed to find any effect of flicker or pattern adaptation. We can think of one possible reason for the enhancement produced by stationary and flickering grattings. Both contain spatial contrast which, after prolonged viewing, lowered the apparent contrast of the test gratings. Several authors (Kulikowski and McCana, 1978; von Grunau, 1978; Keck et al., 1976; Green, 1983) have reported results which indicate that high apparent contrast decreases motion sensitivity. For example, Kulikowski and McCana (1978) found that adaptation to a stationary grating enhanced motion perception. However, this cannot be the entire explanation since it is not clear why stationary gratings fail to enhance motion at short SOA's. At the moment, we can think of no convincing explanation for this difference.

Direction-Specific Threshold Elevation (DSA)

Adaptation to a drifting grating failed to produce a direction specific threshold elevation of the stroboscopically presented test patches. This result contrasts with the DSA effect (e.g., Sekuler and Ganz, 1963) obtained with real motion. The conclusion from the DSA experiment appears to be that contrast of stimuli in real and apparent motion are not detected by the same mechanism.

Before accepting this conclusion, we considered one alternate interpretation. Threshold elevation effects are greatest when the adaptation and test gratings drift at similar velocities (Pantle and Sekuler, 1968). It is possible that the "velocity" of the apparent motion was sufficiently different from the adaptation grating that no DSA effect occurred. Unfortunately, this possibility is difficult to evaluate. While it is true that threshold elevation magnitude depends somewhat on similar velocities, there are no studies magnitude depends somewhat on similar velocities, there are no studies which test the effect of velocity on the size of DSA. However, we tend to doubt this sort of explanation because it fails to explain the disparity between velocity aftereffects and DSA: the same conditions of ISI which produced a marked velocity aftereffect produced no DSA.

Our results seem to conflict with those of Barbur (1981) who measured luminance detection summation between two spots, one that was in real motion and a second which was present in apparent motion. He found that when the spots moved in the same direction, threshold was lowered while opposite motion produced no summation. It was concluded that real and apparent motion are processed by the same mechanism. He posited and then rejected the possibility that spatial and temporal summation could account for his results. We are not convinced that this explanation can be ruled out.

Conclusion

Results from our velocity aftereffect experiment support a unitary mechanism model while the DSA data are more consistent with a two-process model. The proper conclusion seems to be that real and apparent motion are processed by related but not identical mechanism (presumably the short and long range systems). The major constraint on motion processing models supported by our data is that there are two motion processing mechanism, separated at the level of detection but related (or unitary) at the level of conscious motion perception. Doubtless many models could be invoked to describe possible relationships. Exner (1875), for example, suggested that real motion is directly sensed while apparent motion is "inferred" from the change in position of stationary objects (see also Bonnet, 1977 for a similar idea). This kind of model is consistent with our data, since it suggests that the earliest stages of real motion processing are direction selective while the initial stage of apparent motion is not directionally selective. At some later level, however, the information from the two detection systems must merge, perhaps through a mutual inhibition (Pantle and Picciano, 1976).

References

Barbur, J. Subthreshold addition of real and apparent motion, Vision Research, 1981, 21, 314-322.

Barlow, H.B. and Levick, W.R. The mechanism of directionally selective units in the rabbit's retina. Journal of Physiology, 1965, 173, 377-407.

Bonnet, C. Visual motion detection models: features and frequency filters, Perception, 1977, 6, 519-527.

Carlson, V. Adaptation in the perception of visual velocity: Journal of Experimental Psychology, 1962, 64, 192-197.

Clatworthy, J.L. and Frisby, J.P. Real and apparent movement: evidence for unitary mechanism. Perception, 1973, 2, 161-164.

Clymer, A. The effect of seen motion on the speed of subsequent test velocities: speed tuning of movement. A. Es. Ph.D. Thesis, Columbia University, 1973.

Exner, S. Uber des Sehen von Bewegung und die Theorie des zusammengesetzten Auges. Sitzber., Akad. Wiss. Wien, 1875, 72, 156-190.

Gibson, J.J. Visual perception of objective motion and subjective movement. Psychological Review, 1954, 61, 304-311.

Green, M. (1982) Visual masking by flickering surrounds. Vision Research, in press.

Green, M. Inhibition and facilitation of apparent motion by real motion. Vision Research, in press, 1983.

Grunau, M.W. von. Interaction between sustained and transient channels: form inhibits motion in the human visual system. Vision Research, 1978, 18, 197-202.

Kaufman, L., Cyrulnick, I., Kaplowitz, L., Melnick, G., and Stof, D. The complimentarity of apparent and real motion. Psychologische Forschung, 1971, 34, 343--348.

Keck, M., Palella, T. & Pantle, A. Motion aftereffect as a function of the contrast of sinusoidal gratings. Vision Research, 1976, 16, 187-192.

Kolers, P. Some differences between real and apparent visual movement. Vision Research, 1963, 3, 191-206.

Kolers, P. & Grunau, M.W. von. Fixation and attention in apparent motion. Quarterly Journal of Experimental Psychology, 1977, 29, 389-395.

Kulikowski, J., & McCana, F. Is the antagonism between pattern and movement detection? Journal of Physiology, 19 , 231, 22P.

Pantle, A.J. & Picciano, L. A multistable movement display: evidence for separate motion systems in humans. Science, 1976, 193, 500-502.

Pantle, A. & Sekuler, R. Velocity-sensitive elements in human vision: initial psychophysical evidence. Vision Research, 1968, 8, 445-450.

Rock, I. Introduction to Perception. Published by MacMillan PUblishing Co.: New York, 1975.

Scott, T., Jordan, A., & Powell, D. Does the visual aftereffect of motion add algebraically to objective motion of the test stimulus? Journal of Experimental Psychology, 1963, 66, 500-505.

Sekuler, R.W. & Ganz, L. Aftereffect of seen motion with a stabilized retinal image. Science, 1963, 139, 419-420.

Sekuler, R., Pantle, A., & Levinson, E. Physiological basis of motion perception. In R. Held, W.W. Leibowitz and H.-L. Teuber (Eds.), Handbook of Sensory Physiology: Vol. VIII: Perception, Berlin: Springer Verlag, 67-96, 1978.

Smith, K.R. Visual apparent motion in the absence of neural interaction. American Journal of Psychology, 1948, 61, 73-77.

Thompson, P. Velocity aftereffects: The effect of adaptation to moving stimuli on the perception of subsequently seen moving stimuli. Vision Research, 1981, 21, 337-345.

Wertheimer, M. Experientalle Studien uber das Sehen von Bewegung, Zeitschrift fur Psychologie, 1912, 61, 161-265.

COHERENT GLOBAL MOTION PERCEPTS FROM STOCHASTIC LOCAL MOTIONS

D. W. Williams and R. Sekuler

Departments of Psychology,

Neurobiology & Physiology,

and Ophthalmology,

Cresap Neuroscience Laboratory

Northwestern University

Evanston, Illinois

ABSTRACT

A percept of global, coherent motion results when many different localized motion vectors are combined. We studied the percept with dynamic random dot kinematograms in which each element took an independent, random walk of constant step size. Directions of displacement from frame to frame were chosen from a uniform distribution. The tendency to see coherent, global flow along the mean of the uniform distribution varied with the range of the distribution.

Psychometric functions were obtained with kinematograms having various step sizes (0.1 to 1.4 degrees) and element densities (0.2 to 1.6 dots per square degree). Results fall into two categories, depending on whether the step size is larger of smaller than 1.0 degree. For step sizes greater than 1.0 degree changes in dot density altered the psychometric function. No change was found if the step size was less than one degree. These changes in the psychometric function with step size and density are consistent with Ullman's "minimal map theory" of motion correspondence.

Motion: Representation and Perception
N.I. Badler and J.K. Tsotsos, Editors
Published by Elsevier Science Publishing Co., Inc.

For the smaller step sizes, the constancy of the results over a large range of dot densities suggests that spurious directions of displacement due to the interference of random walks for different dots are not important. That is, only the directions of local motion determined by the predefined distribution of directions significantly contribute to the percept. We also found that although temporal summation occurred in a nonlinear manner over frames, it depended only on the set of directions present from frame to frame, not on the spatial relationships between local motion vectors over time. Taken together, these two results suggest that directions of the individual steps are independently detected and that these responses are pooled over time and space to generate the perception of coherent motion.

Optic Flow

OPTICAL FLOW

Bernd Neumann
University of Hamburg
Hamburg

1. *INTRODUCTION*

A large number of contributions to this workshop is concerned with computing or making use of optical flow. This is the term now commonly used for an intermediate representation of time-varying imagery where each pixel is assigned a velocity vector describing its temporal displacement in the image plane or - for human vision - in the retinal field. Optical flow can be consciously experienced by human observers (e.g. when travelling in a car) and has early been recognized as a valuable source of information pertaining to the motion and 3D characteristics of a scene (GIBSON 50). Thorough quantitative analyses, however, have only become available during the last five years, when an increasing number of vision researchers turned to motion problems. As can be seen from this workshop, interesting results on how to exploit optical flow are still being uncovered.

Before making use of optical flow it must be computed - unfortunately. As it turns out, no computational theory has get been offered which promises satisfactory results for unrestricted real-world images. Nevertheless, considerable progress has been made in certain restricted situations. This is also documented by several contributions to this workshop. In this introductory survey I shall try to point out the major differences in the approaches taken so far.

Fig. 1 gives a rough sketch of the representations and the processing connected with optical flow. Much of the variety of the research contributions is due to certain assumptions about the visual world. These will be discussed in the following section. The visual world is projected yielding intensity arrays from which optical flow computation per se

Motion: Representation and Perception
N.I. Badler and J.K. Tsotsos, Editors
Published by Elsevier Science Publishing Co., Inc.

proceeds. Three rather distinct directions of processing have been proposed. As a first possibility, optical flow is directly computed from the intensity array. The result is usually a dense flow field. Alternately, descriptive elements like prominent points or edges may be computed first. Points usually give rise to a sparse flow field after correspondence is established. Edges lead to a quite different flow compentation due to the remaining degree of freedom. In section 3 these distinctions are elaborated in some more detail. Finally, I shall briefly review ways of extracting useful information from optical flow.

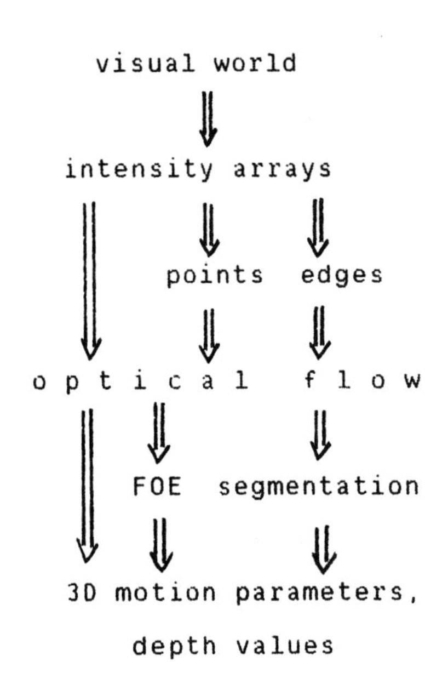

Fig. 1: Alternate ways of computing and exploiting optical flow.

2. ASSUMPTIONS ABOUT THE VISUAL WORLD

Intensity arrays are the result of a physical process which is known to involve an <u>observer</u> (i.e. an optical sensor), <u>objects</u> of the real world, and sources of <u>illumination.</u> In this section some useful restrictions or assumptions regarding these three components will be pointed out. Depending on the choice of restrictions different optical flow problems requiring different solutions may arise.

OBSERVER

For a given scene the complexity of the optical flow field may depend heavily on the motion characteristics of the observer. The following cases have been distinguished so far:

a) stationary

b) translating

c) rotating

d) translating + rotating

While a) is only interesting in connection with moving objects, the other cases are also interesting in a stationary environment. Note that a moving observer in a stationary world is usually not equivalent to a stationary observer facing a single moving object, since the latter case usually also involves stationary background. The distinction between pure translation, pure rotation and the general case is very important for extracting 3D information from optical flow. As shown by PRADZDNY 80 as well as NAGEL and NEUMANN 81, rotation (about the optical center) does not provide depth information. It only complicates the nice properties of optical flow generated by a translating observer. In fact, several approaches to 3D reconstruction from optical flow attempt to "derotate" the images before entering the 3D analysis proper (RIEGER and LAWTON 83). Unfortunately, the full complexity involving both rotation and translation appears to be required for the important case of a translating human observer, since his/her gaze is generally held fixed on a point not straight ahead.

Several subcases of specialized observer motion can also be distinguished. For example, if the direction of an observer translating via stationary environment is known, the focus of expansion (FOE) and hence the direction of the optical flow vectors can be easily determined (JAIN 83). Certain simplifications also arise if the rotation axis is assumed fixed or perpendicular to the direction of translation (BONDE 79).

OBJECTS

There are numerous assumptions and restrictions which may be imposed on the objects of a scene. The following is a list compiled from research dealing with special cases.

a) single / multiple

b) 3D / flat

c) rigid / deforming

d) independent / jointed

e) translating

f) rotating

g) translating + rotating

While most of the difficulties of optical flow computation and subsequent 3D analysis can be studied with a single moving object (particularly if it moves in a non-uniform stationary environment), it is also useful to consider multiple object scenes because of the more complicated segmentation. Single motion concepts like the FOE may sometimes be carried over as shown by JERIAN and JAIN 83.

Another special case arises if objects are essentially flat, e.g. the earth viewed from high altitude. Optical flow computation is far less involved in this case because depth discontinuities do not occur. The mathematical analysis is also easier (TSAI and HUANG 81+82). It is interesting that Gibson considered the approach to a (not necessarily flat) surface fundamental for human spatial perception (GIBSON 50).

Objects need not be rigid. The assumption of rigidity is an essential simplification for a quantitative motion analysis - it amounts to assigning identical 3D motion parameters to all parts of an object. The assumption has not get been brought to bear an optical flow computation, however, which is usually conceived as a process independent of subsequent 3D interpretations. In fact, the 2D smoothness criterion which has been proposed by HORN and SCHUNCK 81 gives rise to nonrigid flow vectors if applied to certain rigidly moving edge shapes (HILDRETH 83). Instead of complete rigidity one may also postulate partial rigidity, in particular jointed motion. Several results pertaining to the analysis of this type of motion are available (WEBB and AGGARWAL 80), all of which assume that the 2D joint positions can somehow be recovered from the flow field. It is not yet clear how this should be done.

Assumptions about the motion characteristics of objects play a similar role for motion analysis as assumptions about observer motion. Depending on the type of restriction, simplified motion equations apply and less complex procedures can be employed. When considering pure rotation it is important to be explicit about the position of the axis. While observer rotation is naturally described about an axis through the optical center, object rotation is sometimes specified w.r.t. other reference systems, e.g. an axis through the object's centroid. It is well known that for a given motion the translation vector (and not the rotation matrix) depends on this choice of reference, hence motion may have a translation component in the object-centered reference system but may be pure rotation if the axis is taken through the optical center. In the latter case it can be recognized immediately that motion stereo will not work for this particular motion. Hence, if 3D analysis is an issue, an observer centered specification of rotation is preferable. For other issues, e.g. the decomposition of motion into "natural" components, other reference systems may be useful.

ILLUMINATION

Time varying illumination gives rise to time varying imagery - a fact which has not enjoyed much attention so far (CORNELIUS and KANADE 83 is an exception). Yet most real-life scenes show changes due to illumination, e.g. darkening from clouds or shadows cast by moving objects. It is not clear what optical flow should be assigned to such phenomena. For most purposes one would probably want to separate illumination effects from physical motion. None of the procedures for optical flow computation does this, however. There appears to be a tacit agreement on leaving illumination problems to higher-level processing. On the other hand, most of the higher-level algorithms proposed for this matter are severely impaired if fed with such optical flow data. This area surely deserves further research.

3. COMPUTING OPTICAL FLOW

In this section I shall try to point out the major issues in computing optical flow as raised by the approaches taken so far.

QUANTIZATION

If one considers quantization part of the computational procedure, the first point in order concerns the temporal and spatial resolution required for a particular approach. Very often, interimage displacements must not exceed a certain amount if the algorithm is to perform well. For example, in gradient-based optical flow algorithms (e.g. HORN and SCHUNCK 81) the displacements must be small enough to permit a linear approximation of the image intensity function over that distance. Token-based schemes, i.e. algorithms which track prominent image elements may be devised to cope with much larger displacements, depending on the correspondence computation. Some algorithms may deteriorate earlier than others if displacements become too large, e.g. Ullman's minimal mapping (ULLMAN 79) as opposed to Radig 's structural matching (RADIG et al. 80).

It is useful to distinguish two phases of optical flow computation: local displacement computation and motion integration. In the first phase, local evidence is evaluated to obtain local displacement information. In the second phase, global measures or constraints are applied to yield a globally consistent flow field. The remainder of this section deals with the two phases in order.

LOCAL DISPLACEMENT COMPUTATION

There are essentially two ways of computing displacements from local information. The first way is directly based on the intensity array. Displacement information is obtained by explaining a temporal intensity change in terms of the displacement of a sloping intensity function. The original version of this approach (due to HORN and SCHUNCK 81) determines only one component of each displacement vector locally. The other component follows from applying a global smoothness constraint. The local ambiguity is not inherent to this approach, however, as can be seen from NAGEL and ENKELMANN 82 and NAGEL 83.

In the second type of approach prominent image features are extracted as a basis for optical flow computation. Further processing depends on whether these features are edge-like or point-like. Edge displacements can usually only be determined with one remaining degree of freedom, hence global criteria must be brought to bear for disambiguation (HILDRETH 83). Point features give rise to the correspondence problem which is disambiguation of displacements in another disguise. It is interesting to observe that in spite of the superficial variety of point and edge operators, the idea is generally the same: edges are locus curves of steepest intensity slope, points are high curvature locations on edges. Nagel's contribution to this workshop is a valuable step towards a unified treatment of edges and points.

MOTION INTEGRATION

What criteria are used to obtain a globally consistent flow field from local measurements? The criterion most frequently encountered is 2D-spatial and temporal smoothness which can be achieved, for example, by minimizing the sum of the squared local velocity differences (HORN and SCHUNCK 81). It is clear that any criterion based on properties of the uninterpreted image will yield a flow field which is in general physically inconsistent, i.e. flow vectors do not always conform with the displacements of physical surface elements. This is particularly evident at object boundaries where the smoothness criterion is not even approximately true. One can cope with this to some extent by keeping track of the smoothing error. In areas where it is large, boundaries may be suspected and a different approach can be taken (CORNELIUS and KANADE 83). Note also the weighting matrix proposed in NAGEL 83. It enforces the smoothness criterion in areas of high ambiguity while preserving the local measurements if based on prominent features (corner points). A similar idea has been followed in YACHIDA 81 where flow vectors are interpolated between prominent point displacements.

It is interesting to compare the computational approaches with psychophysical evidence, e.g. as presented in ADELSON and MOVSHON 83. Within certain limits humans tend to resolve local motion ambiguity following the rule of "common fate". This is in agreement with the smoothness criterion discussed above.

In cases of restricted motion it may be possible to bring to bear a much stronger consistency criterion. If, for example, observer and object translate w.r.t. each other (without rotation), the optical flow field must exhibit a focus of expansion (FOE). Hence all displacement vectors associated with that object must intersect in one point. It should be kept in mind, however, that this is only true for the displacement of physical surface elements. Errors from tracking equal intensity values instead of surface elements may make it difficult to find a FOE.

4. *EXPLOITING OPTICAL FLOW*

In Fig. 1 two directions for exploiting optical flow are indicated: segmentation and 3D analysis. Both are major areas of research which cannot be adequately covered in this introductory survey. Suffice it to point out some important aspects of these tasks.

The key element of motion-based segmentation is flow discontinuity at potential object boundaries or equivalently homogeneity within boundaries. This is about as valid as the analogous assumption in intensity-based segmentation. Both types of errors - missed and false boundaries - are bound to occur in situations of unrestricted motion. Yet for a subsequent 3D analysis most computational procedures require single-object data. NEUMANN 80 investigates a way of grouping flow vectors into independently moving objects based on 3D interpretability. The Hough-transform approach of O'ROURKE 81 is based on the same idea but may be more suitable for a dense flow field.

Over the last few years the mathematics of motion stereo have been clarified considerably. This is also exemplified by a contribution to this workshop (YEN and HUANG 83) which offers a geometrical interpretation for a formal approach developed earlier. Most theoretical results are based on a certain number of points observed in a certain number of images, e.g. 5 points in 2 images. As such the analysis is not well adapted to a dense optical flow field. Overconstrained sets of equations have been formulated to incorporate a larger number of points. The resulting optimization problem is nontrivial due to a potentially high noise sensitivity (FANG and HUANG 83). When extending 3D analysis to deal with a dense flow field instead of isolated points, one has to bear in mind how the field has been computed. One cannot gain much by incorporating flow vectors which have resulted from interpolation.

Theoretical formulations differ depending on whether a central or parallel projection model is adopted. Proponents of parallel projection usually enjoy less complex mathematics and argue that motion stereo should not be based on perspective effects anyway since these effects may be very small for distant objects. In view of the small number of experiments which have been carried out with real-life data and the scarcety of theoretical results pertaining to noise sensitivity, it is difficult to see decisive advantages with one or the other model.

This concludes the brief survey of approaches to computing and exploiting optical flow. Only a small fraction of the relevant work has been cited. The idea was to give at least one reference - preferably a paper presented at the workshop - for each issue addressed in the discussion. It is hoped that the problems which have been treated so far as well as some open questions have become apparent.

5. REFERENCES

IWMPR = International Workshop on Motion: Representation and Perception, Toronto, Canada, April 3-5, 1983

Adelson and Movshon 83
The Perception of Coherent Motion in Two-Dimensional Patterns
E.H. Adelson, J.A. Movshon
IWMRP 11-16 (1983)

Bonde 79
Untersuchungen zur dreidimensionalen Modellierung bewegter Objekte durch Analyse von Formveraenderungen der Objektbilder in TV-Aufnahmefolgen
T. Bonde
Diplomarbeit, FB Informatik, Universitaet Hamburg, 1979

Cornelius and Kanade 83
Adapting Optical-Flow to Measure Object Motion in Reflectance and X-Ray Image Sequences
N. Cornelius, T. Kanade
IWMRP 50-58 (1983)

Fang and Huang 83
Some Experiments on Estimating 3-D Motion Parameters of a Rigid

Body from Two Consecutive Image Frames

J.Q. Fang, T.S. Huang

(to appear in IEEE-PAMI)

Gibson 50

The Perception of the Visual World

J.J. Gibson

Houghton Mifflin Co., Boston/MA 1950 reprint by Greenwood Press, Westport/CT, 1974

Hildreth 83

Computing the Velocity Field Along Contours

E. Hildreth

IWMRP 26-32 (1983)

Horn and Schunck 81

Determining Optical Flow

B.K.P. Horn and B.G. Schunck

Artificial Intelligence 17 (1981) 185-203

Jain 83

Complex Logarithmic Mapping and the Focus of Expansion

R. Jain

IWMRP 42-49 (1983)

Jerian and Jain 83

Determining Motion Parameters for Scenes with Translation and Rotation

C. Jerian, R. Jain

IWMRP 71-77 (1983)

Nagel 83

On the Estimation of Dense Displacement Vector Fields from Image Sequences

H.-H. Nagel

IWMRP 59-65 (1983)

Nagel and Enkelmann 82

Investigation of Second Order Greyvalue Variations to Estimate Corner Point Displacements

H.-H. Nagel, W. Enkelmann

ICPR-82 768-773 (1982)

Nagel and Neumann 81

On 3D Reconstruction from Two Perspective Views

H.-H. Nagel and B. Neumann

IJCAI-81, 661-663

Neumann 80

Motion Analysis of Image Sequences for Object Grouping and Reconstruction

B. Neumann

ICPR-80, 1262-1265

O'Rourke 81

Dynamically Quantized Spaces Applied to Motion Analysis

J. O'Rourke

JHU-EE 81-1 (January 1981) Electrical Engineering Department The Johns Hopkins University, Baltimore/MD

Prazdny 80

Egomotion and Relative Depth Maps from Optical Flows

K. Prazdny

Biol. Cyb. Vol. 36, 87-102 (1980)

Radig et al. 80

Matching Symbolic Descriptions for 3-D Reconstruction of Simple Moving Objects

B. Radig, R. Kraasch, W. Zach

ICPR-80 1081-1084 (1980)

Rieger and Lawton 83

Determining the Instantaneous Axis of Translation from Optic Flow Generated by Arbitrary Sensor Motion

J.H. Rieger, D.T. Lawton

IWMRP 33-41 (1983)

Tsai and Huang 81

Estimating Three-Dimensional Motion Parameters of a Rigid Planar Patch, I

R.Y. Tsai, T.S. Huang

IEEE-ASSP-29, 1147-1152 (1981)

Tsai and Huang 82

Estimating Three-Dimensional Motion Parameters of a Rigid Planar Patch, II

R.Y. Tsai, T.S. Huang

IEEE-ASSP-30, 552-534 (1982)

Ullman 79

The Interpretation of Visual Motion
S. Ullman
MIT Press, 1979

Webb and Aggarwal 80
Observing Jointed Objects
J.A. Webb, J.K. Aggarwal
ICPR-80, 1246-1250 (1980)

Yachida 81
Determining Velocity Map by 3-D Iterative Estimation
M. Yachida
IJCAI-81 716-718 (1981)

Yen and Huang 83
Determining 3-D Motion Parameters of a Rigid Body: A
Vector-Geometrical Approach
B.L. Yen, T.S. Huang
IWMRP 78-90 (1983)

COMPUTING THE VELOCITY FIELD ALONG CONTOURS

Ellen C. Hildreth
MIT Artificial Intelligence Laboratory

ABSTRACT

In this paper, we present a computational study of the measurement of motion. Similar to other visual processes, the motion of elements is not determined uniquely by information in the changing image; additional constraint is required to compute a unique velocity field. Given this global ambiguity of motion, local measurements from the changing image cannot possibly specify a unique local velocity vector, and in fact, may only specify one component of velocity. Computation of the full two-dimensional velocity field generally requires the integration of local motion measurements, either over an area, or along contours in the image. We examine the integration of local motion measurements along contours, using an additional constraint of smoothness of the velocity field. The predictions of an algorithm based on this constraint are compared with human motion perception on a few demonstrations.

1. INTRODUCTION

The study of visual motion is the study of how information about the organization of movement in the changing two-dimensional image can be used to analyze the environment in terms of objects, their motion in space, and their three-dimensional structure. The analysis of motion can be divided into two parts. The first is the measurement of motion; for example, the computation of a two-dimensional velocity field which assigns a direction and magnitude of velocity to elements in the image, on the basis of the changing intensity array. The second is the use of motion measurements; for example, to separate the scene into distinct objects, and infer their three-dimensional structure. For some tasks, it may be sufficient to detect only certain properties of the velocity field, rather than measure it completely and precisely. For example, in order to respond quickly to a moving object, motion must be detected, but not necessarily measured. Other tasks, such as the recovery of three-dimensional structure from motion, require a more complete and accurate measurement of the velocity field [1-5].

In this paper, we present a computational study of the measurement of visual motion. It is a problem which was found to be surprizingly difficult, both in computer vision, and in modelling biological vision systems. Motion measurement poses significant theoretical problems, which the human visual system appears to solve almost effortlessly. The most important of the theoretical problems is the fact that the motion of objects is not determined uniquely from information in the changing image. Figure 1 illustrates two simple examples. In Figure 1a, the solid and dotted lines represent the image of a moving circle, at different instants of time. In the first frame (solid line), the circle lies parallel to the image plane, while in the second frame (dotted line), the circle is slanted in depth. One velocity field consistent with the two frames is derived from pure rotation of the circle about the central vertical axis, as shown to the left in Figure 1a. (The arrows represent a sample of the local velocities.) However, there could also be a component of rotation in the plane of the circle, about its center, as shown to the right in Figure 1a. Both velocity fields correspond to valid motions of the circle. This ambiguity is not particular to circles. In Figure 1b, the solid curve C_1 rotates, translates and deforms over time, to yield the dotted curve C_2. The mapping of points from C_1 to C_2 is much less clear (consider, for example, different possible velocities for the point p). The precise computation of the velocity field in this case is important, when one considers the subsequent computation of structure from motion. Different velocity fields may imply a different three-dimensional structure for the curve in space.

Given this global ambiguity of the velocity field, local motion measurements from the changing image cannot possibly specify a unique local velocity vector, and in fact, may only specify one component of velocity. This local ambiguity of motion measurements is illustrated in Figure 2; Marr and Ullman [6] refer to it as the *aperture problem* (see also [7-12]). If the motion of the edge E is to be detected by operations which examine an area A that is small compared to the overall extent of the edge, the only motion that can be extracted is the component **c** perpendicular to the local orientation of the edge. For example, such operations cannot distinguish between motion in the directions **b**, **c**, and **d**. To determine the motion completely, a

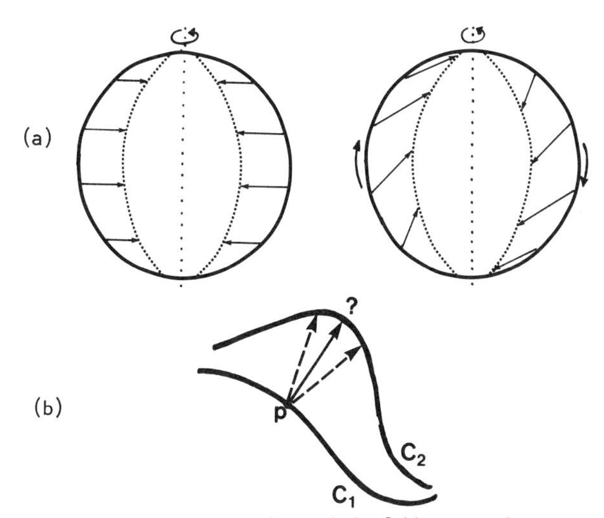

Figure 1. Ambiguity of the velocity field computation.

Motion: Representation and Perception
N.I. Badler and J.K. Tsotsos, Editors
Published by Elsevier Science Publishing Co., Inc.

second stage of analysis is required, which integrates the local motion measurements, either over an area of the image, or along contours. We refer to this as the *motion integration problem.*

Section 2 presents related work in motion analysis. Our work builds primarily on two studies: the first is Marr and Ullman's [6] work on the initial measurement of motion along intensity changes in the image; the second is Horn and Schunck's [8] work on the optical flow computation. In Section 3, we discuss an additional constraint on the velocity field which allows for the general motion of surfaces in space, and leads to a unique velocity field. An algorithm based on this constraint is described in Section 4. We compare predictions of the algorithm with human motion perception, on a few simple demonstrations.

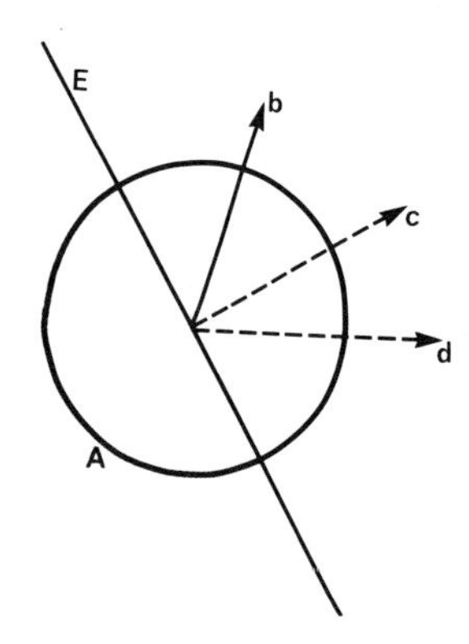

Figure 2. The aperture problem. Motion in the directions **b**, **c**, and **d** can not be distinguished when viewed through the local aperture A.

2. RELATED WORK IN MOTION ANALYSIS

The measurement of motion may be performed at different stages in the processing of an image, utilizing different motion primitives. It is useful to draw a distinction between two main schemes. At the lowest level, motion measurements may be based directly on local changes in light intensity values; these are called *intensity-based* schemes. Examples include cross-correlation, subtraction, and gradient schemes. Alternatively, it is possible to first identify features such as edges and their termination points, corners, blobs, or regions, and then measure motion by matching these features over time, and detecting their changing positions. Schemes of this type are called *token-matching* schemes. Gradient schemes have addressed more directly, the two theoretical problems which we discussed in the introduction. We therefore focus our review on these schemes.

In gradient schemes, initial motion measurements are derived via a comparison between intensity gradients, and temporal intensity changes. A one-dimensional example, illustrating the basic principle, is shown in Figure 3. Consider the intensity profile (intensity I as a function of position x), indicated by the solid curve in Figure 3. At the point p, the profile has a positive slope. If the profile moves to the left, indicated by the dashed curve, the intensity value I at p increases; for a rightward motion, indicated by the dotted and dashed curve, $I(p)$ decreases. The sign of the temporal change in $I(p)$ thus signals the direction of motion, and from the magnitude of the spatial and temporal intensity changes, the speed of motion can be determined.

In principle, measurements of motion may be obtained wherever the image intensity gradient is non-zero; however, the measurements are more reliable at the location of features, such as edges, where steep intensity gradients are induced. Marr and Ullman [6] propose that initial motion measurements in the human visual system are made at the locations of significant intensity changes. These intensity changes give rise to zero-crossings in the output of the convolution of the image with an operator whose shape is the Laplacian of a Gaussian (approximated by the difference of two Gaussians) [13,14]. At the location of a zero-crossing, the spatial gradient of the $\nabla^2 G$ convolution can be combined with its time derivative to compute the motion of the zero-crossing. Davis, Wu and Sun [9] also present a gradient scheme which focuses the motion computation along contours in the image.

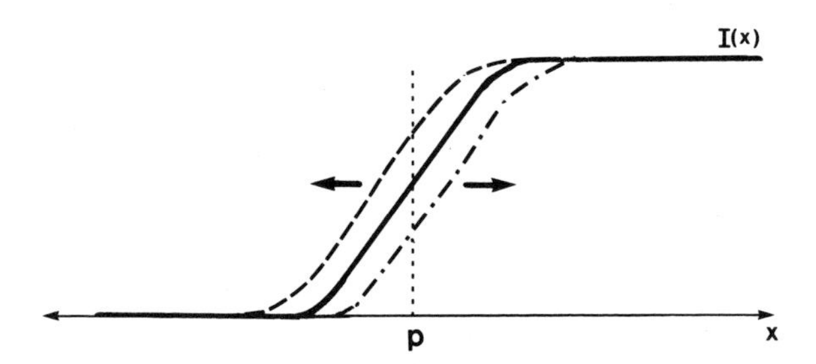

Figure 3. Comparison of the sign of the spatial and temporal derivatives of intensity at the point p yields the sign of direction of motion.

In two dimensions, the spatial and temporal intensity changes alone are not sufficient to determine the local direction and magnitude of velocity [7-12]. For the case of contours, local measurements provide only the component of motion in the direction perpendicular to the orientation of the contour. The component of velocity along the contour remains undetected. More formally, we may express the velocity field along a contour by the function $\mathbf{V}(s)$, where s denotes arclength. $\mathbf{V}(s)$ can be decomposed into components tangent and perpendicular to the contour, as illustrated in Figure 4. $\mathbf{u}^{\top}(s)$ and $\mathbf{u}^{\perp}(s)$ are unit vectors in the directions tangent and perpendicular to the curve, and $v^{\top}(s)$ and $v^{\perp}(s)$ denote the magnitudes of the two components:

$$\mathbf{V}(s) = v^{\top}(s)\mathbf{u}^{\top}(s) + v^{\perp}(s)\mathbf{u}^{\perp}(s) \qquad (1)$$

The component $v^{\perp}(s)$ is given directly by the initial measurements from the changing image; the computation of $\mathbf{V}(s)$ requires the further recovery of $v^{\top}(s)$. In Figure 5, the constraint provided by a single measurement of $v^{\perp}(s)$, is illustrated graphically in *velocity space* (a space in which the x and y axes represent the x and y components of velocity). The true velocity of the corresponding point on the contour

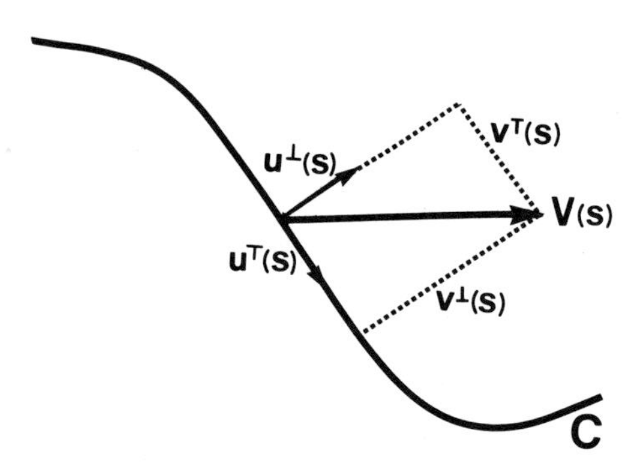

Figure 4. The decomposition of velocity $V(s)$ into tangential and perpendicular components.

must lie along the line l in velocity space which is perpendicular to the vector $v^{\perp}(s)\mathbf{u}^{\perp}(s)$; examples are shown by the dotted arrows. Important aspects of gradient schemes that have been proposed are the way in which local motion constraints are combined to compute the full two-dimensional velocity field, and the additional constraint used to obtain a unique solution.

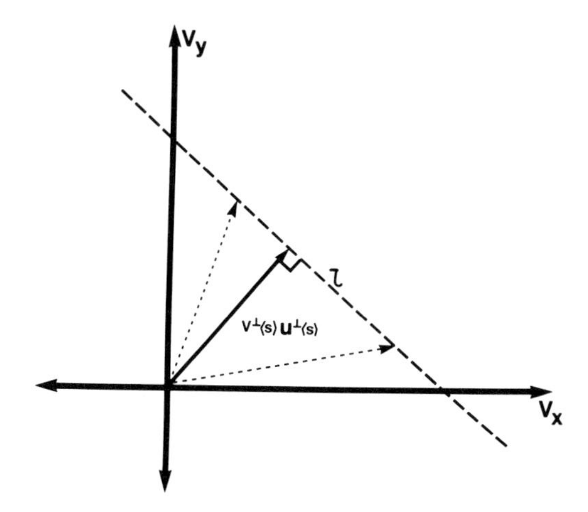

Figure 5. The constraint imposed in velocity space by a single measurement of $v^{\perp}(s)$.

Some gradient schemes assume uniform translation of objects in the image. In this case, the lines of constraint formed in velocity space intersect at a single point. Adelson and Movshon [10] suggest the use of this intersection point for the measurement of motion. In an earlier gradient scheme, Fennema and Thompson [7] proposed the use of the Hough transform for computing the intersection of multiple motion constraints. Marr and Ullman [6] proposed a scheme in which each local measurement restricts the true velocity of a patch to lie within a 180^0 range of directions to one side of a segment of the local zero-crossing contour. A set of measurements taken at different orientations along a zero-crossing contour then further restrict the allowable velocity directions, until a single velocity direction is obtained, which is consistent with all the local measurements.

Other gradient schemes extend the class of motions to rigid rotation and translation in the image plane. Davis, Wu and Sun [9] present an iterative scheme for propagating motion constraints along contours, which begins from points of known velocity, and utilizes the perpendicular component of velocity along the contour. Ullman and Hildreth [12, 15] describe a simple geometrical construction for computing the velocity field in this case. From known directions of velocity at two points along the contour, the direction of velocity can be computed everywhere along the contour. If the perpendicular component of velocity is available, magnitude of velocity may also be obtained.

The above schemes cannot account for the generality of human motion perception; however, they may be useful for the initial detection and rough measurement of motion in the periphery, or analysis of motion during smooth pursuit eye movements, in which stationary objects translate rigidly with respect to the eye. In computer vision, there are restricted applications for these techniques, such as the tracking of objects, or computation of camera motion [16].

A more general constraint was suggested by Horn and Schunck [8] for the optical flow computation. The real world consists predominantly of solid objects, whose surfaces are generally smooth compared with their distance from the viewer. Under motion, the velocity field over these surfaces will generally be smooth as well. Horn and Schunck suggested that a single velocity field solution can be obtained by computing the velocity field which satisfies the constraints derived from the image, and varies as little as possible. In the next section, we apply a similar smoothness constraint along contours in the image.

3. THE SMOOTHNESS CONSTRAINT

In this section we derive an additional constraint on the velocity field, which allows us to analyze the projected motion of three-dimensional objects allowed to move freely in space, and deform over time. The specific analysis assumes that we have measured the perpendicular components of velocity along contours in the image. However, the general constraint may be utilized in other motion measurement schemes as well. The constraint uses the idea of smoothness of the velocity field, as suggested by Horn and Schunck [8]. Intuitively, the motion computation should seek a velocity field which is consistent with the constraints derived from the changing image, and which varies smoothly along a contour. However, there is an infinity of possible solutions satisfying these two properties. A single solution may be obtained by finding the velocity field which varies as little as possible.

To achieve this, we need some means of measuring the variation in velocity along a contour. There are various ways in which this could be done. For example, we could measure the change in direction of velocity as we trace along the contour. Total variation of the velocity field could then be defined as the total change in direction over the entire contour. A second definition involves measuring the change in magnitude of velocity along the contour. This leads to a velocity field solution for which speed is as uniform as possible along the contour. Finally, we could measure the change in the full velocity vector, $\frac{\partial \mathbf{V}(s)}{\partial s}$, incorporating both the direction and magnitude of velocity.

In order to define the variation of the velocity field more formally, first recall the decomposition of velocity into components tangent and perpendicular to the curve (equation 1 and Figure 4). Aside from knowing $v^{\perp}(s)$ everywhere along the curve, there may be points at which the direction and magnitude of velocity, and hence both $v^{\perp}(s)$ and $v^{\top}(s)$, are known. In addition, the direction of velocity alone, and hence the ratio $\frac{v^{\perp}(s)}{v^{\top}(s)}$, may be known at points on the curve. For example, if $v^{\perp}(s) = 0$, the direction of velocity at this point is constrained to lie along the tangent to the curve.

We now consider a more formal means for measuring the variation in the velocity field. Mathematically, this can be accomplished by defining a functional Θ, which maps the space of all possible vector fields (along the contour), $\mathbf{V}$, into the real numbers: $\Theta:\mathbf{V} \mapsto \Re$. This functional should be such that the smaller the variation in the velocity field, the smaller the real number assigned to it. Two candidate velocity fields may then be compared, by comparing their corresponding real numbers. This raises the question of what functional should be used to measure the variation of a velocity field. In the remainder of this section, we evaluate a set of possible functionals, based on the three measures of variation that we previously presented informally: (1) variation in $\mathbf{V}(s)$, (2) variation in the direction of velocity, and (3) variation in the magnitude of velocity, all with respect to the curve.

(1) Variation in V(s)

A scalar measure of the local variation of $\mathbf{V}(s)$ with respect to the curve is given by $|\frac{\partial \mathbf{V}(s)}{\partial s}|$. Two nearby velocity vectors along the image curve in Figure 6a are translated to a common origin in velocity space in Figure 6b, where the vector $\frac{\partial \mathbf{V}(s)}{\partial s}$ is indicated with a dotted arrow. For convenience of notation, we omit the argument to $\mathbf{V}(s)$, writing $|\frac{\partial \mathbf{V}}{\partial s}|$. A measure of the total variation of the velocity field along the curve may then be given by the functional:

$$\Theta(\mathbf{V}) = \int |\frac{\partial \mathbf{V}}{\partial s}| ds$$

We may also consider variations on this functional, involving higher order derivatives, or higher powers, such as:

$$\Theta(\mathbf{V}) = \int |\frac{\partial^2 \mathbf{V}}{\partial s^2}| ds \qquad \text{or} \qquad \Theta(\mathbf{V}) = \int |\frac{\partial \mathbf{V}}{\partial s}|^2 ds$$

(2) Variation in Direction

Let the direction of velocity be given by the angle φ, measured in the clockwise direction from the horizontal, as shown in Figure 6c. In Figure 6d, $\frac{\partial \varphi}{\partial s}$, for two nearby velocity vectors along the image curve, is shown in velocity space. Total variation of direction along the curve could be given by functionals such as the following:

$$\Theta(\mathbf{V}) = \int |\frac{\partial \varphi}{\partial s}| ds$$

or variations involving higher order derivatives, or higher powers.

(3) Variation in Magnitude

Finally, we could measure the change in magnitude of velocity alone, using functionals such as:

$$\Theta(\mathbf{V}) = \int \frac{\partial |\mathbf{V}|}{\partial s} ds$$

Again, we could also consider variations on this measure.

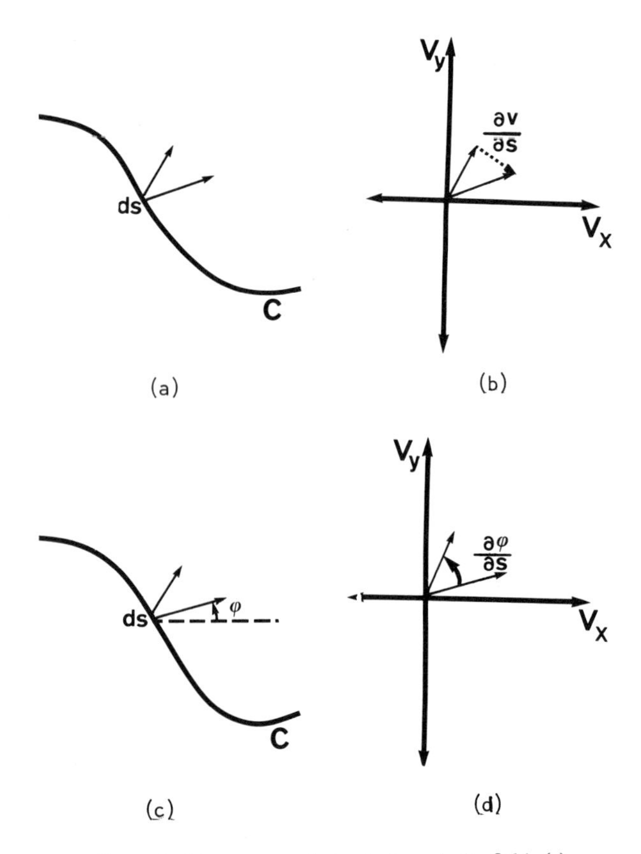

Figure 6. Measuring variation in the velocity field. (a) and (b) Change in the full velocity vector. (c) and (d) Change in direction of velocity.

The functional that we use to measure smoothness may also incorporate a measure of the velocity field itself, rather than strictly utilizing changes in the velocity field along the curve. For example, we could incorporate a term which is a function of $|\mathbf{V}|$. This might be useful if we sought a velocity field which also exhibits the least total motion. In addition, the functional could become arbitrarily complex in its combination of $|\frac{\partial \mathbf{V}}{\partial s}|, |\frac{\partial \varphi}{\partial s}|, \frac{\partial |\mathbf{V}|}{\partial s}$, or higher order derivatives.

We have at least three means of evaluating these measures of smoothness. From a mathematical point of view, there should exist a unique velocity field which minimizes our particular measure of smoothness; this requirement imposes a set of mathematical constraints on our functional. Second, the velocity field computation should yield physically plausible solutions. Finally, if we suggest that such a smoothness constraint underlies the motion computation in the human visual system, this minimization should yield a velocity field consistent with human motion perception.

An examination of these smoothness measures from a physical and mathematical point of view suggests that a measure involving the full velocity vector, such as $\Theta(\mathbf{V}) = \int |\frac{\partial \mathbf{V}}{\partial s}|^2 ds$, is most appropriate for the velocity field computation [12]. Of particular importance are the mathematical properties of this functional. It can be shown that, given a simple condition on the constraints that we derive from the image, there exists a unique velocity field which satisfies our constraints, and minimizes $\int |\frac{\partial \mathbf{V}}{\partial s}|^2 ds$. This condition is almost always satisfied by our initial motion measurements. To obtain this result, we take advantage of the analysis used by Grimson [17] for evaluating possible functionals for performing surface interpolation from stereo data. The basic mathematical question is, what conditions on the form of the functional, and the structure of the space of velocity fields, are needed to guarantee the existence of a unique solution? These conditions are captured by the following theorem (see also [18]):

Theorem: *Suppose there exists a complete semi-norm Θ on a space of functions H, and that Θ satisfies the parallelogram law. Then, every nonempty closed convex set $E \subset H$ contains a unique element v of minimal norm, up to an element of the null space. Thus, the family of minimal functions is*

$$\{v + s \mid s \in S\}$$

where

$$S = \{v - w \mid w \in E\} \cap \mathcal{N}$$

and $\mathcal{N}$ is the null space of the functional

$$\mathcal{N} = \{u \mid \Theta(u) = 0\}.$$

It can be shown that the functional $\{\int |\frac{\partial \mathbf{V}}{\partial s}|^2 ds\}^{\frac{1}{2}}$ is a complete semi-norm, which satisfies the parallelogram law. Second, the space of all possible velocity fields, which satisfies the constraints derived from the image, is convex. It then follows from the above theorem that this space contains a unique element of minimal norm, up to an element of the null space. Since our smoothness measure is non-negative, minimizing $\{\int |\frac{\partial \mathbf{V}}{\partial s}|^2 ds\}^{\frac{1}{2}}$ is equivalent to minimizing $\int |\frac{\partial \mathbf{V}}{\partial s}|^2 ds$. The null space in this case is the set of constant velocity fields, since $\int |\frac{\partial \mathbf{V}}{\partial s}|^2 ds = 0$ implies $|\frac{\partial \mathbf{V}}{\partial s}| = 0$ everywhere, which implies $\mathbf{V}(s)$ constant. As a consequence, the following can be shown [12]: If $v^{\perp}(s)$ is known at two points, for which the orientation of the curve is different, then there exists a unique velocity field which satisfies the known velocity constraints and minimizes $\int |\frac{\partial \mathbf{V}}{\partial s}|^2 ds$. An extended straight line does not yield measurements for two different orientations, but in all other cases, there is sufficient information along a contour to guarantee a unique solution to the velocity field.

We can apply the constraint of least variation and compute a projected two-dimensional velocity field for any three-dimensional surface, whether rigid or non-rigid, undergoing general motion in space. If we measure the variation in the full velocity vector along a contour in the image, using a functional such as $\int |\frac{\partial \mathbf{V}}{\partial s}|^2 ds$, we are guaranteed that there exists a unique solution to the velocity field computation that minimizes this variation. While it is not yet clear that the general smoothness constraint, or the particular measure $\int |\frac{\partial \mathbf{V}}{\partial s}|^2 ds$, is the most appropriate for the motion computation, it is important that this measure satisfies certain essential mathematical requirements, that the other measures do not. For example, the use of a functional incorporating only a measure of velocity direction, which attempts to make the local velocity vectors as parallel as possible, does not yield functionals which are semi-norms, and consequently, does not lead to a unique velocity field solution. For a scheme to underly the motion computation in the human visual system, it is essential that it be mathematically well-founded.

4. AN ALGORITHM TO COMPUTE THE VELOCITY FIELD

The velocity field computation has been formulated as an optimization problem. We seek a solution which satisfies the constraints derived from the image, and minimizes the measure of a discrete correlate to $\int |\frac{\partial \mathbf{V}}{\partial s}|^2 ds$ over the image curve. In general, we are interested in algorithms which are biologically feasible, which suggests that they involve simple, local, parallel operations [1,17,20]. In this section, we present an algorithm which is based on the smoothness constraint.

Our motivation is to examine the plausibility of our general formulation of the velocity field computation. First, we can examine the physical plausibility of the smoothness constraint, through empirical demonstration. While we expect the velocity field generally to be smooth across physical surfaces, it is not clear that the velocity field of *least* variation is still physically plausible. Second, we can examine the plausibility of this constraint for the human visual system, through its predictions on known perceptual demonstrations. We do not propose that the particular algorithm we present here is utilized by the human visual system; our interest is in testing the theoretical aspects of the computation. This section provides only the beginning of such a study. The initial indications, however, offer support for our formulation of the velocity field computation.

To develop an algorithm, we utilize techniques from mathematical programming [21]. First, we express our functional in terms of the x and y components of velocity, V_x and V_y. The continuous functional becomes:

$$\Theta = \int \left[\left(\frac{\partial V_x}{\partial s} \right)^2 + \left(\frac{\partial V_y}{\partial s} \right)^2 \right] ds$$

The general mathematical programming problem can be stated as:

$$\begin{array}{lll} \text{minimize} & f(\mathbf{x}) & \\ \text{subject to} & h_i(\mathbf{x}) = 0 & i = 1, \ldots, m \\ & g_j(\mathbf{x}) \geq 0 & j = 1, \ldots, r \\ & \mathbf{x} \in S & \end{array}$$

where $\mathbf{x}$ is an n-dimensional vector of unknowns, $\mathbf{x} = (x_1, x_2, \ldots, x_n)$. The objective function f, and constraints h_i, $i = 1, \ldots, m$ and g_j, $j = 1, \ldots, r$ are real-valued functions of the variables $x_1, x_2, \ldots, x_n$. The set S is a subset of the n-dimensional space.

In our case, we let:

$$\mathbf{x} = \{V_{x1}, V_{x2}, \ldots, V_{xn}, V_{y1}, V_{y2}, \ldots, V_{yn}\}$$

That is, $\mathbf{x}$ consists of the n x-components and n y-components of velocity, for n points along a contour. For now, we assume that the points are evenly spaced along the contour, but the computation can easily be extended to allow points that are not evenly spaced. For the case of a closed contour, we choose the following discrete formulation of the objective function f:

$$f_c(\mathbf{x}) = \sum_{i=2}^{n} [(V_{x_i} - V_{x_{i-1}})^2 + (V_{y_i} - V_{y_{i-1}})^2] + [(V_{x1} - V_{xn})^2 + (V_{y1} - V_{yn})^2]$$

For the case of an open contour, a different expression for the derivative at the endpoints is used to compute $f(\mathbf{x})$.

With regard to the constraints we have derived from the image, we can either force the velocity field to satisfy the constraints exactly, or satisfy them approximately. In the first case, these constraints are of the form:

$$\mathbf{V} \cdot \mathbf{u}^{\perp} - v^{\perp} = 0$$

The constraint states explicitly that the normal component of velocity for the computed velocity field should be equivalent to the measured normal component. Letting $\mathbf{V} = (V_x, V_y)$ and $\mathbf{u}^{\perp} = (u_x^{\perp}, u_y^{\perp})$, the constraints are simple linear constraints:

$$V_{x_i} u_{x_i}^{\perp} + V_{y_i} u_{y_i}^{\perp} - v_i^{\perp} = 0$$

At this point, we can either set up the computation as a constrained or unconstrained optimization. In the constrained case, we specify the above linear constraints explicitly. To obtain an unconstrained problem, we first use the constraints to express the y-components of velocity in terms of the x-components:

$$V_{y_i} = \frac{v_i^{\perp} - V_{x_i} u_{x_i}^{\perp}}{u_{y_i}^{\perp}}$$

We can now substitute these expressions for the y-components into $f(\mathbf{x})$ and let $\mathbf{x} = \{V_{x1}, V_{x2}, \ldots, V_{xn}\}$. There is no further constraint on the x-components of velocity.

In general, there will be error in the measurements of $v^{\perp}$. From a practical standpoint, it may be advantageous to require that the velocity field only approximately satisfy the image constraints. This can be accomplished by requiring that the difference between $\mathbf{V} \cdot \mathbf{u}^{\perp}$ and the measured $v^{\perp}$ be small. The continuous functional can be extended as follows:

$$\Theta = \int \left[\left(\frac{\partial V_x}{\partial s} \right)^2 + \left(\frac{\partial V_y}{\partial s} \right)^2 \right] ds + \beta \int [\mathbf{V} \cdot \mathbf{u}^{\perp} - v^{\perp}]^2 ds$$

β is a weighting factor, which expresses our confidence in the measured velocity constraints. The second term describes the least squares difference between the computed and measured normal components of velocity. The above functional leads to the following objective function:

$$f_u(\mathbf{x}) = f_c(\mathbf{x}) + \beta \sum_{i=1}^{n} \left[V_{x_i} u_{x_i}^{\perp} - V_{y_i} u_{y_i}^{\perp} - v_i^{\perp} \right]^2$$

The space S is the entire n-dimensional space. Expressed in this way, the problem is an unconstrained optimization. In general, there may be points where V, or the direction of velocity alone, is known. This additional constraint would transform the problem back into a constrained optimization.

For the three examples that we discuss here, we set up the velocity field computation as an unconstrained optimization, in which the image constraints are only approximately satisfied. The conjugate gradient algorithm is used to obtain the solution. This is an iterative algorithm which utilizes the gradient of the objective function to choose an optimal path to follow along the solution surface to the final solution (see Luenberger [21] for details of the algorithm). Our initial velocity field is given by the normal velocity vectors, $v^{\perp}\mathbf{u}^{\perp}$ along the curve. After a number of iterations, the algorithm converges to a unique solution.

In Figure 7, we show a sampling of the true velocity fields for three curves in motion. In Figure 7a and 7b, an ellipse and an arm of a logarithmic spiral are rotating rigidly in the image plane about the point O. In Figure 7c, we have a three-dimensional circular helix, rotating about its central vertical axis. The three-dimensional curve, shown on an imaginary cylinder to the left in Figure 7c, and its velocity field, have been projected onto the image plane, using orthographic projection. The resulting two-dimensional velocity vectors are shown to the right in Figure 7c. These particular examples were chosen because their perceived motion differs from the true motion of the curves. We ask whether the results of our algorithm, based on the additional smoothness constraint, also differ from the true velocity field, in a way that is consistent with our perception.

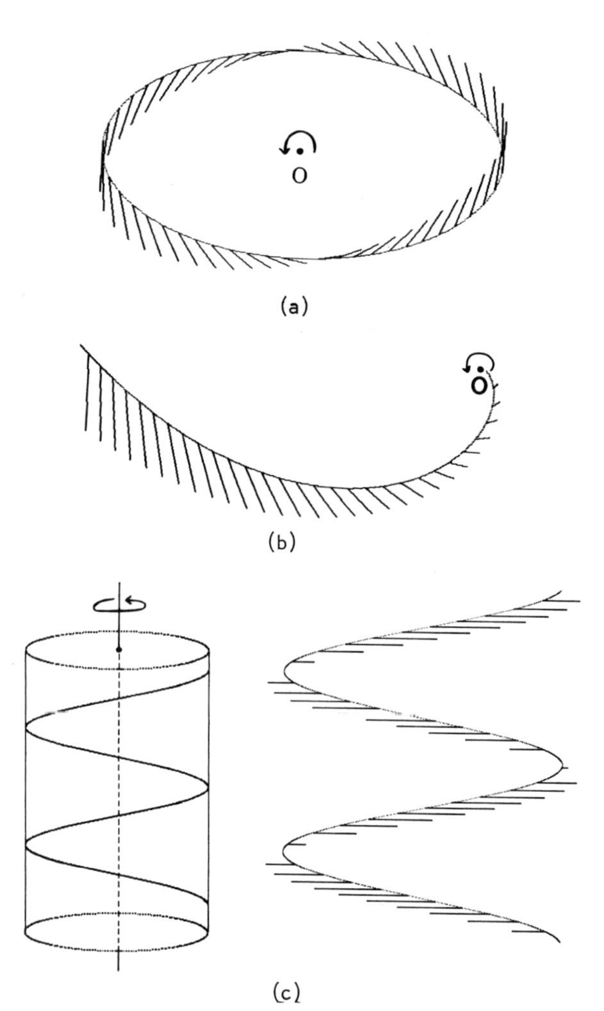

Figure 7. Sample vectors for the true velocity field for (a) an ellipse, rotating rigidly in the image plane about its center, (b) one arm of a logarithmic spiral, rotating about the point O, and (c) a circular helix, rotating about its central vertical axis.

Figure 8 illustrates a sample of the initial measurements of motion along the curve. The curve, and its normal components of velocity, were generated analytically, and therefore represent ideal input data. This data is the only input to the algorithm. Figure 9 shows the results of the conjugate gradient algorithm, again sampling the velocity field along the curve.

In Figure 9a, we can see that for the rotating ellipse, the predicted velocity field is quite different from the true velocity field shown in Figure 7a. In particular, there is less torsion in the computed velocity field. At first glance, one might not consider this a plausible solution. However, in some early perceptual experiments by Wallach, Weisz and Adams [19], they noted that a rigid ellipse does not appear rigid under rotation; it appears to deform continuously as it rotates. In their experiments, they placed simple geometric figures on a rotating turntable. After viewing the ellipse for some time, they noted that it also appears as a circle, rising out of the plane of the turntable. In repeating the experiments ourselves, with a steady fixation of the eyes at the center of the turntable, we observed that during the three-dimensional perception, the circle itself appears rigid, but does not remain in a rigid configuration with respect to the turntable. Rather, the circle appears to roll around the table. The motion of the ellipse, in

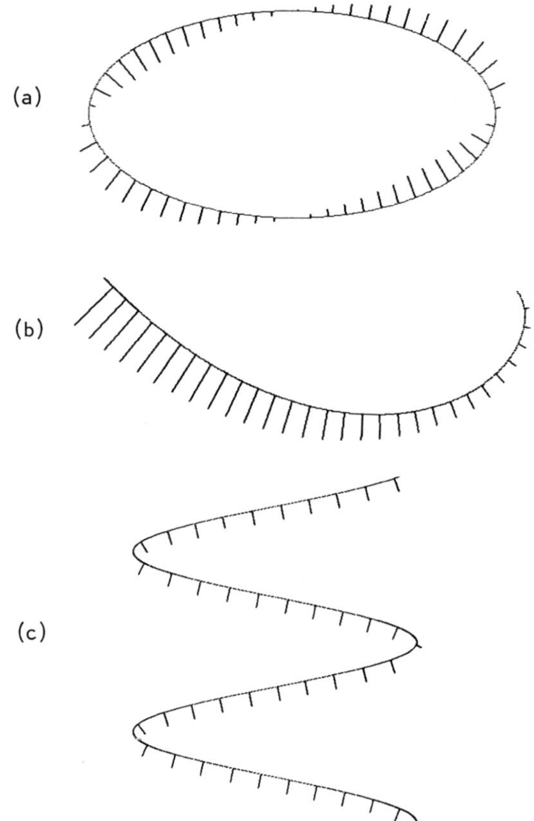

Figure 8. The initial velocity vectors for the (a) ellipse, (b) logarithmic spiral, and (c) circular helix.

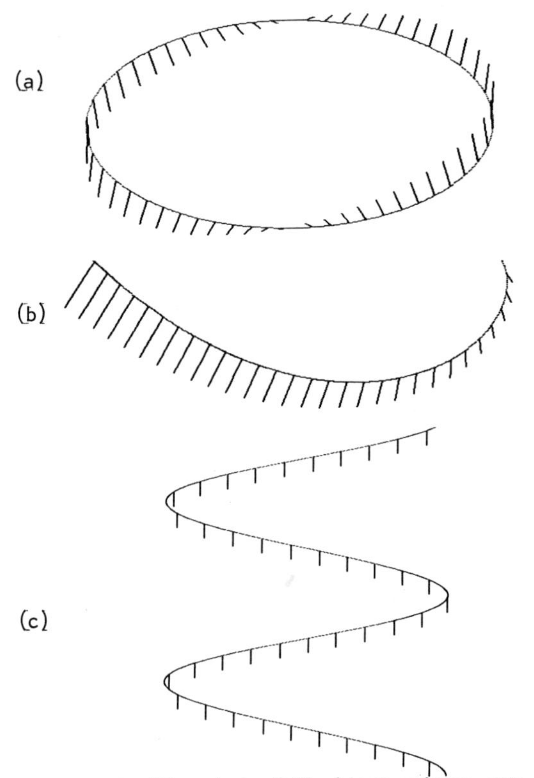

Figure 9. The velocity field of least variation, computed with the conjugate gradient algorithm, for the (a) ellipse, (b) logarithmic spiral, and (c) circular helix.

both the planar and three-dimensional configurations, is qualitatively consistent with the velocity field of Figure 9a. Both perceptions correspond to an underestimate in the torsion of the velocity field.

When we view the logarithmic spiral in rotation, we perceive a strong radial component of motion, which is particularly apparent at the center of the spiral. This suggests that the local velocity vectors are directed outward from the center of the spiral. The results of the algorithm are illustrated in Figure 9b. Toward the center of the spiral, the predicted velocity vectors have a large radial component. This component decreases as we move away from the center. The computed velocity field exhibits a larger radial component than the initial velocity vectors shown in Figure 8b, over most of the curve.

Finally, the third example corresponds to the familiar barberpole illusion. The true motion of the helix gives rise to velocity vectors that are strictly horizontal, yet our perception of the direction of motion of the helix is vertical. This is also consistent with the smoothest velocity field, shown in Figure 9c. The initial velocity measurements would not predict this vertical motion.

From a physical point of view, a possible constraint on the velocity field, which we have not considered explicitly, is rigidity. One might consider rigidity to be an important factor in the physical plausibility of the velocity field, yet demonstrations such as the rotating ellipse and spiral (also see other demonstrations in [19]), suggest that it may not be used as an explicit constraint on the motion computation in the human visual system.

There are two special classes of motion that we should note. First, if a curve undergoes pure translation, the velocity vectors are constant along the curve, so our measure of variation yields a value of zero. Since zero is the minimum value that the measure can obtain, it follows that if there exists a valid solution (consistent with the image constraints) which is consistent with pure translation, then the algorithm would find this solution. Second, when a straight line moves rigidly in space, the velocity field that it generates is also the smoothest one consistent with the constraints provided in its projection onto the image plane, given the particular measure of variation that we have derived here [12]. Thus, for these two special classes of motion, our algorithm, based on the smoothness constraint, always computes the correct physical solution.

5. SUMMARY

To summarize, the main theoretical problem for the velocity field computation is the ambiguity inherent in the information supplied by the changing image. As a result, the computation requires the combination of local constraints, either over an area, or along contours in the image, together with additional constraints. These additional constraints should be derived from properties of the physical world that generally hold true. We explored one particular additional constraint, the smoothness constraint, first proposed by Horn and Schunck [8] for the optical flow computation. This constraint relies on the fact that physical surfaces are generally smooth.

From mathematical considerations, we derived a measure of variation of the velocity field along a contour that led to a unique solution. This measure, $\int |\frac{\partial \mathbf{V}}{\partial s}|^2 ds$, incorporates the change in the full velocity vector along the contour. We presented one example of an algorithm which embodied this constraint, and showed that the predictions of the algorithm were qualitatively consistent with human motion perception, on three perceptual demonstrations. Areas of further study include a continued examination of the physical and perceptual validity of the smoothness constraint, and consideration of alternative algorithms for performing the velocity field computation.

Acknowledgements

This article describes research done in the Artificial Intelligence Laboratory of the Massachusetts Institute of Technology. Support for the laboratory's artificial intelligence research is provided in part by the Advance Research Projects Agency of the Department of Defense under the Office of Naval Research Contract N00014-75-C-0643 and in part by National Science Foundation Grant MCS77-07569.

References

1. S. Ullman: *The Interpretation of Visual Motion* (MIT Press Cambridge and London, 1979)
2. S. Ullman: Proc. Roy. Soc. Lond. B **203**, 405-426 (1980)
3. K. Prazdny: Biol. Cyb. **36**, 87-102 (1980)
4. W. F. Clocksin: Perception **9**, 253-269 (1980)
5. H. C. Longuet-Higgins, K. Prazdny: Proc. Roy. Soc. Lond. B **208**, 385-397 (1981)
6. D. Marr, S. Ullman: Proc. Roy. Soc. Lond. B **211**, 151-180 (1981)
7. C. I. Fennema, W. B. Thompson: Comp. Graph. Image Proc. **9** 301-315 (1979)
8. B. K. P. Horn, B. G. Schunck: Artif. Intel. **17** 185-203 (1981)
9. L. S. Davis, Z. Wu, and H. Sun: In *Proc. ARPA Image* Understanding Workshop, ed. by L. S. Baumann (Science Applications Inc., Arlington, VA, 1982)
10. E. H. Adelson, J. A. Movshon: "Phenomenal Coherence of Moving Visual Patterns", Nature **300**, 523-525 (1982)
11. H. Wallach: "On Perceived Identity: 1. The Direction of Motion of Straight Lines," In *On Perception*, ed. by H. Wallach, (Quadrangle, N.Y., 1976)
12. E. Hildreth: In *Proc. Workshop on Rep. and Control.* (IEEE Computer Society Press, Los Angeles, 1982)
13. D. Marr, E. C. Hildreth: Proc. Roy. Soc. Lond. B. **207**, 187-217 (1980)
14. D. Marr, T. Poggio: Proc. Roy. Soc. Lond. B. **204**, 301-328 (1979)
15. S. Ullman, E. C. Hildreth: In *Proc. Workshop on 'Physical* and Biological Processing *of Images'*, Roy. Soc. Lond., (Springer-Verlag, Heidelberg, 1983). also appears as: (MIT AI Memo 699, 1982)
16. A. Bruss, B. K. P. Horn: *Passive Navigation* (MIT AI Memo 661 1981)
17. W. E. L. Grimson: *From Images to Surfaces. A Computational* Study of the Human Early *System* (MIT Press, Cambridge and London, 1981)
18. W. Rudin: *Functional Analysis*, (McGraw-Hill Co., New York 1973)
19. H. Wallach, A. Weisz, P. A. Adams: Am. J. Psych. **69**, 48-59 (1956)
20. D. Marr: *VISION* (W. H. Freeman Co., San Francisco, 1981)
21. D. G. Luenberger: *Introduction to Linear and Nonlinear* Programming (Addison-Wesley Co., Reading, MA., 1973)

Determining the Instantaneous Axis of Translation from Optic Flow Generated by Arbitrary Sensor Motion

J. H. Rieger and D. T. Lawton
Computer and Information Science Department
University of Massachusetts
Amherst, Massachusetts

Abstract

This paper develops a simple and robust procedure for determining the instantaneous axis of translation from image sequences induced by unconstrained sensor motion. The procedure is based upon the fact that difference vectors at discontinuities in optic flow fields generated by sensor motion relative to a stationary environment are oriented along translational field lines. This is developed into a procedure consisting of three steps: 1) locally computing difference vectors from an optic flow field; 2) thresholding the difference vectors; and 3) minimizing the angles between the difference vector field and a set of radial field lines which correspond to a particular translational axis. This method does not require *a priori* knowledge about sensor motion or distances in the environment. The necessary environmental constraints are rigidity and sufficient variation in depth along visual directions to endow the flow field with discontinuities. The method has been successfully applied to noisy, sparse, and low resolution flow fields generated from real world image sequences. Experiments are reviewed which indicate that the human visual system also utilizes discontinuities in optic flows in determining self-motion. In addition, due to the computational simplicity of the procedure, hardware realization for real-time implementation is possible.

1. Introduction

The motion of an observer/sensor is in general composed of a translation and a rotation. It generates an optic flow field in the image plane of the sensor due to changes of visual directions of details in the environment over time (Gibson *et. al.* 1955). The instantaneous translatory velocity of the sensor induces a radial image velocity field with the intersection of the translational axis and image plane as its center. Likewise the rotational velocity of the sensor induces a rotational velocity field in the image that is purely direction dependent. In general an optical velocity field is the vector sum of translational and rotational fields (footnote 1).

The translational component (and its spatial and temporal derivative fields) contains, e.g., information about the shape of objects (Koenderink and van Doorn 1977), about the relative depth properties of the environment (Lee 1980, Prazdny 1980), or about motion parameters for navigating along curved trajectories (Rieger 1983). The rotational component, on the other hand, contains no environmental depth information and needs to be separated from the translational component. The main computational step toward extracting the translational field from optic flows containing translational and rotational components is to determine the instantaneous axis of sensor translation.

There are several problems in using these and related formulations for determining camera motion parameters and environmental information from real world image sequences. The inference techniques generally require high resolution image displacements as input and are sensitive to the noise and errors that current techniques for determining image motions typically produce. They can also involve solving complex equations and require significant computation.

We show that the recovery of camera motion parameters can be simplified and performed robustly from noisy, low resolution, and sparse displacement fields by analyzing the image displacements at image positions where environmental depth changes occur. This procedure utilizes an observation made by Longuet-Higgins and Prazdny (1980) that details in the environment located in the same direction from an observer/sensor (i.e. along the same ray of projection), but are at different depths, will differ in their image velocity vectors by the difference of their translational components only.

footnote: we refer to both the *optical velocity field* generated by continuous sensor motion and the *optical displacement field* generated by discrete sensor motion as *optical flow fields*. Which is being referred to should be obvious by context.

Motion: Representation and Perception
N.I. Badler and J.K. Tsotsos, Editors
Published by Elsevier Science Publishing Co., Inc.

This is because the rotational components of optic flows are purely direction dependent. The axis of sensor translation is obtained from the intersection of radial fieldlines which are determined by such difference vectors. In cluttered environments we find details being located in the same visual direction but being separated in depth at occluding edges. Figure 1 shows an optical flow field induced by a sensor translating and rotating relative to two surfaces that are separated in depth. We can see that the difference vectors of the vector pairs at the edge point toward the direction of translation indicated in the figure.

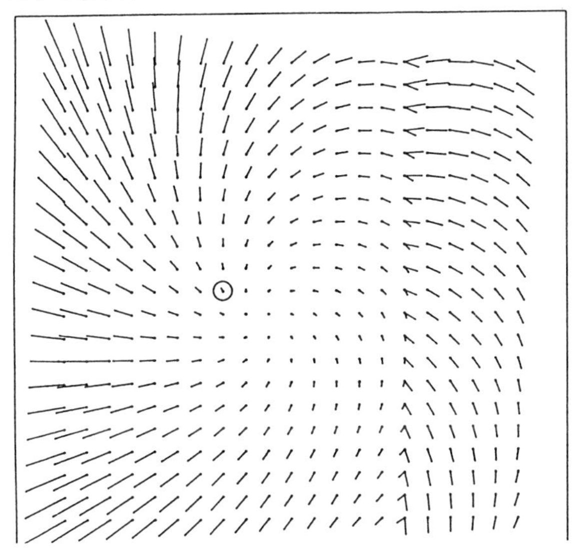

Figure 1. Optic Flow at an occluding edge (from Rieger (1983)).

There are significant difficulties in applying this observation to actual image sequences. From images formed at discrete, successive instants we obtain image displacements and not instantaneous optic velocities. Thus the computation must be expressed in terms of discrete sensor motions. Flow fields computed from actual image sequences are not arbitrarily dense and are in fact generally sparse so there will not be two distinct flow vectors positioned at the same image point. Thus it is necessary to perform the computation using difference vectors determined from image displacement vectors which are spatially separated. Also, flow fields computed from actual images are errorful, especially near occlusion boundaries because of the changes in image structure that occur there. Thus the procedure must be robust to such distortions in the determined difference vectors.

These problems are addressed in this paper. In section 2 we analyze the error introduced by computing difference vectors from spatially separated pairs of vectors. This will lead to a simple method of determining sensor translation from sparse image displacement fields in section 3. In section 4 the results of applying the procedure to simulated data will be compared with its predicted behavior. Results obtained using displacement fields generated from actual image sequences will be described in section 5. Finally, we will discuss in section 6 the relation between this procedure and psychophysical experiments.

2.1. Components of difference vectors between spatially separated optic flow vectors

In this section we decompose a difference vector formed from spatially separated image velocity vectors into a signal component oriented along the correct translational field line and a noise component. The signal component increases for difference vectors formed at image locations where large depth changes occur in the corresponding environmental positions. It also increases with increasing distance between the difference vector and the intersection of the translational axis with the image plane. To the extent that these conditions are satisfied for an optic flow field, its difference vector field will approach the corresponding set of correct translational field lines. Note that this does not require knowledge about the location of occlusion boundaries or of image areas corresponding to large visual slant.

Let us consider a difference vector $\overrightarrow{\Delta F}$ at a point $\tilde{P}_1$ in the image that has been obtained by subtracting the image velocities at $\tilde{P}_1$ and at some adjacent image position $\tilde{P}_2$, where $\tilde{P}_1$ and $\tilde{P}_2$ are separated by (d_x, d_y). (See figure 2 and the appendix for a more detailed derivation of the following equations).

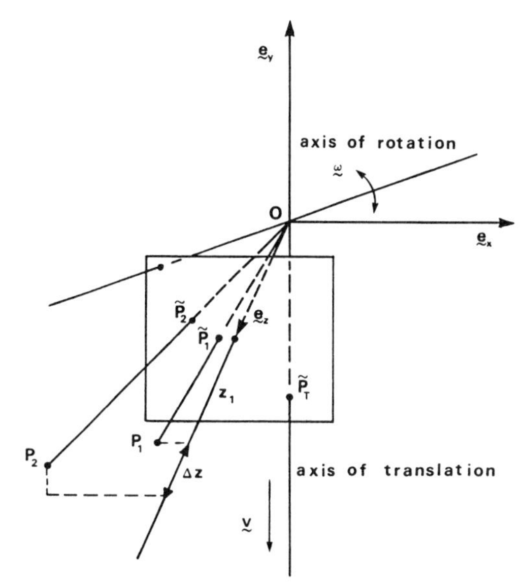

Figure 2. Coordinate system employed in the text: $\tilde{P}_1$ and $\tilde{P}_2$ are the images of the environmental details P_1 and P_2.

We obtain a component $\overrightarrow{\Delta F_R}$ that is due to changes in the rotational component field [the field induced by the rotational velocity (ω_x , ω_y , ω_z) of the sensor]

$$\overrightarrow{\Delta F_R} =$$

$$[d_y\omega_z - d_x(2\tilde{x}_1 + d_x)\omega_y + (\tilde{x}_1 d_y + \tilde{y}_1 d_x + d_x d_y)\omega_x]\underset{\sim}{e}_x$$
$$+ [-d_x\omega_z - (\tilde{x}_1 d_y + \tilde{y}_1 d_x + d_x d_y)\omega_y + d_y(2\tilde{y}_1 + d_y)\omega_x]\underset{\sim}{e}_y$$

If we let $\tilde{P}_T$ denote the intersection of the translational axis with the image plane, V_Z the translational velocity of the sensor along the z-axis, Z_1 the depth of P_1, and ΔZ the depth difference between P_1 and P_2, the translational component of the difference vector reads

$$\overrightarrow{\Delta F_T} = \frac{V_Z}{Z_1 + \Delta Z} \left\{ \overrightarrow{\tilde{P}_1\tilde{P}_2} - \frac{\Delta Z}{Z_1} \overrightarrow{\tilde{P}_T\tilde{P}_1} \right\} .$$

We can rewrite $\overrightarrow{\Delta F}$ as consisting of a component along a translational fieldline and a noise component

$$\overrightarrow{\Delta F} = \left[\frac{-V_Z \Delta Z}{Z_1 Z_2} \overrightarrow{\tilde{P}_T\tilde{P}_1} \right]_{Signal} + \left[\frac{V_Z}{Z_2} \overrightarrow{\tilde{P}_1\tilde{P}_2} + \overrightarrow{\Delta F_R} \right]_{Noise} .$$

For difference vectors with sufficient angular separation from the translatory axis and separation in depth $\overrightarrow{\Delta F}_{Signal} >> \overrightarrow{\Delta F}_{Noise}$.

2.2. Global behavior of difference vector field, possible filter operations

From the previous discussion it is now possible to make some predictions about deviations between the orientations of the difference vector field and the corresponding translational field. The orientation of the noise component of each difference vector is affected by the relative positions of its orginal vector pair ($\overrightarrow{\tilde{P}_1 \tilde{P}_2}$). Therefore, if the velocity/displacement vectors of the original optic flow field are randomly distributed in the image plane, the difference vector field fits the correct set of translational field lines better as the number of difference vectors is increased since $\overrightarrow{\Delta F}_{Signal}$ and $\overrightarrow{\Delta F}_{Noise}$ are additive. An additional factor affecting the fit of the difference vector field to the correct translational field lines is the length of the difference vectors. If a difference vector is small compared to the local average magnitude of the difference vector field, it is more probable that its orientation will be different from the correct translational field line. We therefore filtered the difference vector fields in the experiments described in the next sections by thresholding on difference vector length. As was expected the fit of the difference vectors improved up to a certain magnitude of the length threshold and then detoriated again due the smaller number of difference vectors. So far the value for the threshold has been found by experimenting with different values for several different flow fields and corresponding motions and environments. It may be useful to adjust the threshold automatically to an optimal value given information about the difference vector field like density, average magitude and dispersion of vectors or to refine it after an initial determination of motion parameters.

3. Implementation

Before describing the optimization procedure, it is necessary to develop the calculation in terms of discrete sensor motions and describe how image difference vectors are determined.

If we are dealing with displacements of details in images formed over discrete time intervals instead of an instantaneous optic velocity field, we have to be careful to describe all quantities with respect to the same reference system. Suppose two environmental points lie along the same ray of projection in an image at time t. Translating and rotating the sensor will displace the projections of these points to new positions in the image at time t+1. In the image at time t+1, the image points will be separated due to the translational component of the sensor motion (unless they are located on the translational axis). The separated image points and the intersection of the translational axis with the image plane will be collinear at time t+1. This is the discrete analog of the fact that difference vectors at discontinuities of an instantaneous optic velocity field are oriented along translational field lines.

3.1. Determining Difference Vectors from Sparse Flow Fields

Given image displacements D1 and D2 at positions P1 and P2, the difference vector between points 1 and 2 is obtained by subtracting D2 from D1 and positioning the resulting vector at P1 + D1. Two thresholds are used in evaluating difference vectors. The separation threshold determines the maximal allowable distance between displacement vectors in determining difference vectors. The neighborhood of a given displacement vector contains all other displacement vectors which lie within a distance determined by the separation threshold. The length threshold determines the minimal allowable length for a difference vector.

Since it is not necessary to determine occlusion boundaries before forming the difference vectors, the procedure is applied homogenously with respect to a displacement field. Each vector determines a set of difference vectors in its neighborhood which are of sufficient length. The resulting set of difference vectors determined for all the displacement vectors are then used by the optimization procedure.

3.2. Optimization Procedure

The procedure used to determine a translational axis from a set of difference vectors is similar to that used in Lawton (1982) to determined a translational axis from a noisy displacement field generated by translational motion. It involves finding a translational axis and the corresponding set of radial field lines which minimizes the measure

$$\sum_{i=1}^{n} (1.0 - abs(\cos\theta_i))$$

where θ_i is the angle between the ith difference vector and the radial field line at that position in the image.

The error measure is defined on a unit sphere with each point corresponding to a possible translational axis. The sphere has significant advantages over the image plane as the domain since it allows for a uniform, global sampling of the error function. Using a spherical coordinate system (r , θ , ϕ) where

$$x = r \sin\phi \sin\theta$$

$$y = r \cos\phi$$

$$z = r \sin\phi \cos\theta$$

each axis of translation is defined by some point (θ , ϕ) or ($\theta + \pi$, $\phi + \pi$) on the unit sphere. Due to this redundancy, we restrict the error function to the hemisphere determined by the bounds $-\pi/2 \le \theta \le \pi/2$ and $0 \le \phi \le \pi$.

The search process consists of a global sampling of the error measure to determine its rough shape followed by a local search to find a minimum. The global search is an instance of a generalized Hough Transform (Ballard 1980, O' Rourke 1981) in which each difference vector votes against a particular translational axis by the term in the error measure above. The results of the global search are stored in a global error histogram which is indexed by positions seperated by regular intervals on the unit sphere. In the experiments below the global sampling was performed densely with respect to the unit hemisphere to display the error function. The local search utilizes steepest descent with diminishing step-size and is inititialized where the minimum value is determined by the global sampling. Below, the local search was defined with respect to pixel coordinates in the image plane with the finest resolution set to one pixel. Subpixel interpolation was not used but could be without difficulty.

4. Experiments with Simulated Data

Several experiments have been performed with simulated displacement fields to understand the effects of such factors as resolution, neighborhood size, noise, and environmental depth variance. Two are presented here. The first shows the effects of using low resolution displacement fields. The second shows the behaviour of the procedure as environmental depth variance is increased.

4.1. Experiment 1

The flow field in figure 3a shows image displacements positioned at pixel positions having coordinates which are multiples of 8 from a 128x128 pixel field. The components of the displacement vectors were stored as 8 bit integers. Fewer bits were actually required since the maximal displacement was less than 16 pixels in length. The environment consisted of a surface at depth of 10 units along the z axis and a background surface at a depth of 30 units along the z axis. The obvious discontinuities in the flow field in figure 3a indicate the boundary of the nearer surface. The sensor motion consisted of an initial rotation of 0.1 radians about the (1,1,1) axis followed by a translation of 2 units along (0,0,1). The separation threshold was set to 1 pixel since there was a displacement vector at each pixel location. The length threshold was set to 3 pixels. The resulting error function is shown in figure 3b (Darker in the figure corresponds to less error; also recall that this is a plot of a hemisphere in θ , ϕ coordinates and not the image plane). As can be seen, it is strongly unimodal. The minimum in the global histogram corresponded to the image position (60.28, 60.28). The local search determined the minimum to be at (63, 63). The correct, subpixel, position was (63.5, 63.5).

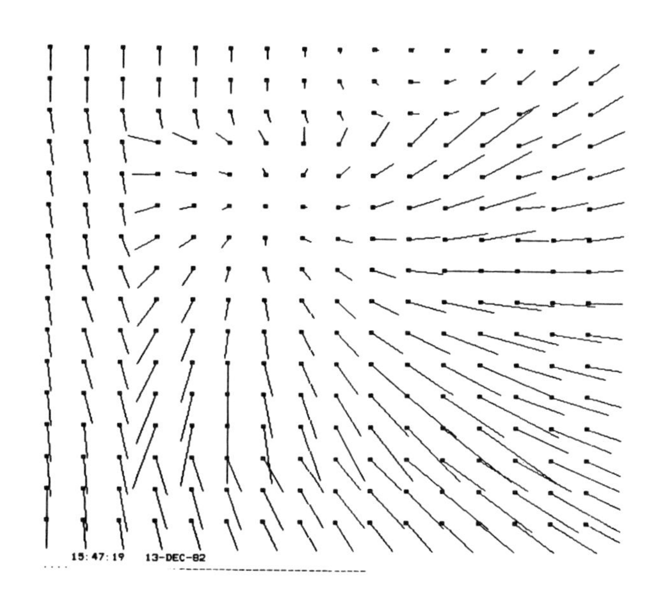

Figure 3a

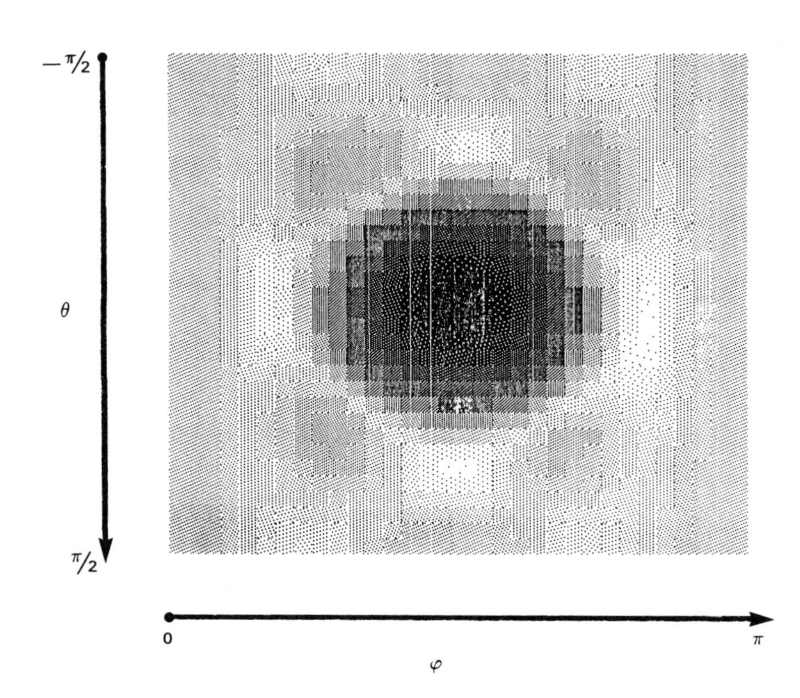

Figure 3b

4.2. Experiment 2

The flow fields in figures 4a and 5a are 32x32 pixel fields using 16 bit real numbers for their components. They were generated using the same motion as in experiment 1 except the translational displacement along the z-axis was only 1 unit. To see the effects of depth variance, the field in figure 4a was produced by moving relative to a (x,y)-plane at 20 units along the z-axis (no depth variance) while the field in figure 5a was produced by moving relative to environmental points with random depths between 20 and 120 units. The separation threshold was set to one pixel because of the density of the field and the length threshold was set to 0. The associated error functions are shown in figures 4b and 5b (the error function plots were normalized independently; the error values in figure 5b are smaller than those in figure 4b). For the case of no depth variance (figures 4a and 4b), the minimum in the global histogram was determined to correspond to the image position (19.97, 7.897). The local search determined the minimum to be at (20, 9). The correct position was (15.5, 15.5). As is expected from the discussion in section 2, the minimum is incorrect and the error function unsharp. For the case of increased depth variance (figures 5a and 5b), the minimum in the global histogram was determined to be at (16.29, 14.71). The local search determined the minimum to be at (17,15), showing the expected improvement with increased variance in environmental depth.

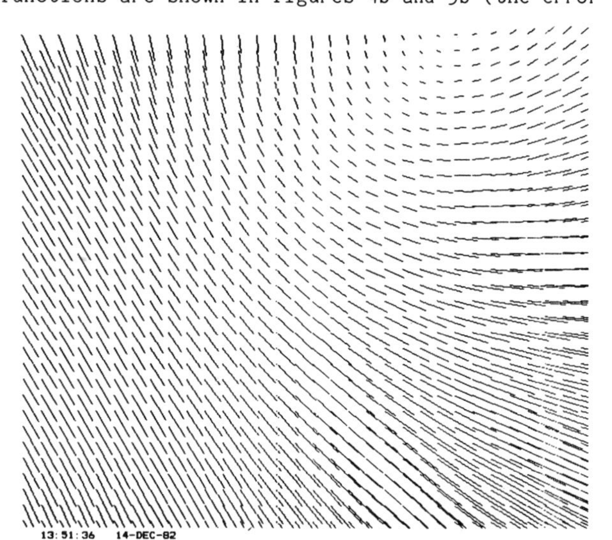

Figure 4a

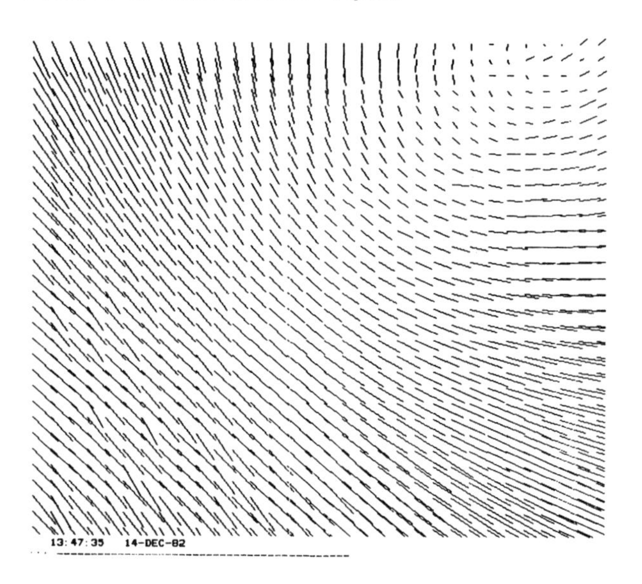

Figure 5a

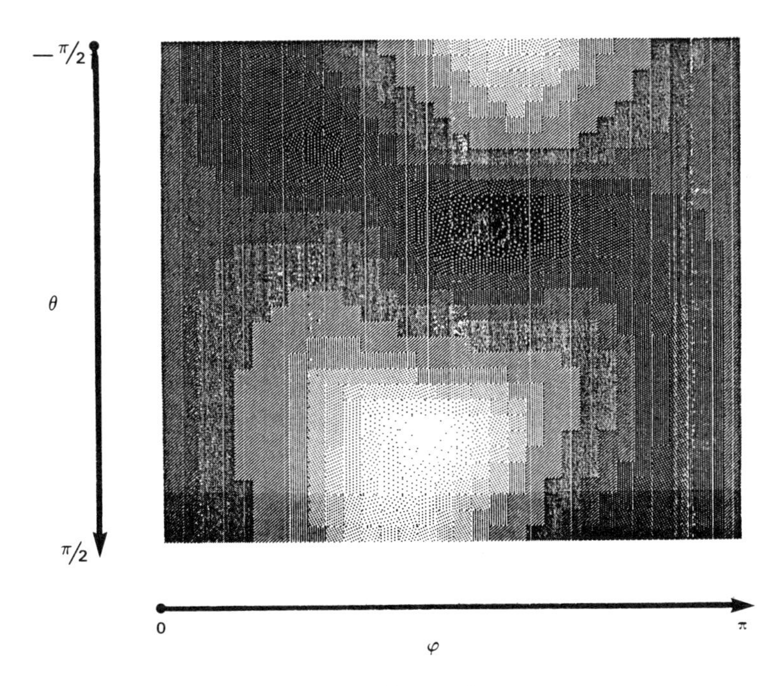

Figure 4b

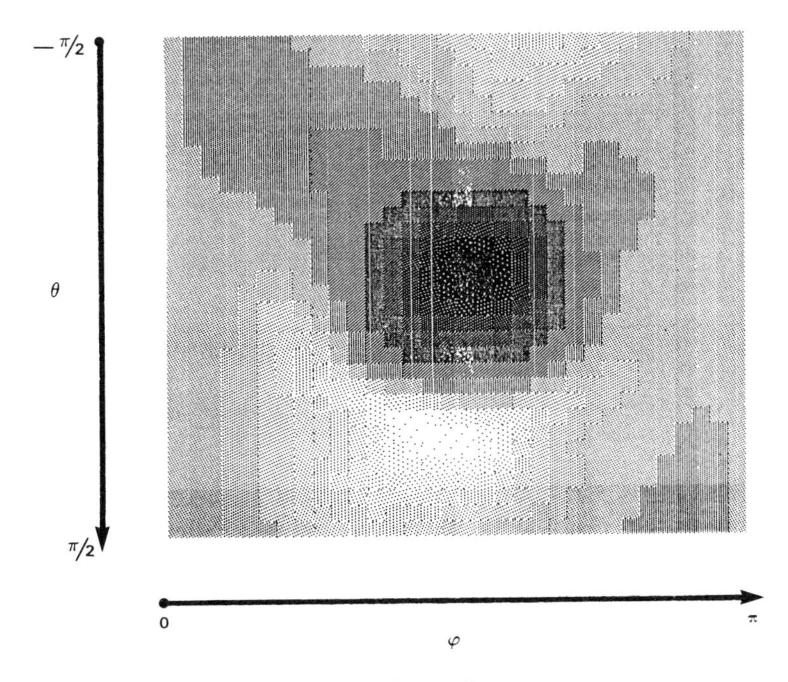

Figure 5b

5. Real Data

Figures 6a and 6b are 128x128 pixel images with 256 intensity levels taken from a GE TN2200 solid state camera. The camera was displaced roughly in the general direction of its z-axis between two textured objects towards a textured background and then rotated about its y axis a few degrees. Figure 6c shows the displacements determined for a set of interesting points extracted from the image in figure 6a using the interest operator described in Lawton (1983). The displacements were found by correlating 5x5 pixel windows centered at these positions in the first image with 5x5 pixel windows positioned at locations within +/- 15 pixels in the x and y directions in the succeeding image. Displacements for points within 10 pixels of the image boundary were ignored.

The separation threshold was set to 10 pixels and the length threshold was set to 3 pixels. A plot of the error function produced using these threshold values is shown in figure 6d. The local search found a minimum at (52, 75). The correct position of the intersection of the translational axis with the image plane for the second image was determined to be at (57.97, 74.58). Since the focal length was rather long, the determined translational axis was well within 5 degrees of the actual one.

Increasing the separation threshold smoothed the error function and made it more stongly unimodal. It also moved the determined intersection of translational axis and the image plane away from the correct position towards the center of the image. Increasing the length threshold significantly decreased the number of difference vectors.

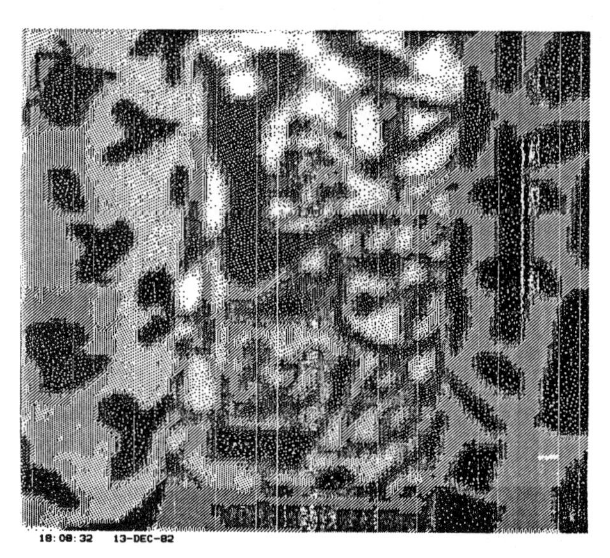

Figure 6a

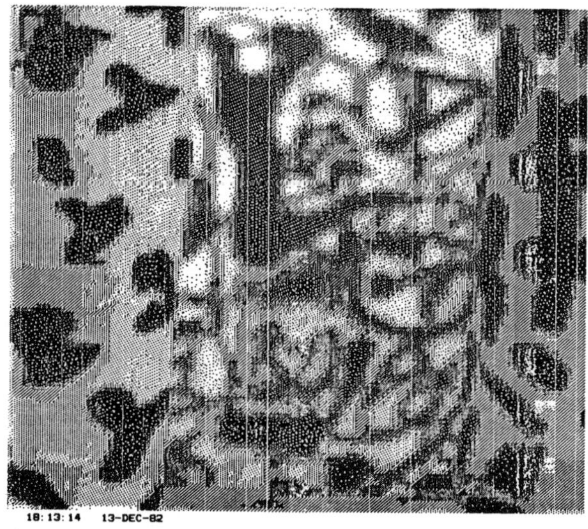

Figure 6b

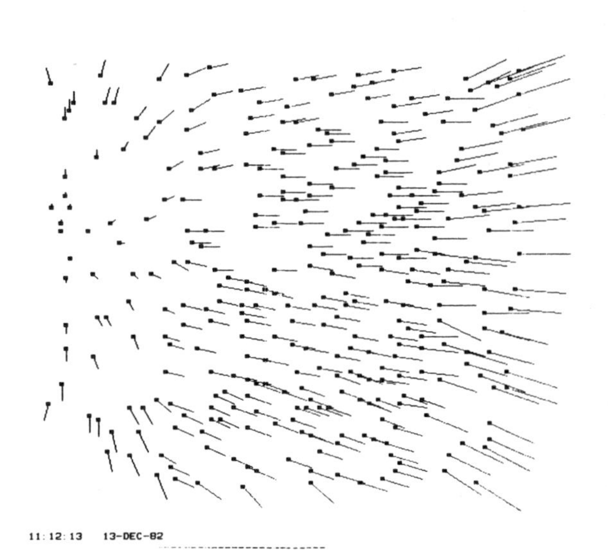

Figure 6c

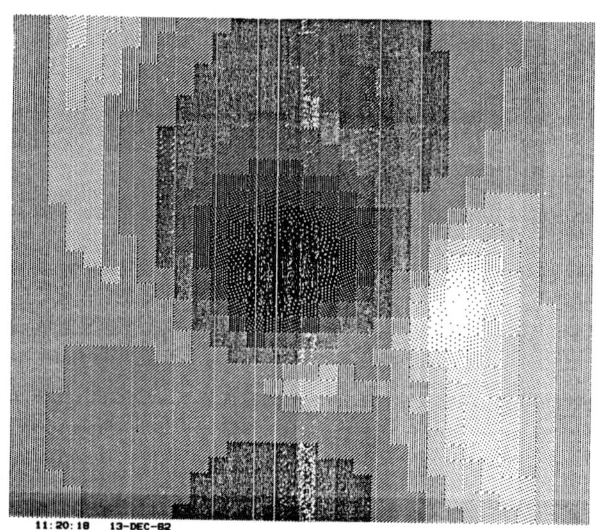

Figure 6d

6. Discussion

We have developed a procedure that determines the translational axis of sensor motion from sparse optic displacement fields given some variation in depth of the surroundings. The simplicity of the procedure makes it attractive for utilizing special hardware architectures for real-time implementation. The inference of camera motion parameters is considerably simplified once the translational axis has been determined. Given the translational axis, the sensor rotations between the frames can be obtained utilizing the fact that the components of displacement vectors perpendicular to the translational field lines are induced by the rotation. Given the sensor rotation between each pair of successive images, the computation of the translational and rotational component fields is straightforward.

6.1. Determining direction of motion from optic flow: results from human vision

Studies of the accuracy with which humans can determine their direction of translation from optic flow were initiated by Gibson, who was the first to point out the possible importance of the 'focus of expansion' (see footnote) for navigation (Gibson et. al. 1955, Gibson 1958). In these and subsequent studies errors of about 10 degrees of visual angle were found (Llewellyn 1971, Johnston et. al. 1973, Regan and Beverley 1982). This led many researchers to conclude that optic flows do not, at least in humans, play an important role in navigation. On closer inspection all three experiments have in common the following: they all simulated an approach toward a plane perpendicular to the axis of translation, and the performance of the observers detoriated as the separation of the directions of gaze and translation increased (i.e. the rotational component of the optic flow increased -- in the first two experiments this was not an independent variable but a side effect of increasing deviation of translatory axis from the screen center). These errors in judging the direction of translation are as one would expect from the procedure described in this paper. An indication of the importance of depth seperation comes from Warren (1976) who simulated the case of an observer moving along an axis of translation parallel to the ground, i.e. $\Delta Z/Z_1 \neq 0$, and found somewhat smaller errors (in the average 5 degrees of visual angle). Only recently Cutting did an experiment with systematic variation of the separation of details in depth along visual directions and of the angle between directions of gaze and translation (Cutting submitted). Indeed, the performance of human observers changed dramatically as $\Delta Z/Z_1$ changed: for $\Delta Z/Z_1 = 0$ angles between axes of translation and gaze of nearly 20 degrees could not be distinguished -- in contrast: for $\Delta Z/Z_1 \approx 1.5$ the differentiable visual angle was 37.5 minutes of arc! However, in this experiment the observers had to fixate some detail located on the second of three transparent planes. Thus deviations of the axes of gaze and translation are also indicated by a sign reversal of image velocities between details in front and behind the fixated one. It is therefore not possible to tell if the observers relied in their judgements on local discontinuities or on sign reversals of the optical flow. An experiment that isolates the two kinds of information could be designed easily. If human observers had to rely on the information given by sign reversals of image velocities, they could not determine the translational axis immediately -- instead they had to change their direction of gaze until it coincided with the translational axis, in which case the sign reversal vanished.

footnote: The 'focus of expansion' is the center of the radial fieldlines of the translational component field of optic flows. However, it does in general not coincide with a focus of expansion of a flow field in the mathematical sense, namely a maximum of the divergence of the field (Koenderink and van Doorn 1981).

APPENDIX

Consider an observer/sensor O moving relative to a static environment. As in Fig. 2 $\tilde{P} = (\tilde{X}, \tilde{Y}) = (X/Z, Y/Z)$ is the image of an environmental detail $P = (X, Y, Z)$. We obtain the image velocity $\vec{F}$ at $\tilde{P}$ by differentiating with respect to time

$$(1) \quad \vec{F} = Z^{-1}\left\{(\dot{X} - \tilde{X}\dot{Z})\, e_X + (\dot{Y} - \tilde{Y}\dot{Z})\, e_Y\right\}.$$

Letting $V = (V_X, V_Y, V_Z)$ and $\omega = (\omega_X, \omega_Y, \omega_Z)$ denote the translational and rotational velocities of O the relative motion of P becomes

$$(2) \quad (\dot{X}, \dot{Y}, \dot{Z}) = -V - \omega \times (X, Y, Z).$$

Eliminating $\dot{X}$, $\dot{Y}$, and $\dot{Z}$ between (1) and (2) gives the translational and rotational components of the optic flow

$$(3) \quad \vec{F}_T = Z^{-1}\left\{(\tilde{X} V_Z - V_X)\, e_X + (\tilde{Y} V_Z - V_Y)\, e_Y\right\}$$

$$(4) \quad \vec{F}_R = (-\omega_y + \tilde{y}\omega_z - \tilde{x}^2\omega_y + \tilde{x}\tilde{y}\omega_x)\, e_x + (-\tilde{x}\omega_z + \omega_x - \tilde{x}\tilde{y}\omega_y + \tilde{y}^2\omega_x)\, e_y$$

Two image points $\tilde{P}_1$ and $\tilde{P}_2$ that are separated by $\overrightarrow{\tilde{P}_1\tilde{P}_2} = (d_X, d_Y)$ differ in their rotational flow vectors by

(5) $\overrightarrow{\Delta F_R} =$

$$[d_y\omega_z - d_x(2\tilde{x}_1 + d_x)\omega_y + (\tilde{x}_1 d_y + \tilde{y}_1 d_x + d_x d_y)\omega_x]\underset{\sim}{e}_x$$

$$+ [-d_x\omega_z - (\tilde{x}_1 d_y + \tilde{y}_1 d_x + d_x d_y)\omega_y + d_y(2\tilde{y}_1 + d_y)\omega_x]\underset{\sim}{e}_y$$

If $P_T = (V_X / V_Z, V_Y / V_Z)$ denotes the intersection of the translational axis with the image plane we can rewrite (3) as $\overrightarrow{F_T} = V_Z\overrightarrow{P_T\tilde{P}} / Z$. Then the difference vector of the two translational flow vectors at separated image positions P_1 and P_2 becomes

$$\overrightarrow{\Delta F_T} = \overrightarrow{F_{T2}} - \overrightarrow{F_{T1}} = V_Z\left\{\overrightarrow{\tilde{P}_T\tilde{P}_2}/Z_2 - \overrightarrow{\tilde{P}_T\tilde{P}_1}/Z_1\right\} =$$

(6) $$\frac{V_Z}{Z_1 Z_2}\left\{Z_1(\overrightarrow{\tilde{P}_T\tilde{P}_1} + \overrightarrow{\tilde{P}_1\tilde{P}_2}) - Z_2\overrightarrow{\tilde{P}_T\tilde{P}_1}\right\}$$

$$= \frac{V_Z}{Z_1 + \Delta Z}\left\{\overrightarrow{\tilde{P}_1\tilde{P}_2} - \frac{\Delta Z}{Z_1}\overrightarrow{\tilde{P}_T\tilde{P}_1}\right\}$$

where $\Delta Z = Z_2 - Z_1$ is the depth separation of the environmental details P_1 and P_2 that correspond to $\tilde{P}_1$ and $\tilde{P}_2$ in the image.

ACKNOWLEDGEMENTS

We would like to thank Steve Epstein for introducing us and Kate Greenspan for feeding us. The image sequences we have been working with could not have been obtained without the UMASS Robotics group, Bill Nugent, and especially Gerry Pocock. Jeff Walker rigged up an interesting camera mount. This research was supported by DARPA grant N00014-82-K-0464 and NIH grant 5 R01 NS14971-04.

BIBLIOGRAPHY

Ballard, D. H., "Parameter Networks: Towards a Theory of Low-Level Vision", Proc. of 7th IJCAI, Vancouver, British Columbia, pp. 1068-1078, 1981.

Cutting, J. E., "Motion Parallax and Visual Flow: How to Determine Direction of Locomotion", submitted, 1982.

Gibson, J. J., Olum, P., and Rosenblatt, F., "Parallax and Perspective During Aircraft Landings", Am. Journ. Psychol., vol. 68, pp. 372-385, 1955.

Gibson, J. J., "Visually Controlled Locomotion and Visual Orientation in Animals", Br. Journ. of Psychol., vol. 49, pp. 182-194, 1958.

Johnston, I. R., White, G. R., Cumming, R. W., "The Role of Optical Expansion Patterns in Locomotor Control", Am. Journ. Psychol., vol. 86, no. 2, pp. 311-324, 1973.

Koenderink, J. J., and van Doorn, A. J., "How an Ambulant Observer can Construct a Model of the Environment from the Geometrical Structure of the Visual Inflow", Kybernetik 1977, G. Hauske and E. Butenandt, editors, R. Oldenbourg Verlag, Munich, 1978.

Koenderink, J. J., and van Doorn, A. J., "Exterospecific Component of the Motion Parallax Field", J. Opt. Soc. Am., vol. 71, pp. 953-957, 1981.

Lawton, D. T., "Motion Analysis via Local Translational Processing", IEEE Workshop on Computer Vision: Representation and Control, pp. 59-72, 1982.

Lawton, D. T., "Processing Translational Motion Sequences", Computer Graphics and Image Processing, in press, 1983.

Lee, D. N., "The Optic Flow Field: the Foundation of Vision", Phil. Trans. R. Soc. Lond. B., vol 290, pp. 169-179, 1980.

Llewellyn, K. R., "Visual Guidance of Locomotion", Journ. Exper. Psychol., vol. 91, no. 2, pp. 245-261, 1971.

Longuet-Higgins, H. C. and Prazdny, K., "The Interpretation of a Moving Image", Proc. R. Soc. Lond. B., vol 208, pp. 385-397, 1980.

O'Rourke, J., "Motion Detection Using Hough Techniques", Proceedings of PRIP. pp. 82-87, 1981.

Prazdny, K., "Egomotion and Relative Depth Map from Optical Flows", Biol. Cybernet., vol. 36, pp. 87-102, 1980.

Regan, D. and Beverly, K. I., "How do We Avoid Confounding the Direction We are Looking and the Direction We are Moving?", Science, Vol. 215, pp. 194-196, 1982.

Rieger, J. H., "Information in Optical Flows Induced by Curved Paths of Observation", J. Opt. Soc. Am., vol. 73, in press, 1983.

Warren, R., "The Perception of Egomotion", Journ. Exper. Psychol: Human Percep. and Perform., vol. 2, no. 3, pp. 448-456, 1976.

COMPLEX LOGARITHMIC MAPPING AND THE FOCUS OF EXPANSION *

Ramesh Jain

Department of Electrical and Computer Engineering

The University of Michigan

Ann Arbor MI

Abstract

Complex logarithmic mapping has been shown to be useful for the size, rotation, and projection invariance of objects in a visual field for an observer translating in the direction of it's gaze. Assuming known translational motion of the observer, the ego-motion polar transform was successfully used in segmentation of dynamic scenes. By combining the two transforms one can exploit features of both transforms and remove some of the limitations which restrict the applicability of both. In this paper we show that by using complex logarithmic mapping with respect to the focus of expansion rather than the center of the visual field perfect projection invariance and better size and rotation invariance may be obtained for any arbitrary motion of the observer.

1. INTRODUCTION

Recently many researchers [BGT82, CaP77, CAV78, SaT80, SAW72, SCH77, SCH80, SCH81, SCH82] have suggested application of complex logarithmic mapping for the size and rotation invariance of objects in a visual field. Schwartz [SCH77,SCH80,SCH81,SCH82] suggested that the retino-striate mapping can be approximated by complex logarithmic mapping for most visual systems and is responsible for size, rotational, and projection invariance. Cavanagh [CAV78] suggests that a combination of Fourier transform and complex logarithmic mapping may be used for translation, size, and rotational invariance. He argues [CAV81] that the mapping alone results in the invariance only in very limited cases. Many approaches have been proposed for obtaining size and rotation invariance in computer vision systems using either transformations based on complex logarithmic mapping [CaP77, SaT80, SAW77, SWC81, ChW79] or using sensors for acquiring images which have complex logarithmic property.

For the segmentation of dynamic scenes in stationary and nonstationary components of the scene obtained using a moving observer, Jain [JAI82a, JAI82b] suggested the use of an Ego-Motion Polar (EMP) transform. This transformation uses the known location of the Focus Of Expansion (FOE) as the origin of the polar coordinate system and converts the original image into another rectangular image whose abscissa and ordinates are r and ϑ, respectively. He developed an algorithm for the segmentation using this transform. The experience with this algorithm [JAI82b] shows that if the observer motion is known then the scene can be sucessfully segmented using the EMP transforms.

The EMP transform has many similarities with the logarithmic mapping. It appears that by combining these two concepts we may be able to understand the scope of the CLM in vision systems better, particularly for projection invariance. Another advantage may be the possibility of extracting more information from a scene sequence describing the motion of the observer.

In this paper, first, we briefly review the CLM and the EMP and then suggest a modified EMP transformation. We discess some mathematical aspects of this transform and present results of our experiments to study the efficacy of this transform.

2. COMPLEX LOGARITHMIC MAPPING

The application of complex logarithmic mapping (CLM) has been suggested for implementing a roation and size invariant mechanism in pattern recognition [CaP77, SAW72]. Schwartz [SCH77, SCH80, SCH81 SCH82] has shown that the mapping from retina to the striate cortex can be approximated by CLM and is responsible for size and rotational invariance in human visual system. Cavanagh [CAV78] suggested that a composite of spatial frequency mapping and CLM would provide a translation, rotation, and size invariant mechanism for human vision. Cavanagh's suggestion was certainly influenced by the work of Casasent and Psaltis [CaP77] in pattern recognition. Schwartz challenged the validity of the Cavanagh's suggestion arguing that application of the Fourier transform leads to certain problems in human visual systems and can not be justified using anatomical arguments also. He posited [SCH81] that the CLM leads to rotation and size invariance and is the mechanism found in most retino-striate

*This research was supported by NSF under Grant No. MCS-8219739.

Motion: Representation and Perception
N.I. Badler and J.K. Tsotsos, Editors
Published by Elsevier Science Publishing Co., Inc.

mappings. He further showed that in addition this mapping also results in projection invariance, by which he meant the sequence of projective changes that a stimulus fixed in the environment would undergo as an organism approached a fixation point. In a series of papers [SCH77, SCH80, SCH82] Schwartz has shown the importance of the CLM in many tasks of the human visual system.

As argued by Cavanagh [CAV81], the CLM results in invariances only under severe restrictions. The size and rotation invariance is obtained only for those cases when the measurements are made with respect to the origin of the CLM. For the human visual system the CLM represents the retino-striate mapping only in peripheral areas and CLM leads to problems in case of a shift in position of objects. Moreover, the projection invariance claimed by Schwartz is true only when the direction of motion and the axis of gaze are colinear.

In this paper our aim is not to address the relevance of the CLM to the human visual system; but to show that if the FOE is known then the limitation of the CLM in case of projection invariance can be removed and the modified mapping may be used for the extraction of information in the dynamic scenes acquired using a controlled moving camera. Here we discuss some mathematical properties of the transform and in a later section show that by modifying the CLM only slightly it can be made very useful, at least, in the case of a moving observer. The relevance of the proposed modification to the human vision remains to be studied.

2.1. Mathematical Aspects of CLM

Let us consider a point P in an image. If we consider the center of the image as the origin then the cartesian coordinates (x, y) and the polar cordinates (r, ϑ) of the point are related by:

$$r = \sqrt{(x^2 + y^2)} \tag{1}$$

$$\vartheta = \tan^{-1}(y/x) \tag{2}$$

If it is desired to map a circular region onto a rectangular region, as in case of the human visual system, then it is well known from the theory of complex variables and conformal mapping that an appropriate mapping is:

$$w = log(z) \tag{3}$$

where

$$z = x + iy \tag{4}$$

and

$$w = u(z) + iv(z) \tag{5}$$

are the complex variables. The mapping given by equation 3 simplifies to:

$$u(r, \vartheta) = \log(r) \tag{6}$$

$$v(r, \vartheta) = \vartheta \tag{7}$$

The selection of the above mapping for modelling the retino-striate structure was strongly influenced by the fact that the magnitude of the cortical magnification factor is approximately proportional to retinal eccentricity. It can be easily verified [SCH80] that the proposed mapping satisfies the above relationship as

$$\frac{dw}{dz} = \frac{d(log(z))}{dz} = \frac{1}{r} \tag{8}$$

Another attractive feature of the above mapping is the fact that CLM is the only analytic function which maps an annular region to a rectangular region. The analytic nature guarantees that the mapping preserves the direction and magnitude of the local angles. These properties may be very useful in the extraction of information from the transformed space.

As shown by several researchers [CaP77, SCH80] the rotation of surfaces in the z space becomes vertica displacement in the w space; the magnification of a surface in z space becomes a horizontal shift in the w space. From dynamic scene analysis view point, a very interesting property of this mapping is that if the observer is moving towards its fixation point then the projection of stationary surfaces remain of the same size and show only horizontal displacements.

3. EGO-MOTION POLAR TRANSFORMATION

The intersection of the 3-D vector representing the instantaneous direction of the observer motion, and the projection plane is called the Focus Of Expansion (FOE) The retinal velocity vectors due to stationary points meet at the FOE. It has been shown [CLO80, GIB79 LEE80, PRA80] that the FOE plays a very important role in the extraction of information from the optical flow.

If the camera motion is known then the FOE can be computed and the fact that the flow vectors for the stationary points meet at the FOE can be used to classify points in an image into stationary and moving points After computing the optical flow, a test for colinearity of the FOE and the flow vector for the point may be used for the classification. The difficult part in this approach is the computation of the optical flow and applying colinearity test to every point in the image.

If the observer continues its motion in the same direction then the FOE the remains same and the image of a point moves along the radial line originating at the FOE [GIB79, BaB82]. If we transform the original image into another two dimensional rectangular image such that ϑ is along the conventional Y axis and r is along the conventional X axis, then due to the observer motion the image of a stationary point will display motion only along r in the transformed image. Note that the motion of all stationary points in the origina image is in assorted directions but in the transformed image all points show motion in one direction. The surface coherence property and the rigidity assumption allows us to assume that regions in an image representing a surface will show the same motion in the r-ϑ plane for the translational motion of the observer and the surface. Thus, if the observer has only a translational component to his motion then we can classify all the regions that show only horizontal velocity in the EMP space as due to stationary surfaces. The regions having vertica component of velocity are due to nonstationary surfaces.

Note that the segmentation of a dynamic scene into its stationary and nonstationary components can now be performed by detecting the presence of vertical component of motion in the EMP space. There is no need to compute optical flow and determine the colinearity of the vectors with the FOE for the segmentation. The velocity components for a region may be determined using techniques used for the stationary camera case in the EMP space. In [JAI82a, JAI82b] the efficacy of this approach was demonstrated considering several scenes containing stationary and moving objects. In Fig.1 we show three frames from a sequence used in [JAI82b]. The first frame of these frames is shown in EMP form in the Fig.2 and the segmentation obtained using the proposed approach is shown in Fig.3.

Figure 1 Three frames of a sequence acquired using a moving camera. The connecting rod was moving to the left, the yoke to the right and all other objects in the scene were stationary. The camera was moving on a rail.

A

B

C

Figure 2 The EMP picture of the frame shown in Figure 1a. Note the distortion in the objects. It should be mentioned here that several issues related to the implementation of the transform are not well understood yet.

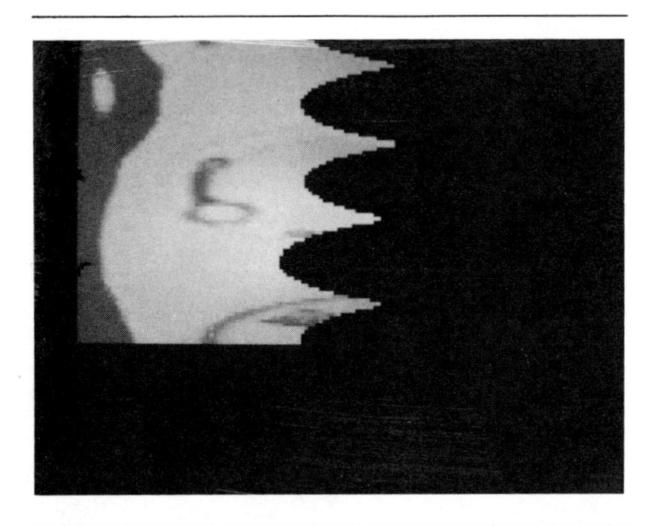

Figure 3 The stationary and moving components of the frame shown in Figure 1a. The brighter objects are the one classified as moving objects and the darker as the stationary object.

4. MODIFICATIONS

Let us consider a stationary point P in the environment whose real world coordinates with respect to the observer at a time instant are (X,Y,Z). The projection of this point on the image plane, assuming that the projection plan is parallel to XY plan and is at $Z=1$, is at (x',y') given by

$$x' = \frac{X}{Z} \tag{9}$$

$$y' = \frac{Y}{Z} \tag{10}$$

If the observer moves then the relationship between the image plane distance of the projection of the point from the FOE and the distance Z of the point from the observer is

$$\frac{dr}{dZ} = \frac{d(\sqrt{(x'^2+y'^2)})}{dZ} = -\frac{r}{Z} \tag{11}$$

where r is the distance from the center of the image.

The projection invariance shown by Schwartz is based on the fact that for a moving observer the projection of a point at the distance Z from the observer satisfies the following relations in the W plane:

$$\left|\frac{dr}{dZ}\right| = \left|\frac{r}{Z}\right| \tag{12}$$

The complex logarithmic mapping results in

$$\frac{du}{dZ} = \frac{du}{dr} * \frac{dr}{dZ} \tag{13}$$

and

$$\frac{dv}{dZ} = \frac{dv}{d\vartheta} * \frac{d\vartheta}{dZ} \tag{14}$$

since from equation 6, we get

$$\frac{du}{dr} = \frac{1}{r} \tag{15}$$

and

$$\frac{d\vartheta}{dZ} = \frac{d(\tan^{-1}\frac{y'}{x'})}{dZ} = 0 \tag{16}$$

in CLM; we have

$$\frac{du}{dZ} = -\frac{1}{Z} \tag{17}$$

$$\frac{dv}{dZ} = 0 \tag{18}$$

Thus for a stationary point P the displacement along v is zero for a translating observer in the direction of it's fixation point. Moreover, if the stationary point is far away from the observer it's displacement in the CLM image may be considered uniform for the uniform tranatory motion of the observer. If the observer moves in the direction of it's gaze, then the displacement component (dx,dy,dz) is such that $dx=dy=0$ resulting in the FOE at the center of the visual field. As argued by Cavanagh [CAV81], if the direction of the motion of the observer is not in the direction of its gaze then the above relationship will be no longer correct. In this case the FOE will be at the point $(dx/dz,dy/dz)$ and the above equations are satisfied if the CLM is obtained considering the FOE as the origin for the transformation.

The EMP transformation considers the direction o motion of the observer by using the FOE as the origin but does not use CLM. By using CLM with respect to the FOE rather than the center of the image we can combine features of both approaches.

Another important point to be noted is that the displacement of the projection of the point in the transformed space is inversely proportional to the depth of the point. The rate of change of the displacement is thus

$$\frac{d^2u}{dZ^2} = \frac{1}{Z^2} \tag{19}$$

Clearly, the displacement and all it's derivatives of the projection of a point in W space are governed by the 3-D distance of the point in the real space. Thus for a planar surface perpendicular to the direction of the gaze the displacement of every point in the transformed space will be same; for other surface the displacements of different points will depend on the nature and the orientation of the surface. This fact may be exploited for the determination of the orientation of a planar surface. It appears possible to obtain the nature of the non-planar surfaces by determining the displacement component at every point of the surface and then using techniques similar to those used in the photometric stereo.

5. RESULTS

We simulated motion of points at different distances using the perspective transformation. The CLM with respect to the center of the image and with respect to the FOE for the uniform observer motion for several frames are given in Table 1. It can be clearly seen that the angle remains constant when the CLM is taken with respect to the FOE; but varies in the other case. Our experiments on IBM PC showed that in CLM the departure in the angle depends on the relative values of X, Y, Z and dx, dy, dz. From the table we see that the departure was more for points having lower values of X and Z; since $dy = 0$, the Y component has no effect. In the modified transform, labelled EMP in the table, the angle remained constant.

Another interesting observation to be made from this table is the fact that points at the same depth, i.e., equal Z, show equal amount of displacement in the EMP space; not in the CLM space. The DCR and DER columns in the table show the frame-to-frame displacement of points in the CLM and EMP spaces, respectively. Thus, if it is desired to extract the surface information using the known motion of the observer then EMP space retains the information; the CLM introduces noise. Of course in depth extraction using motion stereo one can control camera motion, such that the camera moves in the direction of it's optical axis resulting in the FOE at the center of the image, then CLM and EMP will be same. In more general cases, however, use of the observer motion information is desirable.

In Figure 4 we show motion of the observer with respect to a simulated surface. The FOE is (20,0) from the center of the image of size 128x128. The figure shows the surface in the CLM and the modified EMP space and also the superimposed images of the surface after several frames. Note that image plane motion of the surface in the modified EMP space is horizontal, in the CLM space is not. This figure demonstrates that the projection invariance is obtained in the EMP space, not in the CLM space.

A

B

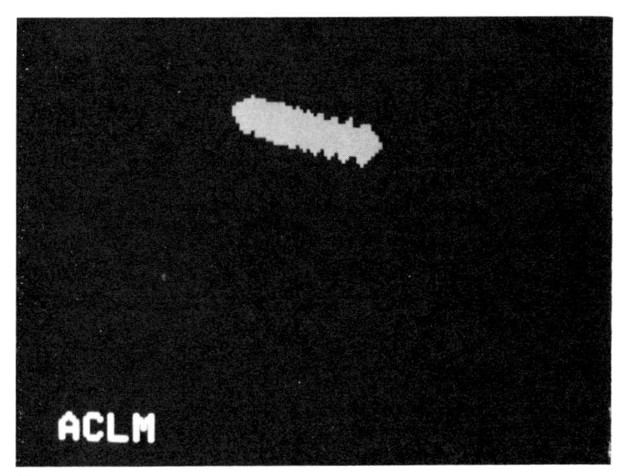

C

Figure 4. The motion of a surface in the CLM and EMP spaces due to the motion of the observer. The FOE is (20,0) from center in the image of the size 128x128.

Table 1

FOR THIS RUN NO OF POINTS WAS 4

THE 3-D COORDINATES OF POINTS WERE

1	150	300	15
2	750	300	15
3	2000	3000	50
4	400	700	50

OBSERVER MOTION 3 0 1

THE FOCUS OF EXPANSION IS 3 0

FOR POINT NO. 1

FRAME	CLM_R	CLM_T	DCR	DCT	EMP_R	EMP_T	DER	DET
	--FOR COMPLEX LOG MAPPING			----	---FOR EGO-MOTION LOG POLAR--			
1.00000	3.17232	1.11518	0.00000	0.00000	3.12250	1.23412	0.00000	0.00000
2.00000	3.24250	1.12328	0.07018	0.00810	3.19661	1.23412	0.07411	0.00000
3.00000	3.31867	1.13144	0.07617	0.00816	3.27666	1.23412	0.08004	0.00000
4.00000	3.40186	1.13966	0.08319	0.00822	3.36367	1.23412	0.08701	0.00000
5.00000	3.49340	1.14794	0.09154	0.00828	3.45898	1.23412	0.09531	0.00000
6.00000	3.59504	1.15629	0.10165	0.00835	3.56434	1.23412	0.10536	0.00000
7.00000	3.70917	1.16470	0.11413	0.00841	3.68212	1.23412	0.11778	0.00000
8.00000	3.83910	1.17317	0.12993	0.00847	3.81565	1.23412	0.13353	0.00000
9.00000	3.98971	1.18170	0.15061	0.00853	3.96980	1.23412	0.15415	0.00000
10.00000	4.16855	1.19029	0.17884	0.00859	4.15212	1.23412	0.18232	0.00000

FOR POINT NO. 2

FRAME	CLM_R	CLM_T	DCR	DCT	EMP_R	EMP_T	DER	DET
	--FOR COMPLEX LOG MAPPING			----	---FOR EGO-MOTION LOG POLAR--			
1.00000	4.05177	0.38189	0.00000	0.00000	4.00235	0.40232	0.00000	0.00000
2.00000	4.12242	0.38328	0.07064	0.00139	4.07646	0.40232	0.07411	0.00000
3.00000	4.19899	0.38469	0.07657	0.00140	4.15650	0.40232	0.08004	0.00000
4.00000	4.28251	0.38610	0.08353	0.00141	4.24352	0.40232	0.08701	0.00000
5.00000	4.37433	0.38752	0.09182	0.00142	4.33883	0.40232	0.09531	0.00000
6.00000	4.47619	0.38896	0.10186	0.00143	4.44419	0.40232	0.10536	0.00000
7.00000	4.59046	0.39040	0.11427	0.00144	4.56197	0.40232	0.11778	0.00000
8.00000	4.72046	0.39185	0.13001	0.00145	4.69550	0.40232	0.13353	0.00000
9.00000	4.87108	0.39332	0.15062	0.00146	4.84965	0.40232	0.15415	0.00000
10.00000	5.04986	0.39479	0.17878	0.00147	5.03197	0.40232	0.18232	0.00000

FOR POINT NO. 3

FRAME	CLM_R	CLM_T	DCR	DCT	EMP_R	EMP_T	DER	DET
	--FOR COMPLEX LOG MAPPING ----				---FOR EGO-MOTION LOG POLAR--			
1.00000	4.29795	0.98349	0.00000	0.00000	4.27569	1.01821	0.00000	0.00000
2.00000	4.31811	0.98418	0.02016	0.00069	4.29631	1.01821	0.02062	0.00000
3.00000	4.33870	0.98487	0.02059	0.00069	4.31736	1.01821	0.02105	0.00000
4.00000	4.35974	0.98557	0.02105	0.00069	4.33887	1.01821	0.02151	0.00000
5.00000	4.38126	0.98626	0.02152	0.00070	4.36085	1.01821	0.02198	0.00000
6.00000	4.40328	0.98696	0.02201	0.00070	4.38332	1.01821	0.02247	-0.00000
7.00000	4.42581	0.98766	0.02253	0.00070	4.40631	1.01821	0.02299	0.00000
8.00000	4.44888	0.98835	0.02307	0.00070	4.42984	1.01821	0.02353	0.00000
9.00000	4.47252	0.98905	0.02364	0.00070	4.45394	1.01821	0.02410	0.00000
10.00000	4.49675	0.98975	0.02423	0.00070	4.47863	1.01821	0.02469	0.00000

FOR POINT NO. 4

FRAME	CLM_R	CLM_T	DCR	DCT	EMP_R	EMP_T	DER	DET
	--FOR COMPLEX LOG MAPPING ----				---FOR EGO-MOTION LOG POLAR--			
1.00000	2.79870	1.05489	0.00000	0.00000	2.71928	1.22777	0.00000	0.00000
2.00000	2.81748	1.05814	0.01878	0.00325	2.73990	1.22777	0.02062	0.00000
3.00000	2.83671	1.06140	0.01922	0.00326	2.76096	1.22777	0.02105	0.00000
4.00000	2.85640	1.06467	0.01969	0.00327	2.78246	1.22777	0.02151	0.00000
5.00000	2.87656	1.06795	0.02017	0.00328	2.80444	1.22777	0.02198	0.00000
6.00000	2.89723	1.07125	0.02067	0.00330	2.82691	1.22777	0.02247	0.00000
7.00000	2.91842	1.07456	0.02119	0.00331	2.84990	1.22777	0.02299	0.00000
8.00000	2.94016	1.07788	0.02174	0.00332	2.87343	1.22777	0.02353	-0.00000
9.00000	2.96247	1.08121	0.02231	0.00333	2.89753	1.22777	0.02410	0.00000
10.00000	2.98539	1.08455	0.02292	0.00334	2.92223	1.22777	0.02469	0.00000

6. DISCUSSION

In this paper we showed that by using the CLM with respect to the FOE the applicability of the CLM in dynamic scenes can be significantly increased. We considered a very simple simulation to illustrate the enhancements in the projection invariance using the modified CLM (or modified EMP). In more realistic situations involving complex motion of observer and planar and curved surfaces in the scene we intend to compute *some* properties of neighborhood of points to extract information about the nature of surfaces.

The plausibility of the modified transformation in the human visual system has not been investigated here. We are investigating the possibility of a simple transform based on the ego-motion to be applied to the CLM to obtain the modified transform. If such a transform can be obtained then the presence of such a transform from the striate cortex to the higher levels should be investigated.

REFERENCES

[BaB82]
Ballard, D.H. and C. M. Brown, *Computer Vision*, Prentice Hall, 1982.

[BGT79]
Braccini, C., G. Gamberdella, and V. Tagliasco, "A model of the early stages of human visual system," *Biological Cybernetics*, 44, 1982, pp.47-80

[CaP77]
"New optical transforms for pattern recognition," Casasent, D. and D. Psaltis, *Proc. of IEEE*, vol. 65, pp.77-84, 1977.

[Cav78]
Cavanagh, P., "Size and position invariance in the visual system," *Perception*, vol. 7, pp.167-177, 1978.

[Cav81]
Cavanagh, P., "Size invariance: reply to Schwartz," *Perception*, col. 10, pp.469-474, 1981.

[ChW79]
Chaikin, G. and C. Weiman, "Log spiral grids in computer pattern recognition", *Computer Graphics and Pattern Recognition*, vol.4, pp.197-226, 1979.

[Clo80]
Clocksin, W.F., "Perception of surface slant and edge labels from optical flow: A computational approach," *Perception*, vol. 9, 1980, pp.253-269.

[Gib79]
Gibson, J.J., *The ecological approach to visual perception*, Houghton Mifflen, Boston, 1979.

[Jai82]
Jain, R., "Segmentation of moving observer frame sequences," *Pattern Recognition Letters*, vol. 1 , pp. 115-120, 1982.

[Jai83]
Jain, R. "Segmentation of frame sequences obtained by a moving observer," *General Motors Research Publication*, no. 4247, Jan. 1983.

[Law82]
Lawton, D. T., "Motion analysis via local translational processes," *Proc. Computer Vision Workshop, 1982, pp. 59-72.*

[Lee80]
Lee, D.N., "The optic flow field: The foundation of vision," *Phil. Trans. Royal Society of London*, vol-B290, 1980, pp. 169-179.

[Pra80]
Prazdny, K., "Egomotion and relative depth map from optical flow," *Biological Cybernetics*, vol. 36 1980, pp. 87-102.

[SaT80]
Sandini, G and V. Tagliasco, "An anthromomorphic retin-like structure for scene analysis," *Computer Graphics and Image Processing*, vol.14, pp.365-372 1980.

[SAW72]
Sawchuk, A. A., "Space-variant image motion degradation and restoration", *Proc. IEEE*, vol. 60 pp.854-861, 1972.

[SCH77]
Schwartz, E. L., "The development of specific visua connections in the monkey and goldfish: Outline of a geometric theory of receptotopic structure", *J. Theooretical Biology*, vol 69, pp.655-683.

[Sch80]
Schwartz, E. L., "Computational anatomy and functional architecture of striate cortex: a spatial mapping approach to coding," *Vision Research*, 20 1980, pp.645-669.

[Sch81]
Schwartz, E. L., "Cortical anatomy, size invariance and spatial frequency analysis," *Perception*, vol.10 pp.455-468, 1981.

[Sch82]
Schwartz, E.L., "Columnar architecture and computational anatomy in primate visual cortex: Segmentation and feature extraction via spatial frequency coded difference mapping," *Biological Cybernetics*, vol. 42, pp.157-168, 1982.

[SWC81]
Schenker, P.S., K.M. Wong, and E.G.Cande "Fast adaptive algorithms for low-level scene analysis Application of polar exponential grid (PEG) representation to high-speed, scale-and-rotation invariant target segmentation", *Proc. SPIE*, Vol 281, Techniques and Applications of Image Understanding, pp.47-57, 1981.

Adapting Optical-Flow to Measure Object Motion in Reflectance and X-ray Image Sequences

Nancy Cornelius

Department of Electrical Engineering

and

Takeo Kanade

Department of Computer Science

Carnegie-Mellon University
Pittsburgh, Pa.

Abstract

This paper adapts Horn and Schunck's work on optical flow[3] to the problem of determining arbitrary motions of objects from 2-dimensional image sequences. The method allows for gradual changes in the way an object appears in the image sequence, and allows for flow discontinuities at object boundaries. We find velocity fields that give estimates of the velocities of objects in the image plane. These velocities are computed from a series of images using information about the spatial and temporal brightness gradients. A constraint on the smoothness of motion within an object's boundaries is used. The method can be applied to interpretation of both reflectance and x-ray images. Results are shown for models of ellipsoids undergoing expansion, as well as for an x-ray image sequence of a beating heart.

Introduction

Interpreting the motion of objects from a sequence of images is difficult because image changes may be due to a number of factors. First, image changes may be due to object translations or rotations, or to relative motion of one object such that it occludes another. Second, changes may occur when non-rigid objects change shape or size. Third, parts of an image need not change even though they correspond to a moving object; for example, regions of an image corresponding to flat surfaces of constant reflectance may exhibit no change if the object undergoes only translation. Fourth, changes may result from motion of the observer. Thus, effective algorithms that measure object motion from sequences of images should do two things:

- They should distinguish between image changes due to motion of objects, due to deformation of objects, and due to occlusion.
- They should determine whether regions of an image that exhibit no apparent brightness changes correspond to moving surfaces.

This paper develops methods for assigning velocities to image points by examining changes in brightness at each point in a sequence of images. While many of the techniques may be applicable to environments where the observer is moving, the emphasis will be on interpreting image sequences where the observer is stationary and only objects move. In general, we must notice that motion analysis from images cannot be solved without making assumptions about the underlying motion of objects represented in the image sequence.

Horn and Schunck [3] addressed a problem of computing optical flow from an image sequence. They define optical flow as "the distribution of apparent velocities of movement of brightness patterns" in a sequence of images. Usually optical flow refers to the flow of the imaged world across the retina as a biological observer moves continuously through the world. However, if we assume a stationary viewer and assume there are no changes in the brightness patterns as a result of the motion, then Horn and Schunck's definition of optical flow gives the velocities of objects projected onto the image plane. To say that there are no changes in the brightness patterns means that the image brightness corresponding to a single physical point on an object is the same from one frame to the next. This restriction permits only translation of objects parallel to the image plane and does not allow arbitrary rotations or perspective transformations. In order to compute optical flow, Horn and Schunck assumed that the velocities varied smoothly over the entire image. This assumption has limited utility in real images where object boundaries are usually places of discontinuous velocity for both the case of a moving object and for an observer moving with respect to a static scene.

Our approach also involves computing velocities at the points in an image, but our method differs from Horn and Schunck's in two important ways. First, the the velocity smoothness constraints are applied only within regions that are separated from the rest of the image by recognizable boundaries. Velocities are free to change abruptly across these boundaries. Second, changes in the brightness patterns are allowed so that velocities more closely represent the arbitrary motions of objects projected onto the image plane. For example, gradual shading changes that occur with rotation relative to the light source may be accommodated.

The methods developed are applied to models of ellipsoids undergoing expansion and to x-ray image sequences of a beating heart. In the latter case the pattern changes of interest are those that occur when the heart changes shape in a direction perpendicular to the image plane.

Motion: Representation and Perception
N.I. Badler and J.K. Tsotsos, Editors
Published by Elsevier Science Publishing Co., Inc.

Problem Statement

If there is no a priori knowledge about the structure of objects in a scene, then measurement of velocity relies on local information about temporal and spatial gradients of image brightness. This local information provides only one constraint, the change in brightness at a given point, while the velocity of a point in an image has two components. In simple situations, where moving objects only undergo translation parallel to the image plane without changing their pattern in the image, this constraint determines the component of velocity parallel to the brightness gradient. When the brightness gradient is zero in the direction of motion (eg. flat region of an object with constant reflectance or a stripe pattern in the direction of motion), then there is no local velocity information. In all cases additional constraints must be imposed to determine the two components of velocity in the image plane as well as to determine the changes in the image pattern.

Let the image brightness projected by a point on a moving object at a time t be given by $I(x,y,t)$. At a later time $t+dt$ the same object point has moved so that its projected position in the image plane is given by $(x+dx,y+dy)$. The brightness of this point may have changed to a value $I(x+dx,y+dy,t+dt)$. Such a change occurs when lighting and shading change as an object rotates or when the object itself changes shape. The total rate of change of brightness dI/dt is given by:

$$\frac{dI}{dt} = \frac{\partial I}{\partial x}\frac{dx}{dt} + \frac{\partial I}{\partial y}\frac{dy}{dt} + \frac{\partial I}{\partial t} \tag{1}$$

where $\partial I/\partial x$ and $\partial I/\partial y$ are the x and y components of the spatial brightness gradient and $\partial I/\partial t$ is the temporal brightness change measured at the point (x,y). The three variables that are to be determined are the x and y components of velocity, i.e. dx/dt and dy/dt, respectively, and the brightness change dI/dt. To simplify the notation, we introduce the abbreviations I_x, I_y and I_t for the partial derivatives of brightness with respect to x, y and t and the abbreviations v_x and v_y for the x and y velocity components. Equation (1) can then be rewritten in the following way:

$$\frac{dI}{dt} = I_x v_x + I_y v_y + I_t \tag{2}$$

To solve this equation for the velocities (v_x,v_y) and the rate of brightness change (dI/dt), other constraints must be applied that restrict the allowable motions. For example, the assumption can be made that the velocity and pattern changes are constant or that they change smoothly within a region. It could also be assumed that the velocities and patterns vary in a constrained manner over time [4].

In the next section, we review Horn and Schunck's method for computing optical flow, and identify problems with it. The remaining sections propose a set of modifications and extensions to cope with those problems. First, we present a technique that permits velocity flow discontinuities at boundaries. Then we suggest a way to accommodate some of the changes in brightness patterns that occur as a result of motion. The final section presents results obtained by applying the modifications to a model of an expanding ellipsoid and an example that incorporates all of these techniques to analyze heart motion from a sequence of x-ray images.

Horn and Schunck's Method for Computing Optical Flow

Horn and Schunck [3] assumed no pattern change in the image so that the brightness change with time corresponding to a single physical point dI/dt is equal to zero, i.e.:

$$I_x v_x + I_y v_y + I_t = 0 \tag{3}$$

This assumption severely limits the allowable motions. Rotations, translations in depth and deformations often result in changes in the image brightness pattern and violate this assumption. Horn and Schunck made the additional assumption that neighboring points have similar velocities. To implement this smoothness constraint, they constrained the local change in velocity by minimizing the square of the magnitude of the spatial gradient of the velocity components:

$$\varepsilon^2 = \left(\frac{\partial v_x}{\partial x}\right)^2 + \left(\frac{\partial v_x}{\partial y}\right)^2 + \left(\frac{\partial v_y}{\partial x}\right)^2 + \left(\frac{\partial v_y}{\partial y}\right)^2 \tag{4}$$

In order to solve for the optical flow v_x and v_y, Horn and Schunck combined the two assumptions (the zero brightness change and the smoothness constraint) by minimizing the following function:

$$\int \left[\left(\frac{dI}{dt}\right)^2 + \alpha^2(\varepsilon^2)\right] dxdy \tag{5}$$

where the integral is over the entire image and α^2 is a weighting factor that depends on the noise in the gradient measurements. The following iterative formulae provide the solution for the flow velocities that minimizes equation (5):

$$v_x^{k+1} = \bar{v}_x^{\,k} - \frac{I_x(I_x\bar{v}_x^{\,k} + I_y\bar{v}_y^{\,k} + I_t)}{\alpha^2 + I_x^2 + I_y^2} \tag{6}$$

$$v_y^{k+1} = \bar{v}_y^{\,k} - \frac{I_y(I_x\bar{v}_x^{\,k} + I_y\bar{v}_y^{\,k} + I_t)}{\alpha^2 + I_x^2 + I_y^2}$$

where $\bar{v}_x^{\,k}$ and $\bar{v}_y^{\,k}$ denote local averages of the velocity components computed at the kth iteration. A region where there is no apparent local velocity information (eg. flat region of constant reflectance) will derive its velocity from the surrounding region, because during the iterative process, velocities will tend to propagate and fill in these regions.

There are three primary problems with this technique. The first two involve the boundaries. First, the technique does poorly when there are discontinuities in the velocity field or in the brightness gradients, because of the smoothness assumption. The discontinuities occur at object boundaries. Second, the same property that allows velocities to propagate within an object tends to extend erroneous velocities outside the area of an object. The problem is most conspicuous for the case where an object is moving against a uniform background. In this case it is not possible to distinguish the velocity of the object from the velocity assigned to the uniform background. Third, motion is constrained to be parallel to the image plane because of the assumption that an object does not change the way it appears in the image from frame to frame.

Boundary Constraints

The previous section suggests that discontinuities in velocity which occur at object boundaries must be explicitly accounted for in order to accurately determine velocities within the boundaries. We propose to allow for these discontinuities by applying the smoothness constraint separately to regions on either side of an image boundary. This can be done once the projection of the object boundaries have been located in the image. As we see next, implementation does not require that the image be segmented into regions corresponding to objects, rather only that the location of *possible* object boundaries be determined.

Image boundaries occur when one object moves in front of another: these are called occluding boundaries. Image boundaries can also occur due to the painted patterns or non-occluding edges on the object: these are non-occluding boundaries. (There are also boundaries due to object shadows, but these are not explicitly dealt with here.) In terms of motions across them, there is an important difference between occluding and non-occluding boundaries. A non-occluding boundary has consistent motion on both sides -- there is no velocity discontinuity. The regions on both sides of an occluding boundary can have different velocities. We must process velocity flow data at a boundary differently according to the type of boundary. The smoothness constraint is enforced across non-occluding boundaries, but not across occluding boundaries. This procedure permits spatial discontinuities in flow velocity to occur when one object moves in front of another.

To apply this method, we need not predetermine whether a boundary is occluding or non-occluding. First, the nearby velocities are computed based on an assumption of non-occlusion; the smoothness constraint is applied across the boundary. Next, the velocities are recalculated assuming occlusion; the smoothness constraint is not enforced across the boundary. Finally, the result that best satisfies the equation for dI/dt (equation (2)) and the smoothness constraint is retained. In this way, the boundary types can be locally determined without explicit segmentation of the image into object regions. This test is repeated with each iteration.

Pattern Changes

A pattern change refers to the change in image brightness of the same physical point on an object from one frame to the next. A pattern change will occur when points on the object are obscured or revealed in successive image frames. This type of change causes discontinuities in the velocity across occluding boundaries. These changes have been accommodated by the method in a previous section.

There is also another type of pattern change. For example, when an object rotates and the lighting hits the object in a different way, it results in different shading. For a Lambertian surface the shading change of a given physical point on an object is given by:

$$\frac{dI}{dt} = \frac{d}{dt}(-k\cos(\tau)) = -k\sin(\tau)\frac{d\tau}{dt} \tag{7}$$

where τ is the angle between the incident light and the the surface normal, and k is a constant. If the surface orientation is known, then dI/dt gives a measure of the change in orientation.

Here we propose to allow for such pattern changes in the image by constraining them to vary smoothly within boundaries. We can think of the pattern change (dI/dt) as another velocity component. While dI/dt is not strictly a velocity, we constrain the variations in dI/dt to vary smoothly within object boundaries, just as was done for the velocity components. Thus we can define a smoothness measure of change in brightness variation:

$$\varepsilon_B^2 = \left[\frac{\partial}{\partial x}\left(\frac{dI}{dt}\right)\right]^2 + \left[\frac{\partial}{\partial y}\left(\frac{dI}{dt}\right)\right]^2 \tag{8}$$

New Algorithm

Now we can present our new algorithm which incorporates the considerations on boundaries and pattern changes. To summarize, this algorithm assumes: (a) the brightness changes of a single physical point can be described by the first order expansion, equation (2); (b) velocity changes in a neighborhood are similar, unless the neighborhood contains an occluding boundary; (c) the rate of pattern change (dI/dt) is also similar in a neighborhood. To impose these assumptions we define an error factor for each.

The first factor is the error in satisfying equation (2), i.e.:

$$\varepsilon_I^2 = \left(\frac{dI}{dt} - I_x v_x - I_y v_y - I_t\right)^2 \tag{9}$$

Since we allow dI/dt to be nonzero, it is included in ε_I^2. The second error factor is a measure of the departure from a spatially smooth velocity field, ε_S^2, which is the same as equation (4). The third error factor is given by equation(8) and measures the departure from a spatially smooth pattern change.

We minimize the sum of these error factors computed over the image:

$$\text{minimize} \sum_i \sum_j \varepsilon_I^2(i,j) + \alpha^2 \varepsilon_S^2(i,j) + \beta^2 \varepsilon_B^2(i,j) \tag{10}$$

An iterative form of the solution is found for the velocities at the $(k+1)$ iteration in terms of the spatial and temporal brightness gradients and the neighboring velocities at the k-*th* iteration:

$$v_x^{k+1} = \bar{v}_x^k - \frac{\beta^2 I_x\left[I_x\bar{v}_x^k + I_y\bar{v}_y^k + I_t - (\overline{dI/dt})^k\right]}{\alpha^2 + 2\alpha^4\beta^2 + \alpha^2\beta^2 I_x^2 + \alpha^2\beta^2 I_y^2} \tag{11}$$

$$v_y^{k+1} = \bar{v}_y^k - \frac{\beta^2 I_y\left[I_x\bar{v}_x^k + I_y\bar{v}_y^k + I_t - (\overline{dI/dt})^k\right]}{\alpha^2 + 2\alpha^4\beta^2 + \alpha^2\beta^2 I_x^2 + \alpha^2\beta^2 I_y^2}$$

$$\left(\frac{dI}{dt}\right)^{k+1} = \left(\frac{\overline{dI}}{dt}\right)^k + \frac{\left[I_x\bar{v}_x^k + I_y\bar{v}_y^k + I_t - (\overline{dI/dt})^k\right]}{\alpha^2 + 2\alpha^4\beta^2 + \alpha^2\beta^2 I_x^2 + \alpha^2\beta^2 I_y^2}$$

where $\bar{v}_x^k$ and $\bar{v}_y^k$ denote averages of the neighboring velocities at the k-*th* iteration and $(\overline{dI/dt})^k$ denotes the average pattern change at the k-*th* iteration. This iterative procedure is applied everywhere in the image, but points in the neighborhood of a boundary are treated differently. Boundaries are located by finding zero crossings in the Laplacian of brightness [1] in each of a sequential pair of images and forming a union of such zero crossings. The size of a neighborhood is determined by the size of the region over which the smoothness constraint ε_S^2 is computed. Velocities are computed separately using points in the neighborhood on one side of the boundary and again using points in the neighborhood that span the boundary. This yields two different estimates for the velocity. The estimate that minimizes $\varepsilon_I^2 + \alpha^2 \varepsilon_S^2 + \beta^2 \varepsilon_B^2$ is used.

Results

Model of Expanding Ellipsoid

The algorithm described in this paper was tested with a sequence of images generated by modelling an ellipsoid that expands uniformly in all directions. The ellipsoid is assumed to have Lambertian surface properties and to be illuminated with a distant source perpendicular to the image plane. The image is resolved to 64 by 64 pixels and quantized to 256 brightness levels (see Figure 1A). The maximum velocity of any point in the image is approximately 0.5 pixels per frame. The background is uniform and therefore provides no information about motion. The actual velocity vectors for the expanding ellipsoid are shown in Figure 1B. These are the results we would like to obtain using our algorithm.

Figure 2 shows the results of applying Horn and Schunck's optical flow technique (equation 6) to the expanding ellipsoid. Here the smoothness constraint is applied across object boundaries. While the velocities are determined fairly accurately within the object, they are propagated erroneously beyond the object boundaries. The total error over the entire image is approximately 15%. When velocity discontinuities are taken into account as outlined above, a more accurate estimate of velocities is obtained as in Figure 3. We see that use of boundary information results in a clear demarcation of velocities within and without the object. The residual errors do not extend substantially beyond the boundaries of the object. The total error is 5%. However, the algorithm tends to overestimate the actual velocities in the vicinity of the boundary. Such inaccuracies are expected, because of the discontinuities in the brightness gradient that occur at the border between one object and another. One way to avoid this problem and possibly improve the flow velocity estimates throughout a region is to determine velocities at the boundaries of the region using another technique (see for example [2]). Such velocities at the boundary provides intitial conditions and remain fixed in the iterative procedure. To see this effect, the actual velocities at the boundary were supplied as initial conditions and remained fixed for the iterative procedure. The result is shown in Figure 4. The total error is less than 3%. For the case of the expanding ellipsoid, the velocities inside the boundary region were close to the correct values whether or not the initial boundary velocities were specified. However, there are probably other cases when a good initial guess of velocity at the boundary will substantially improve the velocity estimates inside the bounded region.

Application to X-ray Images

Though these techniques have been developed for objects imaged in visible light, we have begun to explore application of these techniques to x-ray images. Our goal is to use them to analyze motion of the heart from cine angiograms.

When optical flow techniques are applied to x-ray images, the results have a different meaning. At each point in an x-ray image, the brightness depends on the amount and density of the mass between the x-ray source and the film. Because brightnesses depend on object densities instead of reflectance, the velocities found by this method no longer apply to single physical points on the surfaces of objects. For simplicity, we assume that the density does not change in time and that the brightness changes therefore represent depth changes. In angiograms where radio-opaque dye is injected into the bloodstream, the primary x-ray attenuators are the dye and calcified bone. For this case, the assumption that pattern changes reflect changes in the depth of the heart is accurate since the dye filled heart is the primary source of motion.

X-ray images have two advantages over reflectance images of opaque objects. First, depth information is available. Second, objects are not totally occluded. The disadvantage is that a point in the image does not generally correspond to a single point on a single object. Thus the flow velocities take on a different meaning for x-ray images, as described above. To understand what this difference means, consider a reflectance image and an x-ray density image of the same object, an expanding sphere. (See Fig. 5.) In the reflectance image the brightness due to a single physical point on the sphere is the same in successive images, because it has the same surface orientation. Therefore, to determine velocity at a given point, we need only find a point of matching brightness that satisfies the global smoothness constraint. If we look at brightness along one dimension of the image, then the velocities are found by matching points of similar brightness in successive frames. (See Figure 5A.)

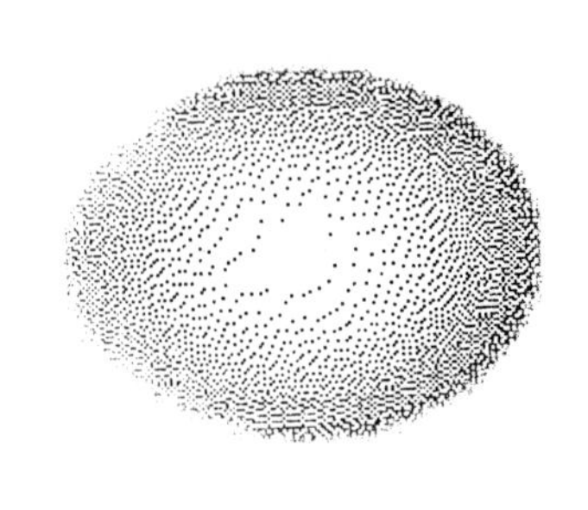

A

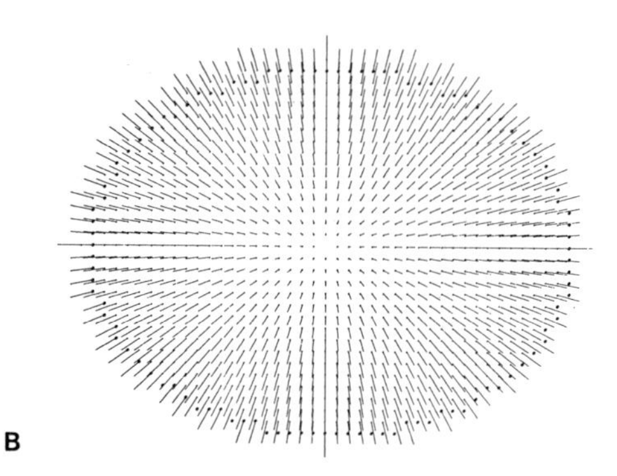

B

Figure 1: An image sequence was obtained by modelling an ellipsoid that expands uniformly in all directions. One frame of the sequence is shown in (A). At each point in the image we have computed the magnitude and direction of the local image velocity. The velocity vectors at each point in the image are plotted here as short line segments representing magnitude and direction. The correct velocity flow pattern determined from the model is shown in (B).

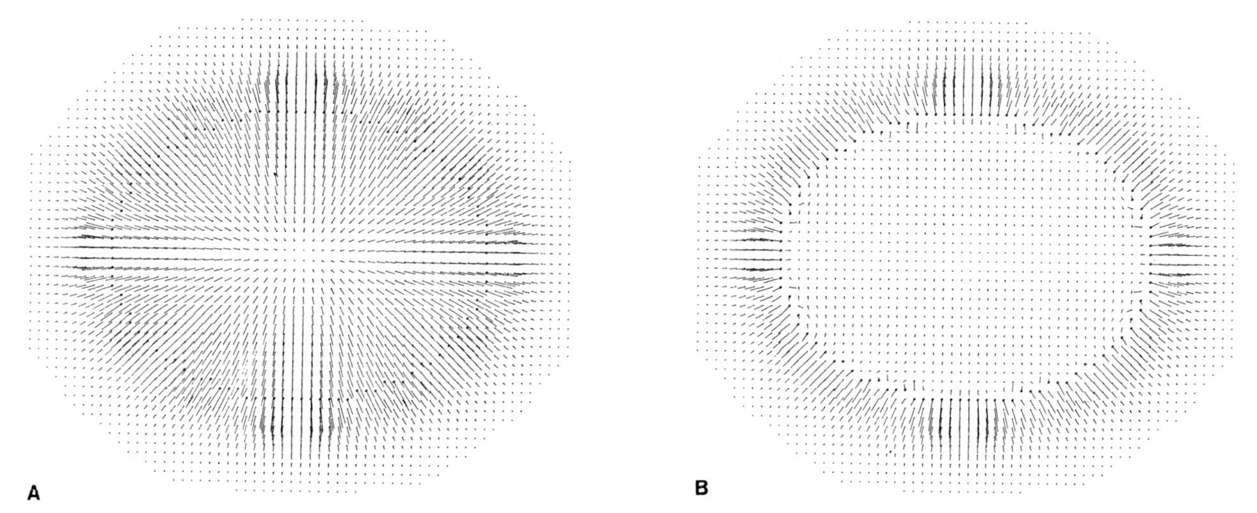

Figure 2: The velocity vectors for the expanding ellipsoid were calculated using Horn and Schunck's optical flow algorithm. The resulting flow pattern is shown in (A). No boundary constraints were imposed, so that the velocity smoothness constraint was applied across the boundaries. The result is that the velocities propagated outside the boundary. A vector plot of the velocity errors is shown in (B). Initial velocities were set to zero. The results are shown after thirty-two iterations.

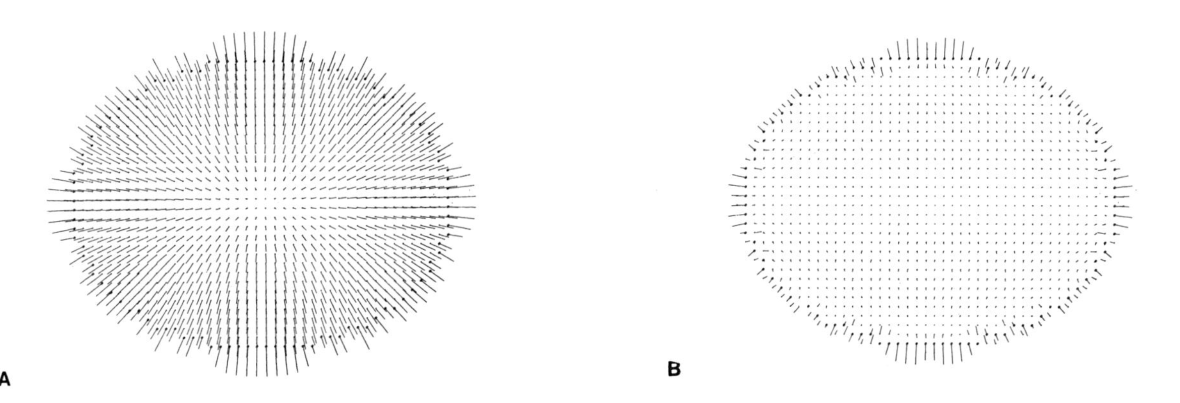

Figure 3: The velocity flow pattern in (A) was calculated for the expanding ellipse assuming that flow discontinuities could occur at the boundaries. The boundaries used are indicated by heavy black dots at the base of some of the velocity vectors. Velocity errors are shown in (B). The velocities computed at the boundaries are substantially greater than the actual velocities, however, the velocities inside the ellipse are very close to the actual velocities. Again, the initial velocities were set to zero and the results are shown after thirty-two iterations.

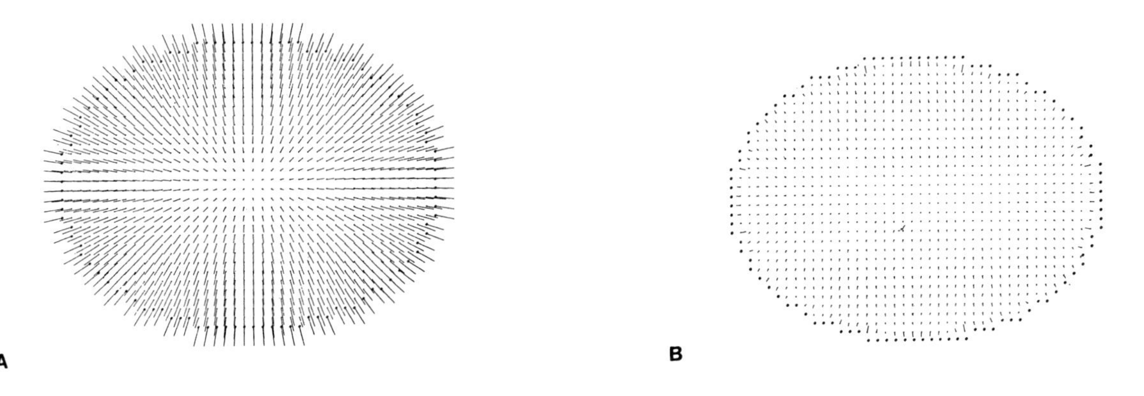

Figure 4: The velocity flow pattern in (A) was calculated for the expanding ellipse with the velocities at the boundaries set to the actual values. As in Figure 3, discontinuities in velocity flow were permitted at the boundary. The boundaries used are indicated by heavy black dots. The plot in (B) shows the velocity errors. The total error is less than 3% after thirty-two iterations.

In an x-ray density image which records the z height as the brightness, such a simple matching of similar brightnesses frequently does not yield sensible velocities. In fact, there may be many points in one image for which there is no matching brightness in successive images. As shown in Figure 5B, matching points in successive frames of the x-ray image of a sphere based on similar brightness values, yields very large velocities near the densest part of the imaged sphere (i.e. center) where the actual velocities are small. A meaningful description of the motion from the density image would be obtained by taking the rate of brightness change (*dI/dt*) into account. This can be interpreted as a change in depth perpendicular to the image plane. (See Figure 5C.) For x-ray images of a beating heart, the brightness at a point in the image is dependent on the depth of the heart cavity perpendicular to the image plane. Thus the pattern changes will reflect the expansion or contraction movement of the heart in the direction perpendicular to the image plane.

Model of X-ray Image of Expanding Ellipsoid

We show an example for a sequence of images generated by modelling an ellipsoid that expands with time. The brightness is proportional to the depth of the ellipsoid perpendicular to the image plane. The ellipsoid is expanding in all directions so that the size and brightness change as a function of time. The velocities projected on the image plane are the same as for the case of the reflectance image. We expect the brightness changes to be proportional to the actual brightness or depth. Figure 6 shows these anticipated velocities and brightness changes. Figure 7 shows the velocity field which is computed by the Horn and Schunck method (i.e., with the assumption that there are no pattern changes, *dI/dt=0*). As expected, we obtain a large flow discontinuity at the center of the image of the ellipsoid. Figure 8 shows the result of our method in which pattern changes are allowed. The velocities and pattern changes are very close to the expected results.

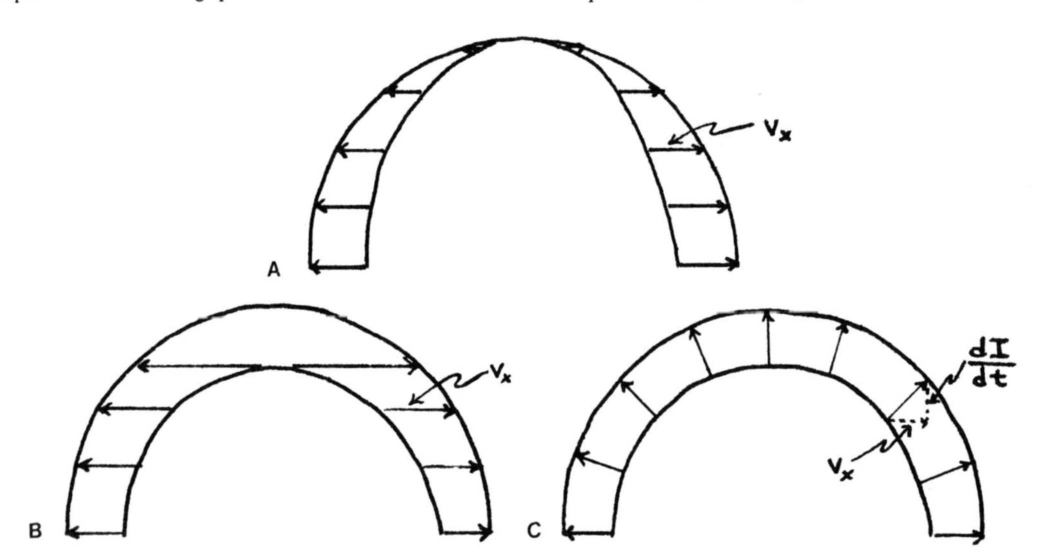

Figure 5: When a sphere expands in a reflectance image, the profile of brightnesses in a cross-section parallel to the x-axis changes as shown in (A). (Assume a distant light source perpendicular to the image plane.) The brightness of points on the surface do not change as the sphere expands, so surface motion can be measured by matching points of similar brightness that satisfy the smoothness constraint. When a sphere expands in an x-ray density image, the brightness of each surface point increases as shown in (B) and (C). It is no longer possible to determine velocities by matching brightnesses. We expect the velocities to be the same as for the reflectance case. However, if we simply match brightnesses as in (B),we obtain very large velocities near the center of the imaged sphere where we expect very small velocities, as well as a large flow discontinuity at the center of the sphere. If, as in (C), we allow for smoothly varying brightness changes (*dI/dt*) in addition to the motion in the plane parallel to the image, then velocities in the image plane are as expected.

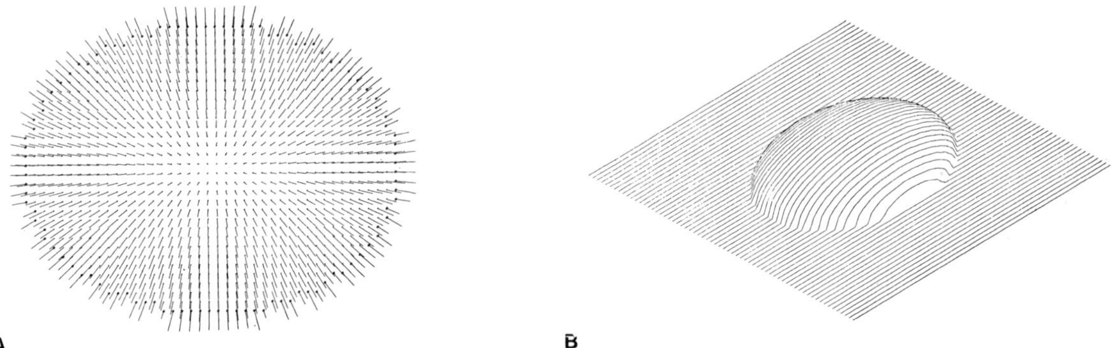

Figure 6: The expanding ellipsoid was modelled again, but the brightness at each point in the image now corresponds to the depth of the ellipsoid measured perpendicular to the image plane. The result is similar to an x-ray image of an ellipsoid. The velocities calculated from the model are shown in (A) and the rate of pattern change (*dI/dt*) is shown in (B). The pattern changes can be thought of as changes in depth.

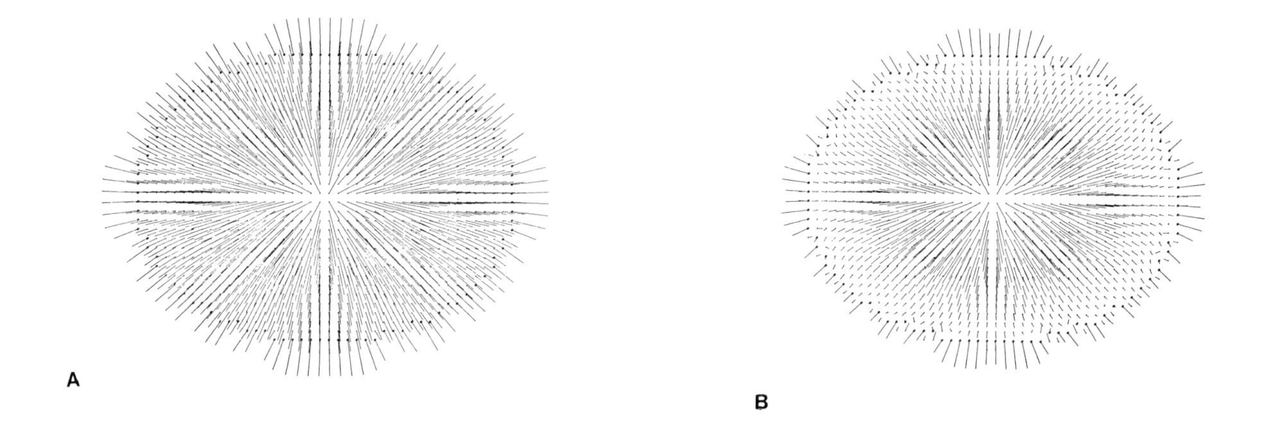

Figure 7: (A) The computed velocities for density images of the expanding ellipsoid assuming no pattern changes, i.e., $dI/dt=0$. Note the large discontinuity in velocity flow at the center of the imaged ellipsoid and the very large velocities near the center. (B) The errors in the computed velocities. Results are shown after 32 iterations.

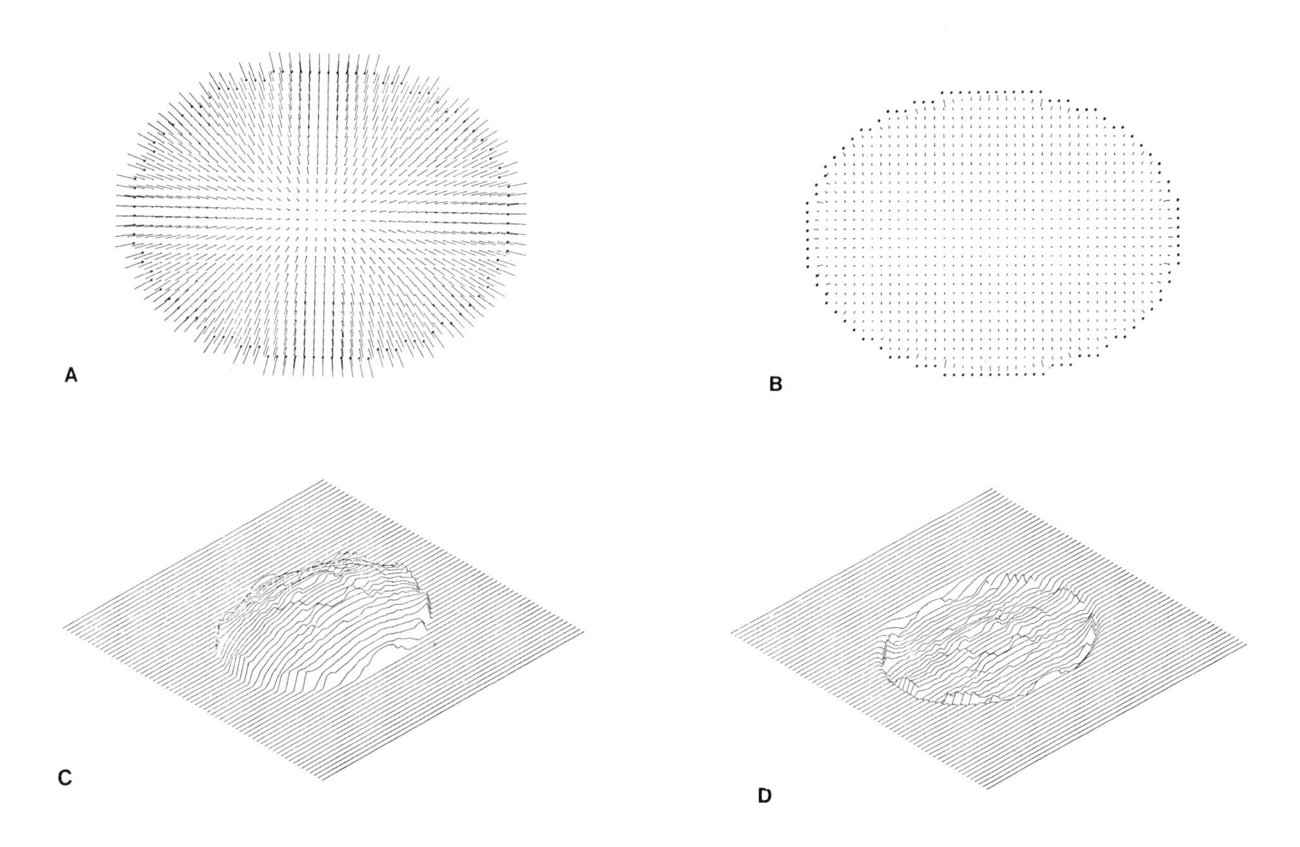

Figure 8: The velocities were computed for density images of the expanding ellipsoid allowing for velocity discontinuities at boundaries and allowing for pattern changes. The velocities were preset to the actual values at the boundaries. All other velocities were initialized to zero. The boundaries used are indicated by heavy black dots. Results are shown after 32 iterations. (A) The velocity flow pattern computed for the image plane. (B) The velocity errors in (A). (C) The computed pattern changes or depth changes. (D) The errors in (C).

Experimental Results for Heart Images

We have applied the methods described in this paper to x-ray images of a dog's heart taken on film at 60 frames a second. Figure 9 shows an example of a single frame of the cine angiogram. A radio-opaque dye was injected into the pulmonary artery just before the image sequence was taken. The dye can be seen filling the left ventricle, the aorta and some of the coronary arteries. The other obvious structures in the images are a couple of catheters left over from some previous injections. The film was digitized with 8 bits per pixel and resolved to 100 x 100 pixels.

The velocities were computed using equations (11). Pattern changes caused by the expansion and contraction of the heart perpendicular to the image plane were permitted and discontinuities in the velocity flow were accommodated at image boundaries as described above. The image boundaries were located at the zero crossings of the Laplacian of a smoothed version of a pair of sequential images. The computed velocities are shown in Figure 10. To verify these results, the computed motion description is used to predict the brightness in a subsequent image from the brightness in the previous image. A comparison of the predicted and actual images shows an error of less than 0.5%. While this does not show that the motion description is actually a good one, it does show that the algorithm is working as expected. In order to get a subjective opinion of the validity of the motion description, we have generated a movie of the velocity vectors for an entire heart cycle and shown that it coincides well with the apparent motion seen in the actual cine angiogram. While the motion description obtained from the analysis of x-ray images may be useful, it does not provide explicit information about motion of object surfaces. This sort of information might be obtained by using additional views of the object from different angles, or by considering a priori information about the object's shape or symmetry.

Summary

This paper extends the work of Horn and Schunck on optical flow. Their velocity smoothness constraint is relaxed at boundaries to permit discontinuities in estimated velocity where there are occluding boundaries. It is not necessary to segment the image into objects in order to use boundary information. Rather it is only necessary to locate possible boundaries which can be done by locating zero crossings in the Laplacian of the smoothed image brightness. Images of an expanding ellipsoid were used to test the resulting iterative algorithm. The results showed that discontinuities in the velocity flow could be accommodated, but that the velocity at the actual boundary may be inaccurate. It is possible to estimate the velocities at the boundary using another technique and then to use these estimates as input to the iterative algorithm.

Though the techniques were originally developed for reflectance images, we have begun to apply them to x-ray density images. To do this it has been necessary to relax Horn and Schunck's restriction that the appearance of an object not change from one image to the next. This was done by assuming that the pattern changes in the image vary gradually within object boundaries. Velocity flow patterns for x-ray images of the expanding ellipsoid were obtained in this way. The results showed that the computed velocities parallel to the image plane were very close to those obtained for reflectance images and the pattern changes were very nearly proportional to the velocities perpendicular to the image plane. The technique was also used to analyze x-ray cine angiograms of a beating heart and produced a subjectively good description of the motion.

Acknowledgements

We would like to thank Bob Selzer and the Biomedical Imaging Lab at the Jet Propulsion Labs in Pasadena, California for providing us with the digitized cine angiogram data.

References

[1] Hildreth, Ellen C.
Implementation of a Theory of Edge Detection.
Technical Report AI-TR-579, Artificial Intelligence Laboratory, MIT, April, 1980.

[2] Hildreth, Ellen C.
The Integration of Motion Information Along Contours.
In *Proceedings of the Workshop on Computer Vision Representation and Control*, pages 83-91. IEEE, August, 1982.

[3] Horn, Berthold B. K. and Schunck, Brian G.
Determining Optical Flow.
Artificial Intelligence 17:185-203, 1981.

[4] Yachida, Masahiko.
Determining Velocity Map by 3-D Iterative Estimation.
In *Proceedings of Seventh IJCAI*, pages 716-718. Vancouver, B.C., Canada, August, 1981.

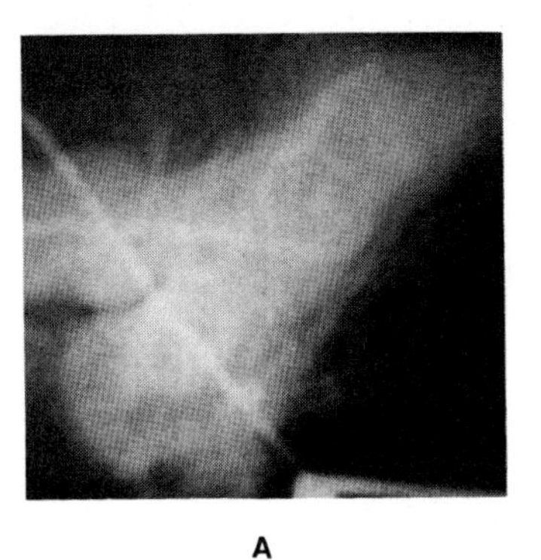

A

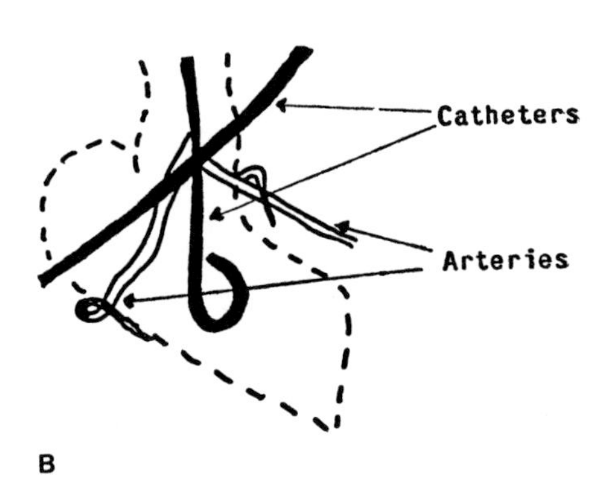

B

Figure 9: Experimental results were obtained for a sequence of x-ray images of a dog's heart injected with radio-opaque dye. (A) An example of a single frame in the sequence. (B) A line drawing identifying the structures.

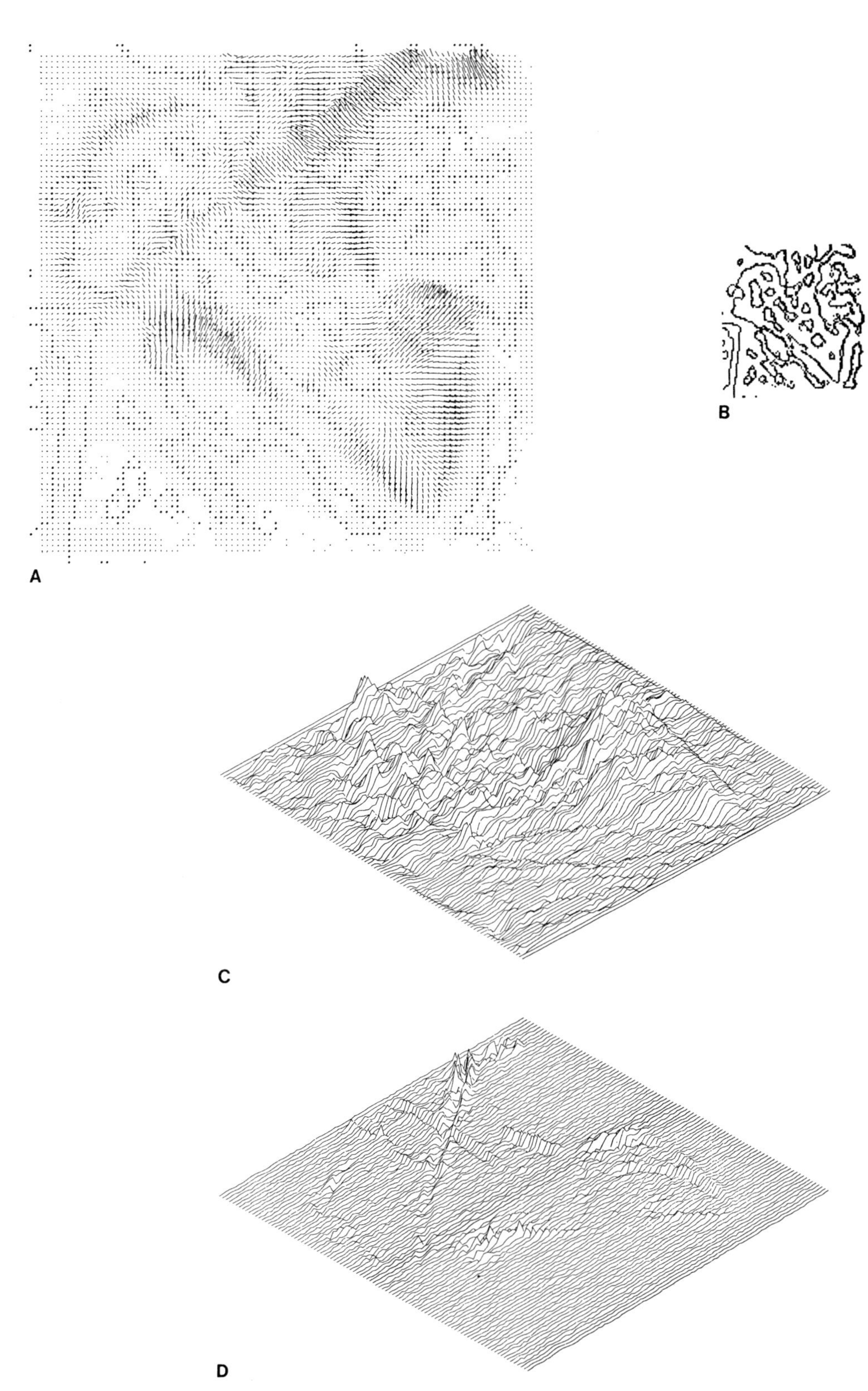

Figure 10: A sequence of x-ray images of a dog's heart were processed to obtain velocity information. Velocity discontinuities were permitted at image boundaries. (A) The velocity flow vectors. The boundaries used are indicated by heavy black dots in the figure. (B) The boundaries alone. (C) The depth changes or pattern changes. (D) The error in predicting the subsequent frame from the previous frame given the motion description in (A) and (C). The total error is less that 0.5%. Initial velocities were set to zero. The results are shown after twenty iterations.

ON THE ESTIMATION OF DENSE DISPLACEMENT VECTOR FIELDS FROM IMAGE SEQUENCES

H.-H. Nagel*
Fachbereich Informatik
Universität Hamburg
Schlüterstr. 70
2000 Hamburg 13
FR Germany

Abstract

Based on recent experimental as well as theoretical investigations, a generalization of previously published approaches towards the estimation of displacement vector fields is formulated. The calculus of variation allows to transform this approach into a set of two partial differential equations for the two components of the displacement vector field. Some simplifying assumptions facilitate the derivation of an iterative solution approach which can be studied in closed form.

1. Introduction

A recent contribution by NAGEL 82 [1] developed a local approach to determine both components of the displacement vector for "gray value corners". An implementation of this approach yielded encouraging results - see NAGEL and ENKELMANN 82a [2].

The same contribution [1] developed an iterative approach to refine such displacement estimates and to extend them into the environment of gray value corners - provided this environment exhibited sufficient gray value variation. An implementation of this iterative estimation procedure encouraged further studies in this direction [3].

Especially interesting appeared the question how such estimates could be extended into areas with predominantly one-dimensional or rather weak gray value variations. In [1] , an idea had been formulated to combine the local approach with a requirement of smooth displacement vector variation which had been first investigated by HORN and SCHUNCK 81 [4]. Based on the experience accumulated during the experiments described in [2,3], this idea is discussed and criticized in section three of the current contribution. A modification of this idea will be described in section four. Some - at least partially justifiable - simplifications facilitate a closed form analysis of this approach which will be presented in section five.

The next section outlines the ideas and notation introduced in [1] to the extent which is required for the subsequent exposition of new developments.

2. Displacement estimation at "gray value corners"

"Corners" in a gray value distribution can be characterized by the following requirement [1]: It is the point of maximum planar curvature in the locus curve of steepest gray value slope. Let the gray value distribution g1(x,y) in frame 1 be approximated by a bivariate second-order polynomial in x and y:

---

*new address
Fraunhofer-Institut für
Informations- und Datenverarbeitung-IITB
Sebastian-Kneipp-Straße 12-14
7500 Karlsruhe 1 / FR Germany

Motion: Representation and Perception
N.I. Badler and J.K. Tsotsos, Editors
Published by Elsevier Science Publishing Co., Inc.

$$g1(x,y) \cong f1_o + f1_x x + f1_y y + \frac{1}{2} f1_{xx} x^2 + f1_{xy} xy + \frac{1}{2} f1_{yy} y^2 \qquad (1)$$

Let the local coordinate system be positioned at a gray value corner, with its axes oriented in the principal curvature directions (i.e. $f1_{xy}=0$). The above requirement will be reflected by:

$$f1_x = \text{maximum} \neq 0 \qquad f1_y = 0 \qquad (2a)$$

$$f1_{xx} = 0 \qquad f1_{yy} = \text{maximum} \neq 0 \qquad (2b)$$

Let $\vec{U}=(u,v)^T$ represent the displacement estimate for all points $\vec{X}=(x,y)^T$ in a small environment around the origin of the coordinate system. The displacement will be determined by minimizing the expression

$$\sum_{\vec{X}} [\, g2(\vec{X}) - g1(\vec{X}-\vec{U}) \,]^2 \implies \text{Minimum} \qquad (3)$$

with the solution (see [1])

$$u = -\left(\overline{g2} - \overline{g1} - \frac{1}{2} f1_{yy} v^2 \right) / f1_x \qquad (4a)$$

$$v = - f2_y / f1_{yy} \qquad (4b)$$

Suppose the displacement estimate $\vec{U}_o = (u_o, v_o)^T$ differs from the true displacement $\vec{U}$ by a small correction $\vec{DU} = (Du, Dv)^T$

$$\vec{U} = \vec{U}_o + \vec{DU} \qquad (5)$$

This correction has to be estimated by minimizing

$$(\vec{dG})^T W_G (\vec{dG}) \qquad (6a)$$

$$\text{with} \quad dG_i = g2(\vec{X}_i) - g1(\vec{X}_i - \vec{U}_o - \vec{DU}) \qquad (6b)$$

W_G represents a weight matrix taking into account our uncertainty about the gray value measurements $g1(\vec{X}_i)$ and $g2(\vec{X}_i)$ at the locations $\vec{X}_i$ in the chosen environment. In order to make the dependance of $\vec{dG}$ on $\vec{DU}$ explicit, we introduce a bivariate polynomial $f1(\vec{X})$ which approximates $g1(\vec{X})$. Since the components Du and Dv are assumed to be small, we may drop all contributions containing higher than first order terms in components of $\vec{DU}$.

Let

$$B_{i,Du} = \frac{\partial dG_i}{\partial Du} = f1_x + f1_{xx}(x_i - u_o) + f1_{xy}(y_i - v_o)$$
$$= h1_x + f1_{xx} x_i + f1_{xy} y_i \qquad (7a)$$

$$B_{i,Dv} = \frac{\partial dG_i}{\partial Dv} = f1_y + f1_{xy}(x_i - u_o) + f1_{yy}(y_i - v_o)$$
$$= h1_y + f1_{xy} x_i + f1_{yy} y_i \qquad (7b)$$

with

$$h1_x = f1_x - f1_{xx} u_o - f1_{xy} v_o \qquad (8a)$$

$$h1_y = f1_y - f1_{xy} u_o - f1_{yy} v_o \qquad (8b)$$

where $(h1_x, h1_y)^T$ denotes the gradient approximation for g1 at location $\vec{X}-\vec{U}_o$. Using this notation, the estimate $\vec{DU}$ is given by

$$\vec{DU} = -(B^T W_G B)^{-1} B^T W_G \{ g2(\vec{X}) - f1(\vec{X}-\vec{U}_o) \} \qquad (9)$$

It has been shown in [1] that for the special case of "gray value corners" the expression (9) yields the closed form solution given in equations (4).

The matrix $B^T W_G B$ can be written in the following form - see [1]:

$$B^T W_G B = \frac{N}{2\sigma^2} \begin{pmatrix} h1_x^2 + f1_{xx}^2 \overline{x^2} + f1_{xy}^2 \overline{y^2} & h1_x h1_y + f1_{xx} f1_{xy} \overline{x^2} + f1_{xy} f1_{yy} \overline{y^2} \\ h1_x h1_y + f1_{xx} f1_{xy} \overline{x^2} + f1_{xy} f1_{yy} \overline{y^2} & h1_y^2 + f1_{xy}^2 \overline{x^2} + f1_{yy}^2 \overline{y^2} \end{pmatrix} \qquad (10)$$

where σ^2 denotes the variance of the gray value measurements and N denotes the number of pixels in the environment used to estimate the displacement. (We employed mostly a 5 x 5 environment in a 192 row x 256 column condensed halfframe, i.e. N = 25.)

The components of this matrix can be understood best if the determinant is discussed. It has the form

$$\det(B^T W_G B) = (N^2/4\sigma^4)\,\{\, \bar{x}^2\bar{y}^2(f1_{xx}f1_{yy} - f1_{xy}^2)^2 + \bar{x}^2(f1_{xx}h1_y - f1_{xy}h1_x)^2 + \bar{y}^2(f1_{xy}h1_y - f1_{yy}h1_x)^2 \,\} \qquad (11)$$

The first term corresponds to the square of the product of the principal curvatures, the next two terms correspond to the squared x- and y-component, respectively, of the gradient weighted curvature in the direction perpendicular to the gradient. This can be seen most easily by writing

$$\sqrt{(h1_x^2+h1_y^2)}\,\frac{1}{\sqrt{(h1_x^2+h1_y^2)}}\begin{pmatrix} f1_{xx} & f1_{xy} \\ f1_{xy} & f1_{yy}\end{pmatrix}\begin{pmatrix} h1_y \\ -h1_x \end{pmatrix} \qquad (12)$$

where the vector $(h1_y, -h1_x)^T / \sqrt{(h1_x^2 + h1_y^2)}$ is a unit vector perpendicular to the gradient.

The matrix $B^T W_G B$ has properties which make it rather interesting for our further discussion.

Proposition: $B^T W_G B$ is positiv semidefinite

Proof: The determinant (11) which can be expressed as the product of the two eigenvalues of $B^T W_G B$ consists of a sum of squared terms, i.e. it can not become less than zero. It follows from this observation that both eigenvalues must have the same sign if both differ from zero. The trace of $B^T W_G B$ again consists of a sum of squared terms. Therefore, the sum of the eigenvalues cannot drop below zero, either. These two observations imply that both eigenvalues of $B^T W_G B$ cannot become less than zero.

Proposition: $\det(B^T W_G B)$ does not vanish at gray value corners

Proof: Let the coordinate axes be aligned with the principal curvature directions whereby $f1_{xy} = 0$. If the requirements expressed by equations (2) are inserted, and the equivalence between $(f1_x, f1_y)^T$ and $(h1_x, h1_y)^T$ is noted, the proposition follows.

As a consequence, $B^T W_G B$ becomes singular only in image areas where the gray value remains constant in at least one direction. $B^T W_G B$ retains its full rank at locations where one principal curvature crosses zero which implies an extremum for the gradient in the associated direction.

Another observation will be important. Since the squared second-order derivatives of the gray value are multiplied by $\bar{x}^2$ or $\bar{y}^2$, respectively, all components of $B^T W_G B$ have the dimension of a gradient squared. The factors $\bar{x}^2$ or $\bar{y}^2$ explicitly indicate that the matrix $B^T W_G B$ expresses a structural property of an entire window, not only of a point. In the latter case, $\bar{x}^2$ and $\bar{y}^2$ would be zero and the matrix $B^T W_G B$ would be reduced to the outer product of the vector $(h1_x, h1_y)^T$ with itself, i.e. the rank of $B^T W_G B$ would only be one rather than two.

3. Smoothness requirement for displacement vector fields.

HORN and SCHUNCK 81 [4] noted that the requirement

$$\nabla g * \vec{U} = - \partial g/\partial t \qquad (13)$$

provides only one constraint equation for the two components of $\vec{U}$. They suggested to supplement this constraint by the additional requirement that the displacement vector field varies smoothly. They formulated this requirement by demanding that

$$\iint dxdy\, \{(u_x^2 + u_y^2) + (v_x^2 + v_y^2)\} \Longrightarrow \text{Minimum} \qquad (14)$$

in addition to

$$\iint dxdy\, (\nabla g * \vec{U} + \partial g/\partial t)^2 \Longrightarrow \text{Minimum} \qquad (15)$$

The subscript x or y in equation (14) should indicate the partial derivative of the components of $\vec{U} = (u,v)^T$ with respect to x or y. This approach suffers from the deficiency that the smoothness requirement is applied indiscrimantly to all parts of the image unless external knowledge is applied to restrict the integration in equations (14) and (15) to the image area corresponding to a single rigid object. Even then difficulties cannot be avoided if this object rotates in space and partially occludes itself.

To alleviate this difficulty, NAGEL 82 [1] suggested to demand smooth variation within a displacement vector field only for image areas with at most linear gray value variations as a function of image plane coordinates. This requirement had been formulated in the following manner. Let

$$M = \iint dxdy \; \{ \; [g2(\vec{X}) - g1(\vec{X}-\vec{U})]^2 + \beta^2 \begin{pmatrix} u_x \\ u_y \end{pmatrix}^T \begin{pmatrix} \gamma+f1_{xx} & f1_{xy} \\ f1_{xy} & \gamma+f1_{yy} \end{pmatrix}^{-1} \begin{pmatrix} u_x \\ u_y \end{pmatrix} + \beta^2 \begin{pmatrix} v_x \\ v_y \end{pmatrix}^T \begin{pmatrix} \gamma+f1_{xx} & f1_{xy} \\ f1_{xy} & \gamma+f1_{yy} \end{pmatrix}^{-1} \begin{pmatrix} v_x \\ v_y \end{pmatrix} \; \} \qquad (16)$$

be a function of the gray values $g2(\vec{X})$ as well as $g1(\vec{X})$ and the vector function $\vec{U}(\vec{X})$. Even considerable spatial variation of u or v would contribute only a small part to the integral M which has to be minimized provided that such variations occur in image areas with significant nonlinear gray value transitions, i.e. large second-order terms in the Taylor series of $g1(\vec{X})$. The summand γ in the diagonal of the matrices in equation (16) should prevent these matrices from becoming singular in image areas with linear sloping gray values. If all gray value distributions are restricted to at most linear variations, the formulation (16) specializes to the approach of HORN and SCHUNCK 81 [4].

Although equation (16) accomodates an attractive suppression of the smoothness requirement across significant gray value transitions - i.e. edges -, closer inspection showed some flaws. First of all, the matrix of second-order partial derivatives of the gray value function is not positive semidefinite, i.e. it cannot be used in this form as a weight in a function to be minimized. Moreover, it becomes singular at gray value corners where one principal curvature crosses zero. In addition, this form of the weight does not express the fact that in areas with large gray value gradients the displacement vector component along the gradient direction can be determined and this need not be subjected to the smoothness requirement.

Due to such objections, an alternative to the inverse matrix of second-order partial derivatives of the gray value function appeared desirable as a spatially variable weight. Experiences accumulated during the investigation of other approaches developed in [1] resulted in a suggestion which appears attractive.

4. A modified weight function for the smoothness requirement.

The properties of the matrix $B^T W_G B$ as outlined in section two let it appear feasible to use some weight matrix of similar structure. This suggestion will be investigated now.

The objections raised against the equation (16) would no longer apply in the case of the following formulation

$$M = \iint dxdy \; \{ \; [g2(\vec{X}) - g1(\vec{X}-\vec{U})]^2 + \beta^2 \begin{pmatrix} u_x \\ u_y \end{pmatrix}^T A^{-1} \begin{pmatrix} u_x \\ u_y \end{pmatrix} + \beta^2 \begin{pmatrix} v_x \\ v_y \end{pmatrix}^T A^{-1} \begin{pmatrix} v_x \\ v_y \end{pmatrix} \; \} \qquad (17)$$

where the matrix A is given by

$$A = \begin{pmatrix} A_{xx} & A_{xy} \\ A_{xy} & A_{yy} \end{pmatrix} \qquad (18)$$

with

$$A_{xx} = f1_x^2 + f1_{xx}^2 \bar{x}^2 + f1_{xy}^2 \bar{y}^2 + \gamma \qquad (18a)$$

$$A_{xy} = f1_x f1_y + f1_{xx} f1_{xy} \bar{x}^2 + f1_{xy} f1_{yy} \bar{y}^2 \qquad (18b)$$

$$A_{yy} = f1_y^2 + f1_{xy}^2 \bar{x}^2 + f1_{yy}^2 \bar{y}^2 + \gamma \qquad (18c)$$

The entities $\bar{x}^2$ and $\bar{y}^2$ are introduced to obtain dimensional equivalence between the terms containing first and second partial derivatives. These values should be determined according to the mask size of the operators used to obtain the derivatives from the tesselated digitized image function. The problem now can be stated in the following form: minimize the squared difference between the gray value at location $\vec{X}$ in frame two and at location $\vec{X}-\vec{U}$ in frame one simultaneously with the requirement that the squared first partial derivatives of the vector function $\vec{U}(\vec{X})$ weighted by A^{-1} are minimized. According to the calculus of variation, this problem can be recast as a system of two coupled partial differential equations:

$$M_u - \frac{d}{dx}M_{u_x} - \frac{d}{dy}M_{u_y} = 0 \tag{19a}$$

$$M_v - \frac{d}{dx}M_{v_x} - \frac{d}{dy}M_{v_y} = 0 \tag{19b}$$

We thus obtain

$$0 = [g2(\vec{X}) - g1(\vec{X}-\vec{U})]\ \{f1_x + f1_{xx}(x-u) + f1_{xy}(y-v)\}$$
$$- \beta^2\{A^{-1}_{xx,x}u_x + A^{-1}_{xy,x}u_y + A^{-1}_{xy,y}u_x + A^{-1}_{yy,y}u_y$$
$$+ A^{-1}_{xx}u_{xx} + A^{-1}_{xy}u_{xy} + A^{-1}_{xy}u_{xy} + A^{-1}_{yy}u_{yy}\} \tag{20a}$$

$$0 = [g2(\vec{X}) - g1(\vec{X}-\vec{U})]\ \{f1_y + f1_{xy}(x-u) + f1_{yy}(y-v)\}$$
$$- \beta^2\{A^{-1}_{xx,x}v_x + A^{-1}_{xy,x}v_y + A^{-1}_{xy,y}v_x + A^{-1}_{yy,y}v_y$$
$$+ A^{-1}_{xx}v_{xx} + A^{-1}_{xy}v_{xy} + A^{-1}_{xy}v_{xy} + A^{-1}_{yy}v_{yy}\} \tag{20b}$$

Here, the third subscript of components of A^{-1} should indicate the derivative with respect to the corresponding variable. The notation indicates that we use the coefficients of the Taylor expansion for $g1(\vec{X}-\vec{U})$ at the location $\vec{X}$ in order to represent the corresponding derivatives $\partial g1(\vec{X}-\vec{U})/\partial u$ and $\partial g1(\vec{X}-\vec{U})/\partial v$, respectively. This implies that terms of higher than second order in the components of $(\vec{X}-\vec{U})$ can be neglected.

5. Analysis of the resulting system of partial differential equations for $\vec{U}$

In order to study properties of the solutions for the equations (20), some simplifying assumptions appear necessary. First, all terms containing derivatives of A are dropped. This is partially consistent with the previous procedure to drop all terms of higher than second order in the Taylor expansion of $g1(\vec{x})$. The matrix A, however, comprises first partial derivatives as well as second partial derivatives of $g1(\vec{x})$. Therefore, dropping $A^{-1}_{xx,x}$, $A^{-1}_{xy,x}$, $A^{-1}_{xy,y}$, and $A^{-1}_{yy,y}$ is not quite consistent. The consequences of this simplification have to be studied. The sum $A^{-1}_{xx}u_{xx} + 2A^{-1}_{xy}u_{xy} + A^{-1}_{yy}u_{yy}$ can be written in the form

$$A^{-1}_{xx}u_{xx} + 2A^{-1}_{xy}u_{xy} + A^{-1}_{yy}u_{yy}$$
$$= \text{trace}\left\{ \begin{pmatrix} A^{-1}_{xx} & A^{-1}_{xy} \\ A^{-1}_{xy} & A^{-1}_{yy} \end{pmatrix} \begin{pmatrix} u_{xx} & u_{xy} \\ u_{xy} & u_{yy} \end{pmatrix} \right\} \tag{21}$$

with an analogous expression for the sum containing second derivatives of v. One thus obtains

$$0 = [g2(\vec{X}) - g1(\vec{X}-\vec{U})]\ \{f1_x + f1_{xx}(x-u) + f1_{xy}(y-v)\}$$
$$- \beta^2\ \text{trace}\left\{ \begin{pmatrix} A^{-1}_{xx} & A^{-1}_{xy} \\ A^{-1}_{xy} & A^{-1}_{yy} \end{pmatrix} \begin{pmatrix} u_{xx} & u_{xy} \\ u_{xy} & u_{yy} \end{pmatrix} \right\} \tag{22a}$$

$$0 = [g2(\vec{X}) - g1(\vec{X}-\vec{U})]\ \{f1_y + f1_{xy}(x-u) + f1_{yy}(y-v)\}$$
$$- \beta^2\ \text{trace}\left\{ \begin{pmatrix} A^{-1}_{xx} & A^{-1}_{xy} \\ A^{-1}_{xy} & A^{-1}_{yy} \end{pmatrix} \begin{pmatrix} v_{xx} & v_{xy} \\ v_{xy} & v_{yy} \end{pmatrix} \right\} \tag{22b}$$

Assume that an initial estimate $\vec{U}_o$ for $\vec{U}$ has been obtained, for example from preceding frames. Using this initial estimate, the equations (22) will be transformed into a system of equations for a correction vector function $D\vec{U}(\vec{X})$ with

$$\vec{U}(\vec{X}) = \vec{U}_o(\vec{X}) + D\vec{U}(\vec{X}) \tag{23}$$

The correction $\vec{DU}$ is assumed to be small so that higher than linear terms in the components of $\vec{DU} = (Du, Dv)^T$ can be neglected. We can write

$$g1(\vec{X}-\vec{U}_o-\vec{DU}) \approxeq g1(\vec{X}) + f1_x(x-u_o-Du) + f1_y(y-v_o-Dv) + \frac{1}{2}f1_{xx}(x-u_o-Du)^2 + \frac{1}{2}f1_{yy}(y-v_o-Dv)^2 + f1_{xy}(x-u_o-Du)(y-v_o-Dv) \quad (24a)$$

or, neglecting higher than first order terms in $\vec{DU}$

$$g1(\vec{X}-\vec{U}_o-\vec{DU}) \approxeq g1(\vec{X}-\vec{U}_o) - B^T*\vec{DU} \quad (24b)$$

with

$$\vec{B} = \begin{pmatrix} f1_x \\ f1_y \end{pmatrix} + \begin{pmatrix} f1_{xx} & f1_{xy} \\ f1_{xy} & f1_{yy} \end{pmatrix}(\vec{X}-\vec{U}_o) \quad (25)$$

The vector $\vec{B}$ represents the gradient of g1 at the location $\vec{X}-\vec{U}_o$.

The system of equations (20) may now be written in the form

$$0 = [g2(\vec{X})-g1(\vec{X}-\vec{U}_o)+B^T*\vec{DU}]B - \beta^2 \begin{pmatrix} \mathrm{trace}\{A^{-1}(\nabla\nabla u)\} \\ \mathrm{trace}\{A^{-1}(\nabla\nabla v)\} \end{pmatrix} \quad (26)$$

or

$$(B^TB)\vec{DU} = -[g2(\vec{X})-g1(\vec{X}-\vec{U}_o)]B + \beta^2 \begin{pmatrix} \mathrm{trace}\{A^{-1}(\nabla\nabla u)\} \\ \mathrm{trace}\{A^{-1}(\nabla\nabla v)\} \end{pmatrix} \quad (27)$$

The outer product matrix B^TB will have rank one rather than two. In analogy to the steps resulting in equations (10) and (9) it is postulated that B^TB must be averaged over a square symmetric neighborhood around $\vec{X}-\vec{U}_o$ of a diameter corresponding to the mask size employed to estimate the derivatives of g1. This averaging process will introduce modifications to the gradient components which depend on the second derivatives of g1 as well as the average squared values of x and y. With other words, this averaging process transforms B^TB into the matrix A given by equations (18). Assuming that A has full rank - compare the discussion in section two and the small terms γ added to the diagonal terms in equations (18a, 18c) for just this purpose - we may write

$$\vec{DU} = -A^{-1}B[g2(\vec{X})-g1(\vec{X}-\vec{U}_o)] + \beta^2A^{-1} \begin{pmatrix} \mathrm{trace}\{A^{-1}(\nabla\nabla u)\} \\ \mathrm{trace}\{A^{-1}(\nabla\nabla v)\} \end{pmatrix} \quad (28)$$

To see the significance of this equation, we assume that the coordinate system is aligned with the eigenvector directions of the matrix A, i.e. A will be diagonal

$$A = \begin{pmatrix} \alpha_1 & 0 \\ 0 & \alpha_2 \end{pmatrix} \quad (29)$$

In this situation we can write

$$\mathrm{trace}\{A^{-1}(\nabla\nabla u)\} = u_{xx}/\alpha_1 + u_{yy}/\alpha_2 \quad (30a)$$

$$\mathrm{trace}\{A^{-1}(\nabla\nabla v)\} = v_{xx}/\alpha_1 + v_{yy}/\alpha_2 \quad (30b)$$

and we obtain

$$\vec{DU} = \begin{pmatrix} Du \\ Dv \end{pmatrix} = -\begin{pmatrix} 1/\alpha_1 & 0 \\ 0 & 1/\alpha_2 \end{pmatrix} B[g2(\vec{X})-g1(\vec{X}-\vec{U}_o)] + \beta^2 \begin{pmatrix} u_{xx}/(\alpha_1\alpha_1) + u_{yy}/(\alpha_1\alpha_2) \\ v_{xx}/(\alpha_1\alpha_2) + v_{yy}/(\alpha_2\alpha_2) \end{pmatrix} \quad (31)$$

The second-order derivatives of u and v could be calculated from the initial values u_o and v_o for the raster points in the environment of the location in question, using the same operators given by BEAUDET 78 [5] which are employed to determine the corresponding derivatives of the gray value as a function of the image plane coordinates.

If the second-order derivatives of g1 vanish

identically within an image area, the eigenvalues of A become equal to the lower limits γ. In this case the smoothness requirement will be felt through the terms with factors β^2/γ^2 which will give large weights to any nonzero values of the Laplacian applied to u and v, with corresponding influence on the components of the correction vector $\vec{DU}$.

As another example, assume that we are at a location with significant second-order gray value variations in both directions as it can be found near gray value corners. In this case, the eigenvalues of A will be proportional to the square of such second-order derivatives and all terms resulting from the smoothness requirement will be smaller by at least a factor $1/\alpha_1$ or $1/\alpha_2$ compared to the contributions from the gray value difference $[g2(\vec{x})-g1(\vec{x}-\vec{U}_o)]$. In this case the smoothness requirement will not have significant influence on the determination of $\vec{DU}$.

The last example addresses the situation of a straight line gray value transition where one eigenvalue - say α_1 - is large whereas α_2 should be equal to γ, i.e. we study a gray value transition along the x-direction which is parallel to the y-direction. In this case the smoothness requirement will contribute to the correction along the y-direction whereas the relative influence on the correction in the x-direction will be much smaller, namely by a factor $1/\alpha_1$.

6. Conclusion

A previous approach to determine displacement vectors locally has been generalized by incorporating a smoothness requirement for the spatial variation of displacement vectors. Deviating from HORN and SCHUNCK 81 [4], the smoothness requirement is not applied indiscriminantly to every location, but it is weighted by a function which depends on the gray value transitions in the image. In this manner, larger variations in the displacement vector can be accommodated automatically across gray value edges. The approach is developped to the point where properties of the solution displacement vector field can be studied in closed form. The simplifications necessary to obtain such a result are pointed out. It has to be investigated whether these simplifications can be avoided and - if not - what effects might be due to them. In addition, it appears rather interesting to study this approach experimentally.

7. Acknowledgements

I gratefully acknowledge many fruitful discussions about this topic with W. Enkelmann. This contribution has been typed under very tight timing constraints by Mrs. U. Bauer whom I want to thank for her help.

8. References

[1] Displacement Vectors Derived from Second Order Intensity Variations in Image Sequences
H.-H. Nagel
FBI-HH/M-97/82 (March 1982)
Fachbereich Informatik, Universität Hamburg
to appear in Computer Graphics and Image Processing (January/February 1983)

[2] Investigation of Second Order Greyvalue Variations to Estimate Corner Point Displacements
H.-H. Nagel and W. Enkelmann
Proc. Int. Conference on Pattern Recognition, Munich, October 19-22, 1982, pp. 768-773

[3] Iterative Estimation of Displacement Vector Fields from TV-Frame Sequences
H.-H. Nagel and W. Enkelmann
Fachbereich Informatik der Universität Hamburg
(submitted for publication, December 10, 1982)

[4] Determining Optical Flow
B.K.P. Horn and B.G. Schunck
Artificial Intelligence 17 (1981) 185-203

[5] Rotationally Invariant Image Operators
P.R. Beaudet
ICPR-78, pp. 579-583

3D Computer Vision

MOTION AND TIME-VARYING IMAGERY

J. K. Aggarwal

Laboratory for Image and Signal Analysis
The University of Texas at Austin
Austin, Texas 78712, USA

1. A Brief History

The analysis of time-varying imagery, image sequence processing and dynamic scene analysis all refer to the young field concerned with the processing of sequences or collections of images with the objective of collecting information from the set as a whole that may not be obtained from any one image by itself. This research area of computer vision, image processing and computer graphics is relatively new. However, the applications are extensive and the issues are fundamental. The applications for analysis of sequences or collections of images covers a broad range of fields including medicine, autonomous navigation, tomography, communications and television, dancing and choreography, meteorology, animation and so on. Motivation for this interest becomes evident if one examines any of the above endeavors in detail. For example, the automatic analysis of the scintigraphic image sequences of the human heart is used to assess motility of the heart and is finding application in diagnosis and supervision of patients after heart surgeries. For the sequences of television images, one is able to reduce the bandwidth necessary for

Motion: Representation and Perception
N.I. Badler and J.K. Tsotsos, Editors
Published by Elsevier Science Publishing Co., Inc.

the transmission of television signals through motion estimation and compensation. The reduction in the necessary bandwidth may enable the transmission of certain classes of television images on existing low bandwidth channels. The processing of sequences of images for the recognition and the tracking of targets is of immense interest to the department of defense of every country. The computation, characterization and understanding of human motion in contexts of dancing and athletics is another field of endeavor receiving much attention. In meteorology, the satellite imagery provides opportunity for interpretation and prediction of atmospheric processes through estimation of shape and motion parameters of atmospheric disturbances. The preceding examples are indicative of the broad interest in motion, time varying imagery and dynamic scene analysis.

This broad interest has been evident since the first workshop in Philadelphia [1]. The workshop was expected to be a meeting of a relatively small number of specialists but it turned into a full scale conference. Since that auspicious beginning in Philadelphia, several additional meetings and special issues have contributed to the exchange of ideas and the dissemination of results. In addition, there are several sessions on motion and related issues at meetings such as the IEEE Computer Society Pattern Recognition and Image Processing Conference (now known as the Computer Vision and Pattern Recognition Conference). The list of workshops and special issues devoted exclusively to motion and time-varying imagery includes three special issues

[2-4], a book [5], a NATO Advanced Study Institute [6], proceedings of this ACM workshop [7], a European meeting on time-varying imagery [8] and a host of survey papers [9-12]. This list is incomplete at best. A better gauge of the breadth and depth of interest is provided by the table of contents of the book published to document the proceedings of the NATO-ASI [14].

2. At the Workshop

The three issues that dominate any discussion of dynamic scene analysis are: segmentation, occlusion and the computation of three-dimensional information from two-dimensional images. Segmentation refers to the process of determining features of interest together with distinguishing interesting changes from uninteresting changes, and establishing correspondence between the features and components in one image to those of the succeeding images. Occlusion analysis includes deriving structural changes due to projection perspective, and the appearance, occlusion and disappearance of objects. The computation of three-dimensional information entails constructing structural models and describing three-dimensional motions through the analysis of two-dimensional image information. These issues have been discussed in considerable detail in earlier review papers [11,12,15]. Therefore, the present discussion is confined to a synopsis of the issues discussed at the workshop. A session consisting of four presentations was devoted to 3-D computer vision. All of the four papers presented new results and stimulated discussion from the audience.

Moving Light Displays

The paper by Jenkins considers the tracking of three-dimensional moving light displays from a sequence of stereo views of moving lights. He investigates two problems simultaneously - namely the correspondence problem, i.e. which point in the right view corresponds to a given point in the left view, and the reconstruction problem, i.e. what are the three-dimensional coordinates of the point given the right and left views. This is done by constructing three-dimensional points corresponding to all pairs and eliminating spurious pairs based on a smoothness assumption. This assumption entails that the position and velocity of the given point would be relatively unchanged from one frame to the next frame. The assumption is exploited by choosing a suitable penalty function based upon position and velocity. The experimental results show that the developed tracking algorithm in fact tracks when the smoothness assumption is satisfied. In certain ways, the present results generalize the work of Rashid [16]. The necessary consideration of a large number of possible combinations is the obvious disadvantage of the method.

Determining 3-D Motion Parameters

The paper by Yen and Huang considers the determination of 3-D motion parameters of a rigid body from point correspondences over two time sequential images. Several earlier researchers have considered many variations of the same problem, for example, Ullman [17], Roach and Aggarwal [18], Tsai and Huang [19] and Nagel [20]. In general, the determination of motion parameters entails nonlinear equations and linear methods in special cases. The present paper presents results based upon a vector geometrical approach including alternative derivations of earlier results [21].

Combining Velocity and Position Information

The paper by Bobick presents results on the computation of structure from motion of rigid objects rotating about an axis fixed in direction. The input information consists of positions and instantaneous direction vectors of three points in orthographic projections in two frames. The author presents both analytical and simulation results. The present results complement the earlier work of Webb and Aggarwal [22,23]. Ultimately, the above problem reduces to fitting an ellipse to given data, whether the data is given in terms of the position alone or position and direction of the tangent. This problem has ramifications based on the position of points around the ellipse and other similar considerations as discussed in detail by Agin [24]. The accuracy of the results will depend significantly upon the choice of data, and if noisy data is confined to a small portion of ellipse, inaccuracies are bound to result.

Hypothesized Motion Parameters

Jerian and Jain consider the problem of finding rotation and translation parameters for a moving camera in a stationary scene. Using the earlier work of Jain [25] and Prazdny [26], the paper shows the viability of using hypothesized motion parameters to find correspondence and better motion parameters to the usual methods of finding correspondence or optical flow first and then solving for motion parameters.

References

1. J. K. Aggarwal and N. I. Badler (Eds.), Abstracts for the Workshop on Computer Analysis of Time-Varying Imagery, University of Pennsylvania, Moore School of Electrical Engineering, Philadelphia, PA., April 1979.

2. J. K. Aggarwal and N. I. Badler (Guest Eds.), Special Issue on Motion and Time-Varying Imagery, IEEE Trans. on PAMI, Vol. PAMI-2, No. 6, November 1980.

3. W. E. Snyder (Guest Ed.), Computer Analysis of Time-Varying Images, IEEE Computer, Vol. 14, No. 8, August 1981.

4. J. K. Aggarwal (Guest Ed.), Motion and Time Varying Imagery, Computer Vision, Graphics and Image Processing, Vol. 21, Nos. 1 and 2, January, February 1983.

5. T. S. Huang, Image Sequence Analysis, Springer-Verlag, New York, 1981.

6. NATO Advanced Study Institute on Image Sequence Processing and Dynamic Scene Analysis, Advance Abstracts of Invited and Contributory Papers, June 21-July 2, 1982, Braunlage, West Germany.

7. Siggraph/Siggart Interdisciplinary Workshop on Motion: Representation and Perception, Toronto, Canada, April 4-6, 1983.

8. European Conference on Time-Varying Imagery, Florence, Italy, 1982.

9. W. N. Martin and J. K. Aggarwal, "Dynamic Scene Analysis: A Survey," Computer Graphics and Image Processing 7, pp. 356-374, 1978.

10. H.-H. Nagel, "Analysis Techniques for Image Sequences," in Proc. IJCPR-78, Kyoto, Japan, November 1978, pp. 186-211.

11. J. K. Aggarwal and W. N. Martin, "Dynamic Scene Analysis," in the book Image Sequence Processing and Dynamic Scene Analysis, edited by T. S. Huang, Springer-Verlag, 1983, pp. 40-74.

12. J. K. Aggarwal, "Three-Dimensional Description of Objects and Dynamic Scene Analysis," Technical Report 83-1-20, Laboratory for Image and Signal Analysis, The University of Texas, Austin, TX., January 1983.

13. H.-H. Nagel, "What Can We Learn from Applications?" in the book Image Sequence Analysis, Edited by T. S. Huang, Springer-Verlag, 1981, pp. 19-228.

14. T. S. Huang (Editor), Image Sequence Processing and Dynamic Scene Analysis, Proceedings of NATO Advanced Study Institute at Braunlage, West Germany, Springer-Verlag, 1983.

15. H.-H. Nagel, "Overview on Image Sequence Analysis," in the book Image Sequence Processing and Dynamic Scene Analysis, edited by T. S. Huang, Springer-Verlag, 1983, pp. 2-39.

16. R. F. Rashid, "Towards a System for the Interpretation of Moving Light Displays," IEEE Trans. on PAMI, Vol. PAMI-2, No. 6, pp. 574-581, November 1980.

17. S. Ullman, The Interpretation of Visual Motion, MIT Press, Cambridge, MA., 1979.

18. J. W. Roach and J. K. Aggarwal, "Determining the Movement of Objects from a Sequence of Image," IEEE Trans. on PAMI, Vol. PAMI-2, No. 6, pp. 554-562, November 1980.

19. R. Y. Tsai and T. S. Huang, "Estimating 3-D Motion Parameters of a Rigid Planar Patch-I," IEEE Trans. on ASSP, Vol. ASSP-29, No. 6, pp. 1147-1152, December 1981.

20. H.-H. Nagel, "Representation of Moving Rigid Objects Based on Visual Observations," Computer, Vol. 14, No. 8, pp. 29-39, August 1981.

21. R. Y. Tsai and T. S. Huang, "Determining 3-D Motion Parameters of Rigid Objects with Curved Surfaces," in Proc. of the IEEE Conf. on PRIP, 1982, pp. 112-118.

22. J. A. Webb and J. K. Aggarwal, "Visually Interpreting the Motion of Objects in Space," IEEE Computer, Vol. 14, No. 8, pp. 40-46, August 1981.

23. J. Webb and J. K. Aggarwal, "Structure from Motion of Rigid and Jointed Objects," Artificial Intelligence, Vol. 19, pp. 107-130, 1982.

24. G. J. Agin, "Fitting Ellipses and General Second-Order Curves," Technical Report CMV-RI-TR-81-5, The Robotics Institute, Carnegie-Mellon University, Pittsburgh, Penn., July 1981.

25. R. Jain, "An Approach for the Direct Computation of the Focus of Expansion," in Proc. of the IEEE Conf. on PRIP, 1982, pp. 262-268.

26. K. Prazdny, "Determining the Instantaneous Direction of Motion from Optical Flow Generated by a Curvilinearly Moving Observer," Computer Graphics and Image Processing. 17, pp. 238-248, 1981.

Acknowledgement: This research was supported in part by the Air Force Office of Scientific Research under Contract F49620-83-k-0013.

Tracking Three Dimensional Moving Light Displays

Michael Jenkin

Department of Computer Science
University of Toronto

ABSTRACT

A method is presented for tracking the three-dimensional motion of points from their changing two-dimensional perspective images as viewed by a nonconvergent binocular vision system. The algorithm relies on a general smoothness assumption to guide the tracking process, and application of the tracking algorithm to a three-dimensional moving light display based on Cutting's[1] Walker program as well as other domains are discussed.

Evidence is presented relating the tracking algorithm to certain beliefs about neurophysiological structures in the visual cortex.

1. Introduction

One of the most basic problems in computer vision is that of determining the correspondence between two or more sets of points. Such a correspondence may be required to follow an object through time, or to locate the same object in different views of a scene. Various algorithms have been suggested as solutions for each of these problems. Rashid [9] tracked points in a moving light display (MLD) by predicting each point's expected location. Marr and Poggio [3] developed an algorithm for determining depth from a grey scale stereo view of a scene. In our case it was required to track a three-dimensional MLD through time by fusing the two views of a scene: In effect we had to solve both problems.

A MLD is defined by Rashid [9] to be, "a sequence of binary images representing points of one or more moving objects in an actual or synthesised scene". A three dimensional moving light display is a MLD in which the lights are located in three dimensions rather than two. A MLD is a useful starting point for analysis. There is a large reduction in the number of data points from grey scale images, and object detection and recognition in a MLD does not generally reduce to frame by frame analysis. Johansson found that human subjects asked to identify the contents of a single frame are frequently unsuccessful [5].

A program was written based on Cutting's Walker program [1] to generate a three dimensional MLD similar to the two dimensional MLD described by Rashid [9]. Instead of generating a two-dimensional perspective view of the synthetic walker, the program was designed to generate a stereo view of the figure by producing two perspective views, as viewed by two eyes a fixed distance apart. A display was produced that could be viewed using a stereo viewer (fig. 1). In the display the lights were joined up to make the object more visible. An algorithm was required which would be able to match up the left and right views of a scene, and which at the same time would track the three-dimensional points through time.

Algorithms using constraints such as Ullman's rigidity constraint [12], Hoffman's planarity assumption [4], and Webb's rotation and translation assumption [13] all restrict the type of object to be tracked to be rigid, or to be composed of rigid parts. We required an algorithm that would not be restricted in either the type of object that could be tracked, nor in the types of motion the object could exhibit. Instead of basing the algorithm on assumptions of the above type, we assumed that the object's motion would obey a general smoothness assumption.

The smoothness assumption was based upon certain beliefs from Gestalt Psychology. The law of proximity, and rule of good continuation[6], suggested that we organize objects into higher groups if they are close together, and that we prefer objects to be smooth rather than to end abruptly. The principal of least action[11] suggested that when an object is perceived as moving we tend to perceive it as moving along a path that in some sense is the shortest, simplest or most direct. A temporal extension of these laws formed the basis of our smoothness assumption.

The tracking algorithm assumed a basic smoothness assumption; a given point would be relatively unchanged from one frame to the next. In particular that

1) The location of a given point would be relatively unchanged from one frame to the next.
2) The scalar velocity, or speed of a given point would be relatively unchanged from one frame to the next.
3) The direction of motion of a given point would be relatively unchanged from one frame to the next.

Notion: Representation and Perception
N.I. Badler and J.K. Tsotsos, Editors
Published by Elsevier Science Publishing Co., Inc.

As these constraints treat each moving point independently they do not restrict either the type of motion exhibited by the object nor do they restrict the type of object to be tracked.

In order to simplify the problem it was assumed that at all times, all of the points were visible by both eyes. This precluded the possibility of occlusion of data points. It was also assumed that the three dimensional position and velocity of the points were known at some initial time.

2. Tracking Algorithm

Let (x,y,z) be a point in a right handed co-ordinate system. If we view this point using a non-convergent binocular vision system whose "eyes" are located at (-e,0,f) and (+e,0,f) looking towards infinity in the -z direction; then the left and right eyes will view a point $P=(x,y,z)$ as images in the x,y plane as

$$x_r = \frac{(x-e) * f}{f-z} \tag{1a}$$

$$y_r = \frac{y * f}{f-z} \tag{1b}$$

$$x_l = \frac{(x+e) * f}{f-z} \tag{2a}$$

$$y_l = \frac{y * f}{f-z} \tag{2b}$$

Where f is the focal length, the system baseline length is 2e, and (x_l, y_l) and (x_r, y_r) are the left and right eye views respectively[2].

Provided that a point (x,y,z) is visible by both the left and right eyes, then the three-dimensional coordinate of the point can be reconstructed as

$$x = \frac{e * (x_r + x_l)}{x_l - x_r} \tag{3a}$$

$$y = \frac{2 * e * y_r}{x_l - x_r} \tag{3b}$$

$$z = f - \frac{2 * e * f}{x_l - x_r} \tag{3c}$$

by solving for x, y and z in equations (1) and (2).

Let L_j, R_k be left and right eye co-ordinates of point j in the left eye, and k in the right eye. We will denote the three dimensional point which can be constructed from L_j and R_k using (3) as [j,k].

Let φ_l and φ_r be functions mapping points in the left and right images respectively from time t to time $t+1$. We seek functions $\varphi_l : n \rightarrow n$ and $\varphi_r : n \rightarrow n$, such that φ_l and φ_r are one-to-one and onto, and that the point P_i at time t has location $[\varphi_l(i), \varphi_r(i)]$ at time $t+1$. We can now state the problem more formally.

Given $P_1(t)$ to $P_n(t)$, the three dimensional location of the points at time t, $V_1(t)$ to $V_n(t)$, the vector velocity of the same n points at time t, and L_1 to L_n and R_1 to R_n (where L_i does not necessarily correspond to R_i), the left and right eye views of the points at time $t+1$, we seek one-to-one and onto mappings φ_l and φ_r such that P_i has location $[\varphi_l(i), \varphi_r(i)]$ at time $t+1$.

Let $w(i,j,k)$ be a disparity function measuring the difference between $P_i(t)$ and the three-dimensional point constructed from L_j and R_k at time $t+1$. Define $w(i,j,k)$ as

$$w(i,j,k) = d(i,j,k) + v(i,j,k) \tag{4}$$

where d and v are

$$d(i,j,k) = \frac{distance(P_i(t),[j,k])}{\sum_{l,r} distance(P_i(t),[l,r])}$$

$$v(i,j,k) = \frac{\left| V_i(t) - \frac{([j,k]-P_i(t))}{time} \right|}{\sum_{l,r} \left| V_i(t) - \frac{([l,r]-P_i(t))}{time} \right|}$$

where the sums over l and r take on all valid combinations for the left and right eye views, *time* is the time between snapshots, *distance* is the Euclidean distance between two points and $|a|$ denotes the length of the vector a.

A point $[l,r]$ is not a valid point if the absolute value of the y coordinate of L_l minus the y coordinate of R_r is larger than a given tolerance, where the tolerance is related to the resolution and noise of the input data. This is clear since from equations (1b) and (2b) it must be true that $y_l = y_r$ for the point $[l,r]$ to be valid.

The d component weights $w(i,j,k)$ by the displacement between P_i and the candidate point $[j,k]$. This corresponds to the first of the basic smoothness assumptions. The v component weights $w(i,j,k)$ by the difference in the vector velocity V_i and the vector velocity over the next time interval if the point $[j,k]$ is chosen. This corresponds to the second and third basic smoothness assumptions.

It remains to determine the correspondence functions φ_l and φ_r given $w(i,j,k)$. We wish these functions to have the property that

$$\Omega = \sum_{i=1}^{n} w(i, \varphi_l(i), \varphi_r(i)) \tag{5}$$

is minimized.

For each point P_i at time t there are at most n^2 possible matchings [l,r] at time $t+1$. This gives rise to $(n!)^2$ total possible matchings. Searching through all possible mappings is not practical. We approximated the solution by applying the greedy heuristic. We chose the n smallest values of $w(i,j,k)$ to construct the mappings φ_l and φ_r, only excluding an entry if it would violate the one-to-one and onto properties of the φ_l and φ_r entries already chosen.

3. Experimental Results

The tracking algorithm was applied to three different sets of input data. The first set of data was that of a synthetic walker (fig. 1). Two other sets of input data were generated. The first produced images of colliding hexagons (fig. 3), while the second produced an image of a translating, rotating disk (fig. 5). In all of these figures certain points were joined in order to make the object more visible. Each figure represents a multiple exposure of some of the frames of the film produced.

A graphic display program was written to show the results of the tracking algorithm. The values of the correspondence function were displayed, as well as the value of Ω for each frame. The results from tracking the synthetic walker (fig 2.), the colliding hexagons (fig. 4), and the rotating, translating disk (fig. 6) were very informative. They demonstrated that the tracking algorithm was capable of tracking objects that obeyed the general smoothness assumption, and that the value of Ω was a useful measure of the uncertainty of the matching found by the algorithm.

The input to the tracking algorithm was generated so that the correct mappings φ_l and φ_r would be the identity mappings. We note that this was the case for the synthetic walker (figs. 1 and 2). It was found that from arbitrary orientation the tracking algorithm was capable of tracking the figure correctly.

The results obtained with the colliding hexagons were as expected. At the point of collision the algorithm fails - the collision violates the smoothness assumption. It was interesting that the value of Ω reaches a maximum for this frame. It was also found that the value of Ω reached a maximum while tracking the synthetic walker when the body parts passed nearest each other.

The final set of input data was that of a rotating translating disk (fig 5.). It was found that for simple rotation below a specific angular velocity the tracking was perfect. At higher angular velocities some points were tracked as rotating backwards, while others were tracked correctly (fig. 6). This is an effect that is found in human perception, quickly rotating objects appear to rotate backwards at certain speeds. Fast acceleration violates the smoothness assumption, and the tracking algorithm fails.

4. Discussion

Orban [7] performed experiments with humans on the perception of moving dots during saccades. On the basis of his experiments he put forward the following hypothesis

1) Two different detection mechanisms exist within the visual system, one dealing with velocity (and direction), the other with amplitude (and direction) of displacement.
2) The input to both of these movement channels is the movement of retinal images.
3) In the displacement channel an evaluation of entering information is made as a function of an expected value.

In later work with cats, Orban [8] discovered that all neurons exhibiting velocity specificity, or amplitude sensitivity or specificity, were also direction specific. As result of this research Orban hypothesised that we perceive velocity or amplitude of motion as a vector quantity. Our tracking algorithm has these properties. In particular we used the result that we perceive velocity as a vector quantity, rather than as a scalar speed and direction, to construct the v term of the weight function.

In Orban's model for the first steps of cortical elaboration of movement perception [8], he proposed that the simple cells of area 18 of the cat's visual cortex receive as input the velocities of moving points on the retina from the Y-neurons of the lateral geniculate nucleus. He identified with these cells the ability to analyze retinal images, at low velocities, in terms of direction, velocity and length of movement. This is exactly the function of our tracking algorithm.

Certain functional properties of these simple cells are also exhibited by our tracking algorithm. Both fail to track at higher velocities, the simple cells do not respond [8], and our algorithm fails. Orban found that the simple cells did not respond to random dot patterns, or moving square patterns [8]. This effect was also exhibited by the tracking algorithm. When presented with a sequence of random dots in three space, the tracking algorithm tracks but produces a large value of Ω.

The tracking algorithm makes no attempt to model the low level structure of the simple cells. Rather we have attempted to model the three dimensional functional equivalent of the simple cells, and we have found that properties exhibited by these cells are also exhibited by our algorithm.

5. Conclusions

The tracking algorithm was used to track the three dimensional moving light displays generated by the programs mentioned above (figs. 1, 3, 5). For those frames in which the general smoothness assumption was satisfied tracking was performed correctly. In developing the algorithm it was found that it was necessary to use both location and vector velocity weighting to find the correct matching. It was found that in the three-dimensional case predictive functions such as Rashid's [9] were unable to track smoothly moving points correctly.

Experimentation with different weightings for the distance and velocity components revealed that the two components were equally important in determining a correspondence. Normalization of the two factors was used to weight the two components equally.

As a side effect the tracking algorithm computes the value of Ω, the total weight per frame of the chosen correspondence. The value of Ω is a measure of how inaccurate a given correspondence is. For example, in the case of the colliding hexagons (fig 2.), there is a sharp increase in the value of Ω during the frame of collision (fig 3.). This value would be very useful to higher level processes. The value of Ω could be compared against a threshold value, and could be used to "turn off" the algorithm whenever the smoothness constraints are violated.

The use of a greedy heuristic to construct the matching has both advantages and disadvantages. As a heuristic it is possible that although the weight function predicts the correct matching, the greedy heuristic will miss it. It will however predict the best matches first. Regan et al. [10] suggested that velocity ratios would be a useful technique for detecting objects moving at velocities greater than that of the rest of the scene. This processing could be performed quickly, allowing fast response to potentially dangerous objects. The greedy heuristic produces similar results but for different reasons. Matches will be found for objects that differ significantly from the rest of the scene. For example, in the process of tracking the walking figure the head and feet were usually tracked first; these features stood out from the rest of the figure. Similarly points which have different velocities from the rest of the image will be matched early in the processing.

6. References

[1] Cutting, James E., A Program to Generate Synthetic Walkers as Dynamic Point-light Displays, Behavior Research Methods and Instrumentation, 10(1), 91-94.

[2] Duda, Richard O. and Hart, Peter E., Pattern Classification and Scene Analysis, John Wiley and Sons, New York, 1973.

[3] Marr, D. and Poggio, T., Cooperative Computation of Stereo Disparity, Science 194, 1976, 283-287.

[4] Hoffman, D. D. and Flichbaugh B. E., The Interpretation of Biological Motion, Massachusetts Institute of Technology A.I. Memo No. 608, 1980.

[5] Johansson, G., Visual Perception and Biological Motion and a Model for its Analysis, Perception and Psychophysics 14, 2, 201-211, 1973.

[6] Koffa, K., Principles of Gestalt Psychology, Harcourt, Brace and Co., New York, 1935.

[7] Orban G. A., et al., Movement Perception During Voluntary Saccadic Eye Movements, Vision Res. 13, 1343-1353, 1973.

[8] Orban G. A., Visual Cortical Mechanisms of Movement Perception, Katholieke Universiteit Leuven, Belgium, 1975.

[9] Rashid, R. F., Towards a System for the Interpretation of Moving Light Displays, IEEE PAMI PAMI-2, November 1980, 574-581.

[10] Regan D. et al., The Visual Perception of Motion in Depth, Scientific American, Vol 241, No. 1, July 1979, 136-151.

[11] Shepard, Roger N., and Cooper, Lynn A., Mental Images and their Transformations, MIT Press, Cambridge, 1982.

[12] Ullman S., The Interpretation of Visual Motion, MIT Press, Cambridge, 1979.

[13] Webb, J. A., Structure from Motion of Rigid and Jointed Objects, Proceedings of the 7th IJCAI-81, Vancouver.

7. Acknowledgements

I would like to thank John Tsotsos and Bernd Neumann for their helpful suggestions and criticisms during the preparation of this paper.

Financial support was gratefully received from a NSERC Summer Research Award and from a contract administered by the Defense Research Establishment Atlantic.

8. Figures

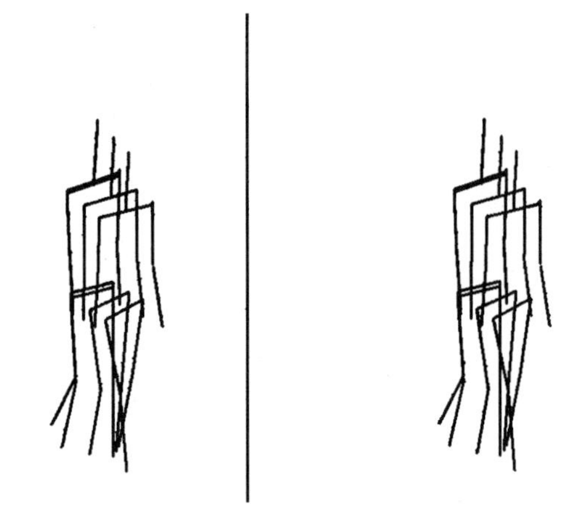

Figure 1

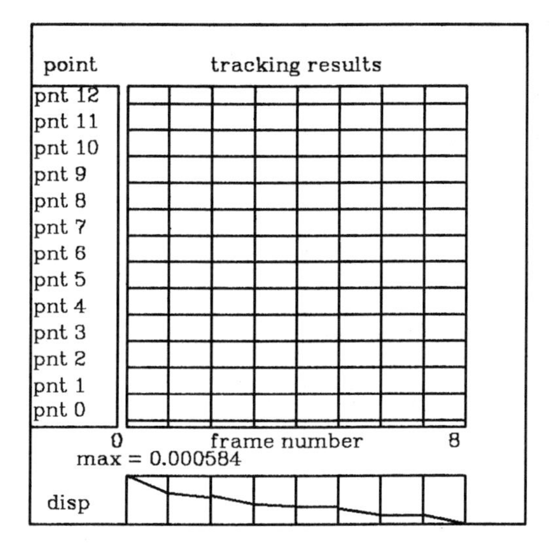

Figure 2

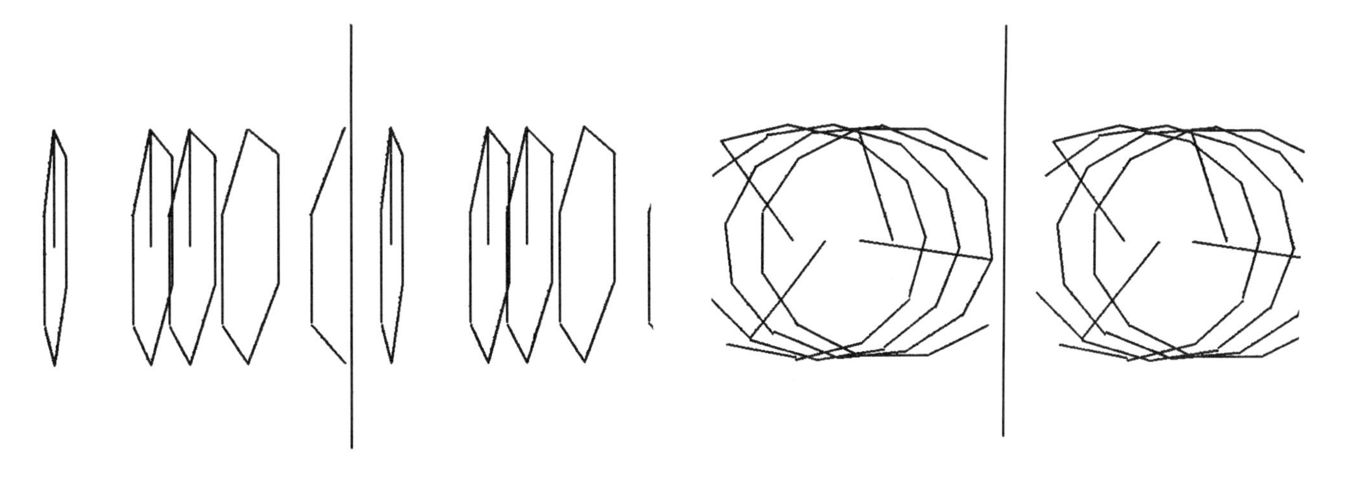

Figure 3

Figure 5

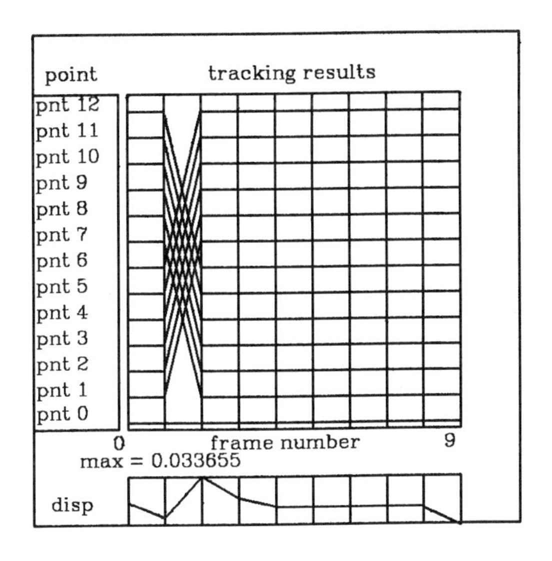

Figure 4

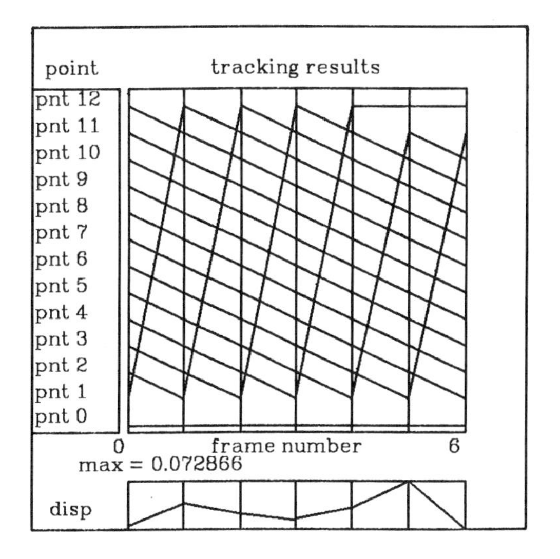

Figure 6

DETERMINING MOTION PARAMETERS FOR SCENES WITH TRANSLATION AND ROTATION

CHARLES JERIAN and RAMESH JAIN

Department of Electrical and Computer Engineering

University of Michigan

Ann Arbor MI

Abstract

A study of methods that determine the rotation parameters of a camera moving through synthetic and real scenes is conducted. Algorithms that combine ideas of Jain and Prazdny are developed to find translational and rotational parameters. An argument is made for using hypothesized motion parameters rather than relaxation labelling to find correspondence.

Introduction

There have been many different approaches to extracting the motion information from sequences of dynamic scenes[Nag81a,Nag81b,ULL79a,ULL79b]. A number of these methods concentrated on determining the FOE or focus of expansion form of the translational parameters. Most of these methods require optical flow vectors or corresponding points in different discrete frames as input. Other methods seek to use the focus of expansion to yield the correspondence[LAW82]. The FOE as the intersection of all optical flow vectors does not exist when the system undergoes rotation, however we feel that the concept of an FOE can still be used in analyzing three-dimensional motion parameters and in solving the correspondence problem for such cases. [PRA79b,PRA80,PRA81a,PRA81b].

One simple method for determining the focus of expansion in scenes where corner points or other matchable entities have been extracted has been proposed by Jain[JAI82a]. This method does not use a correspondence between the frames, rather it uses the triangle inequality to show that in approaching motion the sum of the distances of points in the second frame from the FOE minus the sum of the distances of points in the first frame from the FOE is maximized. This method is difficult to apply directly to real scenes because most corner detectors can lose points from frame to frame and can also find multiple points in both frames. The idea of hypothesizing a likely focus of expansion and verifying it in some way against the image is an important method. In this paper we will suggest other variations of this basic paradigm. In Prazdny's work also optimization of a quality of fit function was proposed to find the rotation parameters from given optic flow vectors and this was tested on synthetic data [PRA81a].

As an alternative to this method a linear solution to rotation and translation parameters have been found by Tsai and Huang [TSA81,TSA82a,TSA82b]. Tsai and Huang have a linear algorithm for finding the direction of motion and the rotation parameters given optical flow Their method is in contrast to the optimization methods in that it uses a system of linear equations. All other systems studied have non-linear equations in their methods.

Their algorithm was implemented in a straight forward way by least squares using subroutines from the *NAAS IMSL* package. The results indicated that there are problems with using low accuracy vectors, vectors for objects which are close in depth or vectors that are spatially close. These problems become worse for objects that are far away. This study showed that their method requires some modification and that the least squares approach might not be the best approach to use. It also shows that getting the rotation from the hyperbolic vs. linear error that occurs locally is next to impossible. If rotation rather than the more genera distortion is desired a problem arises in that their method does not specifically obtain a rotation represented by 3 real numbers but 8 parameters of a 3 by 3 distortion matrix. Only a subset (the orthonorma matrices) actually are rotations. Therefore it is possible that the optimal solution is not constrained to be a rotation because of noise or other errors. In any case it seems that their method will not recover the motion of a synthetic version of Nagel's[DRE81a] car with the optical flow we can obtain using a least squares implementation[CPJ82].

Determining rotation in synthetic scenes using feature points

If a collection of points from a translating rigid body or, equivalently, if the observer translates, the disparity (or optical flow) vectors lie on lines that intersect at a point called the FOE. The FOE gives the direction of motion of the object or observer. Figure 1 shows the disparity vectors for an observer moving toward two boxes. These boxes are at position (0,0,10) and (0,10,25) where (0,0,1) represent the center of the image. The boxes are inclined by a rotation of -.035 radians about the x-axis followed by -0.50 radians about the y- axis The observer is displaced 2 units in the z direction. For this example the focus of expansion is at (0,0) and the direction of motion is (0,0,1).

Motion: Representation and Perception
N.I. Badler and J.K. Tsotsos, Editors
Published by Elsevier Science Publishing Co., Inc.

Imagine a spherical retina whose center is at (0,0,0). Then the lines in this figure would correspond to a collection of great circles of the sphere which pass through a pole at (0,0,1). Now, if the imaging system is rotated before it is translated, a motion component is added to each point on the sphere along a latitude line of the axis of rotation. The projection of such latitude lines on a plane is discussed by Prazdny[PRA81a]. For rotations about axes on the x-y plane, these circles map onto hyperbolic arcs and, for rotations about the z-axis they map into circular arcs. The synthetic images in this section all are created using a $\frac{\pi}{2}$ radians field of view. This gives an aperture ratio of 1. Locally the hyperbolic arcs are almost linear, especially near the center of the cameras field of view. For a camera with a narrower field of view than our synthetic imaging system, they are fairly linear throughout the image. Rotation of the camera causes each point in the plane to be displaced in a way depending on its position on the plane. The actual mapping of image points (X,Y) to rotated points $(NEWX, NEWY)$ under rotation about the y axis by ϑ radians is given by the equations:

$$NEWX = \tan(arctan(X) + \Delta\vartheta)$$

$$NEWY = Y\sqrt{(1+NEWX^2/\,1+X^2)}.$$

Then, translation moves the points along lines passing through the FOE. The effect for most imaging systems, with a reasonable field of view, is almost constant over the entire plane. Note that such rotations can be closely simulated by a translation of an object, at distance r, by $r\vartheta$ in the direction perpendicular to the axis of rotation.

In figure 3, we have the same two objects as figure 1, however the camera is rotated .5 radians about the y-axis before translating 2.5 units along the z-axis. The motion vectors in figure 3 fall on lines (shown in figure 4) that tend to cluster at two intersection points along the x-axis. Any method that finds good possible FOEs will find likely values at these points. The object points that have these values would then be found. This facilitates a segmentation of the scene based on the variation in depth. It is observed that the intersection points all are roughly along a line (in this case the x-axis since the FOE is at (0,0)) and the axis of rotation is always perpendicular to that line. The axis of rotation therefore, can be quickly and easily determined by examining the local FOEs.

Different objects in the image can be at different depths. The points at each depth are observed to intersect near a single point. If the image is pre-segmented into regions that are each part of one object of approximately constant depth, a clear single intersection point for the optical flow vectors of that region would result. This is the converse of using the distinct local peaks to help segment the image. Standard techniques such as finding the least squares intersection point may suffice to obtain reliable results for such a region. Similarly, Jain's method would give a local FOE for such a segmented region. The different local peaks would be fitted with an appropriate curve to find the direction and amount of rotation. The position of the local FOEs would give an ordering of the regions by depth.

Rotational parameters seem to operate orthogonally to the radial direction on which Jain's function depends, so that rotations have little consistent effect on its value. Consider the case where the FOE is at (0,0) and rotation is about the z-axis. The value of the function at the correct FOE (0,0) is not affected by any amount of rotation. The maxima may not be at (0,0) anymore.

From a correct FOE if we draw radial lines to a distinctive feature point, its matching point is found on the same line within a certain distance that depends on the maximum velocity allowed. A smooth error measure can be devised that accounts for points that do not precisely fall on a line by finding the nearest matching point within a range of angles and using the sum of the absolute or square of the error the best matching points. If there is no match, a large amount such as π radians can be added to the sum of the errors. This allows us to effectively find an optimizable function for the FOE that can be tailored to find good peaks along the different clusters or that can average the values to find an average FOE. To find strong peaking the maximum angle for allowing a match is decreased and changes are made to the error function to cause poorer matches to weigh more heavily against a given choice for the FOE. Since rotation of the camera is merely a mapping of the image to itself, the effects of such rotation can be reversed by performing an inverse mapping of the image using equations previously described here or those of Prazdny[PRA81a]. Such transformations are called de-rotation mappings here to distinguish them from three-dimensional rotations. Since such a measure is very sensitive to error in the FOE caused by rotation it can be used to effect controlled de-rotation of the image by intersecting the vectors more tightly in the de-rotated image and reducing the error measure This procedure does not require any a-priori match however if any a- priori matching information is available the routine can selectively examine such matches to select the match that is most consistent with its hypothesized FOE.

Figure 5 shows the same objects as figure 1. The observer translates along the z-axis 2.5 units and rotates 0.5 radians about the z-axis. Figure 5 shows the optical flow vectors that result. Figure 6 shows the lines that extend these vectors to intersection points. As these figures show, the lines intersect on a circle where the FOE of the lines is at the center of the circle The width of the circle is proportional to the amount of rotation and inversely proportional to the translationa velocity. If the rotation angle is zero the circle degenerates to a point. If the angle is π radians, the circle becomes the point at infinity. A circle detector could be used on the local FOE histogram to find the circle and obtain the FOE and amount of z- rotation. If we have only part of the circle is represented strongly because objects are more concentrated in some parts of the image than others, an averaging method would locate the FOE at the center of mass of the points on the circle rather than at the center of the circle. The amount of error would depend on the non-uniformity of the distribution and the radius of the circle. Jain's algorithm for example, tends to find a value near such a center of mass. This leads to increasing inaccuracies if the feature points are non-uniformly distributed. Inaccuracy increases as the radius of the circle increases.

A local approach that would find the FOE in windows would be useful here since we could plot the local FOEs and apply a circle detector to find the FOE and amount of rotation.

As a variation of the above methods, an FOE can be computed by Jain's method using data that has been de-rotated using Prazdny's equations. The computed FOE is used to obtain a correspondence of the feature points. This involves a search in the radial direction from the FOE for a matching feature point. In the synthetic scenes first explored, only position information was present, so no comparison of the feature points was possible. To discern the presence and amount of rotation the following are computed: the distance in the x direction of the matched point, the perpendicular distance, and the difference in angle of the matching points. We only need to explore a few point in the neighborhood of a given point to find the matching point. All these points can be preselected by using a maximum velocity constraint. Additionally, if feature points have feature values, these would be used to further constrain the list to be searched. We de-rotate by moving the position of points in one of the images along hyperbolic paths. For de-rotation purposes rotation about an axis in the x-y plane can be replaced by a rotation about the x-axis followed by a rotation about the y-axis. To simplify the work only rotations about the y-axis were considered, however, everything will still work with an increase in cost for rotations about two axes. The amount of rotation about a given axis is treated as a parameter to vary in an optimization method. The best average FOE is redetermined using Jain's method. It may be necessary to do much re-computation. To avoid this assume that the position of the best FOE varies smoothly and by a small amount with small changes in the amount of rotation. Then only a small amount of additional computation is required to find a new average FOE, given a new angle. In this method we have tested various parameters to determine the amount of error from the correct FOE. Among these is the mean square angular error of all the corresponding points. Other measures tested were the distance of the hypothesized line connecting the corresponded points along the x direction, from the FOE, and from the average of all the x positions. One could also find the least squares value of the intersections and determine the orthogonal distance of each line from the least squares value. A different error measure is the amount of spread of the clusters t the intersections.

Table 1 summarizes the results of an implementation of this theory. The table shows the extracted position of the FOE for different amounts of assumed rotation about the y-axis. The correct values used in generating the data would give an angle of -0.03 radians and locate the FOE at (0.0,-1.00) on the x axis at the extreme left of the image. The column labelled match angular error shows the average of the sum of the squared differences in angular position of vectors from the extracted FOE to pairs of points that are established as corresponding. The correct result yielded the minimum error. The average of the sum of the squares of the angular errors was computed as a measure of the quality of the FOE and is reported in Table 1. Lines were drawn between the matched points to a line going through the computed FOE orthogonal to the x-axis. The actual average position of the intersections of the extended optical flow lines and this line were determined. The x position column reports that position. If there is no systematic error this value should be the same as the x coordinate of the FOE. The variance of the intersections is noted under the variance column. This is an excellent measure of the spread of the intersection cluster and gives the best indication of rotation of all the measures. The FOE variance column reports the variance using the x-coordinate of the FOE rather than the refined x-position obtained from the intersecting vectors. As one can see from examining this table, all of the error measures are quite capable of determining the correct amount of rotation to within .01 radians.

The plot represents the search space for an optimization algorithm that seeks the peak of the surface The function is examined over a coarsely quantized subset of its range. The range is quantized using equa angles of the x and y axes of a polar coordinate sphere This allows for the uniform treatment of cases where the z velocity is much less than the x or y velocities or the rotation is large and the FOE is not on the image. Then a finer search of the area showing the best results in the first search is performed. This gives us a coarse to fine strategy. Figure 9 shows the FOE function for the de-rotated feature points that was generated by the program from the same run whose results are in Table 1 Figure 10 shows the close-up of the function near the correct value.

If there is a non-geometric check on correspondence available, then the quality of correspondence can be evaluated. The possibility of no- match could be included. Correspondence can be used with such equations as those of Tsai and Huang to extract rotation and translation parameters. If correspondence can be verified with a non-geometrical test, then a hypothesize and test paradigm becomes possible. A preliminary correspondence of some points can be found. This can be used to determine some motion parameters. These parameters can then be used to establish a better correspondence for more points. The procedure can be iterated until a sufficiently satisfactory result ensues. If the process fails to converge, new starting values can be hypothesized from the data. A random sampling and consensus algorithm[FIS81] [FIS82] may be useful here.

Real scenes

Since edge points seem to cause problems for the methods described in this paper[CPJ82] corner points were used for work with real scenes. The input image consisted of a pair of 128 by 128 images quantized to 256 grey levels. The images contain a frame surrounding a cube. The observer is approaching the center of the curb with uniform velocity.

A corner detector suggested by Kitchen and Rosenfeld[KIT81] was applied to the two scenes and corner points were extracted. All corner points within a certain fixed distance determined by an estimate of the maximum allowed velocity (in this example 10 pixels) were considered as possible matching corner points The square grey-level differences of 5 by 5 windows was used as an error measure. To more accurately locate the actual matching point a 3 by 3 neighborhood of the point in the previous frame was examined to find the lowest error match. At most five of the best such matching corner points are recorded in an array with the quality of the match for each corner point in frame2.

The space of possible FOEs is searched on a coarse 10 by 10 grid. For a given point in frame2, only those possible matching points determined earlier which are not further away from the FOE in frame1 are tested for approaching motion, for receding motion only points which are not closer would be checked. Of these points, the point with the least difference in angle is selected, if there is such a point. A weighted sum of this angular error and the grey value error measure would be better here. The chosen point is the best match consistent with the FOE. Running totals of angular error and grey value errors are kept.

An attempt was made to minimize the error by changing the FOE. Many choices of the FOE result in an identical correspondence of points in frame 1 and frame 2. These choices yield the same sum total error. The angular error values are not constant for these values but they are less reliable because of quantization errors. One could attempt to more accurately optimize one of these functions but the results would be meaningless unless the disparity vectors are more precisely determined. With a correspondence of points in the two frames a conventional approach to finding the FOE can be used such as a least squares pseudo-intersection of the vectors. Such an answer is also no more accurate than the result that the point is in a region near the center of the image (the image is 128 by 128 with center at 64,64) between (40,40) and (60,60) because of the quantization error. One would obtain a high variance if such a method were applied. To more accurately position the FOE the flow vectors must be determined to a higher degree of precision using interpolation techniques. Since our method gives an excellent correspondence of the points the search costs of interpolation are greatly reduced. The more precise vectors can be intersected to find the FOE with more certainty. Figure 9 shows the disparity vectors generated by this method for the real scene Figure 10 shows the intersecting lines and gives an idea of the quantization errors in angle and the uncertainty of the position of the FOE.

If the points extracted by this method are used with any version of Jain's algorithm, no reasonable answer results. For real scenes, Jain's[JAI82a] algorithm fails to deal with missing points and addition of new points. It also ignores the invaluable information contained in the feature values returned by a corner detector or contained in the actual grey-value window around the feature point.

Quantization problems leads to uncertainty with some of the methods proposed in this paper when they are applied to real scenes. If a point moves one pixel, its direction of motion is only discernable to $\frac{\pi}{4}$ radians precision. To correctly determine the differences between rotation and translation and to precisely fix the FOE information of a much higher resolution is required. To some extent these problems may be ameliorated by the use of image sequences rather than image pairs. Such sequences can be fitted with a curve and the curve used instead of the actual points in the calculations.

Conclusions

In this paper we have shown that methods that seek to use preliminary or hypothesized motion parameters to find correspondence and better motion parameters are viable alternatives to the usual paradigm of attempting to obtain correspondence or optical flow first and then to solve for motion parameters. This will be especially important in a system that has many sources of information that can generate hypothesized motion trajectories and integrate these sources into the dynamic scene problem.

Acknowledgement

We gratefully acknowledge Richard Hill for his endless ideas, discussions and editorial advice.

References

[CPJ82] Jerian,Charles. *Determining Motion Parameters for scenes with Translation and Rotation.* Wayne State University,Department of Computer Science,Masters Thesis, Detroit Michigan,1982.

[DRE82a] Dreschler, L. and H.H. Nagel. "Volumetric Model and 3D- Trajectory of a Moving Car Derived from Monocular TV-Frame Sequences of a Street Scene." *Proceedings IJCAI* (1981) Vancouver, Canada

[DRE82b] Dreschler, L. and H.H. Nagel. "On the Frame-to-Frame Correspondence between Greyvalue Characteristics in the Images of Moving Objects." *Proceedings GI Workshop on Artificial Intelligence*, Bad Honneff, Germany, 1981

[FIS81] Fischler M.A. and Bolles R.C. "Random Sample Consensus: A paradigm for model fitting with applications to image anlysis and automated cartography," *CACM* , Vol. 24(6). pp. 381-395 (June 1981)

[FIS82] Fischler M.A., Barnard S.T., Bolles R.C.,Lowry M., Quam L. Smith G. and Witkin A. "Modelling and Using Physical Constraints in Scene Analysis", in *AAAI-82* pp. 30-35

[GLA81] Glazer,"A simple method for determining optical flow", in *IJCAI-81*

[HIL80] Hill,Richard,"Determining optical flow",Wayne State University Computer Science Department Masters Thesis (1980)

[HOR81] Horn, B.K.P. and B.G. Schunck. "Determining Optical Flow". *Artificial Intelligence* 17 (1981) 185-203

[JAI82a] Jain,R. "An Approach for the Direct Computation of the Focus of Expansion", in *PRIP-82* pp 262-268

[KIT80] Kitchen,L. and Rosenfeld A., "Gray-level Corner Detection", Computer Vision Laboratory, Computer Science Center. University of Maryland College Park, MD 20742 TR-887 April,1980

[LAW82] Lawton D.T. "Motion Analysis via Local Translational Processing" in *Workshop on Computer Vision:Representaion and Control* pp. 59-72

[MOR77] Moravec, H.P. "Towards Automatic Visual Obstacle Avoidance." *Proceedings 5th IJCAI* (August 1977), p. 584.

[MOR79] Moravec, H.P. "Visual Mapping by a Robot Rover." *Proceedings 6th IJCAI* (1979), pp. 598-600.

[MOR80] Moravec, H.P. *Obstacle Avoidance and Navigation in the Real World by a Seeing Robot Rover.* Stanford:~~Stanford University, Department of Computer Science, Ph.D. Thesis, 1980; also Stanford AI Lab Memo AIM-340; also Computer Science Department Report Number STAN-CS-80-813; also Pittsburgh:~~Robotics Institute CMU-RI-TR-3, September 1980.

[NAG81a] Nagel, H.H. "On the Derivation of 3-D Rigid Point Configuration from Image Sequences." *Proceedings IEEE Conference on Pattern Recognition and Image Processing* (1981).

[Nag81b] Nagel, H.H. and B. Neumann. "On 3-D Reconstruction from Two Perspective Views." *Proceedings IJCAI* (1981).

[PRA79b] Prazdny, K. "Motion and Structure from Optical Flow." *Proceedings 6th IJCAI* (1979), pp. 702-704.

[PRA80] Prazdny, K. "Egomotion and Relative Depth map from Optical Flow." *Biological Cybernetics* , 36 (1980), pp. 87-102.

[PRA81a] Prazdny K. "Determining the Instantaneous Direction of Motion from Optical Flow Generated by a Curvilnearly Moving Observer" *CGIP* 17 (1981) 238-248

[PRA81b] Prazdny K. "A Simple Method for Recovering Relative Depth Map in the case of a Translating Sensor" *IJCAI-81* , pp. 698-699

[TSA82a] Tsai R.Y. and Huang T.S. "Uniqueness and Estimation of Three- Dimensional Motion Parameters of Rigid Objects with Curved Surfaces", in *PRIP-82* pp. 112-118

[TSA81] Tsai R.Y. and Huang T.S. "Estimating Three-Dimesional Motion Parameters of a Rigid Planar Patch", in *IEEE Transactions on ASSP* Volume ASSP-29 no. 6 (December 1981) pp. 1147-1152

[TSA82b] Tsai R.Y. and Huang T.S. "Estimating Three-Dimensional Motion Parameters of a Rigid Planar Patch", in *IEEE Transactions on ASSP* Volume ASSP-30 no. 4 (August 1982) pp. 525-534

[ULL79a] Ullman, S. *The Interpretation of Visual Motion* Massachusetts: MIT Press, 1979.

[ULL79b] Ullman, S., "The Interpretation of Structure from Motion, " *Proc. Royal Soc. London* , Vol. B, No. 203, pp. 405-426

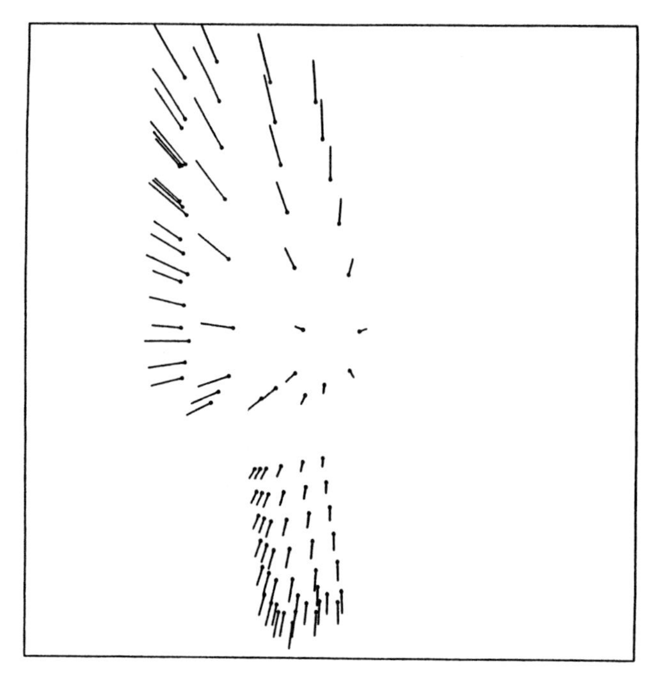

Figure 1 Flow vectors for z translation

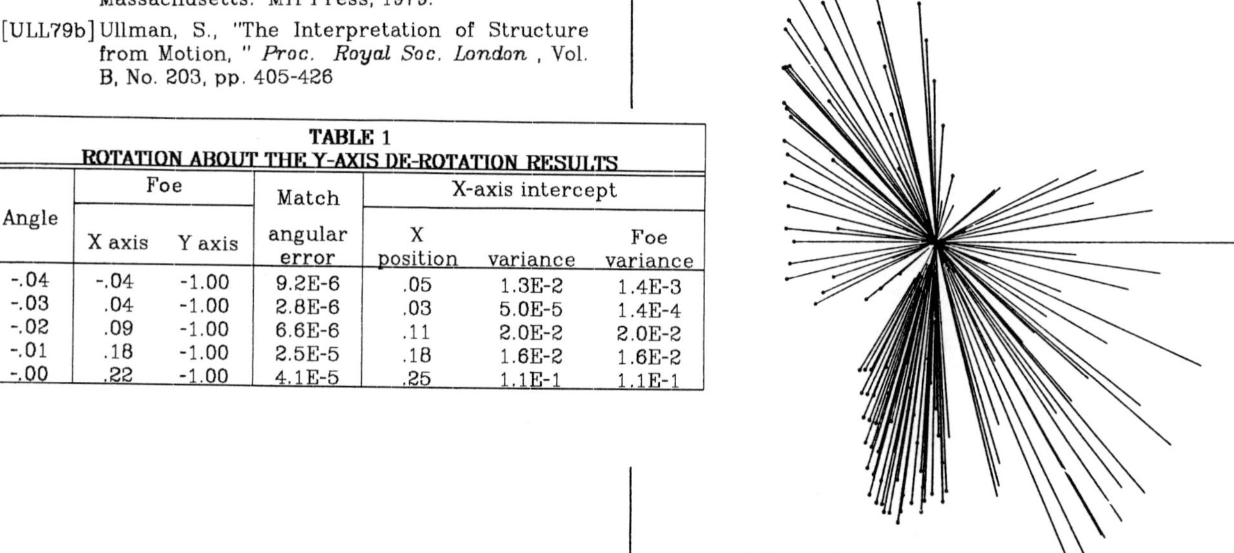

TABLE 1
ROTATION ABOUT THE Y-AXIS DE-ROTATION RESULTS

Angle	Foe		Match	X-axis intercept		
	X axis	Y axis	angular error	X position	variance	Foe variance
-.04	-.04	-1.00	9.2E-6	.05	1.3E-2	1.4E-3
-.03	.04	-1.00	2.8E-6	.03	5.0E-5	1.4E-4
-.02	.09	-1.00	6.6E-6	.11	2.0E-2	2.0E-2
-.01	.18	-1.00	2.5E-5	.18	1.6E-2	1.6E-2
-.00	.22	-1.00	4.1E-5	.25	1.1E-1	1.1E-1

Figure 2 Lines extending flow vectors to meet at foe

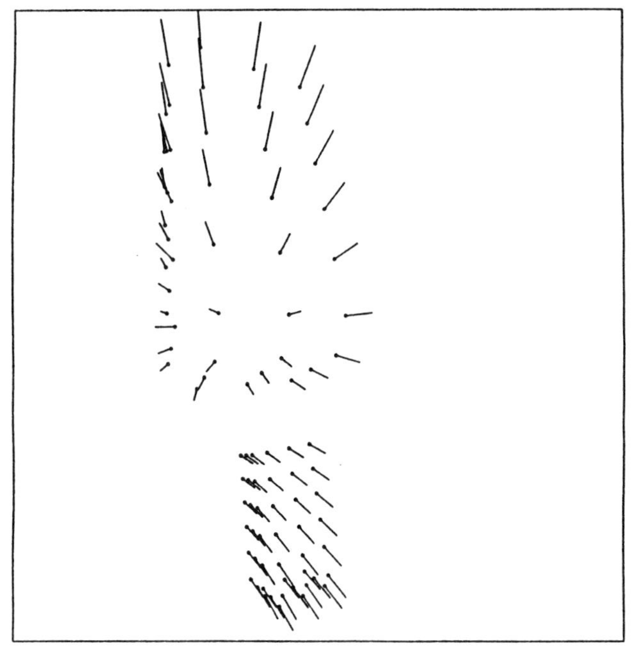

Figure 3 Flow vectors showing y-rotation and translation

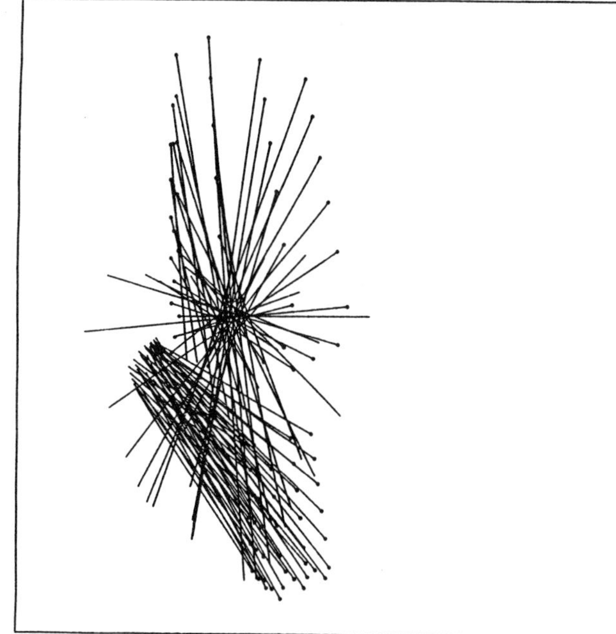

Figure 4 Lines extending flow vectors for y-rotation and translation

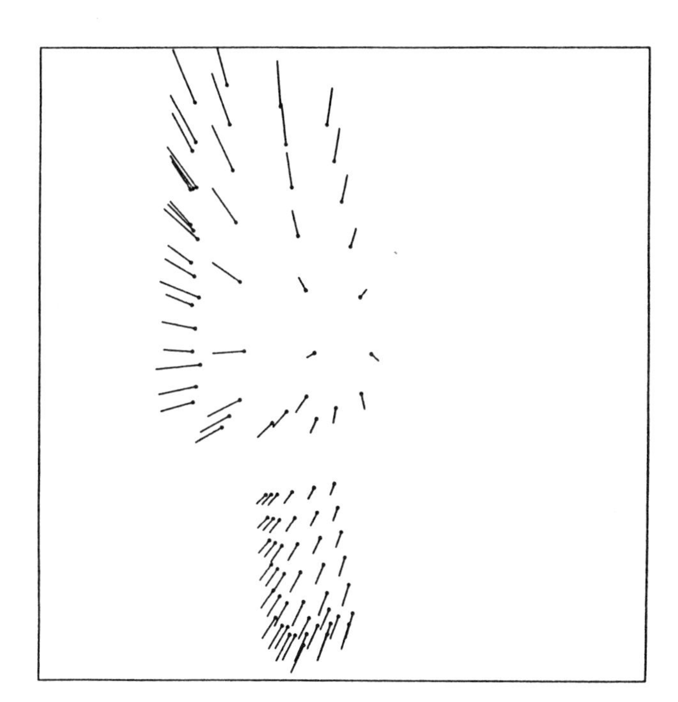

Figure 5 Flow vectors for z-rotation and x and z-translation

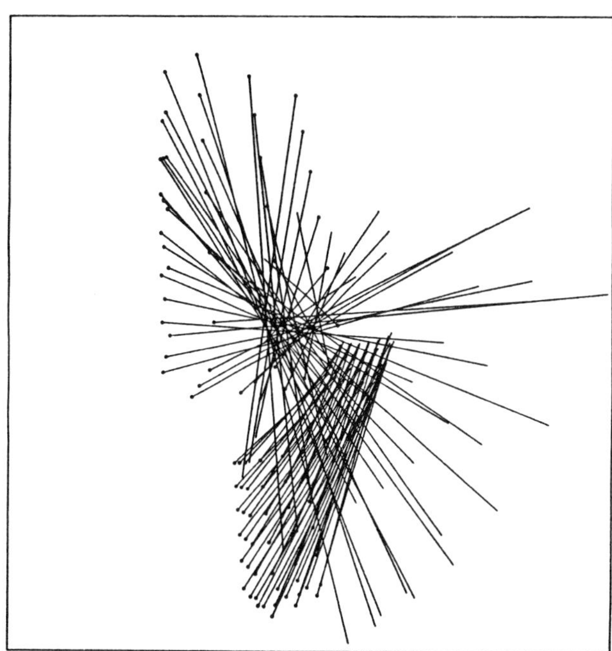

Figure 6 Lines extending z-rotated flow vectors showing a circle pattern

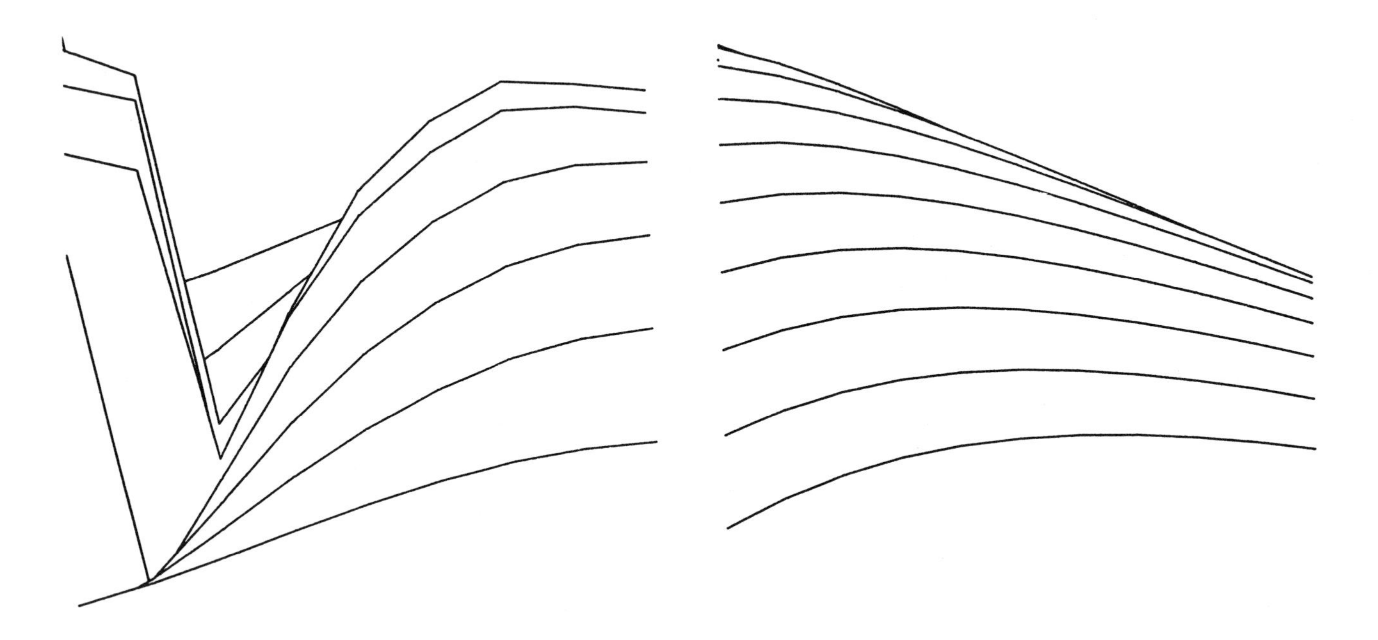

Figure 7 Foe function for foe at 0,-1

Figure 8 Close-up of foe function at 0,-1

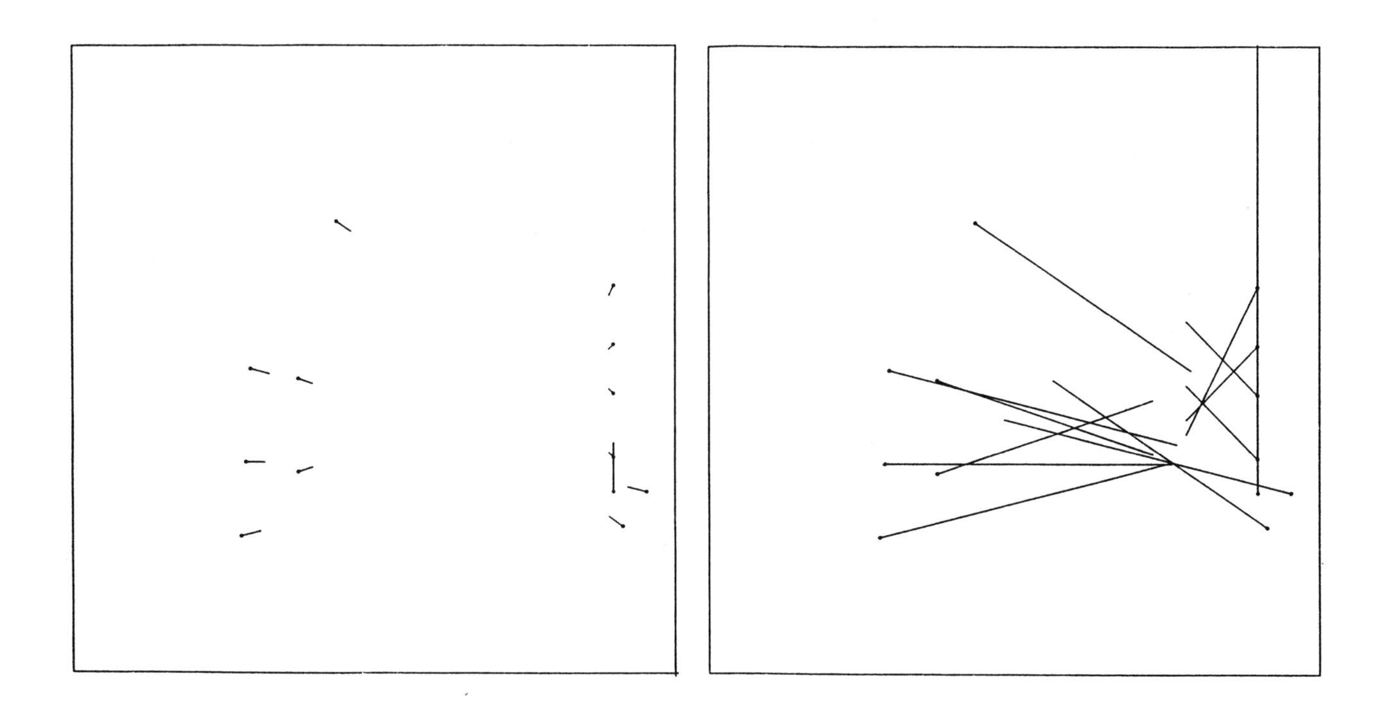

Figure 9 Flow vectors of a cube scene

Figure 10 Lines extending flow vectors of a cube scene

DETERMINING 3-D MOTION PARAMETERS OF A RIGID BODY: A VECTOR-GEOMETRICAL APPROACH

B. L. Yen and T. S. Huang
Department of Electrical Engineering
and Coordinated Science Laboratory
University of Illinois, Urbana, IL

ABSTRACT

A vector-geometrical approach is given for the determination of 3-D motion parameters of a rigid body from point correspondences over 2 time sequential images. The resulting algorithms are similar to existing methods [2-3]. However, the geometrical interpretations provide much valuable insight into the nature of the problem and the uniqueness question.

I. Introduction

The determination of 3-D motion and structure of a rigid body from an image sequence has important applications in many areas, such as target tracking, stereopsis, and robotic vision. A 3-D rigid body motion can be decomposed into a rotation R about an axis through the origin, followed by a translation t. The 3-D motion parameters to be determined are the 3 components of R and the 2 components of $\hat{t}$. 3-D object structure is found as a map of relative depths of object surface points and/or lines.

Methods for determining the 3-D motion and structure of a rigid body from an image sequence have been based on the geometrical/transformation properties of central projections of object surface points and/or straight line segments on the image plane/unit sphere. For general 3-D object motions, correspondences over the image sequence are used. For objects in general, point correspondences (PCs) are used [1-4]; for objects with straight line segments, line correspondences (LCs) can be used [5].

In the past, PC based methods for 3-D motion over 2 frames required the solution of nonlinear equations [1]. Recently, linear methods have been developed [2-3]. In [2], an E matrix is found from 8 PCs. Based on the singular value decomposition of E, a solution to $(R,\hat{t})$ can be found. An alternative method based on the vector analysis of E is given in [3].

This paper deals with the uniqueness/computation of $(R,\hat{t})$ for a rigid body motion over 2 frames. Based on a vector geometrical approach, alternative derivations of previous results [2-3] are given. Given the E matrix, it is shown that there are 4 solution sets to $(R,\hat{t})$. Given the E matrix and a PC, a new method is shown to uniquely determine $(R,\hat{t})$. In addition, further results and geometric interpretations are given.

II. Two-Frame PC Method for Objects with Curved Surfaces

An xyz cartesian coordinate system is established with origin 0 at the focal point of the camera: $\hat{z}$ is aligned with the optical axis, and $\hat{x}/\hat{y}$ aligned with the X/Y axes of the image plane z=F (respectively) (fig. 1). **Define $\bar{p}/\hat{p}$ as the central projection of a 3-D point p on the image plane/unit sphere (respectively).**

$$\bar{p} = \begin{bmatrix} X \\ Y \\ F \end{bmatrix} = \frac{F}{z} p \quad \hat{p} = \begin{bmatrix} \hat{x} \\ \hat{y} \\ \hat{z} \end{bmatrix} = \frac{p}{\|p\|} \quad \text{where } p = \begin{bmatrix} x \\ y \\ z \end{bmatrix} \tag{1}$$

Note that $\bar{p}$ and $\hat{p}$ are parameterized by their 3-D xyz coordinates, i.e. homogeneous coordinates. The notations CO and NCO denotes whether a plane or line contains the origin or not.

Over 2 frames, an object point p at time τ (frame 1) moves to the (corresponding) object point p' at time τ' (frame 2). Associated with the 3-D PC (p,p') are the PCs $(\bar{p},\bar{p}')$ and $(\hat{p},\hat{p}')$ on the image plane and unit sphere (respectively).

The 3-D rigid body motion can be decomposed into a rotation R (by angle θ about axis $\hat{n}$) followed by a translation t.

$$p' = Rp + t \tag{2}$$

where

$$R = [r_{ij}] = \cos\theta \; I + (1 - \cos\theta)\hat{n}\hat{n}^T + \sin\theta \; N$$

Motion: Representation and Perception
N.I. Badler and J.K. Tsotsos, Editors
Published by Elsevier Science Publishing Co., Inc.

$$N = \begin{bmatrix} 0 & -n_3 & n_2 \\ n_3 & 0 & -n_1 \\ -n_2 & n_1 & 0 \end{bmatrix} \quad \hat{n} = \begin{bmatrix} n_1 \\ n_2 \\ n_3 \end{bmatrix} \quad t = \begin{bmatrix} \Delta x \\ \Delta y \\ \Delta z \end{bmatrix}$$

For the following analysis, it is assumed that $t \neq 0$. The case of pure rotation about an axis (CO) is excluded. Note that the case of pure translation ($R=I$ and $t \neq 0$) is included.

The induced motion on the unit sphere can be decomposed into 2 motions (fig. 2) Under R, $\hat{p}$ rotates by angle θ about axis $\hat{n}$ to $R\hat{p}$. Geometrically, $\hat{p}$ and $R\hat{p}$ lie on a circle contained in a plane perpendicular to $\hat{n}$. Under t, $R\hat{p}$ rotates on a great circle (containing $\hat{t}$) to $\hat{p}'$. Geometrically, $R\hat{p}$, $\hat{p}'$, $\hat{t}$ lie on a great circle T. Specifically, $\hat{p}'$ lies on the arc with endpoints $R\hat{p}$ and $\hat{t}$.

From the fact that the points $R\hat{p}_i$, $\hat{p}_i'$, $\hat{t}$ lie on a great circle T_i, a set of n scalar homogeneous equations can be found

$$\hat{t} \cdot (R\hat{p}_i \times \hat{p}_i') = 0 \qquad i = 1, \ldots, n \tag{3a}$$

or

$$\hat{p}_i'^T E \hat{p}_i = 0 \qquad i = 1, \ldots, n \tag{3b}$$

where

$$E = [\hat{t}\times\hat{c}_1 \quad \hat{t}\times\hat{c}_2 \quad \hat{t}\times\hat{c}_3] = GR$$

$$R = [\hat{c}_1 \quad \hat{c}_2 \quad \hat{c}_3]$$

$$C = \begin{bmatrix} 0 & -\hat{\Delta z} & \hat{\Delta y} \\ \hat{\Delta z} & 0 & -\hat{\Delta x} \\ -\hat{\Delta y} & \hat{\Delta x} & 0 \end{bmatrix} \quad \hat{t} = \begin{bmatrix} \hat{\Delta x} \\ \hat{\Delta y} \\ \hat{\Delta z} \end{bmatrix}$$

It is important to note, that the constraint that $\hat{p}_i'$ lies on the arc with endpoints $R\hat{p}_i$ and $\hat{t}$ is not applied, and not reflected in (3a,b). The equation set (3b) can also be derived by elimination [2] or tensor analysis [3]. However, the above derivation is considerably simpler.

III. Uniqueness of Solution to $(R,\hat{t})$

In this section, it is shown that a unique solution to (R,t) for (3a) is impossible.

Theorem 1

Given a solution to $(R,\hat{t})$ for (3a), there exist 3 other solutions

$(R^a,\hat{t}^a)$	(given solution)
$(R^a,\hat{t}^b)$	(alternative solution 1)
$(R^b,\hat{t}^a)$	(alternative solution 2)
$(R^b,\hat{t}^b)$	(alternative solution 3)

where

$$R^b = 2\hat{t}^a(R^{a^{-1}}\hat{t}^a)^T - R^a \qquad \hat{t}^b = -\hat{t}^a$$

and a solution to the PCs $\pm(R^{a^{-1}}\hat{t}^a,\hat{t}^a) = \mp(R^{b^{-1}}\hat{t}^b,\hat{t}^b)$.

Proof:

Assume the existence of a solution $(R^a,\hat{t}^a)$ to $(R,\hat{t})$ for (3a). The existence and form of additional solutions are now analyzed.

Consider additional solutions to $\hat{t}$ in (3a). Given that $\hat{t}^a$ is a solution, $\hat{t}^b = -\hat{t}^a$ is also a solution.

Consider additional solutions to R in (3a). Consider R^b defined by

$$R^b = 2\hat{t}^a(R^{a^{-1}}\hat{t}^a)^T - R^a \tag{4}$$

Note that

$$R^b = R_t^\pi R^a \text{ and } R^a = R_t^\pi R^b \tag{5}$$

where

$$R_t^\pi = 2\hat{t}^a\hat{t}^{a^T} - I \text{ (180° rotation about axis } \hat{t}^a)$$

i.e. R^b is R^a followed by R_t^π (a 180° rotation about axis $\hat{t}^a$). $R^b\hat{p}_i$ is the image of $R^a\hat{p}_i$ under R_t^π and is given by

$$R^b\hat{p}_i = 2(\hat{t}^a \cdot R^a\hat{p}_i)\hat{t}^a - R^a\hat{p}_i \tag{6}$$

From (6),

$$\hat{t}^a \cdot (R^a\hat{p}_i \times R^b\hat{p}_i) = 0 \tag{7}$$

Geometrically, $\hat{t}^a$, $R^a\hat{p}_i$, $R^b\hat{p}_i$ lie on a great circle, where $\hat{t}^a$ bisects $R^a\hat{p}_i$ and $R^b\hat{p}_i$. Given that R^a is a solution, it is now shown that R^b is a solution. By assumption, (3a) is true (with $R = R^a$ $\hat{t} = \hat{t}^a$). (3a) and (7) imply

$$\hat{t}^a \cdot (R^b\hat{p}_i \times \hat{p}_i') = 0 \tag{8}$$

i.e., R^b is a solution. The result can also be obtained by the following geometrical argument. By assumption, $\hat{t}^a$, $R^a\hat{p}_i$, $\hat{p}_i'$ lie on a great circle. $R^b\hat{p}_i$ also lies on this great circle, since it is the image of $R^a\hat{p}_i$ under a 180° rotation about axis $\hat{t}^a$. Thus, $\hat{t}^a$, $R^b\hat{p}_i$, $\hat{p}_i'$ lie on a great circle, i.e. R^b is a solution.

From above, it is clear there are 3 additional solutions $(R^a,\hat{t}^b)$ $(R^b,\hat{t}^a)$ $(R^b,\hat{t}^b)$ to $(R,\hat{t})$. From (4), a solution to the PCs $\pm(R^{a^{-1}}\hat{t}^a,\hat{t}^a)$ is determined by

$$R^{a^{-1}}\hat{t}^a = \frac{1}{2}(R^a + R^b)^T\hat{t}^a \tag{9}$$

Note from (5) that $R^{b^{-1}}\hat{t}^a = R^{a^{-1}}\hat{t}^a$. QED

Note that the actual rotation/translation $(R,\hat{t})$ of the 3-D motion is a solution to (3a). Thus, there are at least 2 solutions to each of R and $\hat{t}$ and at least 4 solutions to $(R,\hat{t})$

$$(R,\hat{t}) \quad (R,-\hat{t}) \quad (R^b,\hat{t}) \quad (R^b,-\hat{t})$$

where

$$R^b = 2\hat{t}(R^{-1}\hat{t})^T - R \tag{10}$$

A unique solution to (3a) is impossible. The best possible case is that there are exactly 2 solutions to each of R and $\hat{t}$ and exactly 4 solutions to $(R,\hat{t})$ given by (10). The next section describes such a case.

IV. Determination of $(R,\hat{t})$ from E

Note that (3b) is homogeneous in the elements of E. Thus, E can be found only to a scale. In this section, it is shown that given E to an unknown scale k as E', there are exactly 2 solutions to each of R and $\hat{t}$ and exactly 4 solutions to $(R,\hat{t})$.

Theorem 2

Given

$$E' = kE = [e_1' \ e_2' \ e_3'] \tag{11}$$

$$= [k(\hat{t}\times\hat{c}_1) \ k(\hat{t}\times\hat{c}_2) \ k(\hat{t}\times\hat{c}_3)] = kGR$$

where k = unknown scale there are 2 solutions each for R, $\hat{t}$, k

$R^a = R$ (actual solution) $\quad R^b = 2\hat{t}(R^{-1}\hat{t})^T - R$ (false solution)

$\hat{t}^a = \hat{t}$ (actual solution) $\quad \hat{t}^b = -\hat{t}$ (false solution)

$k^a = k$ (actual solution) $\quad k^b = -k$ (false solution)

there are 4 solutions to $(R,\hat{t},k)$

$(R^a,\hat{t}^a,k^a)$ (actual solution)

$(R^a,\hat{t}^b,k^b)$ (false solution 1)

$(R^b,\hat{t}^a,k^b)$ (false solution 2)

$(R^b,\hat{t}^b,k^a)$ (false solution 3)

In addition, the PCs $\pm (R^{-1}\hat{t},\hat{t})$ under R are uniquely determined.

Proof:

Consider the solution of $\hat{t}$. It is now shown that $\hat{t}$ can be found to a sign. From (11) and the fact that $\hat{c}_1,\hat{c}_2,\hat{c}_3$ form a right-handed orthonormal system,

$$\begin{aligned} e_1' \times e_2' &= k^2\hat{t}[\hat{t}\cdot(\hat{c}_1\times\hat{c}_2)] = k^2\hat{t}(\hat{t}\cdot\hat{c}_3) \\ e_2' \times e_3' &= k^2\hat{t}[\hat{t}\cdot(\hat{c}_2\times\hat{c}_3)] = k^2\hat{t}(\hat{t}\cdot\hat{c}_1) \\ e_3' \times e_1' &= k^2\hat{t}[\hat{t}\cdot(\hat{c}_3\times\hat{c}_1)] = k^2\hat{t}(\hat{t}\cdot\hat{c}_2) \end{aligned} \tag{12}$$

Since

$$R^{-1}\hat{t} = \begin{bmatrix} \hat{t}\cdot\hat{c}_1 \\ \hat{t}\cdot\hat{c}_2 \\ \hat{t}\cdot\hat{c}_3 \end{bmatrix} \neq 0 \tag{13}$$

$\pm\hat{t}$ can be determined from at least 1 of 3 cross products in (12). Define $\hat{t}^a = \hat{t}$ (actual solution) and $\hat{t}^b = -\hat{t}$.

Consider the solution of R. It is now shown that there are 2 solutions to R. It is a well known fact in vector analysis that a vector is uniquely determined from its dot and cross product with some known vector. This is used for the solution of the column vectors $\hat{c}_1,\hat{c}_2,\hat{c}_3$ of R.

Consider the determination of the dot and cross products of $\hat{c}_i$ i = 1,2,3 with $\hat{t}$. From (11) and (12),

$$\begin{aligned} (\hat{t}\times\hat{c}_1) &= \frac{1}{k}e_1' & (\hat{t}\cdot\hat{c}_1) &= \frac{1}{k^2}\hat{t}\cdot(e_2'\times e_3') \\ (\hat{t}\times\hat{c}_2) &= \frac{1}{k}e_2' & (\hat{t}\cdot\hat{c}_2) &= \frac{1}{k^2}\hat{t}\cdot(e_3'\times e_1') \\ (\hat{t}\times\hat{c}_3) &= \frac{1}{k}e_3' & (\hat{t}\cdot\hat{c}_3) &= \frac{1}{k^2}\hat{t}\cdot(e_1'\times e_2') \end{aligned} \tag{14}$$

Note that k can be determined to a sign by the following. From (11),

$$\begin{aligned} e_1'\cdot e_1' &= k^2[1-(\hat{t}\cdot\hat{c}_1)^2] \\ e_2'\cdot e_2' &= k^2[1-(\hat{t}\cdot\hat{c}_2)^2] \\ e_3'\cdot e_3' &= k^2[1-(\hat{t}\cdot\hat{c}_3)^3] \end{aligned} \tag{15}$$

Adding the 3 equations in (15) and using $\|\hat{t}\|^2 = \|R^{-1}\hat{t}\|^2 = \sum_i(\hat{t}\cdot\hat{c}_i)^2 = 1$, k^2 is found as

$$k^2 = \frac{1}{2}\sum_i e_i'\cdot e_i' = \frac{1}{2}\mathrm{Tr}(E'^TE') \tag{16}$$

Note that the $\pm$ sign ambiguity for k and $\hat{t}$ generates a $\pm$ sign ambiguity in the dot and cross products in (14). Define $k^a = k$ (actual solution) and $k^b = -k$.

It is now shown that $\pm$ sign ambiguity in k and $\hat{t}$ gives 2 solutions to R from (14). From (14), $\hat{c}_1$, $\hat{c}_2$, $\hat{c}_3$ are solved as

$$\begin{aligned} \hat{c}_1 &= \frac{1}{k}(e_1'\times\hat{t}) + \frac{1}{k^2}[\hat{t}\cdot(e_2'\times e_3')]\hat{t} \\ \hat{c}_2 &= \frac{1}{k}(e_2'\times\hat{t}) + \frac{1}{k^2}[\hat{t}\cdot(e_3'\times e_1')]\hat{t} \\ \hat{c}_3 &= \frac{1}{k}(e_3'\times\hat{t}) + \frac{1}{k^2}[\hat{t}\cdot(e_1'\times e_2')]\hat{t} \end{aligned} \tag{17}$$

Note that the $\pm$ ambiguity in the sign of k and $\hat{t}$ affects only the first terms in the 3 equations in (17). The 4 cases for $(\pm k,\pm\hat{t})$ generate 2 solutions for $\hat{c}_1$, $\hat{c}_2$, $\hat{c}_3$. From (17) and (14),

solution a

from (17) for $(k,\hat{t})$ or $(-k,-\hat{t})$

$$\hat{c}_1^a = \hat{c}_1 \quad \hat{c}_2^a = \hat{c}_2 \quad \hat{c}_3^a = \hat{c}_3 \quad \left[R^a = R \ \text{(actual solution)}\right] \tag{18}$$

solution b

from (17) for $(-k,\hat{t})$ or $(k,-\hat{t})$

$$\hat{c}_1^b = 2(\hat{t}\cdot\hat{c}_1)\hat{t} - \hat{c}_1$$
$$\hat{c}_2^b = 2(\hat{t}\cdot\hat{c}_2)\hat{t} - \hat{c}_2$$
$$\hat{c}_3^b = 2(\hat{t}\cdot\hat{c}_3)\hat{t} - \hat{c}_3$$

$$\left[R^b = 2\hat{t}(R^{-1}\hat{t})^T - R \quad (19) \quad \text{(false solution)} \right]$$

From (5), both R^a and R^b are orthogonal of the 1st kind.

In summary, each of R, $\hat{t}$, k has 2 solutions. As a set, $(R,\hat{t})$ k has 4 solutions.

Finally, it is shown that the PCs $\pm(R^{-1}\hat{t},\hat{t})$ under R can be uniquely determined. From (18,19),

$$\pm R^{-1}\hat{t} = \pm \begin{bmatrix} (\hat{t}\cdot\hat{c}_1) \\ (\hat{t}\cdot\hat{c}_2) \\ (\hat{t}\cdot\hat{c}_3) \end{bmatrix} = \pm(R^a)^{-1}\hat{t} = \pm(R^b)^{-1}\hat{t} \tag{20}$$

Note that both rotations R^a, R^b rotate $\pm\hat{t}$ to $\pm R^{-1}\hat{t}$, i.e. $(R^{-1}\hat{t},\hat{t})$ is a PC under both R^a,R^b. Using (20), $\hat{t}^a = \hat{t}$ gives the PC $(R^{-1}\hat{t},\hat{t})$ and $\hat{t}^b = -\hat{t}$ gives the PC $(-R^{-1}\hat{t},-\hat{t})$. Although R is not uniquely determined, $\pm\hat{t}$ and its image under R is uniquely determined.

QED

A geometric interpretation of the results in Theorem 2 is now given. Consider the determination of $\pm\hat{t}$ by (12). The columns e_1',e_2',e_3' of E' are contained in the plane (CO) with normals $\pm\hat{t}$ (fig. 3a). Clearly, $\pm\hat{t}$ can be obtained as the cross product of any 2 non-aligned vectors among e_1',e_2',e_3'.

Consider the determination of $\hat{c}_i$ $i = 1,2,3$ by (18,19). $\hat{c}_i$ can be decomposed as the sum of 2 vectors u_i (perpendicular to $\hat{t}$, contained in the plane (CO) with normals $\pm\hat{t}$) and v_i (aligned with $\hat{t}$, contained in the line (CO) $k\hat{t}$). These 2 vectors are determined in the solution of $\hat{c}_i$ (fig. 3b). Rewriting (18,19),

$$R: \quad \hat{c}_i = u_i + v_i \quad i = 1,2,3 \tag{21a}$$
$$R^b: \quad \hat{c}_i^b = -u_i + v_i \quad i = 1,2,3 \tag{21b}$$

where

$u_i = e_i \times \hat{t} = \sin\phi_i\,\hat{s}$ = projection of $\hat{c}_i$ on π

$v_i = (\hat{t}\cdot\hat{c}_i)\hat{t} = \cos\phi_i\hat{t}$ = projection of $\hat{c}_i$ on δ

ϕ_i = inclusive angle between $\hat{t}$, $\hat{c}_i$

π = plane (CO) with normals $\pm\hat{t}$

δ = line (CO) $k\hat{t}$

The ambiguity in the solution of $\hat{c}_i$ is the result of the $\pm$ ambiguity in the vector u_i, due to the $\pm$ ambiguity in k and $\hat{t}$. It is geometrically clear that $(\pm\hat{t})\cdot\hat{c}_i$ - the component of $\hat{c}_i$ and $\hat{c}_i^b$ along $\pm\hat{t}$ - is the same. Thus, the PCs $\pm(R^{-1}\hat{t},\hat{t})$ are uniquely determined despite the fact that R is not uniquely determined.

The result of Theorem **2** is consistent with the result on the polar decomposition of a singular, complex matrix. From (11),

$$E'' = iE' = G''R$$

where

$$G'' = ikG \quad \text{(hermitian)} \tag{22}$$

From Theorem 2, there are 2 polar decompositions $G''R^a$ and $-G''R^b$ for (singular) E'. This is consistent with the result that the polar decomposition of a complex matrix is unique if and only if it is non-singular [6].

V. Determination of $(R,\hat{t})$ from E with Visibility Constraint

In this section, it is shown that in some cases a unique solution to $(R,\hat{t})$ is possible given E, with the constraint that the rigid body remains in front of the camera over the motion. The following is an analysis of the 2 solutions R, R^b.

Consider the 2 solutions R, R^b. From (18,19),

$$R + R^b = [\,\hat{c}_1+\hat{c}_1^b \;\; \hat{c}_2+\hat{c}_2^b \;\; \hat{c}_3+\hat{c}_3^b\,] = 2[\,(\hat{t}\cdot\hat{c}_1)\hat{t} \;\; (\hat{t}\cdot\hat{c}_2)\hat{t} \;\; (\hat{t}\cdot\hat{c}_3)\hat{t}\,] \tag{23}$$

The rotational axis/angle of rotation pairs $(\hat{n},\theta)/(\hat{n}^b,\theta^b)$ of R/R^b (respectively) are found [4] as

$$\cos\theta + \cos\theta^b = \frac{1}{2}[\,\sum_i(r_{ii} + r_{ii}^b)] - 1$$
$$= \frac{1}{2}[\,2(\hat{t}\cdot\hat{c}_1)\hat{\Delta x} + 2(\hat{t}\cdot\hat{c}_2)\hat{\Delta y} + 2(\hat{t}\cdot\hat{c}_3)\hat{\Delta z}] - 1$$
$$= R^{-1}\hat{t}\cdot\hat{t} - 1 \tag{24a}$$

$$\sin\theta\hat{n} + \sin\theta^b\hat{n}^b = \frac{1}{2}\begin{bmatrix} (r_{32}+r_{32}^b) - (r_{23}+r_{23}^b) \\ (r_{13}+r_{13}^b) - (r_{31}+r_{31}^b) \\ (r_{21}+r_{21}^b) - (r_{12}+r_{12}^b) \end{bmatrix}$$
$$= \frac{1}{2}\begin{bmatrix} 2[\,(\hat{t}\cdot\hat{c}_2)\hat{\Delta z} - (\hat{t}\cdot\hat{c}_3)\,\hat{\Delta y}] \\ 2[\,(\hat{t}\cdot\hat{c}_3)\hat{\Delta x} - (\hat{t}\cdot\hat{c}_1)\,\hat{\Delta z}] \\ 2[\,(\hat{t}\cdot\hat{c}_1)\hat{\Delta y} - (\hat{t}\cdot\hat{c}_2)\,\hat{\Delta x}] \end{bmatrix}$$
$$= R^{-1}\hat{t} \times \hat{t} \tag{24b}$$

From (24b),

$$0 = \sin\theta[\,\hat{n}\cdot(R^{-1}\hat{t} - \hat{t})] + \sin\theta^b[\,\hat{n}^b\cdot(R^{-1}\hat{t} - \hat{t})]$$
$$= 0 + \sin\theta^b[\,\hat{n}^b\cdot(R^{-1}\hat{t} - \hat{t})] \tag{25}$$

Thus, for $\sin\theta^b \neq 0$ and $\hat{t} \neq R^{-1}\hat{t}$ (nonfixed PC $(R^{-1}\hat{t},\hat{t})$), $\hat{n}^b$ (together with $\hat{n}$) is contained in the plane (CO) with normals $\pm(R^{-1}\hat{t} - \hat{t})$.

From (2), $R^{-1}\hat{t}$ is given by

$$R^{-1}\hat{t} = \cos\theta\hat{t} + (1-\cos\theta)\,(\hat{n}\cdot\hat{t})\hat{n} - \sin\theta(\hat{n}\times\hat{t}) \tag{26}$$

From (26),

$$R^{-1}\hat{t}\cdot\hat{t} = \cos\theta + (1-\cos\theta)(\hat{n}\cdot\hat{t})^2 \quad (27a)$$

$$R^{-1}\hat{t}\times\hat{t} = (1-\cos\theta)(\hat{n}\cdot\hat{t})(\hat{n}\times\hat{t}) - \sin\theta[(\hat{n}\cdot\hat{t})\hat{t}+\hat{n}] \quad (27b)$$

Substituting (27a,b) into (24a,b), $(\hat{n}^b,\theta^b)$ are found as

case 1 $\mathrm{mod}_{2\pi}\theta^b = 0$

$\hat{n}^b$ = any unit vector

case 2 $\mathrm{mod}_{2\pi}\theta^b = \pi$

$$\hat{n}^b\hat{n}^{bT} = \frac{1}{2}[\,2(\hat{t}\cdot\hat{c}_1)\hat{t}-(\hat{c}_1-\hat{x})\;\; 2(\hat{t}\cdot\hat{c}_2)\hat{t}-(\hat{c}_2-\hat{y})\;\; 2(\hat{t}\cdot\hat{c}_3)\hat{t}-(\hat{c}_3-\hat{z})] \quad (28a)$$

case 3 $\mathrm{mod}_{2\pi}\theta^b \neq 0,\pi$

$$(\hat{n}^b\cdot\hat{t})\sin\theta^b = -(\hat{n}\cdot\hat{t})\sin\theta \quad (28b)$$

$$\cos\theta^b = (1-\cos\theta)(\hat{n}\cdot\hat{t})^2 - 1 \quad (28c)$$

$$\sin\theta^b\hat{n}^b = (\hat{n}\cdot\hat{t})[(1-\cos\theta)(\hat{n}\times\hat{t})-\sin\theta\hat{t}] \quad (28d)$$

(28a-d) give the false solution $R^b:(\hat{n}^b,\theta^b)$ in terms of the actual solution $R:(\hat{n},\theta)$. The false solution $R^b:(\hat{n}^b,\theta^b)$ with respect to the actual solution $R:(\hat{n},\theta)$ is now analyzed for 4 cases.

Consider the degenerate case where $\mathrm{mod}_{2\pi}\theta=0$ (Case 0) (fig. 4a). Here, R is a full turn (mod 2π) rotation. Note that $\hat{n}$ can be any ("free") unit vector. Some geometrical facts are now established. Consider the PC $(R^{-1}\hat{t},\hat{t})$. Under the identity rotation $R = I$, the PC $(R^{-1}\hat{t},\hat{t})$ is fixed, i.e. $\hat{t} = R^{-1}\hat{t}$. Consider the false solution $(\hat{n}^b, \theta^b)$. Since $\cos\theta = 1$, (28c) implies that $\cos\theta^b = -1$, i.e. $\mathrm{mod}_{2\pi}\theta^b = \pi$. Since $R = I$ (i.e. $\hat{c}_1=\hat{x}$ $\hat{c}_2=\hat{y}$ $\hat{c}_3=\hat{z}$), (28a) implies $\hat{n}^b\hat{n}^{bT} = \hat{t}(R^{-1}\hat{t})^T = \hat{t}\hat{t}^T$, i.e. $\hat{n}^b$ is aligned with $\hat{t}=R^{-1}\hat{t}$.

Consider the general case where $\hat{n}\cdot\hat{t} \neq 0, \pm1$ (Case 1) (fig. 4b). Here, the rigid motion is not a general pure rotation about an axis parallel to $\hat{n}$ and not a screw motion about an axis $\hat{n}$ (CO). Some geometrical facts are now established. Consider the PC $(R^{-1}\hat{t},\hat{t})$. Since R is not a full turn ($\mathrm{mod}_{2\pi}\theta \neq 0$) rotation and $\hat{n}$ is not aligned with $\hat{t}$ ($\hat{n}\cdot\hat{t} \neq \pm1$), the PC $(R^{-1}\hat{t},\hat{t})$ is not fixed. i.e. $R^{-1}\hat{t} - \hat{t} \neq 0$. Consider the false solution $(\hat{n}^b,\theta^b)$. Since $\hat{n}\cdot\hat{t} \neq 0, \pm1$ and $\cos\theta \neq 1$, (28c) implies that $\cos\theta^b \neq \pm1$, i.e. $\mathrm{mod}_{2\pi}\theta \neq 0, \pi$. Since $\sin\theta^b \neq 0$ and $R^{-1}\hat{t}-\hat{t} \neq 0$, (25) implies that $\hat{n}$ and $\hat{n}^b$ are contained in the plane (CO) with normals $\pm(R^{-1}\hat{t}-\hat{t})$. There are 2 subcases. Consider the general subcase where $\mathrm{mod}_{2\pi}\theta\neq\pi$ (subcase a). Here, R is not a half turn. It is now shown that $\hat{n}^b$ is not perpendicular to and not aligned with $\hat{n}$ or $\hat{t}$. Since $\sin\theta \neq 0$, $\sin\theta^b \neq 0$, $\hat{n}\cdot\hat{t} \neq 0$, (28b) implies that $\hat{n}^b\cdot\hat{t} \neq 0$. Taking the dot product of (28d) with $\hat{n}$ and substituting (28b) gives $\hat{n}^b\cdot\hat{n} = (\hat{n}\cdot\hat{t})(\hat{n}^b\cdot\hat{t})$. Since $\hat{n}^b\cdot\hat{t}\neq 0$ and $\hat{n}\cdot\hat{t} \neq 0$, it follows that $\hat{n}^b\cdot\hat{n} \neq 0$. **Since $\hat{n}\cdot\hat{t}\neq\pm1$, it follows that $\hat{n}^b\cdot\hat{n}\neq\pm1$. Assuming $\hat{n}\cdot\hat{t}=\pm1$ implies that $\hat{n}^b\cdot R^{-1}\hat{t}=\hat{n}^b\cdot\hat{t}=\pm1$, i.e. $R^{-1}\hat{t}=\hat{t}$ (contradiction). Hence, $\hat{n}\cdot\hat{t}\neq\pm1$. Consider the subcase where $\mathrm{mod}_{2\pi}\theta=\pi$ (subcase b). Here, R is a half turn. It is now shown that $\hat{n}$, $\hat{t}$, $R^{-1}\hat{t}$ are perpendicular to $\hat{n}^b$. Since $\sin\theta = 0$, (27b) implies** that $\hat{n}\cdot(R^{-1}\hat{t}\times\hat{t}) = 0$, i.e. $\hat{n}$, $R^{-1}\hat{t},\hat{t}$ are contained in a plane (CO). Since $\sin\theta = 0$ and $\sin\theta^b \neq 0$, (28d) implies that $\hat{n}^b\cdot\hat{n} = 0$.

Consider the case where $\hat{n}\cdot\hat{t} = 0$ (Case 2) (fig. 4c). Here, the rigid motion is a general rotation about an axis parallel to $\hat{n}$. Some geometrical facts are now established. Consider the PC $(R^{-1}\hat{t},\hat{t})$. For the same reasons given for Case 1, the PC $(R^{-1}\hat{t},\hat{t})$ is not fixed, i.e. $R^{-1}\hat{t}-\hat{t}\neq 0$. Note that $\hat{n}\cdot R^{-1}\hat{t} = R\hat{n}\cdot\hat{t} = \hat{n}\cdot\hat{t} = 0$, i.e. both $\hat{t}$ and $R^{-1}\hat{t}$ are perpendicular to $\hat{n}$. Consider the false solution $(\hat{n}^b,\theta^b)$. Since $\hat{n}\cdot\hat{t} = 0$ (28c) implies $\cos\theta^b = -1$, i.e. $\mathrm{mod}_{2\pi}\theta^b = \pi$. Since $\hat{n}\cdot\hat{t}=0$ and $\hat{n}\cdot(\hat{c}_1-\hat{x})=\hat{n}\cdot(\hat{c}_2-\hat{y})=\hat{n}\cdot(\hat{c}_3-\hat{z})=0$, (28a) implies that $\hat{n}^b\cdot\hat{n}=0$. It can be verified from (28a) that $\hat{n}^b\cdot(R^{-1}\hat{t}-\hat{t})=0$. Thus, $\hat{n}^b$ is contained in the plane (CO) with normals $\pm\hat{n}$ and bisects the vectors $\hat{t}$, $R^{-1}\hat{t}$. There are 2 subcases. Consider the subcase where $\mathrm{mod}_{2\pi}\theta\neq\pi$ (subcase a). This subcase is analogous to subcase b of Case 1, where $\hat{n}$, $\hat{n}^b$ have been switched. Consider the subcase where $\mathrm{mod}_{2\pi}\theta=\pi$ (subcase b). Here, the PC $(R^{-1}\hat{t},\hat{t})$ is diametrically opposite, i.e. $\hat{t}=-R^{-1}\hat{t}$. $\hat{n}^b$ is therefore perpendicular to both $\hat{t}$, $R^{-1}\hat{t}$. Thus, $\hat{n}$, $\hat{n}^b$, $\hat{t} = -R^{-1}\hat{t}$ are mutually perpendicular.

Consider the case where $\hat{n}\cdot\hat{t}=\pm1$ (Case 3) (fig. 4d). Here, the rigid motion is a screw about an axis $\hat{n}$ (CO). Some geometrical facts are now established. Consider the PC $(R^{-1}\hat{t},\hat{t})$. Since $\hat{t}$ is aligned with $\hat{n}$, the PC $(R^{-1}\hat{t},\hat{t})$ is fixed, i.e. $\hat{n}=\hat{t}=R^{-1}\hat{t}$. Consider the false solution $(\hat{n}^b,\theta^b)$. Since $\hat{n}\cdot\hat{t}=\pm1$, (28c) gives $\cos\theta^b + \cos\theta = 0$, i.e. $\mathrm{mod}_{2\pi}\theta^b=\mathrm{mod}_{2\pi}(\theta+\pi)$. There are 2 subcases. Consider the subcase where $\mathrm{mod}_{2\pi}\theta\neq\pi$ (subcase a). Since $\sin\theta^b=-\sin\theta$ and $\hat{n}\cdot\hat{t}=\pm1$, (28d) implies that $\hat{n}^b\cdot\hat{n}=1$, i.e. $\hat{n}^b$ is aligned with $\hat{n}$. Consider the subcase where $\mathrm{mod}_{2\pi}\theta=\pi$ (subcase b). Since $\cos\theta= -1$ and $\hat{n}\cdot\hat{t} = \pm 1$, (28c) implies that $\cos\theta^b= 1$, i.e. $\mathrm{mod}_{2\pi}\theta^b = 0$. Note that $\hat{n}^b$ can be any ("free") unit vector. This subcase is analogous to Case 0, where $\hat{n}$, $\hat{n}^b$ have been switched.

The fact that the actual solution R cannot be identified from the 2 solutions $R:(\hat{n},\theta)$ and $R^b:(\hat{n}^b,\theta)$ can be understood from the above analysis of the relative geometrical configuration of the 2 rotational axes $\hat{n}$, $\hat{n}^b$ and the vectors of the PC $(R^{-1}\hat{t},\hat{t})$. Given the general case where $\hat{n}$, $\hat{n}^b$ are not aligned and not perpendicular (Case 1a), it is impossible to identify R from R, R^b. Both R and R^b generate the PC $(R^{-1}\hat{t},\hat{t})$. Given the case where $\hat{n}$, $\hat{n}^b$ are perpendicular and the vectors of the PC $(\hat{t},R^{-1}\hat{t})$ are coplanar with one of $\hat{n}$, $\hat{n}^b$ (Cases 1b, 2a), it is impossible to identify R from R, R^b. It is impossible for $\hat{n}$ to be the axis coplanar with (Case 1b) or perpendicular to $\hat{t}$, $R^{-1}\hat{t}$ (Case 2a). Given the case where one of $\hat{n}$, $\hat{n}^b$ is aligned with the vectors of the fixed PC $(R^{-1}\hat{t} = \hat{t},\hat{t})$ and the other is "free" (Cases 0,3b), it is impossible to identify R from R, R^b. It is possible for $\hat{n}$ to be the axis aligned (Case 3b) or not aligned (Case 0) with $\hat{t} = R^{-1}\hat{t}$.

Given the case where $\hat{n}^a$, $\hat{n}^b$, $\hat{t} = -R^{-1}\hat{t}$ are mutually perpendicular (Case 2b), it is impossible to identify R from R, R^b. Both R and R^b are half turns, which generate a diametrically opposite PC ($R^{-1}\hat{t} = -\hat{t},\hat{t}$). However, θ is uniquely determined as π. Given the case where $\hat{n}$, $\hat{n}^b$ are aligned with the vectors of the fixed PC ($R^{-1}\hat{t},\hat{t}$) (Case 3a) it is impossible to identify R from R, R^b. Both R and R^b have the same rotational axis, with angles of rotation differing by π. However, the rotational axis $\hat{n}$ is uniquely determined.

The above analysis is a complete decomposition of the cases for the 2 solutions R: $(\hat{n},\theta)$ and R^b: $(\hat{n}^b,\theta^b)$. It is interesting to note that θ is uniquely determined from Case 2b and $\hat{n}$ is uniquely determined from Case 3a. It is now shown that in the context of the situation where the rigid body remains in front of the camera over the motion, that R can be uniquely determined from E' for subcases of 2 specific cases (Case 2b and 3a).

Theorem 3

Given

$$E' = kE = [e_1' \; e_2' \; e_3'] = [k(\hat{t}\times\hat{c}_1) \; k(\hat{t}\times\hat{c}_2) \; k(\hat{t}\times\hat{c}_3)] = kGR$$

where k = unknown scale, with the constraint that the rigid body remains in front of the camera over the motion, there are 2 solutions to R

$R^a = R$ (actual solution) $\quad$ $R^b = 2\hat{t}(R^{-1}t)^T - R$ (false solution)

except for the following 2 cases where either $\hat{n}$ or θ or both (i.e. R) is uniquely determined.

case 2b

$\hat{n}\cdot\hat{t} = 0 \quad \mathrm{mod}_{2\pi}\theta = \pi$

(pure rotation by 180° about an arbitrary axis parallel to $\hat{n}$; θ ($=\pi$) uniquely determined

$n_3 = \pm 1 \Rightarrow (\hat{n},\theta)$ uniquely determined

case 3a

$\hat{n}\cdot\hat{t} = \pm 1 \quad \mathrm{mod}_{2\pi}\theta \neq \pi$

(screw motion about an axis (CO))

$\hat{n}$ uniquely determined

$\hat{n}_3 = 0 \Rightarrow (\hat{n},\theta)$ uniquely determined

and 2 solutions each for $\hat{t}$ and k

$\hat{t}^a = \hat{t}$ (actual solution) $\quad$ $\hat{t}^b = -\hat{t}$ (false solution)

$k^a = k$ (actual solution) $\quad$ $k^b = -k$ (false solution)

There are 4 solutions to $(R,\hat{t},k)$

$(R^a,\hat{t}^a,k^a)$ (actual solution)

$(R^a,\hat{t}^b,k^b)$ (false solution 1)

$(R^b,\hat{t}^a,k^b)$ (false solution 2)

$(R^b,\hat{t}^b,k^a)$ (false solution 3)

except for cases 2b, 3a where there are only 2 solutions (1st two)

Proof:

Given Theorem 2, it is only required to show that either $\hat{n}$ or θ (or both) are uniquely determined for cases 2b, 3a.

1st, it is shown that cases 2b, 3a can be detected. From the above analysis, Case 2b is characterized by the fact that the PC ($R^{-1}\hat{t},\hat{t}$) is diametrically opposite. It is a well known fact that a diametrically opposite PC ($R^{-1}\hat{t},\hat{t}$) is peculiar only to Case 2b and is then a sufficient condition to detect Case 2b. The geometric configuration for Case 3a is characterized by the fact that $\hat{n}$ and $\hat{n}^b$ are aligned. From the above analysis, this fact is peculiar only to Case 3a, and is then a sufficient condition to detect Case 3a.

Suppose Case 2b has been detected. Here, the motion is a pure rotation about an arbitrary axis parallel to $\hat{n}$. θ is uniquely determined as π. Consider the case where $\hat{n}$ is aligned with the optical axis $\hat{z}$ ($n_3=\pm1$). Note that $\hat{n}$ and $\hat{t}$ are perpendicular to $\hat{z}$. The motion associated with R^b:$(\hat{n}^b,\theta^b)$ and $\pm\hat{t}$ takes an object point p in front of the camera ($z > F > 0$), rotates it 180° about $\hat{n}^b$ to behind the camera ($z'' < -F < 0$), and translates it parallel to the plane (CO) z=0 to p'. p' is still behind the camera with ($z' < -F < 0$) - this is physically impossible. Thus, $\hat{n}^b$ with zero z-component can be eliminated as $\hat{n}$ and R is uniquely determined. Consider the case where $\hat{n}$ is not aligned with the optical axis $\hat{z}$($n_3\neq\pm1$). It is possible that some object points in frame 2 could be taken behind the camera by the motion associated with R^b:$(\hat{n}^b,\theta^b)$ and $\pm\hat{t}$. This cannot be determined without knowing 3-D object points.

Suppose Case 3a has been detected. $\hat{n}$ is uniquely determined. Consider the case where $\hat{n} = \hat{n}^b = \hat{t} = R^{-1}\hat{t}$ are perpendicular to the optical axis $\hat{z}$($n_3=0$). The actual motion associated with R:$(\hat{n},\theta)$ and $\pm\hat{t}$ takes an object p in from of the camera ($z > F > 0$), rotates it by θ about $\hat{n}$ in front of the camera ($z'' > F > 0$), and translates it parallel to the plane (CO) z=0 to p' in front of the camera ($z > F > 0$). This requires $\mathrm{mod}_{2\pi}\theta$ to lie in the range $(0,\pi)$, otherwise p' would lie behind the camera ($z' < -F < 0$). The motion associated with R^b: $(\hat{n}^b,\theta^b)$ and $\pm\hat{t}$ clearly takes p' behind the camera, where $\mathrm{mod}_{2\pi}\theta^b = \mathrm{mod}_{2\pi}(\theta+\pi)$ lies in the range $(\pi,2\pi)$. Thus, whichever of the 2 solutions to θ (modulo 2π) in the range $(\pi,2\pi)$ is eliminated as θ and R is uniquely determined. Consider the case where $\hat{n}$ is not perpendicular to the optical axis $\hat{z}$ and not aligned with the optical axis $\hat{z}$ ($n_3\neq 0, \pm1$). It is possible that some object points could be taken behind the camera by the motion associated with R^b:$(\hat{n}^b,\theta^b)$ and $\pm\hat{t}$. This cannot be determined without knowing 3-D object points. Consider the case where $\hat{n}$ is aligned with the optical axis $\hat{z}$($n_3=\pm1$). Given that the actual motion associated with R:$(\hat{n},\theta)$ and $\pm\hat{t}$ moves points in front of the camera, the motion associated with R^b:$(\hat{n}^b,\theta^b)$ and $\pm\hat{t}$ also takes object points in frame 2 in front of the camera. Thus, R^b cannot be eliminated based on the fact that object points in frame 2 lie behind the camera.

For the given subcases of Case 2b and Case 3a, the 2 solution sets involving R^b can be eliminated.

QED

Note that for subcases of the 2 specific cases, only R can be uniquely determined-there are still 2 solutions to $\hat{t}$.

VI. Determination of $(R,\hat{t})$ from E and a PC

In this section, it is shown that a unique solution to $(R,\hat{t})$ can be determined given E and some PC.

Theorem 4

Given

$E'=kE=[e_1'\ e_2'\ e_3']$

$=[k(\hat{t}x\hat{c}_1)\ k(\hat{t}x\hat{c}_2)\ k(\hat{t}x\hat{c}_3)] = kGR$

where k=unknown scale, and a PC

$(\hat{p}_i,\hat{p}_i')$ where $R\hat{p}_i$ is not aligned with $\pm\hat{t}$ then $(R,\hat{t})$ and k are uniquely determined. Also, given a set of 2 or more PCs, then $(R,\hat{t})$ and k are always uniquely determined.

Proof

From Theorem 2, there are 2 solutions to each of R, $\hat{t}$, k. There are 4 solutions to $(R,\hat{t},k)$. It is now shown that from some PC, one solution to R can be eliminated and one solution to $\hat{t}$ can be eliminated. That is, $(R,\hat{t})$ is uniquely determined. This isolates one of the 4 solution sets to $(R,\hat{t},k)$ and thus k is also uniquely determined.

Consider a PC $(\hat{p}_i,\hat{p}_i')$ where $R\hat{p}_i$ (and $\hat{p}_i'$) is not aligned with $\pm\hat{t}$ (fig. 5). As stated earlier, $\hat{p}_i'$ lies on the arc with endpoints $\hat{t}$, $R\hat{p}_i$. From (6), with $R^a=R$, $R^b\hat{p}_i$ is given by

$$R^b\hat{p}_i = 2(\hat{t}\cdot R\hat{p}_i)\hat{t} - R\hat{p}_i \tag{29}$$

Geometrically, $\pm\hat{t}$ bisects $R\hat{p}_i$ and $R^b\hat{p}_i$. The great circle is split into 2 halves by a line joining the points $\hat{t}$ and $-\hat{t}$. On 1 side of the great circle lies $R\hat{p}_i$ and $\hat{p}_i'$ - on the other side lies $R^b\hat{p}_i$. Clearly, R^b can be eliminated from the 2 solutions R and R^b by the fact that $R^b\hat{p}_i$ does not lie on the same side of the great circle with $\hat{p}_i'$. Computationally, this is achieved by

$$R^*\hat{p}_i\cdot s \begin{cases} < 0 & R^* = R^b \\ > 0 & R^* = R \end{cases} \tag{30}$$

where

$$s = \hat{t}\ x\ (\hat{p}_i'\ x\ \hat{t}) = \hat{p}_i' - (\hat{t}\cdot\hat{p}_i')\hat{t}$$

$$R^* \in \{R,\ R^b\}$$

Then, $-\hat{t}$ is eliminated from the fact that $\hat{p}_i'$ lies on the arc with endpoints $R\hat{p}_i$ and $\hat{t}$. Computationally, this is achieved by

$$(\hat{p}_i'\ x\ R\hat{p}_i)\cdot(\hat{t}\ x\ R\hat{p}_i) > 0 \tag{31}$$

In general, $R\hat{p}_i$ is not aligned with $\pm\hat{t}$ for a PC $(\hat{p}_i,\hat{p}_i')$. There are rare situations where $R\hat{p}_i$ is aligned with $\pm\hat{t}$. It is now shown that the required PC exists for sets of 2 or more PCs, guaranteeing a unique solution for $(R,\hat{t})$ and k. Assume there are 2 PCs $(\hat{p}_i,\hat{p}_i')$ and $(\hat{p}_j,\hat{p}_j')$ where $R\hat{p}_i$ is aligned with $\pm\hat{t}$ and $R\hat{p}_j$ is aligned with $\pm\hat{t}$. Either $R\hat{p}_i=R\hat{p}_j$ or $R\hat{p}_i=-R\hat{p}_j$. Since R is a 1 to 1 mapping, $\hat{p}_i=\hat{p}_j$ or $\hat{p}_i=-\hat{p}_j$. It is impossible for 2 3-D points on a line (CO) to be distinguished from their identical projections and points to lie behind the camera. Thus, there exists at least 1 PC $(\hat{p}_i,\hat{p}_i')$ (maybe more) where $R\hat{p}_i$ is not aligned with $\pm\hat{t}$. QED

The following is an analysis of the results of Theorems 2-4. The geometry of the problem (see II) requires that $R\hat{p}_i$, $\hat{p}_i'$, $\hat{t}$ lie on a great circle, where $\hat{p}_i'$ lies on the arc with endpoints $R\hat{p}_i$ and $\hat{t}$. Equation (3a) reflects these constraints, excluding the latter essential fact. From 8 PCs, E is computed to a scale as E'. Given E', there are 2 solutions to each of R and $\hat{t}$. There are 4 solutions $(R,\hat{t})$ $(R,-\hat{t})$ $(R^b,\hat{t})$ $(R^b,-\hat{t})$ to $(R,\hat{t})$-all of which satisfy (3a)(Theorem 2). Given a PC $(\hat{p}_i,\hat{p}_i')$ where $R\hat{p}_i$ is not aligned with $\pm\hat{t}$, the unapplied constraint is applied to give a unique solution to $(R,\hat{t})$ (Theorem 4). It is interesting to note that with the visibility constraint, R can be uniquely determined from E, for subcases of 2 specific cases (Theorem 3).

The same equation set (3a) is obtainable from 4 possible PC sets $(\pm\hat{p}_i,\pm\hat{p}_i')$ i=1,...,n. Note that $(\hat{p}_i,\hat{p}_i')$ i=1,...,n lie in front of the camera. Consider a PC $(\hat{p}_i,\hat{p}_i')$ where $R\hat{p}_i$ is not aligned with $\pm\hat{t}$ (fig. 5). The solution $(R,\hat{t})$ is consistent with the PC $(\hat{p}_i,\hat{p}_i')$ and the solution $(R,-\hat{t})$ is consistent with the PC $(-\hat{p}_i,-\hat{p}_i')$. The solutions $(R^b,\hat{t})/(R^b,-\hat{t})$ are consistent with the PCs $(\hat{p}_i,-\hat{p}_i')/(-\hat{p}_i,\hat{p}_i')$ (respectively) or with the PCs $(-\hat{p}_i,\hat{p}_i')/(\hat{p}_i,-\hat{p}_i')$ (respectively). In summary, for a PC $(\hat{p}_i,\hat{p}_i')$ where $R\hat{p}_i$ is not aligned with $\pm\hat{t}$, there is a 1-to-1 correspondence between the 4 solutions to $(R,\hat{t})$ and 4 possible PC pairs $(\pm\hat{p}_i,\pm\hat{p}_i')$.

The above facts give an alternative viewpoint of how the actual solution set $(R,\hat{t})$ is identified from the 4 solutions $(R,\hat{t})$ $(R,-\hat{t})$ $(R^b,\hat{t})$ $(R^b,-\hat{t})$ given a PC $(\hat{p}_i,\hat{p}_i')$. For the case where $R\hat{p}_i$ is not aligned with $+\hat{t}$, only the **actual solution $(R,\hat{t})$ is consistent with the given PC $(\hat{p}_i,\hat{p}_i')$. The false solutions $(R,-\hat{t})$ $(R^b,\hat{t})$ $(R^b,-\hat{t})$ correspond to PCs where 1 or both of the 3-D points in frame 1 and frame 2 lie behind the camera.**

VII. Conclusion

The main result is that the 3-D motion parameters $(R,\hat{t})$ and the PCs $(\pm R^{-1}\hat{t},\pm\hat{t})$ under R can be uniquely determined by a vector-geometrical method, given E (to a scale) and a PC. Then object structure can be found by the method in [4]. In addition, 3 other results were established. 1st, it was shown that a unique solution to $(R,\hat{t})$ for (3a) is impossible. 2nd, it was shown that there are 4 solution sets to $(R,\hat{t})$ and a unique solution to $\pm(R^{-1}\hat{t},\hat{t})$, given kE. 3rd, it was shown that there is a unique solution to R for 2 special cases, given E (to scale) and the visibility constraint.

References

1. H. Nagel, "On the Derivation of 3D Rigid Point Configurations from Image Sequences," PRIP, pp. 103-108, Aug. 1981.

2. R. Tsai and T. Huang, "Uniqueness and Estimation of Three-Dimensional Motion Parameters of Rigid Objects with Curved Surfaces," Technical Report CSL R-921, Oct. 30, 1981. University of Illinois.

3. H. C. Longuet-Higgins, "A Computer Algorithm for Reconstructing a Scene from Two Projections," Nature, No. 293, pp. 133-135, Sept. 1981.

4. B. Yen and T. Huang, "Determining 3-D Motion and Structure of a Rigid Body Using the Spherical Projection," CSL Technical Report R-970, Nov. 1982.

5. B. Yen and T. Huang, Determining 3-D Motion and Structure of a Rigid Body Continaing Lines, Technical Report CSL R-971, Nov. 1982.

6. F. R. Gantmacher, Theory of Mattrices vol. I, Chelsea, 1960.

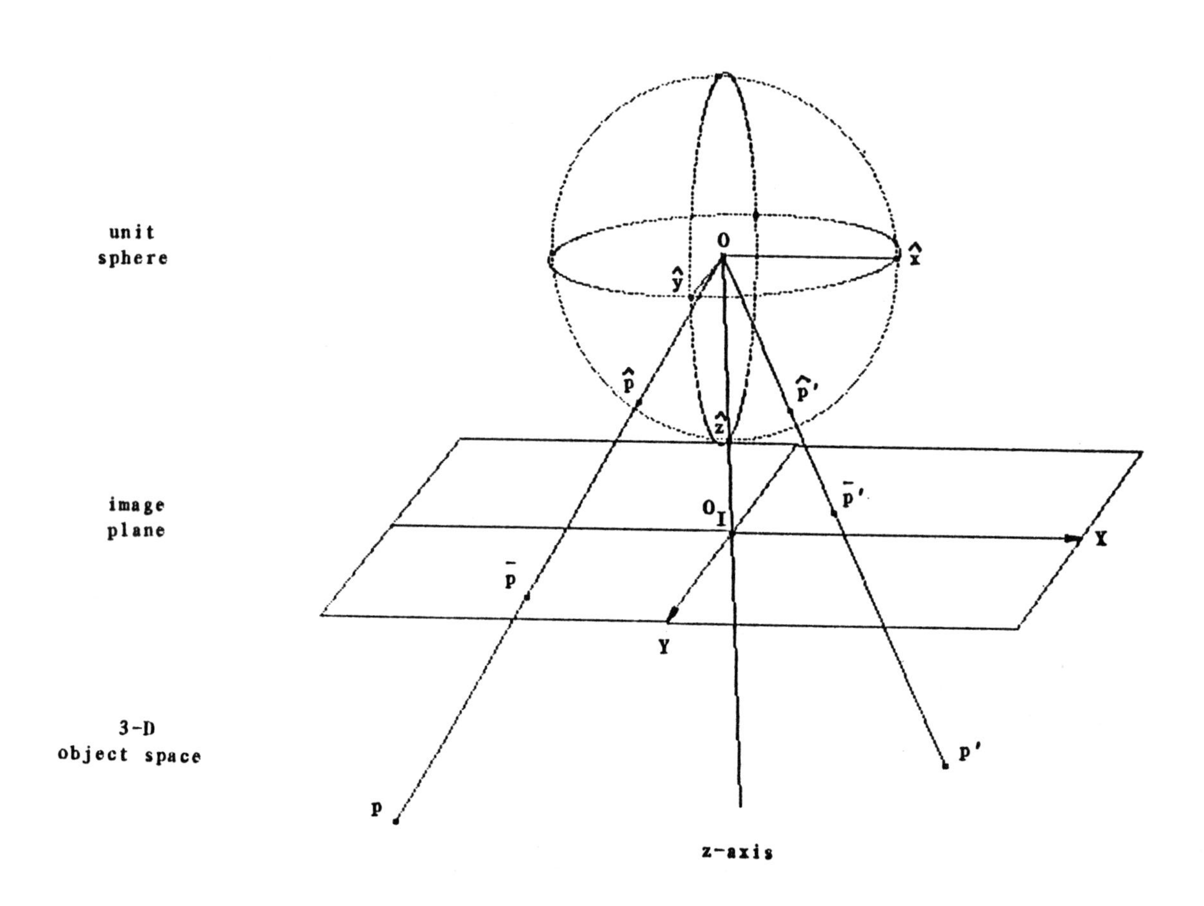

fig. 1 central projection of 3-D PC (p,p') as $(\bar{p},\bar{p}')$ (on image plane) and $(\hat{p},\hat{p}')$ (on unit sphere)

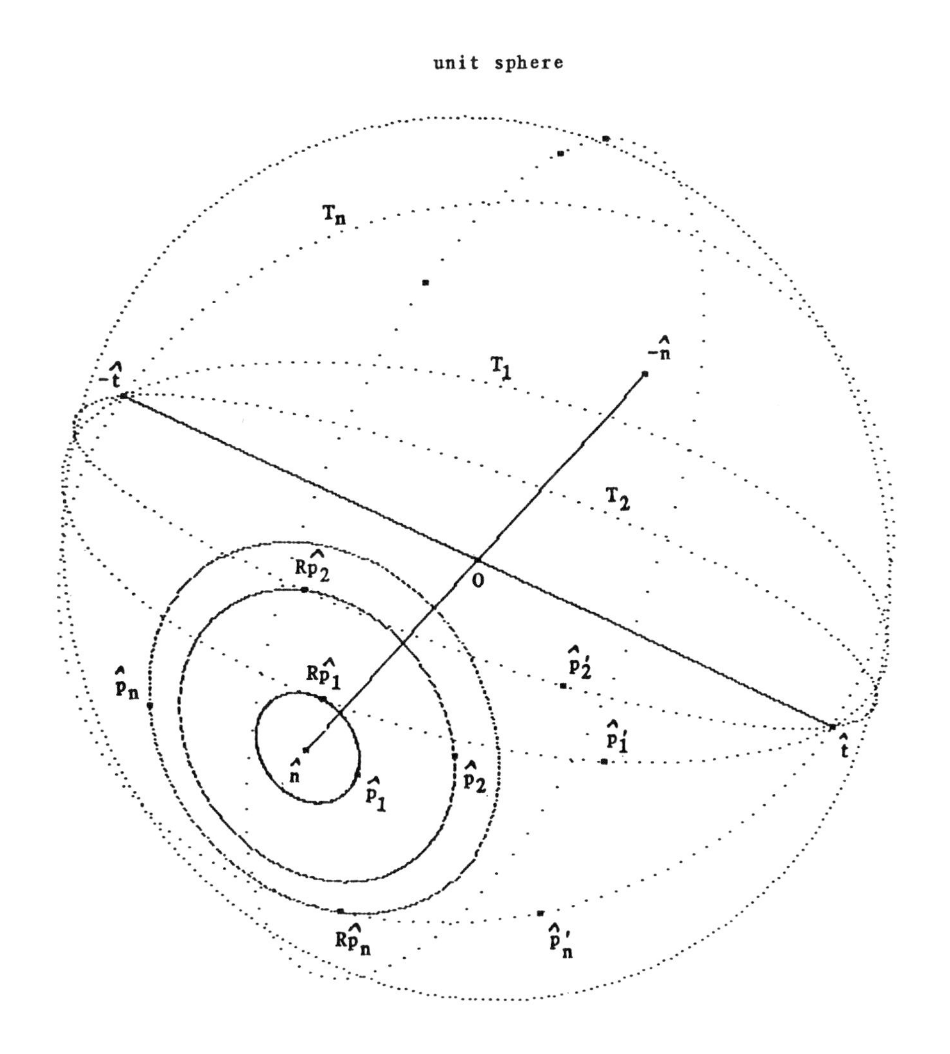

fig. 2 geometry of $\hat{p}_i$, $R\hat{p}_i$, $\hat{p}'_i$ wrt $\hat{n}$, $\hat{t}$

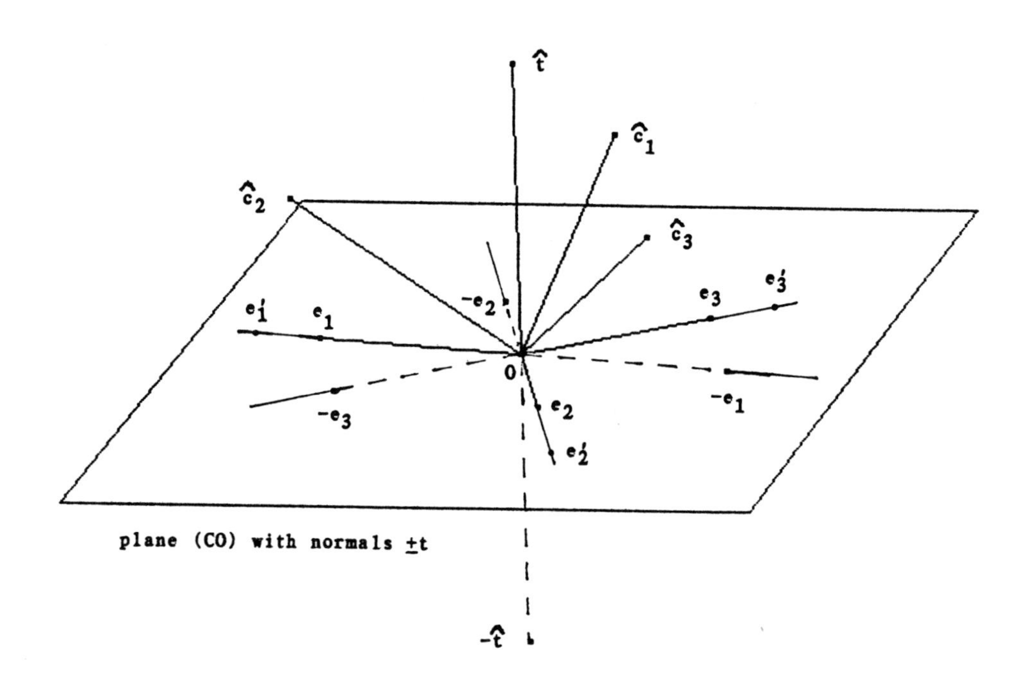

fig. 3a geometry of the determination of $\pm\hat{t}$ from e_i' $i = 1,2,3$

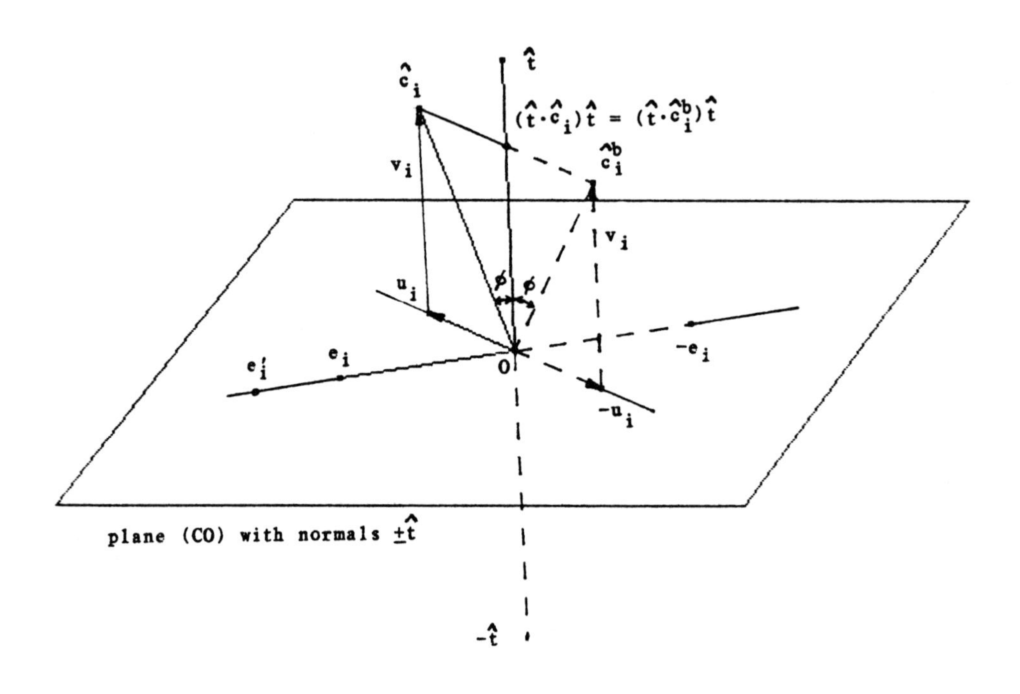

fig. 3b geometry of the determination of $\hat{c}_i$, $\hat{c}_i^b$ $i = 1,2,3$ from $\pm k$, $\pm\hat{t}$, e_i'

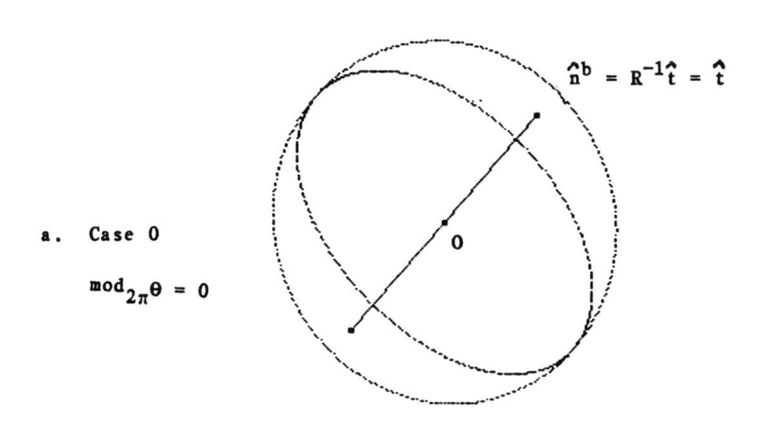

$\mathrm{mod}_{2\pi}\theta^b = \pi$

PC $(R^{-1}\hat{t},\hat{t})$ fixed

$\hat{n}^b \parallel$ to $R^{-1}t = t$ $\quad \hat{n}$ = "free" vector

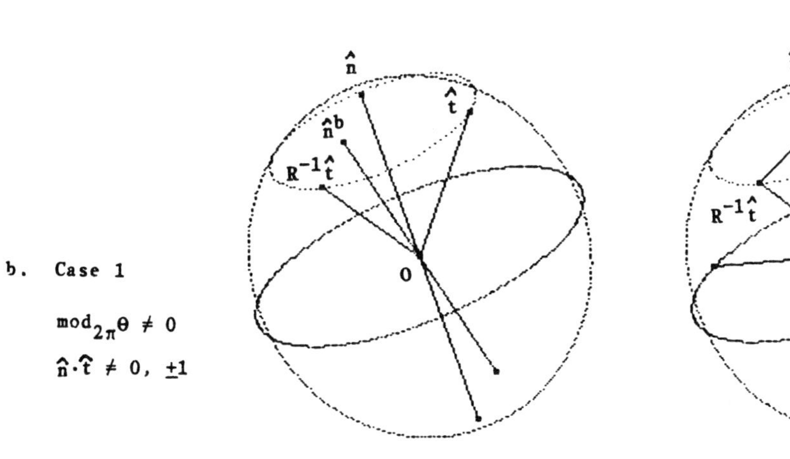

subcase a $\quad \mathrm{mod}_{2\pi}\theta \neq \pi$

$\mathrm{mod}_{2\pi}\theta^b \neq 0, \pi$

PC $(R^{-1}\hat{t}, \hat{t})$ not fixed

$\hat{n}^b$ not $\parallel$ or $\perp$ to $\hat{n},\hat{t}$

subcase b $\quad \mathrm{mod}_{2\pi}\theta = \pi$

$\mathrm{mod}_{2\pi}\theta^b \neq 0, \pi$

PC $(R^{-1}\hat{t}, \hat{t})$ not fixed

$\hat{n}^b \perp$ to $\hat{n}$, $R^{-1}\hat{t}$, $\hat{t}$

fig. 4 geometry of 2 solutions $R:(\hat{n},\theta)$ and $R^b:(\hat{n}^b,\theta^b)$

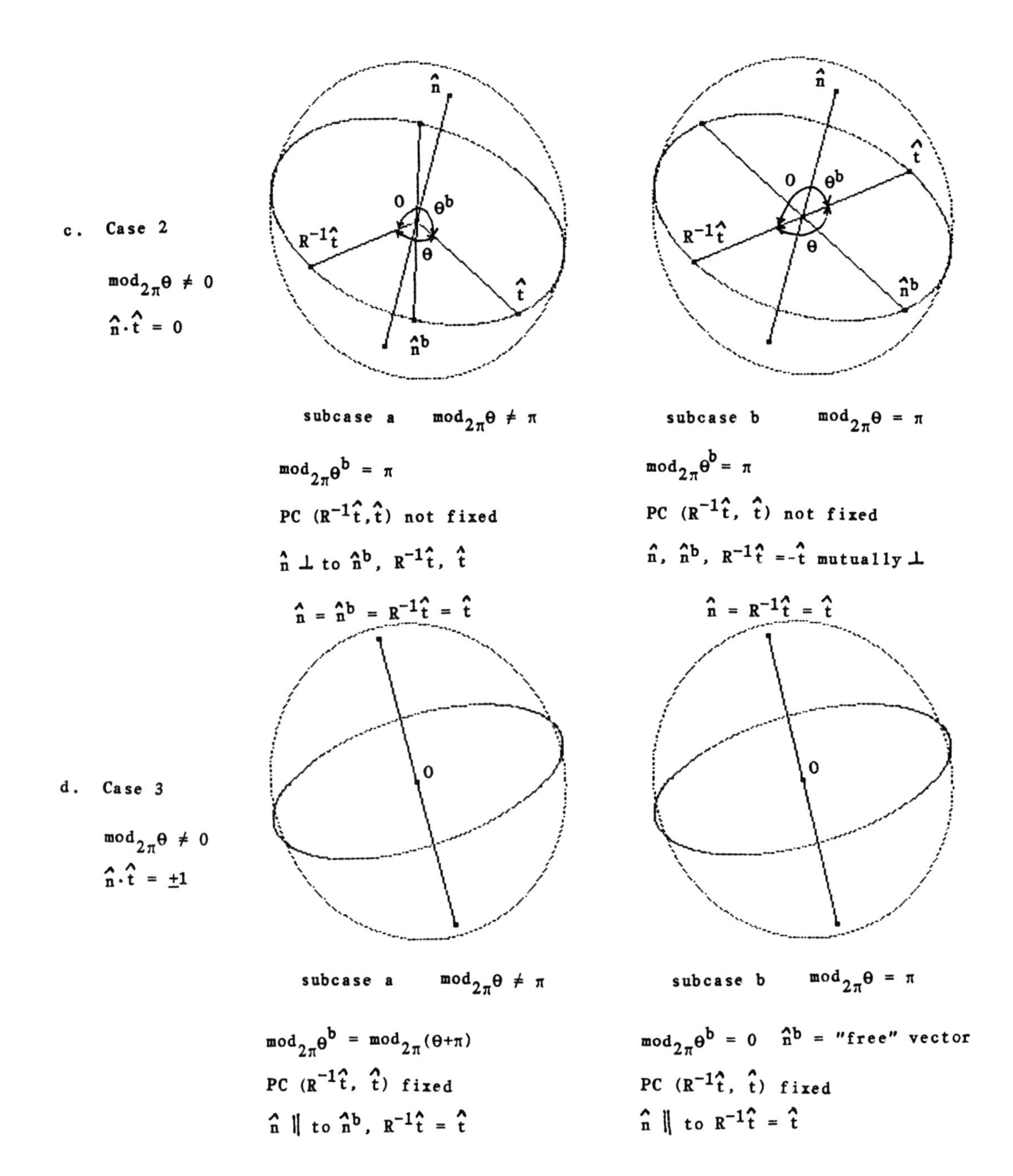

fig. 4 geometry of 2 solutions $R:(\hat{n},\theta)$ and $R^b:(\hat{n}^b,\theta^b)$ (cont'd)

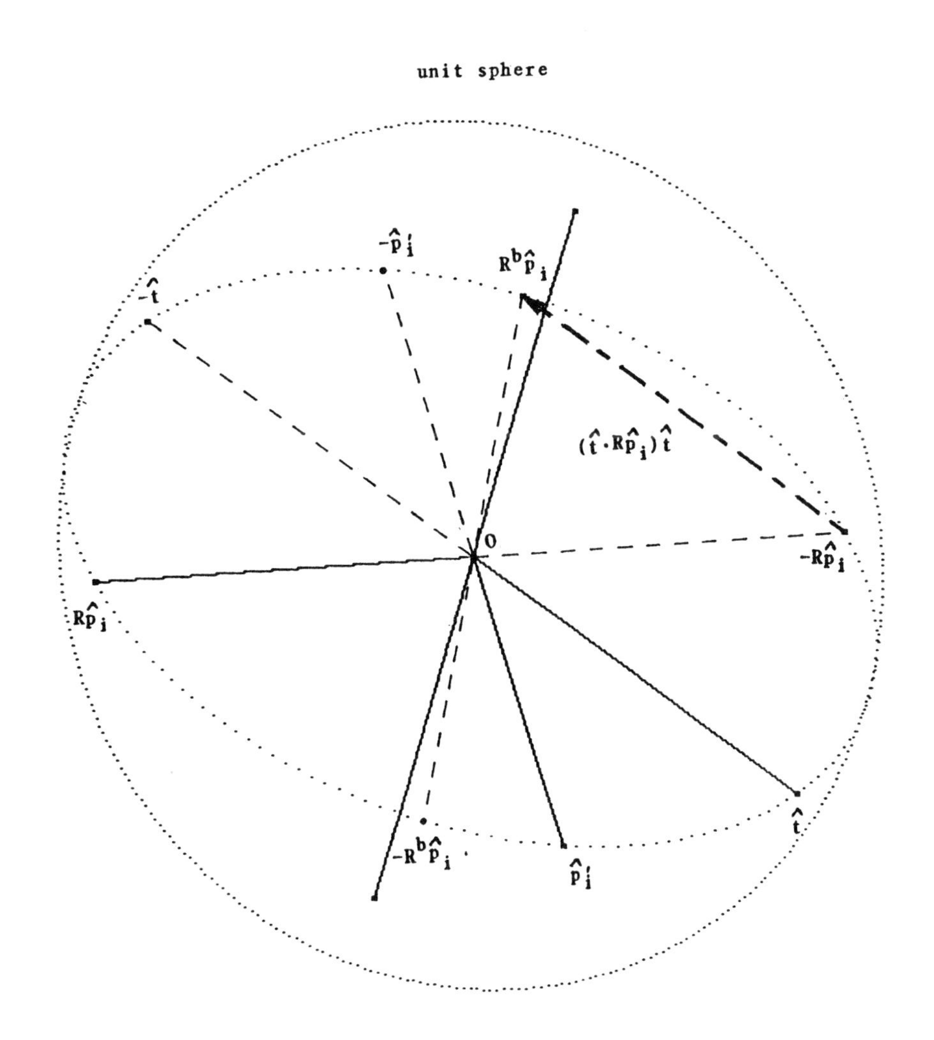

fig. 5 geometry of $\pm R\hat{p}_i$, $\pm R^b\hat{p}_i$, $\pm\hat{p}'_i$, $\pm\hat{t}$ on great circle for PC $(\hat{p}_i, \hat{p}'_i)$ where $R\hat{p}_i$ is not aligned with $\pm\hat{t}$

A Hybrid Approach to Structure-from-Motion

Aaron Bobick
AI Lab
Dept. of Psychology, MIT

Abstract: A method is presented for computing structure from the motion of rigid objects which are rotating about a fixed axis. The input consists of two discrete frames containing the positions and instantaneous direction vectors of three points in orthographic projection. Because only the direction of the velocity vectors and not their magnitudes is needed, the method is insensitive to errors in velocity magnitude estimation. This type of computation could be important in recovering the 3-dimensional structure of objects under dynamic viewing conditions because viewer motion about stationary objects will generate fixed axis rotations.

1. Introduction

When presented with an image generated by a moving structure, we attempt to construct a 3-dimensional, structural interpretation for the display rather than simply perceiving a slightly correlated flow of moving points. This process of determining structure–from–motion has been shown to be a robust psychophysical phenomenon (Wallach and O'Connell,1953; Green, 1961; Johansson, 1964, 1968, 1973; Ullman, 1979a).

In an attempt to explain how our visual system could generate such an interpretation, several theories have been proposed which establish what information is sufficient to make the necessary computations. In their analysis, each theory makes its own particular set of assumptions about the nature of the world and the exact form of the input to the system (Ullman 1979, Hoffman and Flinchbaugh 1982, Longuet-Higgins and Prazdny 1980, Prazdny 1980). For example, one must choose either a perspective projection of the world on to the image plane (where apparent length is scaled by distance), or an orthographic one which is a simple geometric projection. Also there is the choice between using only the position of points as viewed in temporally ordered discrete frames as input versus having an instantaneous velocity available for all points.

This paper presents a "hybrid" approach which uses both the position and the velocity of points in discrete frames. The computation of structure from motion will be shown, however, to require only the *direction* of the velocity vector, not its magnitude (i.e. the "speed"). We

Motion: Representation and Perception
N.I. Badler and J.K. Tsotsos, Editors
Published by Elsevier Science Publishing Co., Inc.

argue that the direction of a velocity vector may be computationally more reliable and more easily measured than its magnitude.

The motion considered is that of a set of points in a rigid configuration which rotates about a fixed axis; the axis itself is allowed to undergo translation. The goal is to provide an analytic solution which generates a provably unique structural interpretation of the given configuration. Since the solution is local in time, this theory could be extended to general tumbling motion as well, given certain constraints.

2. Hybrid Theory Constraints

Positions and Velocities

An important step in formulating a structure-from-motion theory is to decide whether the input should consist of only the positions of points in discrete frames or whether some of the velocity information should be included. Arguments can be made for both, and both schemes have been used.

One of the weaknesses of a theory which uses only discrete frames as input is that there is no inclusion of temporal constraints such as image sequence or smoothness of motion. Consider the prototypical structure-from-motion experiment where three or four static frames are shown to a subject and he is asked what he perceives. According to theories that consider only position (e.g. Ullman 1979b; Hoffman and Flinchbaugh 1982) the order of presentation should be irrelevant. Subjects report, however, that if the order of presentation is not consistent with a smooth velocity, the structural interpretation is not as robust or even absent (Richards and Lieberman 1982). This indicates that velocity information, either implicit or explicit, is useful in structure-from-motion computations.

At the other end of the spectrum are theories which make extensive use of velocity information (Prazdny, 1980; Longuet-Higgins and Prazdny, 1980). In some cases, not only are the instantaneous velocities required, but also their spatial first and second derivatives are needed (Longuet-Higgins and Prazdny, 1980). Unfortunately, either because these theories have been constructed using perspective projection,[1] or because the spatial derivatives of the velocity fields are required, small errors in the measurement of the velocity fields can produce unstable results (Ullman, 1982).

The approach we have chosen is a combination of these two strategies and therefore is a "hybrid" model. We will take as input to the system the 2 dimensional position of points in

[1]The theories which only use velocity information assume perspective projection because under orthographic projection velocity information alone can only determine structure up to a scaling factor (Ullman, 1982).

discrete frames with their associated 2 dimensional *"direction"* vectors — velocity vectors of unknown magnitude. There are two reasons for this choice of input. First, to compute an instantaneous magnitude a system must have very good temporal resolution, a requirement not well suited for biological systems. Second, motion sensitive cells in biological systems have been found to be highly directionally specific (Campbell, *etal.*, 1968; Hubel and Wiesel, 1968), but not very sensitive to differences in speed (Wurtz, 1969).

Orthographic Projection

For the development of the hybrid theory we have chosen an orthographic projection of the 3-dimensional world onto the the 2-dimensional image plane. This choice is dictated by several reasons: First, when a point moving in space undergoes a small change in depth relative to its absolute depth from the observer, the image generated by a perspective projection is locally almost identical to the image generated by an orthographic projection, up to a scaling factor. Second, when the change in depth is small relative to the absolute depth as it is in most common viewing conditions, a formulation using perspective theory can become numerically unstable. This problem will arise if the theory relies on the apparent change in length associated with the small change in distance. Since we are constructing a theory which is to be used in computations, we wish to avoid this type of problem. Finally, orthographic projection simplifies the mathematics, making analytic solutions more feasible (Kender,1982).

Fixed axis motion

Our final constraint upon the hybrid theory was to restrict the motion of the rigid configuration to a fixed axis rotation. Thus the rigid configuration is rotating about an axis whose orientation in space with respect to the image plane is constant, although translation of the axis itself is allowed.

There are several reasons which motivated using the fixed axis constraint. First, fixed axis motion is equivalent to the relative motion of a stationary object generated by a changing viewer position. Specifically, it can be shown that if an observer moves in a plane about an object, the resulting motion in the image is that caused by rotation about a translating axis.[2] Therefore, solving the fixed axis problem quickly will allow the recovery of many 3-dimensional structures under dynamic viewing conditions. This need for a quick computation motivates searching for a temporally local solution.

---

[2]The axis of rotation in this case is not simply the vertical axis even if the observer is moving horizontally. Consider a chandelier hanging from a ceiling. As an observer passes under it, the lamp undergoes an apparent rotation about the horizontal axis.

Second, if a temporally *and spatially* local process could solve the fixed axis problem, this solution could be applied to general motion if one is willing to make the following assumption: the angular velocity of the axis of rotation — $\vec{\omega}$ — is small compared to the angular velocity of the object rotating about $\vec{\omega}$. This assumption would allow making a fixed axis assumption for small periods of time and therefore the fixed axis solution could be applied. The inclusion of the spatially local constraint further allows the theory to partition an image into parts that are moving independently, eliminating the possibility of using any global information.

Finally, there exists psychophysical evidence that the human observer is particularly adept at interpreting displays of fixed axis motion. Green (1961) reports that if a subject is given a display depicting the projection of general tumbling motion, the structural perception is not nearly as strong.

Correspondence

For the hybrid theory we assumed that the correspondence problem has been already solved. That is, all the points in each view are correctly identified with respect to the previous view.

3. Solution of the Fixed Axis Case

A previous attempt to solve the problem of fixed axis rotation was made by Webb and Aggarwal (1981). These authors exploited the fact that any point on an object which is undergoing fixed axis rotation (with no translation) traces a circle in 3 space and hence projects as an ellipse in the image plane. The motion in depth of a point can then be estimated by fitting an ellipse to the path in the image plane and then computing the relative depth. Webb and Aggarwal do not provide any analytic solutions and thus it is not clear how much information is required to generate a unique structural interpretation.

The solution to the fixed axis problem presented here provides an analytic solution local in both time and space and demonstrates that a computationally-feasible, unique and correct structural interpretation is possible provided that certain specified conditions are met.

Although the details of the mathematics have been left for the appendix, here we provide a sketch of the hybrid solution to the fixed axis case. The solution using 2 views of 2 points is given first showing that this information is sufficient to reduce the problem to picking from two possible axis of rotation. As shown in the appendix, another view or another point is necessary to specify the axis uniquely.

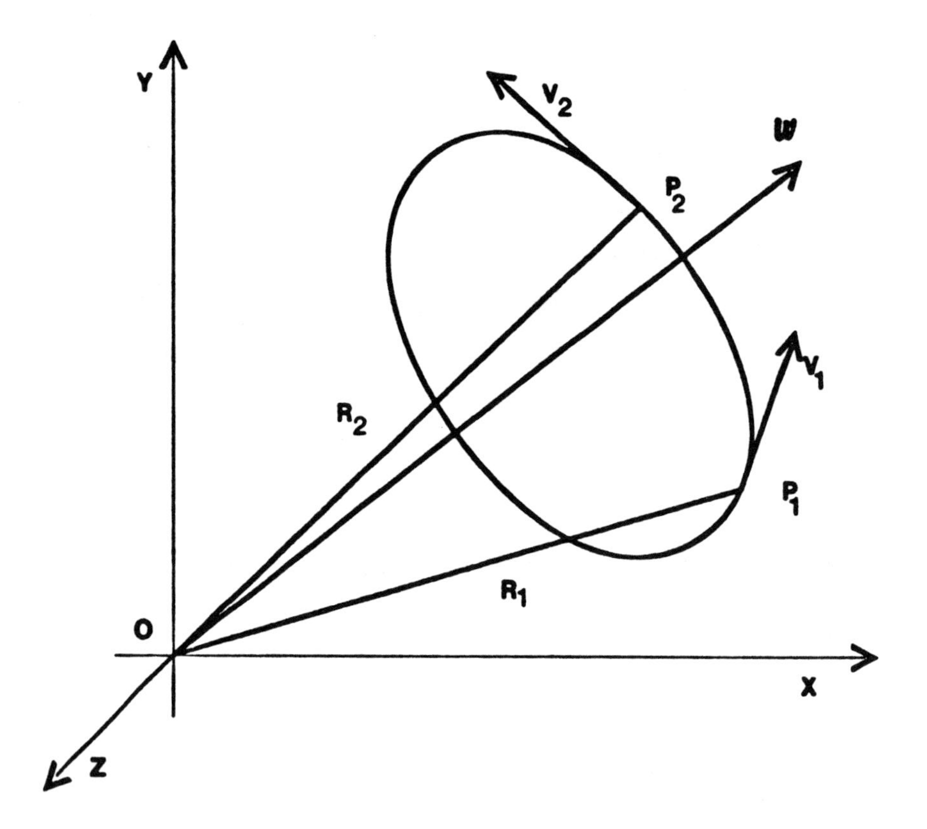

Figure 1. Fixed axis motion.

Two Views of Two Points: The Equations

In Figure 1 $\vec{\omega}$ is a fixed axis in a *translating* coordinate system about which some object is rotating. Assume that on this object there are are two distinguishable points **O** and **P** whose 2 dimensional projections are known. Also, without loss of generality we can assume that **O** is at the origin and therefore **P** is the only point in motion with respect to the translating coordinate system.[3]

Consider the case of 2 views of **O** and **P**. Let $\mathbf{R}_i = (x_i, y_i, z_i)$ be the vector from **O** to **P**, where i is the view index. Correspondingly let $\mathbf{V}_i = (\dot{x}_i, \dot{y}_i, \dot{z}_i)$ be the instantaneous velocity

[3]If **O** is in motion then its velocity vector can be subtracted from both those of **O** and **P**. The resulting conditions are identical in terms of the motion of **P** with respect to **O**, and we can consider the motion of **O** as simply another translational component of the motion of the coordinate system.

vector of **P**. If we assume as given the 2 dimensional projected position, $\mathbf{R}_i^*$, and velocity, $\mathbf{V}_i^*$, of the point **P**, then fixed axis motion allows the use of the constraints and equations in the following table to compute the unknown positions and velocities. Each equation is listed with the constraint which generated it and any new unknowns not introduced by the equations before it..

Constraint	Equation	New Unknowns	
Equal Length	$\|\|\mathbf{R}_1\|\| = \|\|\mathbf{R}_2\|\|$	z_1 , z_2	(1)
Fixed Length	$\mathbf{R}_1 \cdot \mathbf{V}_1 = 0$	$\dot{z}_1$	(2)
Fixed Length	$\mathbf{R}_2 \cdot \mathbf{V}_2 = 0$	$\dot{z}_2$	(3)
Fixed Axis	$\mathbf{V}_2 \cdot [\mathbf{V}_1 \times (\mathbf{R}_1 - \mathbf{R}_2)] = 0$	none	(4)

Note that in each equation in which the velocity vectors appear, the right hand side is equal to zero. Therefore any scalar multiple of the true velocity vectors will also satisfy the equations. Thus, if we are not interested in in solving for the magnitude of the velocities, we can then use the 2-dimensional directions as input, letting $\dot{x}_i = 1$ and $\dot{y}_i = \frac{dy}{dx}$. This allows the hybrid theory to use only the direction of motion without needing the magnitudes.

Although there are as many equations as unknowns using two views of two points, we are not guaranteed a solution unless the equations are independent. We can establish independence by examining the Jacobian matrix which if non-zero guarantees a local solution (Richards, etal. 1981). To compute the Jacobian first we expand the equations from above:

$$(x_1^2 + y_1^2 + z_1^2) - (x_2^2 + y_2^2 + z_2^2) = 0 \tag{1a}$$

$$x_1\,\dot{x}_1 + y_1\,\dot{y}_1 + z_1\,\dot{z}_1 = 0 \tag{2a}$$

$$x_2\,\dot{x}_2 + y_2\,\dot{y}_2 + z_2\,\dot{z}_2 = 0 \tag{3a}$$

$$(\dot{x}_2\dot{y}_1 - \dot{y}_2\dot{x}_1)(z_1 - z_2) + (\dot{z}_2\dot{x}_1 - \dot{x}_2\dot{z}_1)(y_1 - y_2) + (\dot{y}_2\dot{z}_1 - \dot{z}_2\dot{y}_1)(x_1 - x_2) = 0 \tag{4a}$$

The Jacobian matrix, J, is computed by taking the partial differentials of the above equations with respect to the unknown variables z_1, z_2, $\dot{z}_1$, and $\dot{z}_2$:

$$J = \begin{pmatrix} 2z_1 & -2z_2 & 0 & 0 \\ \dot{z}_1 & 0 & z_1 & 0 \\ 0 & \dot{z}_2 & 0 & z_2 \\ c_1 & c_2 & c_3 & c_4 \end{pmatrix} \tag{5}$$

where

$$c_1 = \dot{x}_2 \dot{y}_1 - \dot{y}_2 \dot{x}_1$$
$$c_2 = -c_1$$
$$c_3 = \dot{y}_2(x_1 - x_2) - \dot{x}_2(y_1 - y_2)$$
$$c_4 = \dot{x}_1(y_1 - y_2) - \dot{y}_1(x_1 - x_2)$$

Therefore the determinant of J is:

$$det(J) = 2z_1^2(z_2 c_2 - \dot{z}_2 c_4) + 2z_2^2(z_1 c_1 - \dot{z}_1 c_3) \tag{6}$$

A complete analysis of the above expression is provided in the first appendix. In that appendix the possible degenerate conditions are listed and that list is shown to be exhaustive. There are two classes of conditions in which no solution exists: those caused by the position of the axis of rotation and those caused by viewing the two points at particular points in their path.

Degeneracies

There are two positions of the axis of rotation which are degenerate: *i)* the axis of rotation is parallel to the frontal plane, where the circle traced out in 3-space by the point **P** projects into a line in the image plane;[4] *ii)* the axis of rotation is the line of sight, where the path of **P** is a circle parallel to the image plane. Both of these conditions can be neglected because of a general position argument: we assume that a small change in viewing angle will not qualitatively change the nature of the solution. Also note that in each case the axis of rotation is apparent, only the 3-dimensional structure is ambiguous.

Excluding the above degenerate conditions caused by the position of the axis of rotation, it is still possible for the viewing positions of the points to cause the determinant of the Jacobian matrix to be zero. This condition can occur two ways: *i)* the two views of point **P** occur at the minimum and maximum of depth, corresponding to the minimum points of curvature on the ellipse which is the projection of the path of **P**; *ii)* the two views of point **P** occur at the same depth, $z_1 = z_2$. However, both of these conditions violate the general position condition as a small perturbation in the viewing position will determine whether or not an analytic solution exists.

Finally, it would be possible that there was a "hidden" analytic dependency in the equations whose existence might not be determined in the analysis. However, such dependencies would always be true and would show themselves in any implementation of the solution. As

[4]This is an important special condition for the following reason. If an observer's line of sight is in the same direction as his motion, then any stationary object will appear to rotate about an axis which is parallel to the image plane. However, this case is easily identified from the input data and although the exact depth of the points can not be resolved, it is easy to compute the axis of rotation.

discussed in the next section a successful numerical simulation was executed indicating no such hidden dependencies.

Uniqueness

Since we know that the determinant of the Jacobian matrix is generally non-zero, we are guaranteed to have finite number of solutions to the system of equations $1a - 4a$. However, as mentioned earlier, it is important for a theory which is to recover structure-from-motion to get a unique solution. Therefore, we must solve the system of equations and show whether or not more than one solution is admissible.

When the above system of equations is solved for z_1 (Appendix 2), the resulting equation is of the form:

$$a_1 z_1^4 + a_2 z_1^2 + a_3 = 0 \tag{7}$$

where each a_i is a function of the known inputs of the 2-dimensional positions and velocities. As expected there are no odd powers of z_1; this is because under orthographic projection we cannot distinguish between the two half spaces defined by the frontal plane $z = 0$. However, because equation (7) is a quadratic in z_1^2, we get *two* solutions for z_1^2. If both those solutions are positive, there are two admissible solutions for the depth of z_1, (four, counting reflections) and the solution is not unique.

In fact, it is often the case that both solutions for z_1^2 are indeed positive, leaving an ambiguity to be resolved. This is accomplished by using either another view or another point to constrain the solution. Another point is preferred as the temporal locality is still satisfied and that strategy is the solution chosen in this paper. The exact form of the solution to the above equations as well as a proof of uniqueness using an additional point are left for the appendix. The basic idea presented there is that if three points are used, they can be considered pairwise to eliminate any spurious solutions. Also, by using three points we eliminate the *"false target"* problem: non-rigid points appearing to be moving in a rigid configuration. The introduction of the third point over-constrains the system making the probability of a false target zero.

4. Numerical Implementation and Analysis on Noisy Data

To test the hybrid solution for hidden dependencies as well as to get an estimate of its performance under noisy conditions, a numerical simulation of fixed axis motion was generated. Initially we assumed that the inputs to the system — the 2 dimensional positions and directions of velocity — were exact. As expected, the solution using two views of three points provided an exact solution for the axis of rotation and the relative depths of the moving points. Thus, the analytic solution was confirmed to be valid, with no unexpected dependencies between the variables.

After these initial tests, increasingly larger perturbations of the data were introduced. We chose to perturb the directional component of the input velocity because instantaneous velocities are numerically difficult to compute. The position input was left undisturbed based on the assumption that this information is accurately available.

The perturbations were introduced by adding an angular error to the direction of the 2 dimensional velocity vectors. The errors were normally distributed with a mean of zero and a standard deviation that could be specified for each series of trials. 100 trials were simulated for each selected standard deviation and three statistics were computed: the percentage of trials at which no solution could be found (a square root would be performed on a negative argument); the average angular error between the solved for axis of rotation and the true axis of rotation; the *median* angular error between the solution and the true axis. This last statistic was computed to prevent a few very bad solutions from distorting the apparent performance of the algorithm.

Figure 2 displays the results of the tests. The standard deviation is plotted logarithmically to give a broader view of the data. As shown the theory degrades gracefully both in the number of times in which no solution is found and in the error of those solutions generated. Also, because different sets of points on the same object can be considered independently, the percentage of cases in which no solution is found can be made arbitrarily small by considering additional points. Therefore as long as the standard deviation of error of the measured velocity direction is not very large (less than 15°) reasonably accurate solutions for the axis of rotation can be found. This degree of angular error is within that measured for human vision (Campbell, *etal.*, 1968, Spoerri, *etal.*, 1983).

5. Summary

The computation of the structure of objects from their motion has been investigated. The particular problem solved, fixed axis rotation, has been shown to a useful computation for the human visual system to perform. In order to allow a numerically well behaved implementation, the hybrid theory was generated using numerically stable inputs – position vectors and the direction of velocity vectors – and was successfully tested under noisy conditions. The algorithm proved to be sufficiently robust to make its implementation biologically feasible.

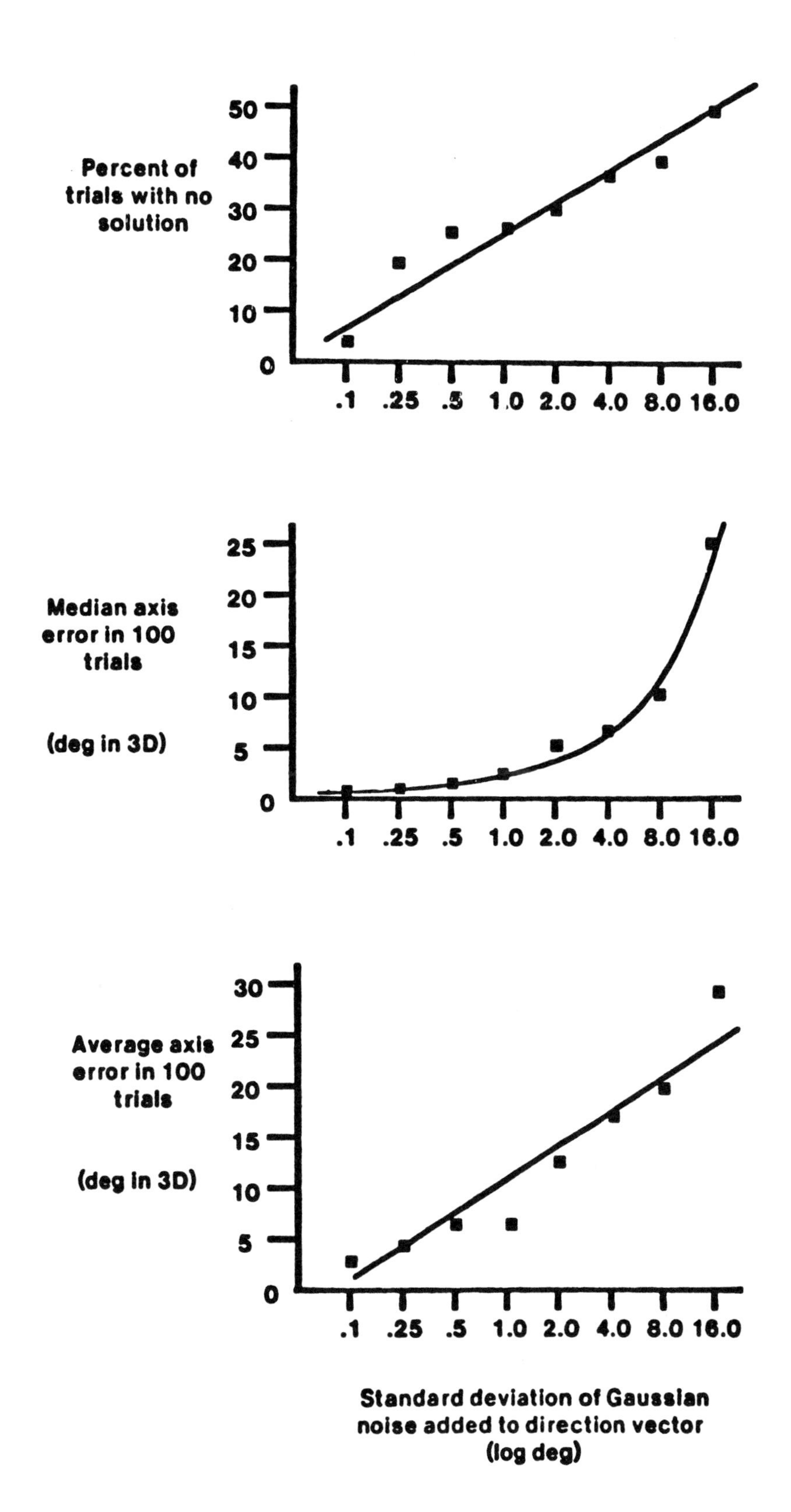

Figure 2. Results of numerical simulations of the hybrid solution of the fixed axis problem. For each standard deviation of noise, 100 trials were conducted to generate the performance data.

Acknowledgments

The author thanks W. Richards for suggesting the problem and providing much aide and incentive. Also thanks to S. Ullman, D. Hoffman, and C. Atkeson for some valuable discussions. This report describes research done at the Artificial Intelligence Laboratory of the Massachusetts Institute of Technology. Support for the laboratory's artificial intelligence research is provided in part by the Advanced Research Projects Agency of the Department of Defense under Office of Naval Research contract N00014-80-C-0505 and by the National Science Foundation and the U.S. Air Force of Scientific Research under grant 7923110-MCS. The author was also supported by NIGMS Training Grant 5T32GM07484

References

Barlow, H.B. and R.W. Levick, "The mechanism of directionally selective units in rabbits retina," *J. Physiology,* (London) Vol. 178, 1965, 477-504.

Bridgeman, B., " Visual receptive fields to absolute and relative motion during tracking," *Science,* Vol. 178, 1972, 1106-1108.

Brown, J.F., "The visual perception of velocity," *Psychological Forsch.,* Vol. 14, 1931, 199-232.

Campbell, F.W., B.G. Cleland, G.F. Cooper, and C.E. Enroth-Cugell, "The angular selectivity of visual cortical cells to moving gratings," *J. Physiology,* (London) Vol. 198, 1968, 237-250.

Green, B.F., "Figure coherence in the kinetic depth effect," *J. Experimental Psychology,* Vol. 62, No. 3, 1961, 272-282.

Hoffman, D.D. and B.E. Flinchbaugh, "The interpretation of biological motion," *Biological Cybernetics,* Vol. 42, 1982, 195-204.

Horn, B.K.P. and B.G. Schunk,*Determining Optical Flow,* MIT A.I. Memo 572, MIT, Cambridge Mass.,1980.

Hubel, D.H. and T.N. Wiesel, "Receptive fields and functional architecture of monkey striate cortex," *J. Physiology* (London), Vol. 195, 1968, 215-243.

Johansson, G., "Perception of motion and changing form," *Scandanavian J. Psychology,* Vol. 5, 1964, 181-208.

Johansson, G. and G. Jansson, "Perceived rotary motion from changes in a straight line," *Perception and Psychophysics,* Vol. 4, No. 3, 1968 , 165-170.

Johansson, G., "Visual perception of biological motion and a model for its analysis," *Perception and Psychophysics,* Vol. 14, No. 2, 1973, 201-211.

Johansson, G., "Visual motion perception,"*Scientific American,* Vol. 232, No. 6, 1975, 76-88.

Kender, J., "Why perspective is difficult: how two algorithms fail," *AAAI Proc. National Conf. on A.I.,* 1982, 9-12.

Longuet-Higgins, H.C., "A computer algorithm for reconstructing a scene from two projections," *Nature,* Vol. 293, 10 Sept., 133-135

Longuet-Higgins, H.C. and K. Prazdny, "The interpretation of a moving retinal image," *Proc. Royal Soc. London,* Vol B, No. 208, 1980, 385-397.

Marr, D. and S Ullman, "Directional selectivity and its use in early visual processing,"*Proc. Royal Soc. London,* Vol B, No.211,1981, 151-180.

Marr, D. and E. Hildreth, "Theory of edge detection," *Proc. Royal Soc. London,* Vol B,1980, 187-217.

Prazdny, K., "Egomotion and relative depth map from optical flow," *Biological Cybernetics,* Vol. 36, 1980, 87-102.

Richards, W.A. and H. Lieberman, "A correlation between stereo ability and the recovery of structure from motion," in press.

Richards, W.A., J. Rubin, and D.D. Hoffman, *Equation counting and the interpretation of sensory data,* MIT A.I. Memo 614, 1981.

Sperling, G., "Movement perception in computer-driven visual displays," *Behavior Research Methods and Instrumentation,* Vol. 8, 1976, 144-151.

Spoerri, A., W.A. Richards, and A. Bobick, " Angular sensitivity for directional selectivity in man," in preparation.

Torre, V. and T. Poggio, "A synaptic mechanism possibly underlying directional selectivity to motion," *Proc. Royal Soc. London,* Vol B, No. 202, 1978, 409-416.

Ullman, S.,*The Interpretation of Visual Motion,* MIT Press, Cambridge and London, 1979.

Ullman, S., "The interpretation of structure from motion," *Proc. Royal Soc. London,* Vol. B. No. 203, 1979, 405-426.

Ullman, S., "Recent computational studies in the interpretation of structure from motion," in Beck and Rosenfeld, eds., *Human and Machine Vision,* Academic Press, 1982, in press.

Wallach, H. and D.N. O'Connell, "The kinetic depth effect,"*J. Experimental Psychology,* Vol. 45, 1953, 205-217.

Webb, J. and J.K. Aggarwal, "Visually interpreting the motion of objects in space," *Computer,* Vol. 14(8), 40-46.

Wurtz, R., "Comparison of effects on eye movements and stimulus movements on striate cortex neurons of the monkey," *J. Neurophysiology,* Vol. 32, 1969, 987 -994.

Appendix 1 - Analysis of the Jacobian Matrix

In order to show when a solution to equations $1a - 4a$ exists, we must be able to completely specify when the determinant of the Jacobian will be zero whenever the fixed axis equations are also satisfied. To do this we will use a geometric interpretation of the coefficients of the equations to show under exactly what conditions they go to zero allowing a degeneracy to occur.

For convenience we repeat equations $1a - 4a$:

$$(x_1^2 + y_1^2 + z_1^2) - (x_2^2 + y_2^2 + z_2^2) = 0 \qquad (1a)$$

$$x_1\,\dot{x}_1 + y_1\,\dot{y}_1 + z_1\,\dot{z}_1 = 0 \qquad (2a)$$

$$x_2\,\dot{x}_2 + y_2\,\dot{y}_2 + z_2\,\dot{z}_2 = 0 \qquad (3a)$$

$$(\dot{x}_2\dot{y}_1 - \dot{y}_2\dot{x}_1)(z_1 - z_2) + (\dot{z}_2\dot{x}_1 - \dot{x}_2\dot{z}_1)(y_1 - y_2)$$
$$+(\dot{y}_2\dot{z}_1 - \dot{z}_2\dot{y}_1)(x_1 - x_2) = 0 \tag{4a}$$

as well as the equation for the Jacobian:

$$det(J) = 2z_1^2(z_2c_2 - \dot{z}_2c_4) + 2z_2^2(z_1c_1 - \dot{z}_1c_3) \tag{6}$$

where

$$c_1 = \dot{x}_2\dot{y}_1 - \dot{y}_2\dot{x}_1$$
$$c_2 = -c_1$$
$$c_3 = \dot{y}_2(x_1 - x_2) - \dot{x}_2(y_1 - y_2)$$
$$c_4 = \dot{x}_1(y_1 - y_2) - \dot{y}_1(x_1 - x_2)$$

If we assume that both z_1 and z_2 are non-zero, then using $2a$ and $3a$ we can substitute for the derivative terms $\dot{z}_i$:

$$det(J) = z_1^2z_2^2(z_2 - z_1)c_1 + z_2^3(x_1\dot{x}_1 + y_1\dot{y}_1)c_3 + z_1^3(x_2\dot{x}_2 + y_2\dot{y}_2)c_4 = 0 \tag{A1}$$

The first condition under which the value of the Jacobian could be zero is if the coefficients of the z_i terms are zero. We can examine the coefficients individually to see what conditions are necessary for them to be zero.

Note that c_1 can be interpreted as the dot product $\mathbf{V}_1^* \cdot \mathbf{V}_2^{*\perp}$. This term is zero only when the two 2-dimensional velocity vectors are in the same direction. Since the path of **P** in 2 dimensions is an ellipse,[5] this occurs only when the two views of **P** occur at exactly opposite points on the path of **P**.

However, c_3 has the geometric interpretation of $-(\mathbf{R}_1^* - \mathbf{R}_2^*) \cdot \mathbf{V}_2{}^{\perp}$ which is never zero, since a cord of an ellipse is never tangent at the point of intersection. Likewise, for c_4. Therefore, for the second and third term of the Jacobian to be zero, the quantities $(x_i\dot{x}_i + y_i\dot{y}_i) = \mathbf{R}_i^* \cdot \mathbf{V}_i^*, i = 1,2$ must be zero. Since the path of point **P** projects into an ellipse in two dimensions, the dot product of the position and velocity vectors are only zero when **P** lies on one of the axis of the ellipse.[6]

[5] If the axis of rotation is parallel to the image plane then the projection of the path of **P** is a line. In this case the velocity vectors are all in the same direction and therefore the coefficients c_1 and c_2 will be zero. Also, the vector $\mathbf{R}_1^* - \mathbf{R}_2^*$ will be in the same direction as the velocities causing coefficients c_3 and c_4 to be zero. Therefore, this is a degenerate condition under which the hybrid theory is not applicable. It is however an easy task to solve for the axis of rotation under such conditions.

[6] This dot product would always be zero if the projection of the path of **P** were a circle instead of an ellipse; this case would arise if the axis of rotation was parallel to the line of sight. However, when that is the case all views of **P** occur at the same depth. As will be shown, if the two views of **P** occur at the same depth, the hybrid solution is degenerate.

Combining the two above conditions, we see that the coefficients of the unknowns z_i of the Jacobian are analytically zero *only* when the two views of **P** occur on the same axis of the ellipse. In three dimensions this corresponds to when either i) one view of **P** is at the maximum depth and the other is at the minimum, or ii) the two views of **P** are at the depth at which z component of the velocity vector is maximum.

If we assume the coefficients of the Jacobian are non-zero, then its value can only be zero if the values of the unknowns z_1, z_2 satisfy equation A1. However, the values of the z_i are also constrained to satisfy equations $1a$ – $4a$. We will show that there is exactly one condition under which all five equations will be satisfied.

Let us re-write equations 1a and 4a as follows using equations $2a$ and $3a$ to substitute for the velocity terms $\dot{z}_i$:

$$z_1^2 - z_2^2 + (x_1^2 + y_1^2 - x_2^2 - y_2^2) = 0 \tag{A2}$$

$$z_1 z_2(z_1 - z_2)c_1 + z_2(x_1\dot{x}_1 + y_1\dot{y}_1)c_3 + z_1(x_2\dot{x}_2 + y_2\dot{y}_2)c_4 = 0 \tag{A3}$$

Combining the above two equations with A1 we have three non-linear equations in two unknowns. In order for all three equations to be satisfied one of the following conditions must be satisfied: *i)* One of the equations is a combination of some multiple of the others;[7] *ii)* The three equations share a common factor; *iii)* There is a degenerate condition under which some of the coefficients become zero causing the equations to be satisfied.

If either condition *i* or *ii* were true, they would *always* be true. That is, there would be no case in which one could generate values from a fixed axis rotation simulation in which the Jacobian would be non-zero. This however cannot be true since successful numerical simulations were computed indicating that the equations have no analytic dependencies.

By examination of equations A1–A3 and based on the above discussion of the zero conditions of the coefficients it is clear that the only case in which the equations can collapse is when $z_1 = z_2$. In this case the terms involving c_1 disappear, c_3 becomes equal to c_4, and $\mathbf{R}_1^* \cdot \mathbf{V}_1^* = -\mathbf{R}_2^* \cdot \mathbf{V}_2^*$. Except for the condition in which the two views of **P** occur at the same depth, all the coefficients of the three equations are non-zero; therefore no other values of z_1 and z_2 which satisfy the fixed axis equations will cause the Jacobian to be zero.

To summarize, there are two classes of conditions in which the Jacobian of the fixed axis equations is zero. The first class contains the degenerate positions of the axis of rotation: *i)* The axis of rotation is parallel to the image plane causing the path of **P** to project into a straight line, making all the coefficients of the Jacobian zero (see footnote); *ii)* The axis

[7]Note that this combination need not be linear since these are non-linear equations.

of rotation is parallel to the line of sight such that all views of **P** are at the same depth. These are the only degenerate axis positions.

The second class of degeneracies are those pairs of viewing positions of **P** for which no solution exists. There are also two of these: *i)* The two views of P occur at the minimum and the maximum depth of the path of **P**; *ii)* The two views of **P** occur at the same depth.

Note that all of the above degeneracies violate a general position argument and occur with probability of zero (a small perturbation of viewing position or the exact location of **P** on its path will allow an analytic solution). Therefore we can conclude that in general the hybrid theory solution to the fixed axis can be applied.

Appendix 2 - Proof of Uniqueness

To solve the fixed axis case, we first solve the set of equations generated by two views of two points. We then show that by using this solution and by considering points pairwise, a unique solution can be generated.

For convenience we repeat the equations of section 3:

$$(x_1^2 + y_1^2 + z_1^2) - (x_2^2 + y_2^2 + z_2^2) = 0 \qquad (1a)$$

$$x_1\,\dot{x}_1 + y_1\,\dot{y}_1 + z_1\,\dot{z}_1 = 0 \qquad (2a)$$

$$x_2\,\dot{x}_2 + y_2\,\dot{y}_2 + z_2\,\dot{z}_2 = 0 \qquad (3a)$$

$$(\dot{x}_2\dot{y}_1 - \dot{y}_2\dot{x}_1)(z_1 - z_2) + (\dot{z}_2\dot{x}_1 - \dot{x}_2\dot{z}_1)(y_1 - y_2) + (\dot{y}_2\dot{z}_1 - \dot{z}_2\dot{y}_1)(x_1 - x_2) = 0 \qquad (4a)$$

Solving these equations for z_1 yields the following expression:

$$z_1^4[k_3^2k_6 - 2k_1k_3k_4 + 2k_2k_3k_5] + z_1^2[k_3^2k_6^2 + 2k_2k_3k_5k_6 + k_2^2k_5^2 - 2k_1k_3k_4k_6 - k_1^2k_4^2] - k_1^2k_4^2k_6 = 0 \qquad (A4)$$

where

$$k_1 = x_1\dot{x}_1 + y_1\dot{y}_1$$
$$k_2 = x_2\dot{x}_2 + y_2\dot{y}_2$$
$$k_3 = \dot{x}_2\dot{y}_1 - \dot{y}_2\dot{x}_1$$
$$k_4 = \dot{x}_2(y_1 - y_2) - \dot{y}_2(x_1 - x_2)$$
$$k_5 = \dot{x}_1(y_1 - y_2) - \dot{y}_1(x_1 - x_2)$$
$$k_6 = x_1^2 + y_1^2 - x_2^2 - y_2^2$$

As previously discussed, the solution is a quadratic in z_1^2. Because we are using orthographic projection, we cannot distinguish between mirror reflections; thus we expect to have no odd powers of z. However, if both solutions for z_1^2 are positive, there are two possible solutions for the depth and the axis of rotation. Therefore we need either another point or another view to determine the structure uniquely. In order to preserve temporal locality and because of the assumed availability of additional points, another point is selected.

Two views of three points

Consider two views of three points (O, P, Q) rotating about a fixed axis $\vec{\omega}$ with O at the origin. If we first consider just the pair of points O and P, we know from the above analysis that *at most* two solutions will be generated for $\vec{\omega}$ (and therefore for the path of **P**). If indeed we get two possible solutions for $\vec{\omega}$, one is correct and the other is "false." Since at this point we have no way of knowing which solution is correct, label the two solutions for $\vec{\omega}$ as $\vec{\omega}_{P_1}$ and $\vec{\omega}_{P_2}$.

Next consider the two views of O and Q. Agai1 we will either get a unique solution, $\vec{\omega}_Q$ (in which case since $\vec{\omega} = \vec{\omega}_Q$ the problem is solved) or two solutions will be generated, $\vec{\omega}_{Q_1}$ and $\vec{\omega}_{Q_2}$ We know that at least one of the $\vec{\omega}$'s generated by P will match one of the axis generated by Q since both of those will correspond to the true axis. In order to show uniqueness, we need to show that the probability that the other pair of axis generated by P and Q will be equal is zero.

To do this, we will show that once the first viewing position of Q is specified, there is at most one other viewing position of Q that will generate a false solution for $\vec{\omega}$ identical to the false solution generated by the two views of P. Finally we will show that there is a probability of zero that the second view of Q occurs in that one position.

Adopt the following notation which is illustrated in Figure 3:

$\vec{\omega}$	$=$	The true axis about which **P** and **Q** are rotating
C_P, C_Q	$=$	The paths traced out in 3 space by **P** and **Q**. C_P and C_Q are circles. Their projections, C_P^* and C_Q^*, are therefore ellipses.
$\mathbf{P}_1, \mathbf{P}_2$	$=$	The two positions on C_P at which the two views of **P** occur.
$\mathbf{Q}_1, \mathbf{Q}_2$	$=$	The two positions on C_Q at which the two views of **Q** occur.
C_{P_f}	$=$	The false path specified by the false solution for $\vec{\omega}$. $C_{P_f}^*$ intersects and is tangent to C_P^* at $\mathbf{P}_1^*$ and $\mathbf{P}_2^*$ since the position and velocities at these points are known.
$\vec{\omega}_f$	$=$	The false axis corresonding to C_{P_f}.

Figure 3 illustrates the case where two views of **P** yielded two solutions for $\vec{\omega}$ and therefore two solutions for the path of **P**: C_P and C_{P_f}, the correct and false solutions respectively. The question which must be answered is "What is the probability that a given $\mathbf{Q}_1^*$ and $\mathbf{Q}_2^*$ on C_Q^* will yield a false solution C_{Q_f} such that we will not be able to distinguish between $\vec{\omega}$ and $\vec{\omega}_f$?"

In order for us not to be able to determine the true axis the following must be true:

(i) The false ellipse, $C_{Q_f}^*$, must have its minor axis lie on $\vec{\omega}_f^*$.

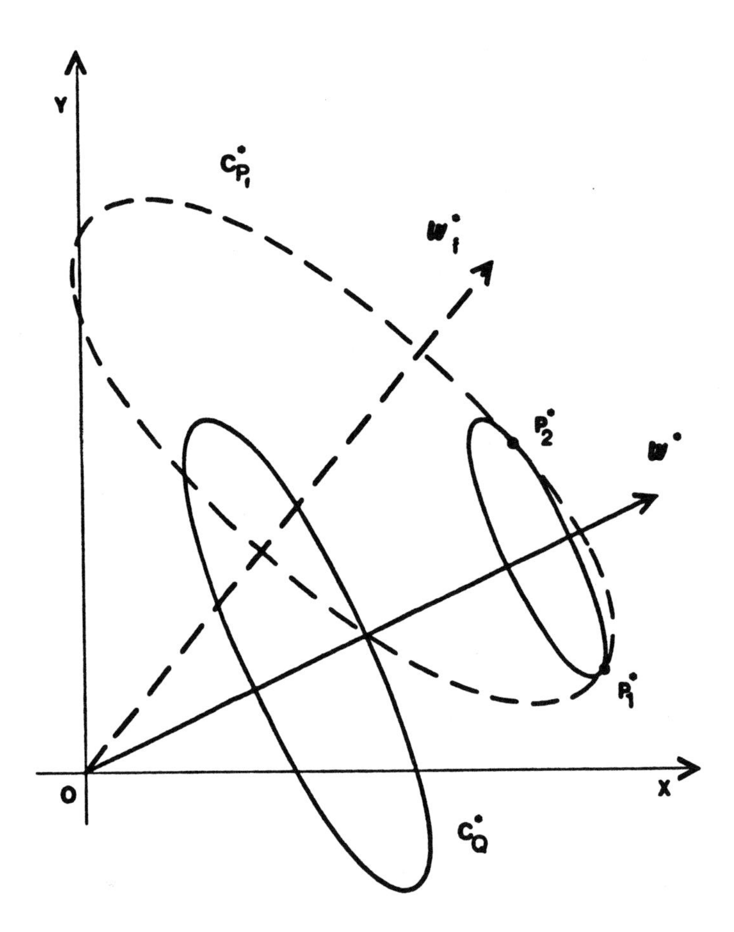

Figure 3. Illustration of when two views of **P** gives two possible interpretations for $\vec{\omega}$.

(ii) The eccentricity of $C^*_{Q_f}$ must equal that of $C^*_{P_f}$. This is true because in the true solution, the ellipses are circles projected at the same slant. For the false solution not to be discriminable the two false ellipses must have the same eccentricity.

(iii) $C^*_{Q_f}$ must intersect and be tangent to C^*_Q at $\mathbf{Q}^*_1$ and $\mathbf{Q}^*_2$.

We can now consider the form that the equation for $C^*_{Q_f}$ (in two dimensions) must have to satisfy the above criteria.

For simplicity, we rotate the coordinate axis such that the new x-axis, x', lies along $\vec{\omega}_f$ (See Figure 4).

In this new coordinate frame (x', y'), the equation for $C^*_{P_f}$ is of the form

$$\frac{y'^2}{a_0^2} + \frac{(x'^2 - h_0)^2}{b_0^2} = 1 \tag{A5}$$

Assume that $C^*_{Q_f}$ has some eccentricity $e_0 = \frac{c_0}{a_0}$ where $b^2 = a^2 + c^2$. Therefore for $C^*_{Q_f}$ to satisfy the first two above conditions, namely lie on the same minor axis and have the same

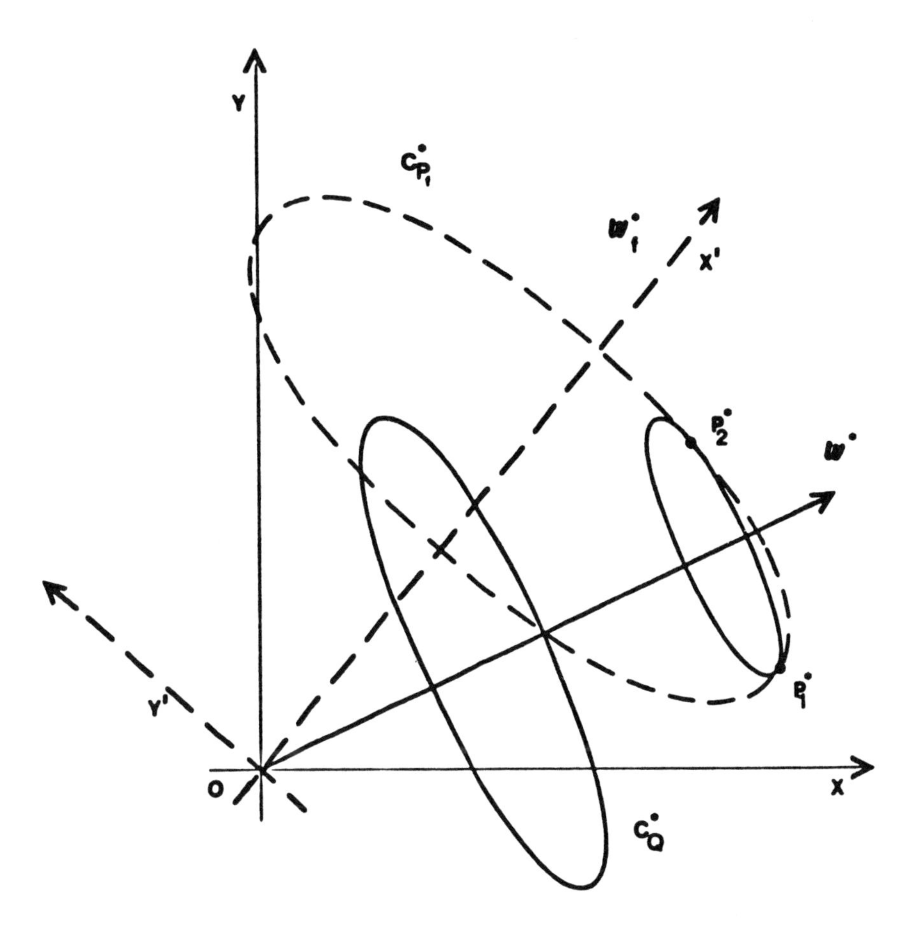

Figure 4. Coordinate axis rotated to align with the projection of the false solution for $\vec{\omega}$.

eccentricity as $C^{*}_{P_f}$, its form must be

$$\frac{y'^2}{a^2} + \frac{(x'^2 - h)^2}{a^2(1 + e_0^2)} = 1 \tag{A6}$$

where the only free parameters are a and h.

Letting $k_0 = (1 + e_0^2)$ we can rewrite equation A6 as:

$$k_0 y^2 + x^2 - 2xh + h^2 = a^2 k_0 \tag{A7}$$

(The prime superscripts have been dropped for clarity; we are still considering the rotated coordinate frame.) Also, we can take the derivative of the equation with respect to x to get:

$$2k_0 \frac{dy}{dx} + 2x - 2h = 0 \tag{A8}$$

At this point, let us consider the 2-dimensional position and velocity of Q in the first view. Let us notate the position as (x_1, y_1) and the velocity as $(\dot{x}_1, \dot{y}_1)$.[8] If we now substitute these

[8]Remember that if we are not concerned with velocity determination we can simply measure the direction of the motion, $\frac{dy}{dx}$, and set $\dot{x}_i = 1$ and $\dot{y}_i = \frac{dy}{dx}$.

values into equations A7 and A8 (using $\frac{dy}{dx} = \dot{y}_1/\dot{x}_1$) we can solve explicitly for a and h:

$$h = k_0 y_1 (\frac{\dot{y}_1}{\dot{x}_1}) + x_1 \tag{A9}$$

$$a = \left[\frac{\left(k_0 y_1^2 + (x_1 - (k_0 y_1 (\frac{\dot{y}_1}{\dot{x}_1}) + x_1))^2 \right)}{k_0} \right]^{1/2} \tag{A10}$$

With a and h now known, the possible false path for Q and its projection $C^*_{Q_f}$ are completely specified. It has been constructed such that C^*_Q and $C^*_{Q_f}$, the projections of the true and false paths of Q respectively, intersect and are tangent at Q^*_1. We also know that there is *at most* one other point on C^*_Q at which the two ellipses could intersect and be tangent. This is true because three points, given position and tangent vectors define a unique ellipse. Therefore, this one point, if it exists, must be Q^*_2 for all three of the above mentioned criteria to be true.

If we now allow the second view of Q to occur anywhere in its path, then the probability that it will be in the one exact position which would be tangent to the false ellipse is zero. However, suppose we require the angular distance between Q_1 and Q_2 to be equal to the angular distance between the two views of P; this is the case if the object is rigid. Then once Q_1 is specified, the position of Q_2 is fixed. We therefore need to know that this constraint does not create any analytic dependencies which will force the second view of Q to be in the one spot which will not allow a determination of the true solution.

At this point we can use the same type of argument provided in the appendix which discusses the Jacobian matrix. If there were any analytic dependencies, then they would always be true. But since the numerical simulations demonstrate the existence of the solution, we know that in general the two views of Q do not yield the same false solution for $\vec{\omega}$ as that generated by the two views of P. Therefore, we can say that in general two views of three points gives a unique solution to the fixed axis problem.[9]

Note that at this point we have also eliminated the *"false target"* problem: non-rigid points in motion appearing to be in a rigid configuration. Recall that if two solutions for $\vec{\omega}$ were generated for each pair of points, (O,P) and (O, Q), we were guaranteed that one of the axis of the first set would match one from the second set, since these would correspond to the true axis of rotation. If however the points O, P, and Q were not rigidly linked, then the probability that there would be an agreement between the two sets of possible axes is equal to the probability that two randomly chosen vectors point in the same direction, namely zero. Thus one could quickly identify such non-rigid cases.

[9]The one thing a proof of this type cannot provide is an exhaustive list of the specific degenerate conditions, special cases in which the solution breaks down. The only such case determined by this author to date is if the three points O, P, and Q are colinear. In this case, the two ellipses generated by the views of Q are just scaled versions of the ellipses generated by P and there would be no way to disambiguate between the two solutions.

Multicomputer Architectures for Real-Time Perception

Leonard Uhr

Computer Sciences Department, Univ. of Wisconsin

Introduction

This paper examines the computing demands that must be met by a system capable of scene description and perception of real-world moving objects. A brief survey is made of the major different kinds of computer systems that have been built, or designed, and of the different sources of potential speed-up of processing that have been exploited. Finally, a number of alternative possible hardware architectures that might be capable of handling real-time perception of moving objects are suggested, and examined.

The Demands on a System for Real-Time Perception of Moving Objects

Real-world objects can move at any speed (up to the speed of light). And arbitrarily large numbers of objects might be moving, in the same scene, in arbitrarily different directions and manners. We might pose the "general problem of the perception of motion" as one of recognizing any number of objects moving in any number of trajectories at any constant or changing speeds. Or we might restrict the problem to only a manageable few objects moving no faster than a fixed maximum speed.

But it seems most reasonable to ask that the system handle those objects, moving at those speeds, that a human being can perceive; for example, a galloping horse but not a bullet. This is not to insist upon modelling human perception, but rather to point out that the "general" problem is potentially infinite in its size, speed, and complexity, and that we must cut it down to reasonable proportions.

The Enormous Computing Burden to Perceive Objects in Motion

Most researchers developing "vision systems" today pose their ultimate problem as one of handling a scene on the order of 250 by 250, 500 by 500, 1,000 by 1,000 or 4,000 by 4,000 "pixels" (picture elements), as resolved by a television camera. For simplicity, let's assume the scene is resolved in a 1,000 by 1,000 array, giving 1,000,000 pixels, each containing an 8-bit value (representing color or grey-scale intensity). This is an extremely large amount of information. But not when compared with the human retina's roughly 10,000,000 cones (for shape and color vision) and 100,000,000 rods (for change, motion, and grey-scale vision).

To date only a few attempts have been made to have a computer program try to recognize and describe the objects in a full television picture (usually on the order of 500 by 500). These programs have taken from 5 or 10 minutes to 10 or 20 hours of CPU time on rather large computers (e.g., Univac 1110; DEC PDP-10). From these times (for programs that are just barely beginning to handle a single static scene) we can extrapolate that several hundreds, or even thousands, of hours of CPU time might be needed to process the 30 television frames that a human being is able to perceive in a single second. Reddy has suggested that for segmentation alone 1,000 processes might be needed per pixel, and therefore from 1 to 10 billion instructions per second (BIPS) are needed to segment a single scene in one second.

Motion: Representation and Perception
N.I. Badler and J.K. Tsotsos, Editors
Published by Elsevier Science Publishing Co., Inc.

Some Possibilities of Speed-Ups Using Parallel Programs

We can use the more parallel(-serial) programs that have been designing with parallel computers specifically in mind, such as the "pyramid" and "cone" systems that I, and a number of other researchers, have been developing (Hanson and Riseman, 1974, 1978; Klinger and Dyer 1974; Levine and Leemet, 1976, Levine 1978; Tanimoto, 1976, 1978; Uhr, 1972, 1974, 1976, 1978; Uhr and Douglass, 1979; Douglass, 1977, 1978; see Tanimoto and Klinger, 1980, Rosenfeld, 1983) to give alternate estimates of the time needed. It appears that from 5 to 50 transforms of the sort that a suitably designed multi-computer network of parallel processors could execute in a single machine instruction might be needed in each of from 10 to 50 layers. This would need a sequence of from 50 to 2500 instructions. Thus, for example, if each processor took 10 microseconds per instruction such a system would need only .50 to 25 milliseconds for an entire (static) scene. (In sharp contrast, large amounts of time can easily be eaten up by the serial processes used by most other kinds of vision systems.)

Empirical Evidence that Human Perception is Highly Parallel

The human perceptual system throws some extraordinarily interesting light on this problem. The "basic meaningful instruction time" for the brain appears to be from 1 to 2 milliseconds, if one accepts the following simple and to my mind rather compelling argument:

The retina is excited by light energy, sending trains of impulses through the neurons of the visual system and the cerebral cortex. The synaptic junctions between neurons compute functions of the impulses firing into them, and fire out their results. That is the underlieing microstructure with which the brain "perceives" and "thinks." Therefore the time needed for a synaptic firing, plus the time needed for a neuron to carry the resulting information to the next synapse, can be thought of as the "basic meaningful instruction time" of the brain.

Now a synapse takes about 1.5 milliseconds to do its job, and an impulse moves along a neuron in a time appreciably smaller than that. So 1.5, or, to be conservative, 1 to 2 milliseconds, is a quite reasonable estimate. Note how slow this is compared to a computer.

Psychologists have also collected extensive data on the time a human being needs to recognize and perceive objects in a wide variety of different scenes. Without bothering to pin this figure down precisely, it lies somewhere between 50 or 100 milliseconds and 1 second.

This gives us a quite amazing result: <u>The human brain perceives with a serial sequence of processes from 25 (or fewer) to at most 1,000 deep</u>!

A television camera typically takes a new picture every 30 milliseconds. This is necessary to give a smooth flowing picture of moving objects, because of the "critical flicker fusion frequency" of the human visual system which (depending upon illumination, resolution, type of object, and a variety of other factors), lies in the range of from 10 to 30 static images per second.

This suggests that each new static picture might have to be processed in 30 milliseconds, which would force us to lower the lower extreme given above to <u>only 15 to 30 serial processes when scenes of moving objects must be perceived</u>. It seems likely, however, that a well-designed visual system (including living visual systems) would not need to process each new scene in its entirety, but rather would use what it had gathered from the moving scene of the most recent past to direct, and drastically cut down on, its processing.

There are at least two additional factors that are almost entirely conjectural:

1) How complex is each individual process in the brain? We know that hundreds, or often many thousands, of neurons synapse on a single neuron. So, conceivably, the functions a neuron computes could be very complex indeed. But somehow it seem intuitively more plausible to assume that much of this is a matter of redundancy and stabilization, and that the functions aren't all that complex - that the degree of parallelness in the function computed by the individual neuron is on the order of 5 or 10, or a few hundred at most. Indeed there is experimental evidence that cortical neurons onto which, typically, 30,000 or so neurons synapse, will fire when only two or three of those neurons fire into them.

2) How much can previous knowledge cut down on the processing needed for each new input frame, and how great a speed-up in processing can this effect? Here very little is known. Estimates

must wait until we have developed computer systems that exhibit some of this speed-up, while performing reasonably well. But it seems unlikely that significantly more than one order of magnitude speed-up could be attained, and it is not unlikely that there might be little or no speed-up at all, since this is hard to achieve in parallel systems.

The Major Types of Parallel Systems Built or Proposed

Clearly, computers must handle scenes much faster than presently possible, to hope to succeed at real-time real-world vision. The alternatives that have been proposed include faster processors, concurrent processors, networks, scanners, pipelines, and arrays. The following examines these briefly:

Speeding up Processors by Brute Speed of Faster Technologies

Today's technology has given us machines that execute meaningful picture-processing instructions in from 100 nanoseconds to 10 microseconds. A serial computer must iterate such an instruction; a parallel array need execute it only once, at each and every cell of the array.

It is conceivable that we will see cryogenic computers with speed-ups on the order of two, or even three, orders of magnitude during the next ten years. But that will be the end, since we will have hit the limiting speed of light. And it seems likely that these will depend upon expensive, bulky and limited (in size) computers, since they will have to be kept within an environment that maintains their temperatures close to absolute zero.

Speeding up Processing by Improving Architecture

We can, in theory, speed up processing to whatever degree we desire, by building special architectures that compute arbitrarily more complex functions in parallel. That is, any serial program (computer) can be converted into a parallel program (computer) - if only by building a system with lookup tables for every possible input state.

But this is an exponentially explosive procedure that quickly grows out of the bounds of reasonable economy and well-balanced architecture. Today we see people building machines capable of parallel fetches of 9 (Duff's CLIP4, 1976), or even 25 (Kruse and Danielsson's PICAP II, 1980) nearest-neighbors (and of course for arithmetic operations 32 or 64 bit numbers are often fetched in parallel).

But it seems unlikely that anybody will ever want to build systems that would fetch thousands, or even a few hundred, pieces of information in parallel in order to compute complex structural functions over them. Thus we can expect to get one order of magnitude increases in speed from parallel input of the separate components of a picture processing function. But it seems unlikely that we can get two, and certainly not more than two.

When arithmetic is appropriate (as in handling grey-scales, intensities, and weights) we can get roughly 10, 30 or 60 fold improvements. (This is already done in serial computers, which fetch and transform up to 64 bits of numerical information in parallel. But the large parallel arrays of today work with only 1 bit of information at a time; otherwise the cost of hardware would be excessive, since thousands of processors are involved.)

By building a special-purpose scanner we can increase speed by 2 to 10 times. Appropriate architectures for the whole system, encompassing the buffer memory that contains the tv image of the scene, the large array (or scanner), and a general-purpose computer, can also increase speed considerably. E.g., Kruse's (1976, 1978) PICAP I may well gain one order of magnitude, or more, in speed from a specially designed scanner, appropriate hardware for picture-processing operations, and a capability to scan over a sample of the large (raw or transformed) image as a function of what the program has found out so far.

Still another very interesting and powerful system can, potentially, gain 100-fold in speed because of its pipeline of 113 processors - Sternberg's Cytocomputer (Sternberg, 1978, 1980). But it seems unlikely that such a long pipeline can be fully exploited unless it has very large and expensive temporary memories associated with it to store intermediate results.

Networks of Concurrent Processors

Rather than on the one hand using one traditional single-CPU serial computer, or on the other hand using what to many people appear to be unacceptably specialized large arrays or scanners, we can try to couple a number of individual computers (each with one CPU and at least a little bit of its own memory) into a network of processors. [Note that all three of these alternatives are actually "general-purpose" in the sense that they all are equivalent to "universal Turing machines" and therefore, potentially, they can all compute exactly the same set of functions, given enough time. But the crucial problem is one of efficiency: Are they fast enough? Are they as cheap as possible?]

Today people have designed or built networks of 5, 16 (e.g., Wittie and van Tilborg, 1980; Despain and Patterson, 1978) or at the most 50 (Swan et al., 1977) or 80 (Manara and Stringa, 1981) processors of this kind. [See Wittie, 1976, 1978; Sullivan et al., 1977, Goodman and Sequin, 1981; Uhr, 1983b for descriptions of networks that might have thousands, or millions, of processors.] A large variety of different interconnection patterns have been proposed that might, potentially, allow thousands, or even millions, of processors to be connected together.

But the problems of developing operating systems, programming languages, and, hardest of all, actual programs that make efficient use of very large numbers of processors of this sort appear to be extremely difficult, and relatively little progress is being made. Results with the small networks already built, like C.mmp (Fuller, 1976) and Cm* (Swan et al., 1977) at Carnegie-Mellon, suggest that 5 or 15 processors will speed up processing only 2 or 4 times, and then only on programs that are highly parallel, where there is virtually no need to pass messages between processors.

These first early results may merely reflect the difficulty of the problems of designing, building and programming networks, rather than revealing any inherent and fundamental limitations. But for the moment there is little reason to count on large increases in speed from large numbers of processors connected in a typical network. One order of magnitude improvement seems possible, two orders of magnitude seems quite unlikely.

Promising Alternative Multi-Computer Architectures

Increases in brute speed of the hardware, increases in the number of bits of information that the individual processor can fetch in parallel, improvements in the architecture of the processors so that they are designed as appropriately as possible for picture processing operations, pipelining, and improvements in the design of the overall system can all effect significant increases in speed. But the only way we can continue to increase speed, with no potential absolute limits (other than the number of particles in the universe) is by adding more processors; and the best way to handle profligate increases in processors appears to be with relatively simple, regular and well-structured architectures.

During the next few years judicious combinations of improvements from this large set of sources of possible speed-up may well prove to be sufficient. Kruse's PICAP is an unusually successful, and instructive, example of how to gain enormous amounts of power and efficiency at a relatively cheap price by careful design not only of the scanning processor but also of the total system. And networking together 10 or 20 much-faster-technology general purpose computers (as Binford and Reddy have suggested doing at Stanford and at Carnegie-Mellon) may give similar improvements of from 2 to 4 orders of magnitude in speed. This suggests that combining all these sources of speed-up might, conceivably, give 4 to 8 orders of magnitude improvements.

But the true parallel array (which will probably have to sacrifice some, but by no means all, of these different subsidiary types of improvements, because their increased costs would become excessive when multiplied by the thousands or millions of processors now involved) appears to be the only way to continue to increase speed at the same time that increasingly larger scenes are processed.

And a pyramid of arrays, or some other 3-dimensional structure of 2-dimensional arrays, appears to give substantial further improvements in coordinating the flow of information.

Very large arrays and pyramids can come in several models, and take advantage of at least some of the other potential sources of speed-up, as follows:

A) A pure array would have one processor assigned to each pixel in the raw input scene, thus giving the most extreme form of parallelism. If we want to contemplate numbers as high as 100,000,000 (to model the 100,000,000 rods in the human visual system) each individual processor will have to be as simple and as cheap as possible, and there would seem to be little opportunity to take advantage of other sources of speed-up from very fast expensive technology, wider parallel input busses, or elaborately designed architectures. But, as we see in the CLIP arrays (Duff, 1976, 1978), near-neighbor logical operations can be built into the hardware at relatively little expense.

It seems unlikely that anybody would seriously contemplate building 100,000,000 computers (unless a technology developed that made this very simple and cheap, e.g., by growing crystals). And since such a system would be inefficient at the later stages of perception it would almost certainly be combined with other types of resources, as discussed above.

More plausible would be a smaller array of this sort that itself was scanned over the larger retinal array. For example, we can today conceive of building a 1,000 by 1,000 CLIP-type array (by 1985 this should be quite possible at a cost appreciably less that $1,000,000). This would need only 100 iterations to scan the larger 100,000,000 rod array. Similarly, even when we want to process smaller pictures (as will usually be the case) it will often be preferable to have a small array scan over the larger picture array. Thus Duff is building a new 500 by 20 CLIP4-type array that will scan over a 500 by 500 television image.

We can assume that this type of array will perform its basic operation in one microsecond or less rather than the 11 microseconds needed by CLIP4. The speed demanded by the particular range of programs that the system will be expected to execute will determine how many such iterations can be tolerated. But only after we have gained experience in writing such programs will we have a clear idea of how deep their serial sequence of processes might be. For now we have only the rough estimates given above - that if we can find the right programs they should have a serial depth of only a few thousand, or even a few hundred, instructions.

B) A large array might be built from conventional off-the-shelf micro-computers, e.g., z80s or M68000s. It seems reasonable to assume that these will be of the same technologies, price and speed as the specially-designed CLIP, DAP or MPP chips. The differences would be that these array chips today have 4 or 8 computers rather than only one. And in 5 or 10 years, since each computer needs only 1,000 or fewer gates, each chip might have 1,024 or even more computers using improved VLSI technologies that are expected to fabricate 10^6 10 or even more transistors on each chip. The CLIP processors are more appropriately designed for picture processing (e.g., with near-neighbor fetches in parallel and appropriate logical operations built into the hardware); but they have much less memory and a much smaller total repertoire of hardware-embodied instructions. In contrast, a traditional computer will always need at least one chip, since to fill its very much larger requirements for memory (see Uhr, 1982, for an examination of these issues). We might therefore expect an 8 to 1000-fold decrease in costs from an array chip with many computers. In addition, the more appropriate design of the array processors would further increase speed and power.

On the other hand, the larger memory and wider range of instructions (especially including arithmetic instructions) with the more conventional micro-processor would, when they were useful, lead to significant improvements on their side. And it might well be that in 5 or 10 years "conventional" micro-processors will be designed to do more and more parallel operations.

C) A somewhat (probably appreciably) smaller array might be built from bit-slice processor chips. These are, today, often used as building blocks for very fast and powerful large CPUs. They are relatively easily configured and constructed into processors, and, because they are micro-coded, they can be given a wide variety of carefully tailored machine language instructions. A typical processor might need roughly 20 chips (PICAP I has about 45), and cost, roughly $200 to $1000 (for raw hardware). If many processors were built the design costs per processor would go down appreciably.

But nobody appears to have built a system of this sort with more than 16 processors, or even contemplated one with more than 1000 or so (Siegel et al., 1979; Bogdanowicz, 1977; Briggs et al., 1979). Probably $500 to $2000 per processor is a reasonable estimate for the cost in an array of from 50 to 200 processors built from off-the-shelf ICs. Such a processor might well be 10 times faster than either of the above alternatives in basic single-processor speed.

This would appear to be a very interesting alternative when the size of the image to be processed is reasonably small. For example, a 16 by 16 array of bit-slice processors might scan a 96 by 96 array of pixels by having each processor scan a 6 by 6 sub-array. This might well be done at virtually the same speed as the 96 by 96 CLIP array. It would probably cost appreciably more. Its main advantage would appear to be its wider range of possible instructions and its great (potential) flexibility (if it can be micro-coded and programmed appropriately) to effect a wide variety of processes, including more traditional MIMD network processes.

D) A very powerful and appropriately designed single scanner would be far slower than the three alternatives described above. Kruse's PICAP is an already built and tested example of such a system. It needs one microsecond per pixel (using a faster technology than that of CLIP4, but that is quite reasonable since only one processor is involved). It will inevitably need increasing amounts of time as the size of the picture to be processed grows. For a 100 by 100 it may well need only 100 times as much time, gaining 100-fold in speed because of the good architecture and overall design features touched on above. But for a 1,000 by 1,000 it would need 10,000 times as much time.

Such a system might be extended by making the single processor into a pipeline of processors (as Sternberg has done in the Cytocomputer). But as the pipeline grows longer the need for more memory for intermediate results grows greater. Sternberg's 113 processors already appear to be hard to use completely, in every instruction. It seems possible that a hundred-fold increase can be got in this way (though possibly only at a very high cost in hardware), but probably little more.

E) An array of several specially designed processors, possibly including pipelines, appears to be an attractive alternative to the more conventional arrays envisioned in C), and especially in B). Here we see obvious trade-offs in terms of the power and cost of each particular processor. To give rough estimates: the PICAP processor might cost $10,000; the Cytocomputer processor might cost $50,000. A more conventional bit-slice processor might cost from $200 to $2,000. There is a need for a great deal of empirical examination of the costs and benefits of different design trade-offs for such processors. Almost certainly a great variety of different individual processors, and total arrays and networks, should be built, and almost certainly there will be no single "right answer."

F) A series of arrays of the sort described above might be built, to serve as a giant 3-dimensional pipeline each of whose processor stages was itself a very large array. This might take one of several forms:

1) The entire system of arrays might be built by simply duplicating the single array. For example, 20 96 by 96 CLIP4s might be put in series.

2) A single array might be programmed to simulate such a series, by decomposing it into a number of sub-arrays of equal size, and shifting information, when appropriate, from one sub-array to the next. (This would be extremely wasteful of time in a conventional CLIP4 array, though changes in the interconnection architecture might overcome many of the problems.)

3) A single array might be used, but one that was re-configurable at the hardware level, under program control. Siegel et al.'s, 1979, proposed 32 by 32 MIMD-SIMD system can be reconfigured into different sub-systems whose sides are powers of 2. Lipovski, 1977, is now building a 16-processor system that is reconfigurable at the hardware level; he hopes to demonstrate that its extra costs for switches and slower message passing can be kept reasonably small.

4) Rather than have all arrays the same size, arrays of different sizes might be used, where appropriate. This is an attractive alternative for picture processing, since the very large array of raw information initially input to the system by the tv camera or other sensing device is successively abstracted and reduced as processing progresses. That suggests starting with a very large array but then transferring information to successively smaller arrays as they become appropriate, giving an overall pyramid structure (Dyer, 1982; Tanimoto, 1982; Uhr, 1983b).

5) Different kinds of processors, and different kinds of arrays, might well be appropriate at different layers of such a system. It seems likely that the early "image processing" operations on near-neighbor cells (for which CLIP for example is especially well designed) can be handled by simpler processors that can be built into larger arrays. "Higher-level" processing might best be done by processors capable of addressing more global sets of information. At the most global level, when information has been greatly reduced, a single powerful processor might be most appropriate (Uhr, 1983a).

G) Each of the above possibilities has its own range of uses for which it is best suited, and its own set of limitations. This strongly suggests that a judicious combination of several of the above alternatives might produce a network of mixed resources that could take advantage, if multi-programmed over a suitable mix of user programs, of each of its sub-system's strengths.

A Network-Array Architecture for Perceptual Systems

The following specifies an architecture for a network-array of processors that appears to be an appropriate system on which to run programs for perception and cognition.

A) Information is input from the sensing device into a large buffer memory, and then processed by a sequence of instructions that are executed by a very large hardware-parallel array of relatively simple processors specially designed with the near-neighbor operations most people feel are appropriate for the first stages of image processing, image enhancement, and scene analysis.

This suggests a cellular array like CLIP4 (Duff, 1976, 1978), DAP (Flanders et al., 1977; Reddaway, 1978, 1980) or the MPP (Batcher, 1980). Today it might only be 128x128. But in five or fifteen years it can be much larger, e.g., 512x512, or more. Therefore it seems attractive to use several successively smaller arrays of this sort, thus building a full, or truncated, pyramid.

B) Next follow instructions executed by appreciably smaller arrays of processors, each of which is appreciably more powerful and has access to an appreciably larger neighborhood of pixels. It seems preferable to have some of these arrays built from the very powerful bit-slice processors, rather than from the cheaper micro-computers on a chip. For example, we have designed a system (Uhr, Thompson and Lackey, 1981) that has a 4 by 4 array of 16 such processors, with 16 slave processors tied to each, giving a 16 by 16 array in toto. This gives a 16 processor MIMD system, where each processor is a 16 processor SIMD system. A fuller system would have several CLIP-like arrays of 1-bit processors, followed by arrays of more powerful N-bit processors.

C) Finally, a general-purpose computer can handle the deepest layers of processing. Probably best is a whole network of general-purpose computers that augments the more specialized pyramid, with a good linking topology within the network (e.g., a Moore graph - see Uhr, 1983b) and between network and pyramid (Uhr, 1983a). Links to this newtwork should make it possible to transfer information to one of its computers at whatever layer the programmer feels is appropriate (because the degree of potential speed-up from parallel processing has become so small as to be relatively unimportant, whereas the advantages of allowing a processor to make global assessments of information no matter where it might be have become sufficiently great).

Summary Discussion and Conclusions

It seems compelling that a general system for the perception of scenes of moving objects must be orders of magnitude faster (and more powerful) than any conceivable single-CPU computer. And it seems much more sensible to build a system that has many processors working in parallel, rather than try to push as much power as possible out of one serial machine. Human perception uses millions of processors arranged in a parallel-serial layered network of arrays, and handles perception in a surprisingly small serial depth of 25 or fewer processes to a few hundred at most.

A variety of different sources for speeding up processing are available to us, and many of these can be combined. But the only way in which we can continue to speed up our systems is to add more physical processors. This paper examines some of these sources of potential speed-ups, and some of the network and array architectures that appear the most promising ways to exploit them.

The use of a parallel-serial pyramid of parallel arrays can, potentially, enormously reduce the times needed to process highly parallel image arrays as they pump through the pyramid (Uhr et al., 1982, Uhr, 1983b). A parallel array gives $O(N^2)$ speed-ups over a serial computer whenever an operation must be iterated over an N^2 array. Parallel W^2 window operations give a further $O(W^2)$ speed-up. In addition a converging pyramid of arrays reduces the $O(N^2)$ diameter of an array (and hence the possibility of $O(N^2)$ message-passing distances) to O(logN).

The most promising total structure appears to be the one with a pyramid of arrays of simple 1-bit computers that link at their higher levels to a similarly structured network of more powerful computers. But a variety of other array-pyramid-network structures appear promising, and worthy of exploration.

Possibly the best way to view such a system is as a network of diverse resources with which to learn how to explore and extend such networks - by simulating, designing, and building them, then programming them to see how well they work. This raises many major problems, in developing operating systems and languages to expedite the smooth use of the total system, handling different kinds of nodes in the total network, and developing appropriate algorithms and programs. But these are extremely interesting, and important, problems in themselves. To the extent they are not solved, or handled well, such a system will still be usable, in its separate components, and in the limited combination of these components for which it has been expressly designed.

References

Batcher, K.E., Architecture of a massively parallel processor, Proc. 7th Annual Symp. on Computer Arch., ACM, 1980, 168-174.

Bogdanowicz, J. F., Preliminary Design of a Partitionable Multimicroprogrammable Microprocessor System for Image Processing, Tech. Report, EE 77-42, School of Electrical Engineering, Purdue Univ., 1977.

Briggs, F., Fu, K. S., Hwang, K. and Patel, J., PM4 - a reconfigurable multimicroprocessor system for pattern recognition and image processing, Proc. AFIPS NCC, 1979, 255-265.

Despain, A.M. and Patterson, D.A., X-tree: a tree structured multiprocessor computer architecture, Proc. Fifth Annual Symp. on Computer Arch., April, 1978, 144-151.

Douglass, R. J., Recognition and depth perception of objects in real world scenes, Proc. IJCAI-4, 1977, 656.

Douglass, R. J., A computer Vision Model for Recognition, Description, and Depth Perception in Outdoor Scenes. Unpubl. Ph.D. Diss., Computer Sciences Dept., Univ. of Wisconsin, 1978.

Duff, M. J. B., CLIP4: a large scale integrated circuit array parallel processor, Proc. 3d Int. Joint Conf. on Pattern Recog., 1976, 4, 728-733.

Duff, M. J. B., Review of the CLIP image processing system, Proc. National Computer Conf., 1978, 1055-1060.

Flanders, P.M., Hunt, D.J., Reddaway, S.F., Parkinson, D., Efficient high speed computing with the Distributed Array Processor, In: High Speed Computer and Algorithm Organization, New York: Academic Press, 1977, 113-128.

Fuller, S. H., Price/performance comparisons of C.mmp and the PDP-10, Proc. Third Symp. on Computer Arch., 1976, 195-202.

Goodman, J.R. and Sequin, C.H., Hypertree, a multiprocessor interconnection topology, submitted for publication, 1981.

Hanson, A. R. and Riseman, E. M., Preprocessing cones: a computational structure for scene analysis, COINS Tech. Rept. 74-7, Univ. of Mass., 1974.

Hanson, A. R. and Riseman, E. M., Visions: A Computer System for Interpreting Scenes, In: Computer Vision Systems, A. R. Hanson and E. M. Riseman (Eds.), New York: Academic Press, 1978, 303-334.

Klinger, A. and Dyer, C., Experiments on picture representation using regular and decomposition, TR Eng. 7497, UCLA, 1974.

Kruse, B. The PICAP picture processing laboratory, Proc. 3d Int. Joint Conf. on Pattern Recog., 1976, 4, 875-881.

Kruse, B. Experience with a picture processor in pattern recognition processing, Proc. National Computer Conf., 1978.

Kruse, B., Danielsson, P.E., and Gudmundsson, B., From PICAP I to PICAP II, In: Special Computer Architectures for Pattern Processing, K.S. Fu and T. Ichikawa (Eds.), New York: CPR Press, 1980.

Levine, M. D., A knowledge-based computer vision system, In: Computer Vision Systems, A. Hanson and E. Riseman (Eds.), New York: Academic Press, 1978, 335-352.

Levine, M. D. and Leemet, J., A method for nonpurposive picture segmentation, Proc. 3d Int. Joint Conf. on Pattern Recog., 1976, 494-498.

Lipovski, J., On a varistructured array of microprocessors, IEEE Trans. Computers, 1977, 26, 125-138.

Manara, R. and Stringa, L., The EMMA system: an industrial experience in a multiprocessor, In: Languages and Architectures for Image Processing M. Duff and S. Levialdi, (Eds.), London: Academic Press, 1981. pp. 215-228.

Reddaway, S.F., DAP - a flexible number cruncher, Proc. 1978 LASL Workshop on Vector and Parallel Processors, Los Alamos, 1978, 233-234.

Reddaway, S.F., Revolutionary array processors, In: Electronics to Microelectronics, W.A. Kaiser and W.E. Proebster (Eds.), Amsterdam: North-Holland, 1980, 730-734.

Rosenfeld, A. (Ed.) Multi-Resolution Pyramids for Image Processing, Berlin: Springer, 1983.

Siegel, H.J., et al., PASM: A Partitionable Multimicrocomputer SIMD/MIMD System for Image Processing and Pattern Recognition, School of Electrical Engineering TR-EE 79-40, Purdue Univ., West Lafayette, 1979.

Sternberg, S.R., Cytocomputer real-time pattern recognition, paper presented at Eighth Pattern Recognition Symp., National Bureau of Standards, April, 1978.

Sternberg, S.R., Language and architecture for parallel image processing, Proc. Conf. on Pattern Recognition in Practice E.S. Gelsema and L.N. Kanal, (Eds.), Amsterdam: North-Holland, 1980.

Sullivan, H., T. Bashkov, and D. Klappholz, A large scale, homogeneous, fully distributed parallel machine, In: Proc. Fourth Annual Symp. on Computer Arch., 1977, 105-124.

Swan, R.J., S.H. Fuller and D.P. Siewiorek, Cm* - A modular, multi-microprocessor, Proc. AFIPS NCC, 1977, 637-663.

Tanimoto, S. L., Pictorial feature distortion in a pyramid, Computer Graphics Image Proc., 1976, 5, 333-352.

Tanimoto, S. L., Regular Hierarchical Image and Processing Structures in Machine Vision, In: Computer Vision Systems, A. R. Hansen and E. M. Riseman (Eds.), New York: Academic Press, 1978, 165-174.

Tanimoto, S. and Klinger, A. (Eds.), Structured Computer Vision, New York: Academic Press, 1980.

Uhr, L., Layered "recognition cone" networks that preprocess, classify and describe. IEEE Trans. Computers, 1972, 21, 758-768.

Uhr, L., A model of form perception and scene description, Computer Sciences Dept. Tech. Rept. 176, Univ. of Wisconsin, 1974.

Uhr, L., "Recognition cones" that perceive and describe scenes that move and change over time, Proc. Int. Joint Conf. on Pattern Recog., 1976, 4, 287-293.

Uhr, L., "Recognition cones" and some test results. In: Computer Vision Systems A. Hanson and E. Riseman, (Eds.), New York: Academic Press, 1978, 363-372.

Uhr, L., Comparing serial computers, arrays and networks using measures of "active resources," IEEE Trans. Computers, 1982, October.

Uhr, L., Pyramid Multi-Computers, and Extensions and Augmentations, In: Algorithmically Specialized Computer Organizations, D. Gannon, H.J. Siegel, L. Siegel and L. Snyder, Eds., New York: Academic Press, 1983. (a)

Uhr, L., Algorithm-Structured Computer Arrays and Networks, New York: Academic Press, 1983. (b)

Uhr, L. and Douglass, R., A parallel-serial recognition cone system for perception, Pattern Recognition, 1979, 11, 29-40.

Uhr, L, Lackey, J. and Thompson, L., A 2-layered SIMD/MIMD Parallel Pyramidal "Array/Net", Proc. Workshop on Computer Architecture for Pattern Analysis and Image Data Base Management, IEEE Computer Society Press, 1981, 209-216.

Uhr, L., Schmitt, L., and Hanrahan, P., Cone/pyramid perception programs for arrays and networks. In: Multi-Computers and Image Processing, K. Preston, Jr. and L. Uhr, (Eds.), New York: Academic Press, 1982. pp. 180-191.

Wittie, L.D., Efficient message routing in mega-micro-computer networks, Proc. Third Annual Symp. on Computer Arch., New York: IEEE, 1976.

Wittie, L.D., MICRONET: A reconfigurable microcomputer network for distributed systems research, Simulation, 1978, 31, 145-153.

Wittie, L.D. and van Tilborg, A.M., MICROS, a distributed operating system for MICRONET, a reconfigurable network computer, IEEE Trans. Computers, 1980, 29, 1133-1144.

3D Motion Perception

Motion from Continuous or Discontinuous Arrangements

Paul A. Kolers

University of Toronto

Previously, the study of motion was directed almost wholly to apparent motion, the compelling sense of motion derived from the properly timed presentation of spatially separated stimuli. Only recently has the study of real motion come somewhat into its own, the result in part of great improvements in the technology of stimulus control. Motion perceived in the cinema is the paradigmatic case of apparent motion--a series of stills presented at the proper rate creates the sense in the observer of a continuous movement of objects. The two kinds of motion, "apparent motion" derived from discontinuous displays, and "real motion" derived from a continuously moving object, are often distinguished only by their means of production; in appearance they can be hard to tell apart.

The similarity of appearance often led people to affirm an identity of mechanism. Apparent motion was sometimes said to be a derivative case of real motion (Gregory, 1966), or real motion was sometimes said to be a limiting case of apparent motion (Ansbacher, 1944). Sometimes it was even claimed that it was unfortunate that anyone had ever sought to make a distinction between the two forms of stimulation (Gibson, 1968); the reason behind this opinion was the belief

Motion: Representation and Perception
N.I. Badler and J.K. Tsotsos, Editors
Published by Elsevier Science Publishing Co., Inc.

that the perception of motion was due wholly to the order and timing of stimulation. The means or mechanisms by which perceptions of motion are created visually remain central questions in the area; I will comment briefly on two aspects, by way of introduction to the papers.

The first concerns equivalence of the two--the claim based on their very similar appearance. In some experiments on motion a few years ago, the interaction of the moving object and visual objects in its path, or upon the path itself, was studied (Kolers, 1963). The findings were that an object in continuous motion influenced objects in its path, whereas an object in apparent motion, despite looking like the object in continuous motion, had no such influence. Other objects interposed into the path could moreover alter the shape of the path of objects in apparent motion, and did so more than they affected objects in continuous motion. A few other conditions were identified that were different for perception of motion from continuous and discontinuous displays and the proposal was made that the mechanisms underlying the perception were different for the two kinds of motion. Especially, motion from discontinuous stimulation seemed to require a filling-in or impletion not required in the perception of continuous motion (Kolers, 1964).

This claim has not been universally accepted. Some people affirm that all forms of perception of motion must derive from a common productive source in the nervous system.

The appeal has been made widely to neurophysiological studies of specific motion detectors as the basis of the claim. Indeed, some investigators in Europe have found that motion detectors in the brain of cat were equally excited by objects in continuous motion and stationary objects whose intermittent flashing created the sense of apparent motion (Gruesser & Gruesser-Cornehls, 1973). A few years ago Braddick (1974) proposed that there were two kinds of motion detectors, one responsible over a short range of retinal space and the other active over a longer range. The short range process was said to operate over spatial separations of less than about 15 min of visual angle (.25 deg), and the long range process thereafter. This view has been widely accepted and praised by many. Braddick (1980), Anstis (1980), and others have used the distinction to specify two different mechanisms of motion.

Now it should hardly need saying that any continuous event has a threshold for discontinuity: two components that cannot be told apart at some degree of separation can be told apart at a larger degree of separation; two lines, for example, or two dots, or two lights. People, therefore, may have some difficulty in understanding the difference between the earlier claim that continuous stimulation produces different effects in the nervous system from those of discrete stimulation, and Braddick's saying that over a very short separation the visual system cannot tell one thing from two but at a wider separation can do so. Moreover I believe that

Braddick introduces an undesirable confusion of description when he alleges that the short range process is directly wired and due to motion detectors, whereas the long range process is interpretive. All perceptual experiences have a source in the nervous system, for one thing; for a second, all perceptual experiences are interpretations or ascriptions or translations of one sort of neural activity to another. I believe it is not informative to say one kind has a physiological basis and the second kind is cognitive, for cognition too has a physiological basis. Finally, it is interesting to notice that multiple instances of a short range process could not be clearly discriminated or interpreted under some conditions (Foster, Thorson, McIlwain, & Biederman-Thorson, 1981), affirming, as it were, the dependence upon interpretation of even the short range process.

A flash presented to a retinal location generates a field effect, a lateral radiation that can be measured by the spatial range of visual masking, by Panum's area, or the like (Kolers, 1972). This field may mark the spatial range of Braddick's short range process, or be the spatial resolution of continuous motion. In the end, it does seem to me that Braddick has merely renamed the difference earlier pointed to as between continuous and discontinuous stimulation in motion perception, and added one or two conditions that help to distinguish them.

I turn now to what Ullman (1979) called the

correspondence problem, and speak of it from two perspectives, definition and data. If we consider two flashes, A and B, presented so as to seem to be in motion, to speak of a correspondence between them is to imply that on the occasion of the first flash something immediately begins to seek pictorial targets or mates, correspondents of some sort. It is not until the second flash has appeared that the correspondent is identified, however, and motion is seen between the location of the first flash and the location of the second. Moreover, the exact location of the second seems to be information of little value in determining the movement relation (Beck, Elsner, & Silverstein, 1977).

A few years ago some experiments on apparent motion made the two flashes different in shape (Kolers & Pomerantz, 1971); most other studies before had used identical flashes as the stimuli. The new finding was that over a large range of conditions, any plane shape would change into any other (Figure 1); indeed, it took some pains before we were able to find ways to retard the motion, and did so with shapes whose texture was dense (Figure 2). The upper pair of Figure 2 are seen in smooth transformation when one of the two is superposed on the second; motion, but not transformation of shape, is seen when the two are side by side. In the latter condition one of the two shapes is seen to replace the other abruptly rather than change into it in a graded fashion. In the lower pair even superposition does not yield a smooth change; rather, flicker is seen. It is not clear that a notion of correspondence would allow smooth change of shape to characterize superposition of the first pair, but motion with replacement to characterize their side by side position. It is not clear in terms of correspondence why the side by side position of the second pair would allow motion with replacement whereas superposition would create flicker.

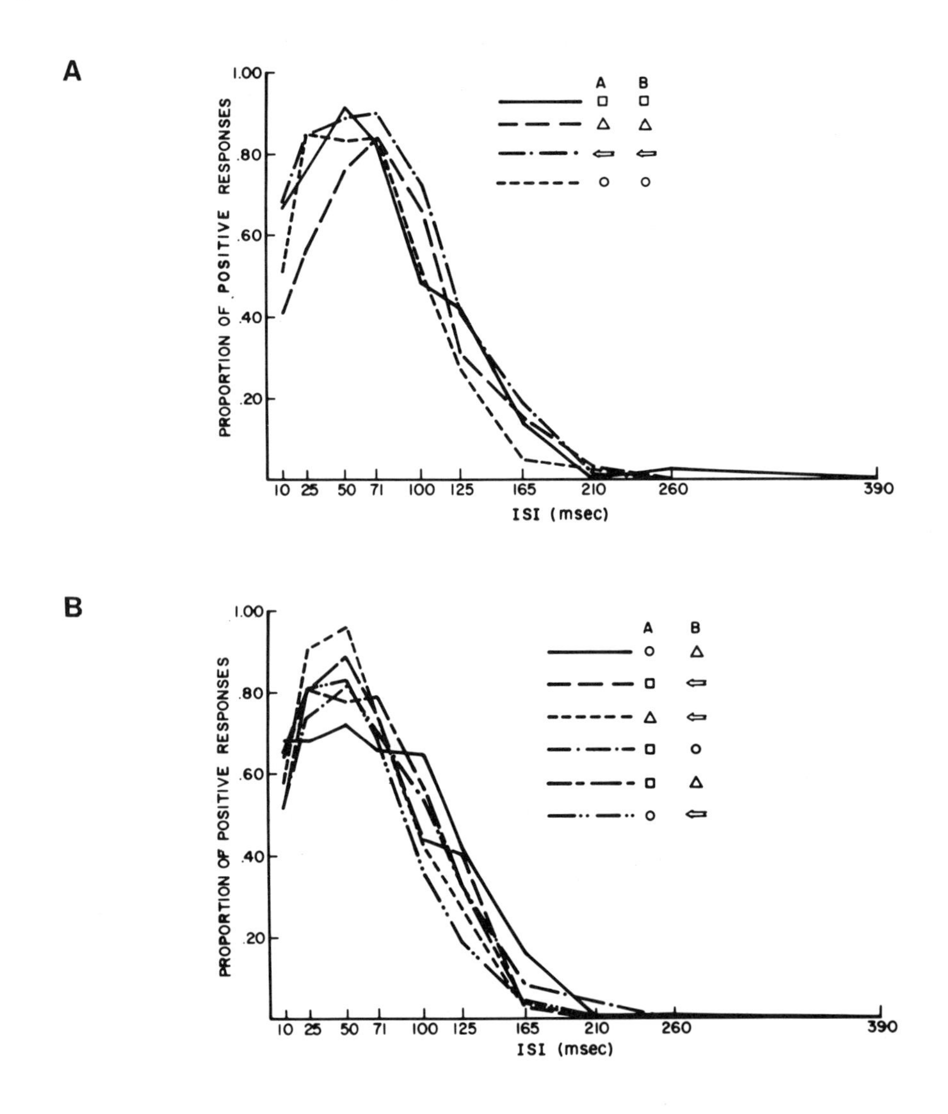

Figure 1. The likelihood of reporting apparent motion between pairs of shapes. The pairs of shapes in A are disparate; the pairs in B are identical. Overall results were about the same. (From Kolers & Pomerantz, 1971)

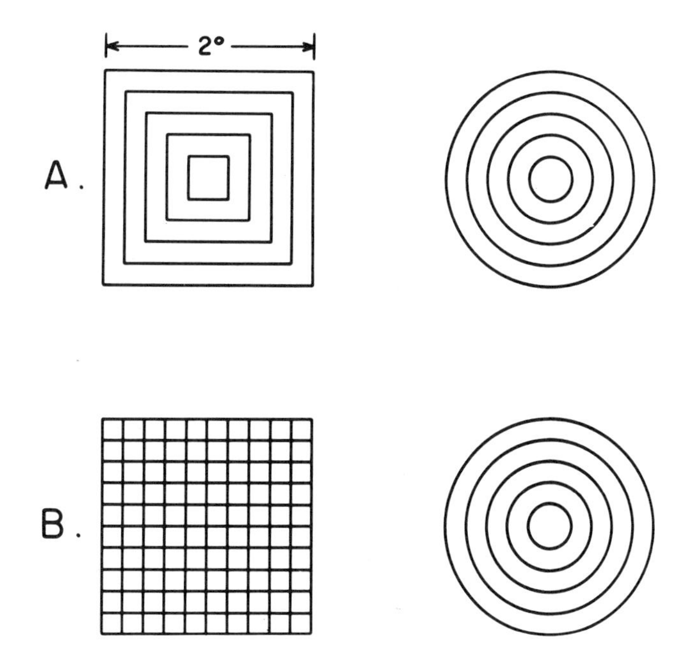

Figure 2. Restrictions on the report of a perception of motion. The pair in A replace each other when presented side by side but change into each other when superposed. The pair in B also replace each other when side by side, but only flicker when superposed. (From Kolers, 1972)

In other tests we tried to force analysis and matching of features by alternating part of one shape with a whole other one (Figure 3). The whole cube on the left was alternated with the parts indicated (and others), one part at a time; the parts represent the outline, vertical lines, oblique lines, and other features as well--both logical and neurological features of the cube. A process of correspondence would presumably match the isolated feature with its mate in the whole cube, and move that common part while the remainder of the cube merely flickered on and off. In perceptual fact, the feature seemed to grow into the cube in one direction of motion, or the cube shrank to the feature in the other direction of motion. There was no correspondence

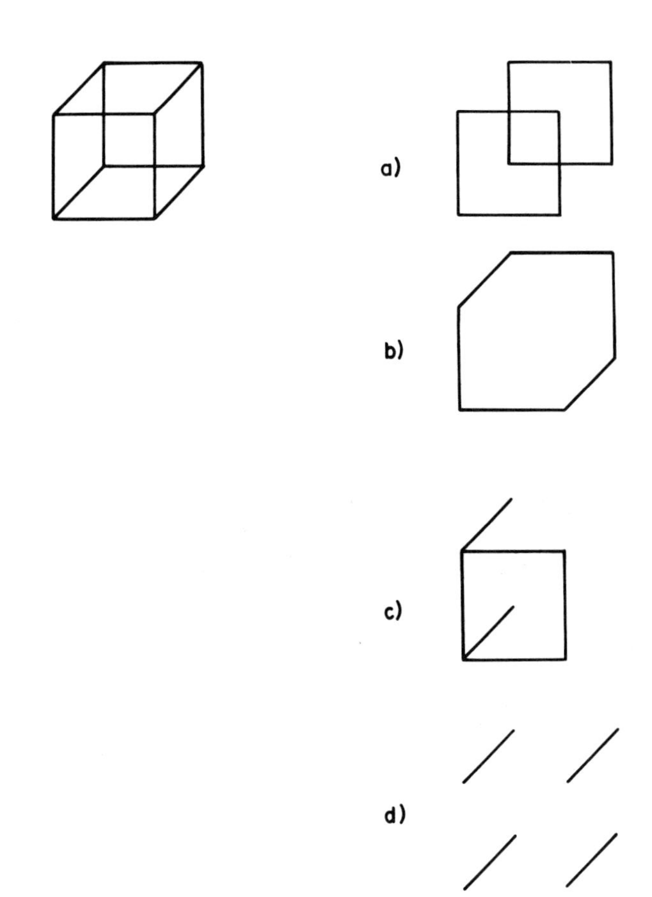

Figure 3. The cube alternated with various of its composing features seems to shrink to the feature in one direction of motion, or the feature grows into a whole cube in the other direction.

of parts or features, but rather movement of an object from one location to another, the object changing its shape to resolve the difference between the appearance at the two locations.

The "tokens" of correspondence in Ullman's work are not well defined; they are neither gray level phenomena nor shapes, but something in between--some attribute of shape but not shape. It seems important to the theory that they be discrete and distinguishable, for they are intended as primitives whose composition would accommodate the perceptual facts (Marr, 1982). Some of the perceptual complexity that causes difficulty for such a conceptualization can be realized in the example of Figure 4. The inscribed triangle or the inscribed circle, when alternated with the square, will seem to exchange identity with the square; square and circle or square and triangle change continuously into each other under proper conditions of timing. Inscribing the triangle within the circle and flashing the two of them simultaneously and in alternation with the square yields a perception of the circle and square exchanging identities while the triangle blinks on and off. A slight extension in size of the triangle, or shrinkage of the circle (not shown), so that the vertices of the triangle contact the circle, changes the perception: circle and triangle exchange identity with the square. Thus, minor connections between parts changes figural composition drastically, and with a marked change in the perceptual outcome (Kolers, 1972).

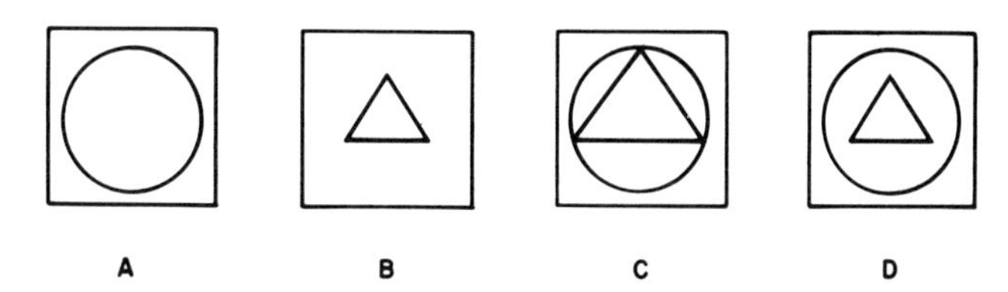

Figure 4. The square was alternated with the circle in A or with the triangle in B, and each seemed to change into the other. When the vertices of the triangle coincided with the circle in C, the whole configuration changed into the square; but when the vertices did not coincide, D, the triangle blinked on and off while the circle and the square exchanged shapes. (From Kolers, 1972)

It is sometimes alleged that a criterion of smoothness is inappropriate as a measure of apparent motion; Ullman stresses this point particularly. I believe that in doing so he may have erred in interpreting some statements to make this point. Relatively few psychological studies of motion perception take smoothness as a dependent variable or try to scale smoothness, except for occasional special purposes. Rather, investigators use as their criterion the description of an object in smooth motion traversing the screen, and do so to distinguish this appearance from the partial motions, jitter, and other perturbations that can characterize incomplete effects. Ullman is no doubt correct in affirming that scaled smoothness does not accommodate certain results found in the perceptual literature, but errs by affirming that smoothness is something more than one other feature of the object seen in motion. He compounds the misinterpretation, I believe, when he asserts that competition is a better criterion by which to assess his parameters of correspondence and affinity. Competition would test to which of two possible correspondents in a second flash movement derived from a first flash would seem to go. It is a fact, however, that competition yields unstable results. Pomerantz and I tried the procedure some years ago with little reliability. For example a circle in one flash followed by a circle and a square in a second flash did not yield motion only between the circles; depending upon distance and retinal position, several different perceptions could be achieved.

Subsequently, von Grunau and I made some observations with color (Kolers & von Grunau, 1976). Among other conditions we alternated different shapes and different colors, trying to test competition and choice. All of the preferences that we studied in this aspect of the investigation were highly unstable, and seemed to change depending upon the part of the eye stimulated, the direction of fixation, the frequency of presentation, the position of the shapes and other variables. I believe consequently that Ullman's reliance on choice and competition as criteria of correspondence and affinity will not yield stable results. The smooth motion of an object seems to me a better index of its movability than that obtained from unstable judgments of choice and competition.

Actually, in concerning himself with correspondence as he does, Ullman seems to be reverting to a classical notion of motion perception--that the perception is due to the discrimination of a single object in different locations at different times. This classical approach has many things wrong with it; some of them have been listed in another place (Kolers, 1972). What we seem to find in the study of motion perception, and apparent motion especially, is that location is more important by far than object; in a large number of studies we have found that the visual system will retain locations and fill them with figures that are made to fit, rather than retain a figure and move it from one location to another, or track it as it moves. It is as if apparent motion were a matter of the visual system taking the object presented at the second location and backfilling the space between it and the first location with a suitably transforming shape. The system so to speak backdates appearances, rationalizing disparities to do so; interaction between locations rather than correspondence between shapes seems to be the driving mechanism. This retrospective filling-in is itself complicated; for example, if a red is alternated with a green, the person sees red moving part of the way across the screen to be replaced by green somewhere near the half-way mark. If it were just a matter of retrospection, why bother with the color change; why not move just one color across the screen? A further complication arises merely from consideration of the timing. If the interval between A and B is made a little different from the interval between B and A, the person sees motion in depth during the longer interval. In many cases, as Ullman remarks, depth is not preserved in line segments of figures; in this case depth is created out of a disparity in timing. It is hard to appreciate what if anything could be the correspondent. And indeed what we had known was that there was little in the way of correspondence as to shape, shape features, colors, or the like underlying the compelling perceptions of apparent motion.

Let me add that I make no claims about the rightness of the mathematical analysis of motion so far as its implementation in a machine is concerned. I have no competence with which to assess that issue, except to say that getting a machine to run can be its own very significant reward. My concern is with the propriety of applying that analysis to human perception of motion, especially when the data are sometimes forced to conform to the characteristics of certain formalisms. I believe that not much is gained in understanding of human perception by an analysis, however elegant, that does not abide by the basic perceptual facts.

Acknowledgment

This work was supported by Grant A7655 from National Sciences and Engineering Research Council Canada. Author's address: Department of Psychology, University of Toronto, Toronto, Canada, M5S 1A1.

References

Ansbacher, H.I. Distortion in the perception of real movement. _Journal of Experimental Psychology_, 1944, 34, 1-23.

Anstis, S. M. The perception of apparent movement. _Philosophical Transactions of the Royal Society London_, 1980, B 290, 153-168.

Beck, J., Elsner, A., & Silverstein, C. Position uncertainty and the perception of apparent movement. _Perception & Psychophysics_, 1977, 21, 33-38.

Braddick, O. J. A short-range process in apparent motion. _Vision Research_, 1974, 14, 519-527.

Braddick, O. J. Low-level and high-level processes in apparent motion. _Philosophical Transactions of the Royal Society London_, 1980, B 290, 137-151.

Foster, D. H., Thorson, J., McIlwain, J. T., & Biederman-Thorson, M. The fine-grain movement illusion: A perceptual probe of neuronal connectivity in the human visual system. _Vision Research_, 1981, 21, 1123-1128.

Gibson, J.J. What gives rise to the perception of motion? _Psychological Review_, 1968, 75, 335-346.

Gregory, R. _Eye and brain_. New York:McGraw-Hill, 1966.

Gruesser, O.-J., & Gruesser-Cornehls, U. Neuronal mechanisms of visual movement perception and some psychophysical and behavioral correlations. In R. Jung (Ed.), _Handbook of sensory physiology: Central processing of visual information, Part A_ (Vol. 7/3A). New York: Springer-Verlag, 1973.

Kolers, P.A. Some differences between real and apparent visual movement. *Vision Research*, 1963, 3, 191-206.

Kolers, P.A. The illusion of movement. *Scientific American*, 1964, 211(4), 98-106.

Kolers, P.A. *Aspects of motion perception*. Oxford:Pergamon, 1972.

Kolers, P.A. & Pomerantz, J.R. Figural change in apparent motion. *Journal of Experimental Psychology*, 1971, 87, 99-108.

Kolers, P.A. & von Grunau, M. Shape and color in apparent motion. *Vision Research*, 1976, 16, 329-335.

Marr, D. *Vision*. San Francisco:Freeman, 1982.

Ullman, S. *The interpretation of visual motion*. Cambridge, Mass.: MIT Press, 1979.

Perception of Rotation in Depth:

The Psychophysical Evidence

Myron L. Braunstein
Cognitive Sciences Group
University of California, Irvine
Irvine, California

1. Introduction

There are a variety of ways in which motion in the environment can provide information about three-dimensional relationships. One transformation that has received increasing attention in both the visual perception literature and in the machine vision literature is rotation in depth. This transformation, which includes any rigid rotation other than a rotation about the line of sight, can provide both a strong impression of depth and specific information about three-dimensional relationships in a rotating object or pattern. Computational theories have been developed concerning the relationships that an observer can potentially extract from the information available in this transformation (Ullman, 1979). If computational theories are to be compared to human performance, a systematic body of data on human perception of rotation in depth is required. Such a body of data has been developing, especially in the last few years. Most of these studies have used computer animation techniques introduced into this area of research by Green (1959, 1961). It is now possible to derive some preliminary conclusions from these data about what information is actually used by observers, what sources of information are dominant when multiple sources are available, and what errors occur in perception that can provide insights into the processes that observers apply to this information.

The objective of this paper is to bring together these empirical findings concerning the ability of human observers to perceive three-dimensional relationships on the basis of rotation in depth. It is intended to systematize and clarify the current state of knowledge in this area. The following sections are organized about a series of conclusions, ranging from general to specific, concerning three major issues in the perception of objects undergoing rotation in depth. The first is the relationship between perceived depth and perceived relative distance in a rotating object. Perceived depth refers to the three-dimensional structure of an object, without regard to the position of the observer (e.g., the perception of a sphere as a sphere rather than as a circle). Perceived relative distance (or depth order) refers to the perception of which parts of an object are closer and which are more distant (e.g., which is the near hemisphere in a transparent sphere). Perceived relative distance in rotating objects is usually measured with direction of rotation judgments. The remaining two issues concern the variables which determine perceived depth and those which determine judgments of relative distance.

2. Depth and relative distance

2.1 Rotation in depth can be perceived even when the direction of rotation is ambiguous.

This observation was recorded as early as 1860 by Sinsteden for distant windmills (Boring, 1942) and was confirmed in experiments by Miles (1931) and Wallach and O'Connell (1953), using shadow projections based on distant light sources. Using true parallel projections generated by computer animation, Braunstein (1966) found that observers could distinguish between two-dimensional and three-dimensional random dot patterns, rotating about horizontal or vertical axes, with 97% accuracy. The perception of depth in parallel projections limits the generality of a "reverse-perspective" theory of depth perception through motion (Johansson, 1977) and provides support for Ullman's (1979) "structure-from-motion" model.

2.1.1 Judgments of depth, coherence and rotation direction are affected differently by perspective and rotation speed.

The ambiguity in the direction of rotation found in parallel projections can be decreased by using a polar projection with a projection point close to the rotating object. Increasing the polar projection distance, in displays of dots rotating about a vertical axis, increases the frequency with which a display is chosen (in pair-comparison judgments) as more coherent or rigid, but decreases the choice of the display as appearing more three-dimensional (Braunstein, 1962). The accuracy of direction of rotation judgments also decreases with increasing projection distance, but the function relating accuracy to projection distance differs from that relating depth ratings to projection distance. Petersik (1980a, b) found a nearly linear decrease in accuracy with projection distance, while depth ratings declined sharply at the shortest projection distance and subsequently were asymptotic. Petersik also found different effects of stimulus-onset asynchrony (which determined both rotation speed and display duration) on accuracy of

Motion: Representation and Perception
N.I. Badler and J.K. Tsotsos, Editors
Published by Elsevier Science Publishing Co., Inc.

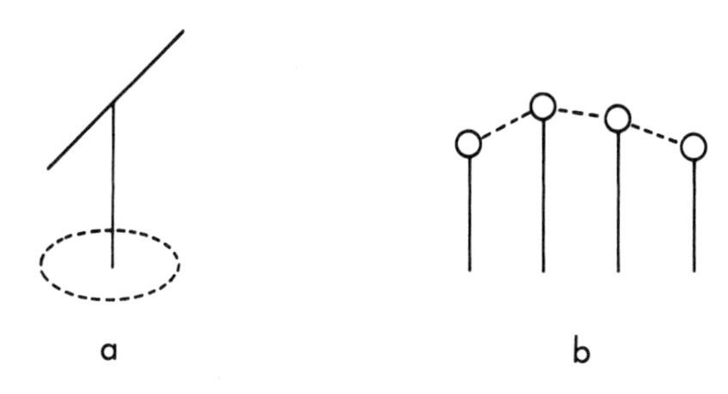

Figure 1. Objects studied by Wallach and O'Connell (1953) that cast shadows with (a) visible or (b) implicit contours changing in length and direction during rotation.

direction judgments and depth ratings. Accuracy peaked at a moderate rotation speed (10 rpm) and declined at both high (100 rpm) and low (.3 rpm) speeds. Depth judgments, on the other hand, were highest at the highest rotation speed and declined steadily with rotation speed. These results indicate that perceived depth, coherence and relative distance are not identical functions of the stimulus variables. Studies of perceived rotation in depth should continue to examine the relationships among these three dependent measures.

3. Variables affecting perceived depth

3.1 Polar perspective is not required for perceived depth.

This conclusion is based on essentially the same body of literature as that which supports the conclusion that rotation in depth can be perceived when the direction of rotation is ambiguous. In the early observations cited above for distant windmills and fans, and in Wallach and O'Connell's (1953) study, subjects observing approximately parallel projections described percepts that were predominantly three-dimensional. Using actual parallel projections of random dot patterns rotating about a vertical axis, Braunstein (1962) found judged coherence to be greater for parallel than for polar projections, although judged depth was greater for polar projections. As noted in section 2.1, two-dimensional and three-dimensional rotating dot patterns can be distinguished with almost perfect accuracy on the basis of parallel projections (Braunstein, 1966).

3.2 One-dimensional changes are sufficient for eliciting reports of perceived rotation in depth.

This conclusion is contrary to the position stated by Wallach and O'Connell (1953), that simultaneous changes in the length and direction of projected contours are necessary for the "kinetic depth effect." These changes could be explicit, as in the case of a rotating tilted line (Figure 1a) or implicit, as in the case of several identical display elements rotating about a vertical axis that are not aligned horizontally (Figure 1b). Braunstein (1977a), however, found that a segmented horizontal line rotating about a vertical axis--which displayed length changes only in its projection--was perceived on most trials as rotating in depth. Four types of displays were studied based on two variables: (1) the waveform of the displayed projection was either sinusoidal or linear (triangular); (2) the waveform was symmetrical, or asymmetrical in accordance with a polar projection. The four display

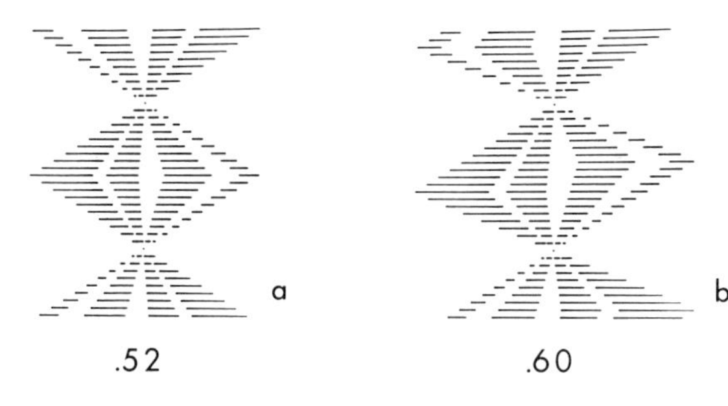

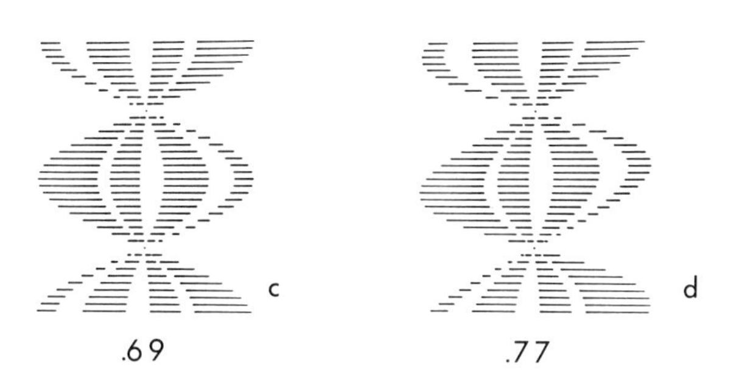

Figure 2. Four waveforms used in displays of a segmented line: (a) linear-symmetrical, (b) linear-asymmetrical, (c) sinusoidal-symmetrical, and (d) sinusoidal-asymmetrical. The mean proportion of rotation in depth responses is shown for each type of display.

types based on these two variables are illustrated in Figure 2. Each line in the figure represents a successive frame in the display. Subjects selected either rotation or expansion-and-contraction as the perceived motion. The results are shown in the figure. Both variables significantly affected judgments of which motion was displayed, but there was no interaction. These results lead to three corollary conclusions:

3.2.1 The probability of reporting rotation in depth is increased by a sinusoidal waveform in the projection.

3.2.2 The probability of reporting rotation in depth is increased by asymmetrical velocities (in the "near" and "far" portions of the rotation cycle) comparable to those found in a polar projection.

3.2.3 The above two effects are additive.

The first of these three conclusions is not surprising, as a projected sinusoidal waveform is consistent with a rigid rotation in depth. The last finding was not expected, however, as asymmetrical velocities are consistent with a rigid rotation only when superimposed on a sinusoidal waveform. Asymmetrical velocities might be expected to increase reports of perceived rotation only for the sinusoidal waveform, but an equal effect was found for the linear waveform.

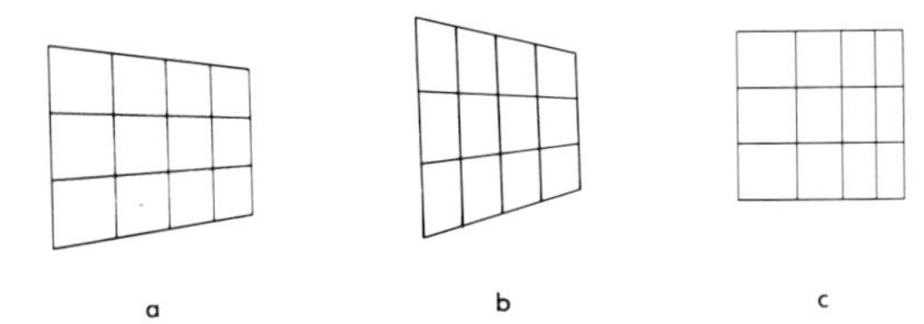

Figure 3. Projections of a rotated grid with (a) both convergence and compression appropriate to a polar projection, (b) convergence only, and (c) compression only.

3.3 Three-dimensional organization can be detected in as few as two successive frames of a rotating pattern.

Perception of a rigid shape rotating in depth can occur with successive presentation of a trapezoid and its mirror image (Kolers & Pomerantz, 1971). Lappin, Donner, and Kottas (1980) presented two frames representing 512 random dots on the surface of a sphere rotating 5.6^{0} about a vertical axis through its center. Two-thirds of the observers spontaneously reported a perception of a sphere rotating in depth. Subjects could discriminate between perfectly correlated pairs of frames and pairs with a small number of uncorrelated dots. Lappin et al. reported that a high perspective level (a projection distance from the center of the sphere of two sphere radii) was required for detecting the three-dimensional structure in these displays. This latter result differs from those of Braunstein (1966) and Petersik (1980b) with longer rotation sequences, suggesting a possible interaction between perspective and either the number of frames or the rate of rotation required for the perception of three-dimensional structure.

4. Variables affecting perceived relative distance

4.1 Relative distance is ambiguous in parallel projections of silhouetted objects (actual or computer-simulated).

Until quite recently, observations of the perception of rotation in depth, based on approximate or true parallel projections, involved actual or computer-simulated silhouettes. In the windmill illusion, the blades of a distant mill and the millhouse must be silhouetted against the sky. Braunstein (1966, 1977b) and Petersik (1980b) used bright filled dots that, like shadows, merged when they overlapped in the projection. It is not surprising that observers in these studies could not exceed chance accuracy in their judgments of relative distance. In a true parallel projection in which uniformly filled shapes merge as they overlap, there is no relative distance information available. There is, indeed, no "correct" direction of rotation that can be assigned to these displays for use in evaluating the accuracy of subjects' responses.

4.2 Increasing polar perspective can increase the accuracy of relative distance perception in shadow projections.

This is obvious as an overall finding, but the specific relationships between perspective and the accuracy of direction of rotation judgments are far from obvious. These relationships, described in the following sections, reveal a great deal about how the visual system uses perspective information.

4.2.1 Convergence strongly dominates compression.

It is useful to subdivide perspective effects in a polar projection in order to make a more specific determination of what information is used by the human observer in judging relative distance. A classification that has been found related to observer performance in a number of studies contrasts perspective effects in the dimension of the axis of rotation to effects in the perpendicular dimension. These effects can be illustrated in a grid pattern rotated about a vertical axis (Figure 3a). Perspective effects in the vertical dimension result in convergence of the horizontal lines in the grid (Figure 3b), while effects in the horizontal dimension result in compression of the space between the vertical lines (Figure 3c). The relative effectiveness of convergence and compression in determining slant judgments in static projections has been studied by Attneave and Olson (1966) and by Gillam (1970). Both investigators found a clear dominance of convergence. Convergence and compression can be easily identified in dynamic displays of regular planar patterns. For convenience in discussion, these terms also will be used to refer to analogous effects in three-dimensional random-dot patterns (shape of the projected path vs. spacing between successive positions in the projection of a rotating dot).

Three experiments provide overwhelming evidence of the dominance of convergence over compression in dynamic displays. Braunstein and Payne (1968), in a study of perspective effects on the rotating trapezoid illusion, included rectangles and trapezoids rotating about a vertical axis with either convergence or compression eliminated. Eliminating compression had little effect on either the high accuracy of direction of rotation judgments for rectangles or the consistent reports of oscillation for trapezoids. With convergence eliminated, however, there was a sharp drop in both the accuracy of the judgments for rectangles and the consistency of the reported oscillations for trapezoids. Judgments for both figures approached chance expectations with convergence eliminated.

The strongest evidence for the dominance of convergence is found in a study intended to show that there are conditions under which compression is dominant (Hershberger, Stewart, & Laughlin, 1976). To obtain such conditions, extreme measures were required to emphasize compression and deemphasize convergence. The figure selected for rotation about a vertical axis consisted of 13 equally spaced vertical lines, 4-mm high. The 13 lines spanned a horizontal extent of 80 mm. Thus, not only was the horizontal extent of the figure 20 times the vertical extent, but the projected compression of 12 equal intervals provided compression information while convergence information was conveyed by variations in the projected lengths of the short vertical lines. Still, when compression and convergence were made to indicate opposite directions of rotation, judgments corresponded most often to the direction indicated by convergence. In order to obtain a condition in which compression was dominant, it was necessary to use a closer polar projection point in computing the horizontal perspective effects than that used in computing the vertical perspective effects. The extreme conditions required to demonstrate a dominance of compression over convergence--a 20:1 ratio in horizontal to vertical extent, 12 horizontal subdivisions, and a closer projection point for computing horizontal effects--provide a clear

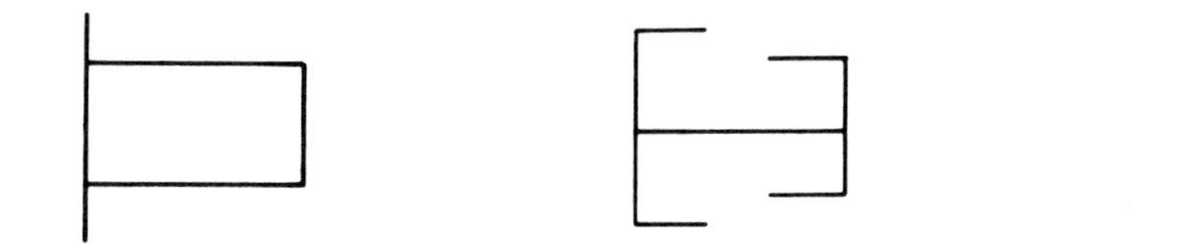

Figure 4. Right-angled figures with unequal vertical sides studied by Börjesson (left) and Braunstein (right).

Figure 5. A figure with equal vertical sides that elicits predictable reports of oscillation when rotated about a vertical axis.

demonstration of the robustness of convergence in determining judgments of direction of rotation.

The dominance of convergence effects over compression in determining direction of rotation judgments has also been found for three-dimensional patterns. Braunstein (1977b) had subjects report direction of rotation for random dot patterns representing spheres rotating about a vertical axis. At the highest perspective level--a projection point three radii from the center of the sphere--the proportion of correct judgments was .80 with both convergence and compression present, .84 with convergence only, and .51 with compression only. This result indicates that the dominance of convergence is not limited to the planar patterns previously studied.

4.2.2 The effectiveness of polar perspective varies with the shape of the rotating object.

Evidence of the effect of shape on the relationship between perspective and the accuracy of direction of rotation judgments comes from two sources. Several studies have varied both projection distance and shape, providing direct comparisons. Other comparisons can be made across studies using different shapes but comparable projection distances. The specific results will be summarized below, but an overall pattern emerges: Certain shapes are especially effective in conveying perspective changes, or, put another way, perspective effects are especially salient for some shapes. For other shapes, there is a general reduction in the accuracy of direction of rotation judgments at all polar perspective levels. There is a third category of shapes in which errors are systematically and predictably related to the shape of the figure.

4.2.2.1 Shapes with right angles provide the highest accuracy.

This conclusion is based on studies in which the axis of rotation is vertical and the sides of the right angle are horizontal and vertical. When the rotating figure is a rectangle, the presence of right angles is confounded with the presence of equal vertical sides. The effectiveness of right angles has been confirmed in figures designed to control for the effects of equal sides (Figure 4) in two studies using different figures (Börjesson, 1971; Braunstein, 1971).

4.2.2.2 Accuracy is markedly lower for circles and spheres.

Braunstein and Payne (1968) obtained direction of rotation judgments for a rectangle 2 units wide and a circle 2 units in diameter, and Braunstein (1977) obtained judgments for a sphere 2 units in diameter. With a projection distance 3 units from the center of each figure, the proportions of correct responses were .99, .58, and .80 for the rectangle, circle, and sphere. (Some of the rectangle's advantage could be related to its greater width at the top and bottom, but even when the projection distance was increased to 9 units the proportion correct for the rectangle was .98.)

4.2.2.3 Systematic errors in direction of rotation judgments occur when opposite vertical sides are enclosed by acute and obtuse angle.

This finding is among the most interesting in the literature on the perception of rotation in depth. It is the principal basis for the rotating trapezoid illusion (Ames, 1951). The effect can be obtained, however, with a similar arrangement of acute and obtuse angles when the enclosed sides are equal in length, as shown in Figure 5 (Braunstein, 1971). In the projection of a trapezoid rotating about an axis parallel to its parallel sides, the angles enclosing the larger side become increasingly acute when that side is approaching. Unless the projection distance is very short (relative to the dimensions of the trapezoid) the angles enclosing the larger side also become increasingly acute when that side is receding. The perceived oscillation of the trapezoid can be predicted from a perceptual heuristic that processes diverging contours as indicating approach of the enclosed side. As changes in the visual angles projected by the vertical sides provide veridical information about the direction of rotation, the proposed heuristic would have to override visual angle information. Contour convergence appears to be the dominant source of information about direction of rotation in this case, just as it is dominant when placed in conflict with compression.

All three of the above findings can be derived from the use of convergence as the principal source of information in determinining direction of rotation. The high accuracy found for figures with right angles can be attributed to the salience of convergence in these figures; the low accuracy for circles and spheres can be attributed to the lack of salient convergence information. Finally, the consistent errors in judgments of direction of rotation for trapezoids that occur in half of each rotation cycle appear to be related to the presence of salient but misleading convergence information.

4.3 Occlusion can provide accurate direction of rotation judgments in parallel projections.

The covering and uncovering of more distant objects by nearer objects, during movement of the objects or the observer, is a potential indicator of relative distance. Braunstein, Andersen, and Riefer (1982) studied two types of occlusion that can occur during rotation in depth: (1) edge occlusion--represented by an opaque sphere with texture elements on its surface that were occluded as they rounded the edge, and (2) element occlusion--represented by a transparent sphere with the nearer

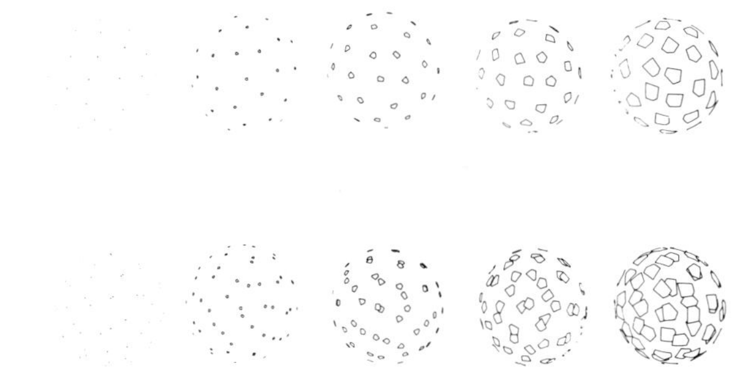

Figure 6. Single frames from edge-occlusion (top) and element-occlusion (bottom) sequences for each level of texture-element size. (The stimulus film showed white figures on a black background.)

elements occluding the more distant elements. The number of element occlusions was determined by element size. There were five size levels (Figure 6). Element density on the surface was constant. All of the stimuli were parallel projections. The accuracy of direction of rotation judgments (Figure 7) was consistently high for edge occlusion, exceeding 90% for the largest element sizes. For element occlusion, accuracy increased from chance for the smallest element size (filled "dots" that did not provide occlusion information) to over 80% at the largest element size.

That experiment did not isolate dynamic occlusion effects for the element occlusion displays, as static interposition was available in single frames of these displays. A second experiment was conducted to isolate dynamic effects in element occlusion for objects rotating in depth (Andersen & Braunstein, 1982). Stimuli were generated in which the contours of the pentagons were not drawn. Instead, dots were randomly located within the area occupied by each pentagon. When one of these implicit pentagons moved in front of another, any dot in the covered area of the more distant pentagon was

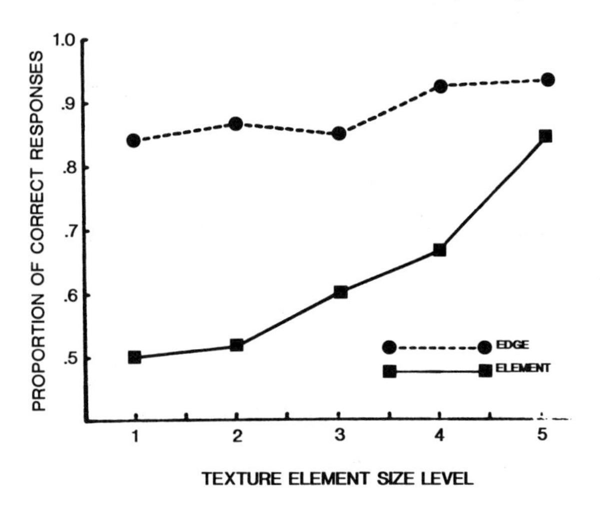

Figure 7. Effects of texture-element size on proportion of correct responses for edge and element occlusion.

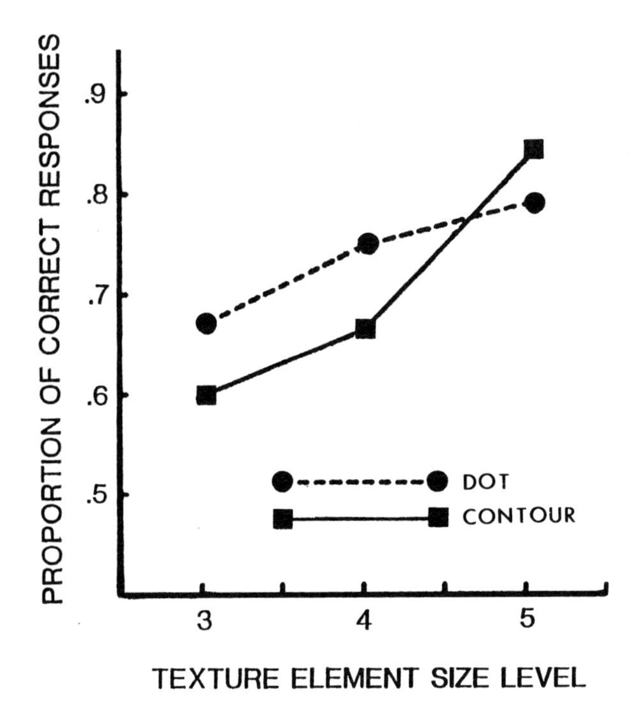

Figure 8. Comparison of proportions of correct responses for dot and contour stimuli displaying element occlusion.

deleted from the display. These displays provided the type of pure kinetic occlusion studied with translating planes by Kaplan (1969) and Rhodes (1980), and by Yonas and Granrud (1982) with infants. As with the contoured elements, the accuracy of direction of rotation responses varied with element size. The levels of accuracy were comparable to those found with contoured pentagons (Figure 8), indicating that the effectiveness of element occlusion does not require the presence of explicit contours--providing static interposition information--but can be obtained with purely kinetic contours.

5. Conclusion

It is clear that rotation in depth provides information to human observers about three-dimensional relationships. Geometric analyses of the projective changes that occur during this transformation are useful in describing the information that is potentially available, but the use, disuse or misuse of this information by the observer is a subject for empirical research. The present paper has described the major findings of this research: Human observers are able to use the information in a parallel projection of an object rotating in depth, possibly the sinusoidal waveform of the projected velocity, in perceiving rigid rotation. Little use is made of compression in making direction of rotation judgments from polar projections when convergence is available as an alternative indicator of relative distance. Contour convergence is misused in judgments of direction of rotation for trapezoids, even when visual angle changes could provide a veridical indication of relative distance. The observer can use alternative indicators of relative distance--the addition of polar perspective or dynamic occlusion--in resolving the relative distance ambiguity in parallel projections.

The emphasis in many of the psychophysical studies summarized above has been on the perception of direction of rotation or on the overall impression of depth or coherence elicited by rotating objects. Computational approaches, on the other hand, have usually emphasized the three-dimensional structure or shape of rotating objects. As both approaches are extended, there should be increasing opportunities to relate perceptual processes suggested by psychophysical data to those proposed in computational models. In doing so, it is useful to note that the processes used by the human observer may not be optimum from either a criterion of assuring veridicality of judgments or efficiency in machine computation. Other criteria may determine the development of these processes--speed in processing biologically important information, bias toward avoiding biologically serious errors, usefulness at earlier stages of evolution, and effectiveness with degraded viewing conditions are some examples. It is important, therefore, that models of movement perception be firmly based on empirical findings if they are to be applied to human observers.

Acknowledgment

I wish to thank George J. Andersen for valuable comments on an earlier draft of this paper.

References

Ames, A. Visual perception and the rotating trapezoidal window. Psychological Monographs, 1951, 67 (7, Whole No. 324).

Andersen, G. J., & Braunstein, M. L. Isolating kinetic occlusion in parallel projections of rotation in depth. Paper presented at the meeting of the Psychonomic Society, Minneapolis, November 1982.

Attneave, R., & Olson, R. K. Inferences about visual mechanisms from monocular depth effects. Psychonomic Science, 1966, 4, 133-134.

Boring, E. G. Sensation and perception in the history of experimental psychology. New York: Appleton-Century-Crofts, 1942.

Börjesson, E. Properties of changing patterns evoking visually perceived oscillation. Perception & Psychophysics, 1971, 9, 303-308.

Braunstein, M. L. Depth perception in rotating dot patterns: Effects of numerosity and perspective. Journal of Experimental Psychology. 1962, 64, 415-420.

Braunstein, M. L. Sensitivity of the observer to transformations of the visual field. Journal of Experimental Psychology, 1966, 72, 683-689.

Braunstein, M. L. Perception of rotation in figures with rectangular and trapezoidal features. Journal of Experimental Psychology. 1971, 91, 25-29.

Braunstein, M. L. Minimal conditions for the perception of rotary motion. Scandinavian Journal of Psychology, 1977, 18, 216-223. (a)

Braunstein, M. L. Perceived direction of rotation of simulated three-dimensional patterns. Perception & Psychophysics, 1977, 21, 553-557. (b)

Braunstein, M. L., Andersen, G. J., & Riefer, D. M. The use of occlusion to resolve ambiguity in parallel projections. Perception & Psychophysics, 1982, 31, 261-267.

Braunstein, M. L., & Payne, J. W. Perspective and the rotating trapezoid. Journal of the Optical Society of America, 1968, 58, 399-403.

Gillam, B. Judgments of slant on the basis of foreshortening. Scandinavian Journal of Psychology, 1970, 11, 31-34.

Green, B. F., Jr. Mathematical notes on 3-D rotations, 2-D perspective transformations, and dot configurations. Group Report No. 58-5, Massachusetts Institute of Technology, Lincoln Laboratory, 1959.

Green, B. F., Jr. Figure coherence in the kinetic depth effect. Journal of Experimental Psychology, 1961, 62, 272-282.

Hershberger, W. A., Stewart, M. R., & Laughlin, N. K. Conflicting motion perspective simulating simultaneous clockwise and counterclockwise rotation in depth. Journal of Experimental Psychology: Human Perception and Performance, 1976, 2, 174-178.

Johansson, G. Spatial constancy and motion in visual perception. In W. Epstein (Ed.), Stability and constancy in visual perception: Mechanisms and processes. New York: Wiley, 1977.

Kaplan, G. A. Kinetic disruption of optical texture: The perception of depth at an edge. Perception & Psychophysics, 1969, 6, 193-198.

Kolers, P. A., & Pomerantz, J. R. Figural change in apparent motion. Journal of Experimental Psychology, 1971, 87, 99-108.

Lappin, J. S., Doner, J. F., & Kottas, B. L. Minimal conditions for the visual detection of structure and motion in three dimensions. Science, 1980, 209, 717-719.

Miles, W. R. Movement interpretations of the silhouette of a revolving fan. American Journal of Psychology, 1931, 43, 392-405.

Petersik, J. T. Rotation judgments and depth judgments: Separate or dependent processes. Perception & Psvchophysics, 1980, 27, 588-590. (a)

Petersik, J. T. The effects of spatial and temporal factors on the perception of stroboscopic rotation simulations. Perception, 1980, 9, 271-283. (b)

Rhodes, J. W. The effect of multiple depth cues in the perception of occlusion. Unpublished doctoral dissertation, University of Tennessee, Knoxville, 1980.

Ullman, S. The interpretation of visual motion. Cambridge, Mass.: MIT Press, 1979.

Wallach, H., & O'Connell, D. N. The kinetic depth effect. Journal of Experimental Psychology, 1953, 45, 205-217.

Yonas, A., Granrud, C. E., & Smith, I. M. Infants perceive accretion/ deletion information for depth. Investigative Ophthalmology and Visual Science, 1982, 22 (3, Supplement), 124.

The Cross-Ratio and the Perception of Motion and Structure

William A. Simpson
Department of Psychology
University of Toronto
Toronto, Ont.

ABSTRACT

Followers of J.J. Gibson have proposed that the cross-ratio, a projective invariant for four collinear points, underlies the perception of objects in motion. Experiment 1 tested this theory by presenting subjects with displays of 3 or 4 dots rotating in depth. Accuracy was equally high in both conditions for motion and structure judgements, so the cross-ratio cannot be necessary. Experiments 2 and 3 tested the cue of lining up, and some evidence for its use was found. The results are consistent with an analysis based on the sinusoidally changing positions of the dots.

INTRODUCTION

Perhaps the most important problem in perception is how the ever-changing two-dimensional image cast on the retina comes to be interpreted as the motion of three-dimensional objects. The problem was of special interest to J.J. Gibson, who thought psychological theories and experiments dealing with static viewers and displays irrelevant since perception is essentially concerned with motion.

Gibson analysed the problem as follows. A moving object delivers a continuous projective transformation to the eye. This transformation contains two sources of information--'the transformation as such is one kind of stimulus information, for motion, and...the invariants under transformation are another kind of stimulus information, for the constant properties of the object' (Gibson, 1966, p.145).

Unfortunately, Gibson was never more precise than this. His followers, though, did try to make the analysis less vacuous. Hay (1966) made a valiant bid to specify what in 'the transformation as such' could be used to determine an object's motion. He had to concede in the end that 'there is far from a direct one-to-one correspondence between the parameters of image transformation and those of object displacement' (p.563). In other words, Hay was unable to say how the parameters relate to the perception of motion.

There have also been attempts to specify what invariants lie behind the perception of structure. Gibson, Owsley & Johnston (1978) and Shaw & Pittenger (1977) propose the cross-ratio. Given a line

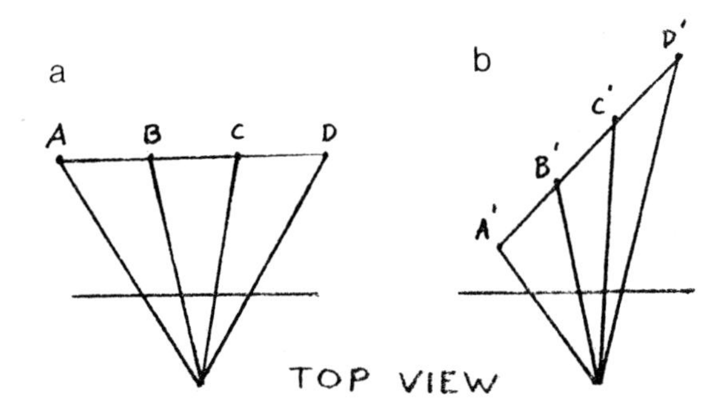

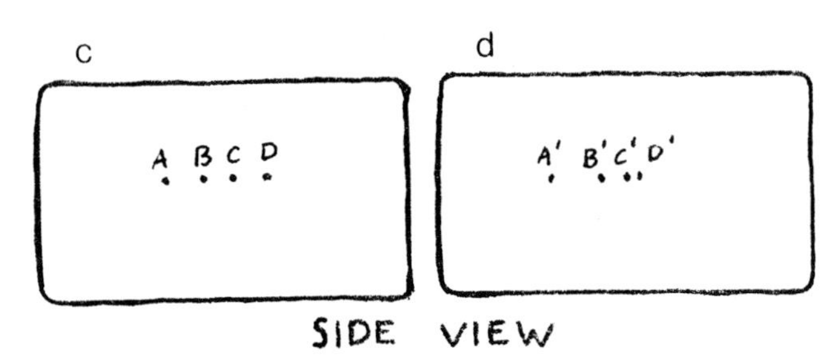

Figure 1. As the dots in (a) are rotated to their position in (b), their projection changes from (c) to (d). The cross-ratio is constant.

Motion: Representation and Perception
N.I. Badler and J.K. Tsotsos, Editors
Published by Elsevier Science Publishing Co., Inc.

divided by four points ABCD, the cross-ratio of the segments

$$(AC/CB)/(AD/DB)$$

is an invariant under projective transformation. As can be seen in Figure 1, the projections of the rotating line ABCD change quite a bit, but their cross-ratios are equal. The cross-ratio is a constant only if the points are collinear; hence it could be used to determine structure.

It is an easy matter to test this theory directly. If viewers can determine the collinearity of three-dot displays (where the cross-ratio is not available), then the cross-ratio cannot be necessary. Experiment 1 tests subjects' performance with three and four dot displays.

If the cross-ratio theory is eliminated, several others await refutation. Johansson (1975) and Borjesson & von Hofsten (1973) argue for perceptual vector analysis into common and relative components. Ullman's polar-parallel scheme requires three distinct views of four noncoplanar points to derive their motion and structure. Longuet-Higgins & Prazdny's (1980) account is based on instantaneous vectors of the optic flow field. Todd's (1982) analysis considers trajectories of moving elements over an extended distance and time period as primitives.

Despite the surfeit of existing theories, yet another was examined in Experiments 2 and 3. The very simple (and situation-specific) cue of 'lining up' was tested. If collinear dots rotating in depth are viewed edge-on, they will all line up (coincide) twice per cycle. If they are not collinear, they will not all line up (see Figure 2).

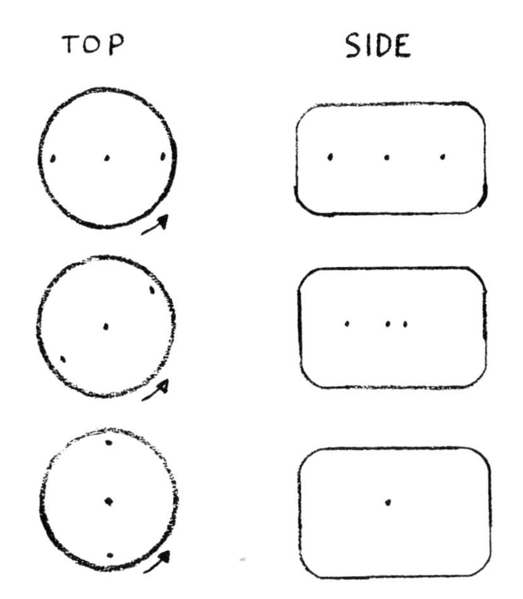

Figure 2. As collinear dots rotate, their projections approach and coincide.

If the lining up cue is used to find the structure of moving displays, removing it should decrement performance. Experiment 2 removed line-ups by blocking the subjects' view of them; Experiment 3 used oscillating displays which either moved through positions that yielded line-ups or did not.

EXPERIMENT 1

The cross-ratio theory of the perception of structure and motion predicts that these judgements are possible only for displays with at least four elements. If subjects can determine collinearity of three-element displays rotating in depth, then the cross-ratio cannot be necessary.

Method

Stimuli. The stimuli were produced by videotaping Q-tips (painted flat black except for the white cotton ends) inserted in a turntable rotating at 12 rpm. The display subtended 11.3 deg of visual angle and the dots were about 1.2 min wide by 3.6 min long. High contrast was used in taping; only the white dots were visible. The dots rotated about the vertical axis and the shooting angle was in the same plane as the rotation (i.e. edge-on). Thus the dots moved to and fro along a horizontal line in the middle of the TV screen.

Four factors were varied in generating the stimuli: collinearity (collinear or not collinear; in the latter case, one dot deviated by 2 deg when the other dots coincided), number (3 or 4 dots), axis of rotation (in the same plane as the dots, or off of it by 2.3 deg), and spacing (even or gradient). Thus there were 32 stimuli in all. Each lasted 15 sec (3 rotations), and the ISI was 5 sec.

Apparatus. Subjects viewed the stimuli on a TV monitor. Viewing was monocular from 1.5 m. This ensured the same projection as that delivered to the camera. The perspectivity (defined as distance from farthest point to camera divided by distance from nearest point to camera) was 1.2.

Procedure. The stimuli were viewed in the same random order by all subjects. Either during or after viewing each one, the subjects described its structure (collinear or not) and motion (in depth or in the plane; rotating, oscillating, nonrigid or random motion). Pains were taken to make clear that the collinearity description was to be based on how the dots were arranged in depth (i.e., how they would look if viewed from above), and not their 2-D arrangement on the TV screen. Examples of collinear and noncollinear Q-tips were shown. The motion descriptions were also demonstrated. Nonrigid motion was described as stretching, twisting, or bending; random motion was described as unintelligible or unpatterned movements.

Subjects. The subjects were 12 students of psychology, all naive as to the rationale underlying the experiments. The same subjects served in all experiments, the order of experiments being counterbalanced.

Results

As expected, subjects were able to determine the structure and motion of three-dot displays. In fact, there was no statistical difference between three- and four-dot performance--in both cases accuracy was near 90% (Structure: 3-dot=89.6%, 4-dot=95.6%; Motion: 3-dot=89.6%, 4-dot=92.7%).

Discussion

The cross-ratio cannot be necessary for the derivation of structure and motion since performance is near perfect when the cross-ratio is not present. Other schemes would also have difficulty with the stimuli used. The Johansson (1975) account is not applicable since the stimuli had no common motion vectors, and thus no partitioning into common and relative motion is possible. Ullman (1979) and Longuet-Higgins & Prazdny (1980) predict difficulties with collinear (coplanar) displays which were not found. Todd's (1982) scheme uses seven parameters, only two of which were really necessary (frequency and phase).

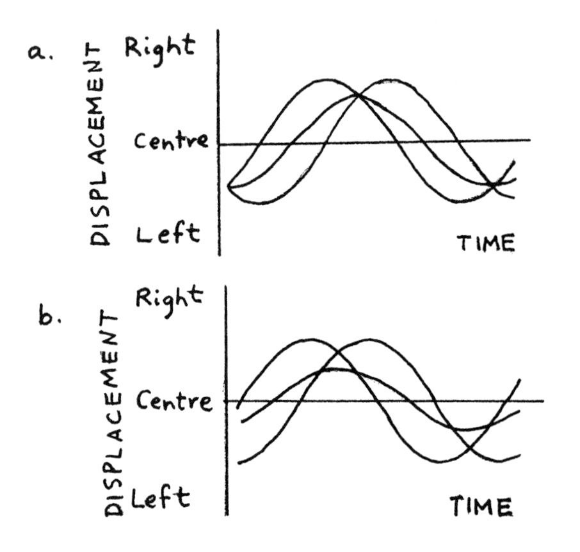

Figure 3. The displacement paths generated by rotating collinear(a) and noncollinear (b) points viewed edge on. Collinear points' paths meet at a common intersection; noncollinear points' paths do not.

EXPERIMENT 2

If a set of collinear dots is viewed edge on, they will all coincide twice per cycle. This is plainly seen when displacement is plotted against time (Figure 3a). Where all curves meet is a line-up. When the dots are not collinear, they never all line up simultaneously (Figure 3b). This lining up cue can be eliminated by blocking the subjects' view of the coincidences.

Method

Stimuli. The stimuli were videotaped as in Experiment 1. In this experiment three factors were varied: collinearity (collinear or not), number (3 or 4 dots), and blocking (line-ups or not). The last factor was varied as follows. Line-

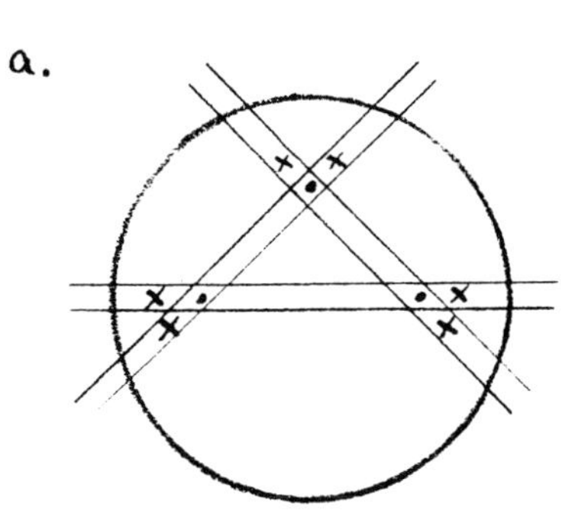

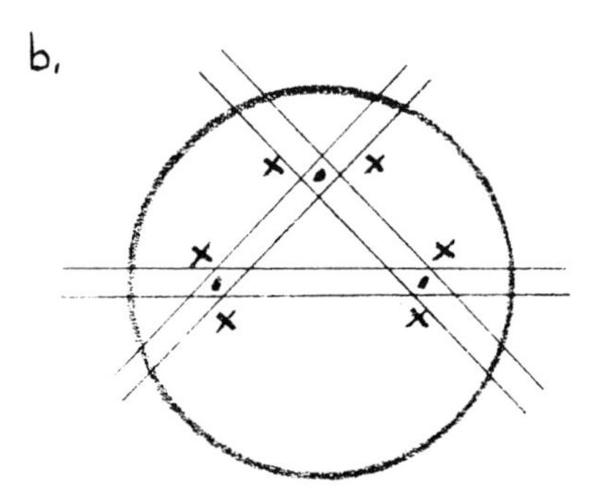

Figure 4. The method used to block line-ups and nonline-ups. Each figure is a top view of the turntable. Line-ups are blocked by placing blocks in the alleys (a).Nonline-ups are blocked by putting them outside the alleys.

ups were blocked by placing 4 min wide cardboard occluders directly in front of the lined up dots on the turntable (see Figure 4a). Each line-up blocked stimulus had a control stimulus with the same number of occluders but placed anywhere except in front of line-ups (Figure 4b). Both gave a flickering appearance, the dots flickering off when occluded.

Apparatus, Procedure, and Subjects. As in Experiment 1.

Results

The ANOVA showed a main effect for blocking on motion judgements, $F\ (1,11)=5.5$, $p<.05$. Fewer correct descriptions were given of the stimuli's motion when the line-ups were blocked from view (85% correct) than when nonline-ups were blocked (94%). There was no effect for structure judgements.

Discussion

The lining up cue seems to play a role in the perception of motion, but not of structure. Perhaps the reason why structure judgements were unaffected by the lack of line-up information is that another source of information was available. When collinear dots' line-ups are blocked, the display is blank for an instant; noncollinear always have at least one dot visible. Whether this cue is used is tested in Experiment 3.

EXPERIMENT 3

A cue other than lining up may have been used to determine collinearity in Experiment 2. If collinear dots' line-ups are blocked, the screen goes blank momentarily; noncollinear displays always have at least one dot visible. This cue can be eliminated by oscillating the dots so they never line up.

Method

Stimuli. The stimuli were videotaped as in the previous experiments. Three factors were varied: collinearity (collinear or not), number (3 or 4 dots), and lining up (present or not present). The dots were oscillated such that the arc through which they moved either included or excluded the line-up positions.

Apparatus, Procedure, and Subjects. As in Experiments 1 and 2.

Results

The ANOVA for structure showed a significant number X line-up interaction, $F\ (1,11)=5.5$, $p<.05$. As Figure 5 shows, accuracy for three and four dot displays was equal when line-ups were present, but when not present, structure was more accurately recovered with three dot displays.

Discussion

Lining up seems to be used in judging structure in four dot patterns but not three dot ones. This result could be attributed to a confound of number of dots and amount of arc through which the display was oscillated. A four dot pattern, since its dots are more closely spaced, must be oscillated through fewer degrees of arc to prevent crossing of the dots' projections (line-ups) than a three dot pattern. However, if this explanation is correct, it is strange that motion judgements were not affected in the same way.

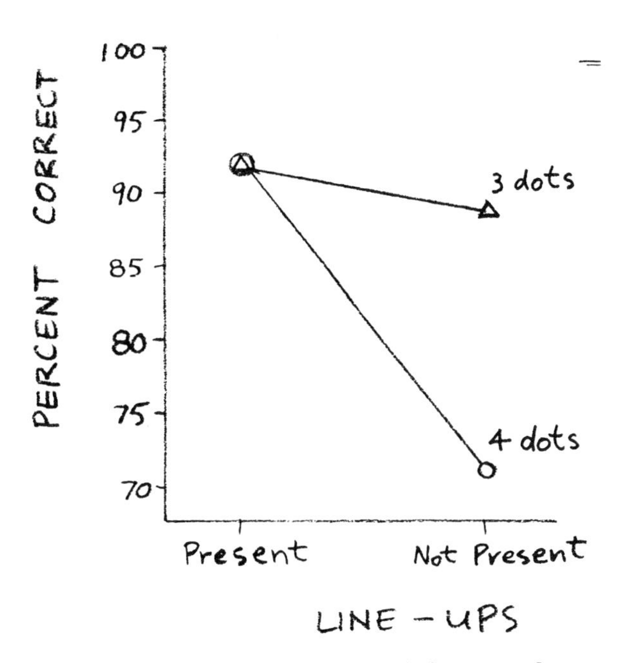

Figure 5. Accuracy of structure judgements for displays with 3 or 4 dots whose oscillations included or excluded line-ups (Experiment 3).

GENERAL DISCUSSION

As a result of these experiments, we can safely conclude that the cross-ratio is not necessary for the perception of structure and motion, and hence cannot be entertained as a general scheme. Some evidence was found for the specific cue of lining up. Overall, the best candidate amongst existing schemes for interpreting structure from motion seems to be Todd's (1982) trajectory analysis.

It might be simpler, however, to consider the displacements of the dots as a function of time Figures 3 and 6). Their sinusoidally varying positions, velocities, and accelerations reflect what physicists call simple harmonic motion. For the simple one-dimensional harmonic motion used in Experiments 1 and 2, derivation of structure would proceed as follows. First, position of the dots is plotted against time. The maxima of the curves represent the distance of the dots from the axis of rotation. The angle separating the dots is obtained from the phase differences between the displacement curves. The motion is given by the shape of the curves--for constant angular velocity rotation, the curves are cosine functions.

The analysis is not restricted to such simple cases. The dots need not rotate at constant angular velocity--if they do not, their phase differences and amplitudes are still readily available. The dots' motion need not be rigid--if their frequencies differ, nonrigid motion is recovered. Translation is also easily dealt with, either alone or in combination with rotation. Most importantly, two-dimensional motions on the projection plane are susceptible to the same rules of interpretation.

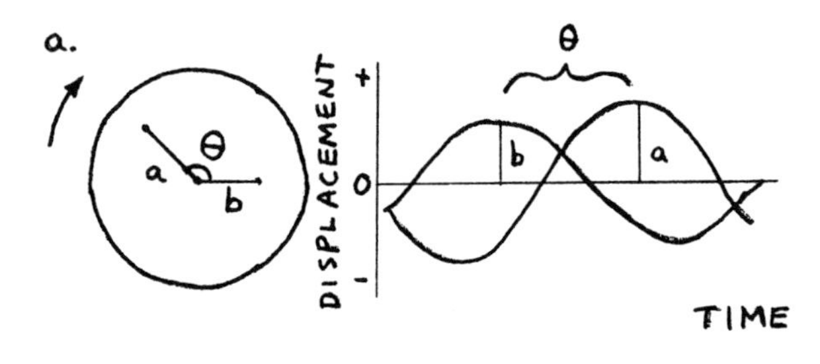

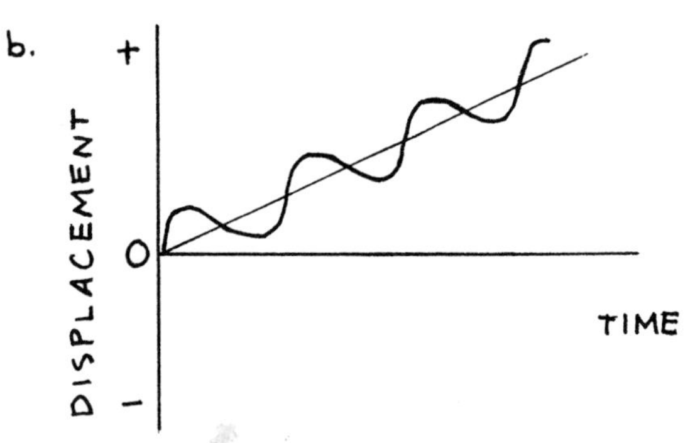

Figure 6. If the dots on the turntable in (a) are viewed edge on, they generate the sinusoidal curves to the right. The distance of dots a and b from the axis of rotation is given by the maxima of the curves; the angle separating them, θ, is given by the phase difference. A dot rotating about a vertical axis and translating from left to right generates the curve in (b). The slope of the curve's zero-crossing gives the translation speed.

Such a scheme based on the positions of the dots over time is very simple, and is immune to difficulties which plague other approaches. In contrast to optic flow schemes (Lee,1974; Koenderink & van Doorn, 1976; Longuet-Higgins & Prazdny, 1980), no complicated mathematics is required, and no difficulties are encountered in separating translational and rotational components. It encounters no problems with planar objects (unlike the schemes of Ullman, 1979, and Longuet-Higgins & Prazdny, 1980) and requires only one point (versus Ullman's 4-5). The number of views required to fit the curve is an empirical question--conceivably, a function could be fit to only one data point (one dot at one time and one position). Unlike many schemes, the rigidity assumption is done away with. It seems that the analysis of motion into its sinusoidal components constitutes a powerful method of recovering structure from motion.

REFERENCES

Borjesson, E. & von Hofsten, C. Visual perception of motion in depth: application of a vector model to three dot motion patterns. Perception & Psychophysics,1973, 13, 169-179.

Gibson, E.J., Owsley, C.J., & Johnston, J. Perception of invariants by five-month-old infants: differentiation of two types of motion. Developmental Psychology, 1978, 14, 407-415.

Gibson, J.J. The problem of temporal order in stimulation and perception. Journal of Psychology, 1966, 62, 141-149.

Hay, C.J. Optical motions and space perception--an extension of Gibson's analysis. Psychological Review, 1966, 73, 550-565.

Johansson, G. Visual motion perception. Scientific American, 1975, 232 (6), 76-88.

Koenderink, J.J. & van Doorn, A.J. Local structure of movement parallax of the plane. Journal of the Optical Society of America, 66, 717-723.

Lee, D.N. Visual information during locomotion. In R.B.MacLeod & H.Pick (Eds.), Perception. Ithaca, N.Y.: Cornell University Press, 1974.

Longuet-Higgins, H.C. & Prazdny, K. The interpretation of a moving retinal image. Proceedings of the Royal Society of London, Series B, 1980, 208, 385-397.

Shaw, R.E., & Pittenger, J. Perceiving the face of change in changing faces: implications for a theory of object perception. In R.E.Shaw & J.Bransford (Eds.), Perceiving, Acting and Knowing. Hillsdale, N.J.: Erlbaum, 1977.

Todd, J.T. Visual information about rigid and non-rigid motion: a geometric analysis. Journal of Experimental Psychology: Human Perception and Performance, 1982, 8, 238-252.

Ullman, S. The interpretation of structure from motion. Proceedings of the Royal Society of London, Series B, 1979, 203, 405-426.

Selective Attention to Aspects of Motion Configurations: Common vs. Relative Motion

James R. Pomerantz and Nelson Toth
Department of Psychology, SUNY at Buffalo
Buffalo, NY

Abstract

The motion of a dot configuration may be described as the sum of its relative (part) and common (whole) motion components. Is either of these two component dimensions extracted before the other in human perception? Reaction time data from selective attention experiments show that neither dimension can be responded to without interference from the other, implying that neither is processed more quickly than or ahead of the other. Following Garner's nomenclature, common and relative motions appear to act as either symmetrically integral or configural dimensions, depending on the particular motion configurations tested.

The experiments to be described below are motivated by the question, Do we see the forest before the trees in motion perception? Many of the puzzles about motion perception that interest psychologists concern the perception of relatively complex configurations that are set in motion. With such configurations, one may attend to the motion trajectory of the configuration as a whole (the forest) or alternatively to the motion of the various parts within the configuration (the trees). The experiments below are somewhat analogous to studies that have been reported in the literature on visual form perception, where attention may similarly be directed to global shapes or to local features (Navon, 1977; Pomerantz, in press).

This research deals with the perception of moving constellations of lights. Many of us are familiar with stimuli of this sort from the classic work of Duncker, Rubin, and Johansson. Johansson's demonstrations make it clear that we do not ordinarily perceive the absolute trajectories of these lights as they move across the screen. Instead, we organize their motion into two vector components. One is the component common to all the lights, or in other words the motion of the forest; the other is the relative motion of lights within the whole, or the motion of the trees. (Of course, even common motion must be perceived as relative to some external frame of reference, but following the customary practice in the literature we restrict the term "relative" to refer to motions that are internal to a configuration of lights.)

To elaborate, consider Johansson's well-known L-shaped motion configuration (Figure 1). At the level of absolute motions of the individual elements, one light moves downward while another moves leftward to form an L. But at the level of the whole constellation, the relative component is the pure approach of the two lights toward their midpoint, whereas the common component is the motion of the imaginary midpoint itself. Frequently, it is these common and relative components that humans perceive: the lights are seen to approach one another along a diagonal path while the whole pair moves along a perpendicular diagonal path. Under these circumstances, observers may fail to realize that the absolute trajectories of the lights follow horizontal and vertical paths.

The question we raise is whether these two dimensions of motion, common and relative, are separable in Garner's (1974) sense and so are capable of being perceived independently of each other; or whether they are mandatorily combined in perception and so are integral or configural dimensions. If they are separable, then either dimension should be capable of being processed and responded to without interference from the other; otherwise, interference should appear. We also ask whether these dimensions are processed symmetrically. If relative and common motion are extracted sequentially or in a hierarchical fashion, then asymmetries should appear, with the primary dimension (the one processed first or receiving higher priority) being separable from the secondary dimension, but with the secondary dimension being integral with the primary.

Cutting and Proffitt (1982) have suggested that relative motion may be processed with a higher priority than common motion in rolling wheel displays, where lights are mounted on the rim and hub of an otherwise unseen, rolling wheel. Once relative motion is determined perceptually, it may be subtracted in vector fashion from the absolute motion to yield the common motion as residual. This scheme may be contrasted with the alternative view, advocated at times by Johansson, that common motion is extracted first, leaving relative as

Motion: Representation and Perception
N.I. Badler and J.K. Tsotsos, Editors
Published by Elsevier Science Publishing Co., Inc.

residual.

Cutting and Proffitt's argument is based not simply on the order of detecting relative and common motions but on the order of minimizing these vectors to their simplest states. Their two-light displays were multistable and could be perceived either as lights on a rolling wheel or as lights mounted on opposite ends of a tumbling stick. According to their reasoning, which of these two organizations should be perceived depends on whether the common or the relative components is minimized first. For example, consider the case of two lights mounted 90 deg apart on the rim of the wheel. If common motion were minimized first, subjects should perceive a rolling wheel, because the common motion would be a simple linear translation of the unseen hub of the wheel; the residual relative motion would be revolution of the two lights about this hub. On the other hand, if relative motion were minimized first, subjects should perceive two two lights as if they were attached to opposite ends of a rigid stick that was spinning about its midpoint, because this circular motion is the simplest description possible for the relative motion of the lights; the residual common motion would follow a more complex path described by a prolate cycloid. Because their subjects perceived the latter organization more frequent than the former, Cutting and Proffitt concluded that relative motion is normally processed first.

Our experiments today approach this same problem from a different perspective. Rather than ask observers to describe or compare their organizations of motion configurations, we ask them to discriminate as quickly as possible between configurations that differ in their relative and common motion components. The speed and accuracy of these discriminations should be determined in part by how common and relative motions are handled in the visual system.

In particular, if relative and common motion can be perceived independently, they should act as separable dimensions in information-processing tasks. Accordingly, subjects should be able to attend selectively to either component and make speeded classification responses to either without interference from the other. Many dimensions of stimuli have been found to be separable, including the color and size of a form, for example.

If common and relative motion are not separable dimensions, they could be asymmetrically integral. This would indicate that the two exist in a nested or hierarchical relationship to one another. In such a case, one dimension could be attended to selectively from the other, but not vice versa. An example of an asymmetrically integral pair is the pitch and place of articulation of a spoken monosyllable. One may attend selectively to the fundamental frequency (pitch) of a speech sound such as "bae" or "gae" while ignoring its place of articulation (the initial phoneme), but not vice versa.

If common and relative motions are neither separable nor asymmetrically integral, they could be either symmetrically integral or configural. Integral dimensions are processed jointly and interactively, and selective attention to either component is difficult. An example is the brightness and saturation of a single color chip. With configural dimensions, on the other hand, the subject appears to be attending to neither of the dimensions used to construct the stimuli; instead, the subject reorganizes the stimulus into a new set of dimensions or emergent features. A good example of configural dimensions are the parenthesis pairs used by Pomerantz & Garner (1973) and by Pomerantz & Schwaitzberg (1975; see Figure 2A).

Integral and configural dimensions may be distinguished in a number of ways, but one major difference is that with configural dimensions, subjects are faster and more accurate in divided attention tasks, which require attending to both dimensions, than in selective attention tasks, where only one dimension need be processed. With the parentheses stimuli, the forest is clearly seen before the trees; responding to the whole configuration is easy, whereas responding to any part is much more difficult.

The moving stimuli we used in the present experiments are shown in Figure 3. The panels of this figure show a constellation of four lights moving with various relative and common motions. The dots always originated in a square configuration and moved with vectors shown by the arrows. In the upper left panel, the four dots move as a unit to the right; in the upper right panel, they move upward as a unit. Thus, these two panels contain only common motion and no relative motion. The next row shows only relative motion, which is horizontal in the left panel and vertical in the right. No common motion exists in these displays, since as a whole they are not going anywhere; they are simply stretching around a stationary midpoint. The third row shows a mixture of simultaneous relative and common motions that are congruent, or move along the same axis. For example, in the left panel the lights move horizontally as a whole while they stretch horizontally at the same time, as indicated by the longer vectors on the righthand lights. The bottom row shows incongruent motion between the common and relative vectors, where the dots move horizontally as a whole while they stretch vertically, or vice versa.

Experiment 1

This experiment measured the success or failure of selective attention with a Stroop-like technique adapted from Navon's (1977) form perception experiments. The subject's task was to decide as quickly as possible about the direction of either the common or relative motion. In one set of trial blocks, they were to attend only to the common motion and ignore any relative motion that might appear. They were to press one button if the common motion were horizontal and another button if it were vertical. In the other set of blocks they were to attend only to the relative motion and make the same horizontal/vertical decision. Failure of selective attention was assessed by the amount of interference (increase in RT) observed when the irrelevant dimension called for the opposite response from the relevant dimension.

The moving lights were presented on the screen of a fast-phosphor (P-11) oscilloscope driven by a computer. The velocity of the relevant common and relative motions were both 2.4 deg/sec, whereas the velocities of the irrelevant motions were varied orthogonally across three different speeds of 1.2, 2.4 and 4.8 deg/sec. Most importantly, these three speeds were identical for common and relative motion, so that neither motion was made easier to perceive or to ignore by virtue of its physical velocity.

The experiment was run both in dim ambient light where the edges of the display device were visible, and in the dark where they were not. Twelve subjects were run in a total of 864 trials each, not counting 144 practice trials. In one-third of these trials (the baseline or control conditions), the lights moved horizontally or vertically along only the relevant dimension (common <u>or</u> relative motion). In another third of the trials (the congruent conditions), motion was present on both the relevant and the irrelevant dimension, but these two were congruent, or in the same direction. In the final third of the trials (the incongruent conditions), these two motions were incongruent or in perpendicular directions. The three types of trials were mixed in a random sequence counterbalanced across subjects. Reaction times (RTs, measured from the onset of motion until a response was made) and error rates were recorded. The instructions to subjects emphasized both speed and accuracy.

<u>Results</u>

The data of principal interest are from trials run in dim ambient light, and they are shown in Table 1. These data were submitted to an analysis of variance involving three factors: Motion Type (common vs. relative), Motion Direction (horizontal vs. vertical) and Irrelevant Motion (baseline control, congruent and incongruent).

Motion Type had no significant main effect on RTs, $F(1,11) < 1$, nor did it enter into any significant interactions. In the baseline conditions where there was no motion on the irrelevant dimension, responses were almost identically fast and accurate to common (623 msec) and to relative (629) motion. That is, the horizontal/vertical decisions were made equally quickly on common and on relative motions that were of equal physical magnitude. This is one bit of evidence that neither relative nor common motion is processed more quickly than the other.

Motion Direction also had no significant effect, $F(1,11) < 1$, nor did it enter into any significant interactions. That is, subjects responded equally quickly regardless of whether the relevant motion was in the horizontal or the vertical direction.

Lastly, the type of Irrelevant Motion had a substantial effect on RTs, $F(2,22) = 16.0$, $p < .001$. RTs were much longer in the Incongruent condition than in either the Congruent or the Control conditions, which in turn were about equal. Most importantly, the absence of a Motion Type by Irrelevant Motion interaction, $F(2,22) = 1.19$, $p > .20$, indicates that common and relative motion behaved in a similar, symmetrical fashion in the three conditions of the experiment. Table 1 does suggest a performance difference between common and relative in the Congruent conditions, but as the above analyses showed, this difference was not significant.

Thus, neither dimension could be attended to selectively without interference from an incongruent direction of motion on the other. This clearly indicates that the dimensions of relative and common motion are not separable. Further, this interference was almost exactly symmetrical for relative (263 msec) and common (260) motions. If either of these two dimensions were processed more quickly than (or ahead of) the other in stimuli containing both, one would have expected an <u>asymmetry</u>, with the slower or less attention-demanding dimension being easier to ignore. In sum, our results indicate that neither dimension enjoys a natural priority or precedence in perceptual processing.

Table 2 shows the corresponding results from the dark conditions, where the edges of the display screen were not visible. The main result here is that responding to common motion becomes quite difficult in the dark, even in the baseline conditions. This is not surprising, because here the common motion has no external reference frame, and so the task is one of detecting an absolute motion. Similarly, the interference from incongruity on the irrelevant dimension is asymmetric, with relative motion being much harder to ignore than common (806 vs. 407 msec interference). This result follows directly from the poorer performance on common motion in the baseline conditions. Note that incongruency from common motion produced substantial interference, although this interference was smaller than that from relative motion. Finally, congruent common motion had a slight (but not significant) facilitating effect when relative motion was being judged, but congruent relative motion actually produced substantial interference (compared with no relative motion) when common was being judged. This effect deserves further exploration, since it is rare in the human information processing literature for an irrelevant but congruent source of information to interfere with performance. The result probably reflects a masking of the less perceptible common motion by the more salient relative motion.

To summarize, in the dark, relative motion gets in much more quickly than common. A large asymmetry is present, with relative interfering more with common than vice versa (although even the latter interference effect is substantial). But this asymmetry is fully explained by the baseline difference in discriminability and tells us little about the ordering of priorities assigned to relative and common motion in human vision except to reconfirm the well-known fact that absolute motions are difficult to perceive.

Table 3 shows the effect of varying the velocity of the irrelevant dimension (for dim ambient light conditions only) when its direction was incongruent with the relevant dimension. As the data show, increasing these velocities increased the interference effect monotonically. This implies that by selecting velocities of relative or common motion appropriately, one can create an asymmetry of interference in either direction. This table contains no interaction even approaching statistical significance, and so there is no detectable asymmetry present. Thus, the important point is when these two velocities are physically equal and an external frame of reference is visible, performance on the two dimensions is symmetrical, and so neither dimension commands precedence.

Experiment 2

Our second experiment replicated the first with the addition of a divided attention task. In this task, subjects had to indicate whether the directions of motion on the two dimensions were congruent or incongruent. This is a condensation (or same-different) task, which requires that both dimensions be processed before a response can be made. Ordinarily, divided attention is harder than selective attention, indicating that it is more difficult to process two dimensions than just one. However, with configural dimensions such as the parentheses, the reverse is true; divided attention, as measured by performance on the condensation task, is easier than selective attention. This indicates that wholistic or emergent properties of the stimulus can be accessed more quickly than can the nominal dimensions (left and right parenthesis) along which the stimuli vary.

The present experiment was run only in the light and used a new set of 12 subjects, but otherwise it followed closely the design and procedure of Experiment 1.

Results

The results are shown in Table 4. The data replicate the pattern of results from Experiment 1 in showing approximately equivalent baseline performance on common and relative motion. The data show that subjects opted for speed over accuracy on the relative baseline condition; but the RT difference was not significant, and given the results of Experiment 1, we can safely conclude that the two baselines are equal. Similarly, symmetric amounts of interference from the irrelevant dimensions were obtained: 77 msec from relative vs. 81 from common. Thus, neither relative nor common motion was easier to ignore. Overall, the RTs were shorter in Experiment 2 than in 1, reflecting a greater number of trials and so larger amounts of practice in the second experiment.

Most important, performance on the divided attention (condensation) task was both faster (by 52 msec) and more accurate (by 1.9 percentage points) than on the selective attention tasks. (Faster RTs for divided attention were shown by 9 of 12 subjects.) This shows that for these stimuli, the dimensions of common and relative motion are neither separable nor integral. Instead, they appear to be configural, meaning that subjects are attending to novel, emergent features of the whole motion pattern. In sum, neither common nor relative motion is normally processed ahead of the other, since subjects can respond more quickly to the joint status of the two than to either one alone. An equally significant conclusion is that subjects may not be processing these two nominal dimensions at all but instead may be attending to some as yet undefined emergent feature of these moving configurations.

Experiment 3

The first two experiments have relied on a measure of selective attention that is based on the well-known Stroop phenomenon (wherein subjects exhibit difficulty in naming the color of ink in which a conflicting color name is printed, such as the word RED printed in green ink). To use this measure, stimuli are presented that vary on two dimensions (here, common and relative motion) which may be congruent or incongruent, and poorer performance with the incongruent stimuli indicates a failure of selective attention to the relevant dimension. One potential shortcoming is that the interference so measured could arise in any stage of processing required by the task, ranging from sensory interactions to competition in response processes; indeed, the basic Stroop effect with colored words is usually attributed to response competition. Interference effects that are due to processing near the response end may be uninformative about any genuinely perceptual interactions of common vs. relative motion, and so it would be valuable to determine if interference (symmetrical or asymmetrical) occurs with other measures of selective attention that are not as sensitive to response effects as the Stroop measure is.

To examine this question, a new experiment was conducted using stimuli whose common and relative motion components did not call for identical response categories and that used an entirely different measure of selective attention. The stimuli used are shown in Figure 4. As in the previous experiments, they consist of four luminous dots that originate in a square configuration. Once motion begins, the configuration as a whole moves either upward or downward (common motion). Independently of this common motion, the dots move either outward or inward (relative motion). The subject's task is to respond to either the common or the relative motion in a two-alternative forced-choice RT task. In contrast to the previous experiments, the two categories into which the stimuli are classified are different: upward vs. downward for common motion, and outward vs. inward for relative motion. Clearly, none of the common-relative pairings involve congruity or incongruity, and so no Stroop-like interference can arise with these stimuli. Instead, failure of selective attention is assessed by the amount of interference generated when the irrelevant dimension varies randomly from trial to trial compared with when it remains constant (Garner, 1974).

This experiment involved nine different conditions. Four of these are baseline control conditions in which subjects discriminate between just two stimuli that differ only in the common motion or only in their relative motion, with the irrelevant dimension held fixed. For example, the subject would receive a series of 40 trials consisting of a random alternation of the up-and-in and the down-and-in patterns, presented one at a time. Two additional conditions, called correlated tasks, require discriminating between just two patterns that differ in both their relative and common motion. An example would be discriminating the up-and-in pattern from the down-and-out. These conditions test for the presence of a redundancy gain (i.e., improved performance relative to the control conditions), which is a defining characteristic of integral dimension.

Two more conditions, called filtering tasks, present the subject with all of the four patterns, shown one at a time in a random sequence. The task is to discriminate the stimuli on the dimension that is indicated as relevant on that trial, and to ignore the variation on the irrelevant dimension. Poorer performance on the filtering tasks than on the control tasks indicates a failure of selective attention, which is a defining characteristic of both integral and configural dimensions. (Note that variation on the irrelevant dimension does not automatically produce interference effects; with truly separable dimensions such as the parenthesis pairs shown in

Figure 2B, performance does not suffer at all in the filtering tasks.) Finally, there is the condensation task, which again involves presenting the subject with all four stimuli. The task requires that one response be made to the up-and-in and to the down-and-out patterns; the other response is to be made to the up-and-out and to the down-and-in patterns. As stated above, this task demands divided attention to both the relative and common motion. Poorer performance on this task than on the filtering tasks is a defining characteristic of both integral and separable dimensions; but equal or superior performance on the condensation relative to the filtering tasks is a defining characteristic of configural dimensions.

In other respects, the design of this experiment parallels that of the preceding ones. Nine subjects all served in 9 full replications of all nine conditions (40 trials per replication), for a total of 3240 trials per subject. The stimuli were presented in the center of the viewing screen (where the subject was fixating) in this experiment; in another experiment we have completed, the patterns were presented to the visual periphery, but the same basic pattern of results was obtained.

Results

The results are shown in Table 5. First, RTs on the baseline (control) tasks were virtually identical for the relative (397 msec) and the common (400 msec) motion conditions; error rates were identical for the two. This indicates that these two dimensions were equally easy to perceive when the irrelevant dimension did not vary.

Second, RTs in the filtering tasks increased by an average of 45 msec over the controls, indicating that variation on the irrelevant dimension could not be ignored, $F(1,64) = 7.62$, $p < .005$. Third, this failure of selective attention was almost perfectly symmetrical for the relative and common motion tasks (48 and 43 msec interference, respectively; $F < 1$). Thus, the two dimensions were equally difficult to ignore, which implies that neither dimension demanded precedence in perceptual processing. Fourth, performance in the correlated dimensions tasks was superior to that on the control tasks, indicating that a redundancy gain was enjoyed. This pattern of results suggests that the two dimensions are symmetrically integral instead of separable or configural, neither of which would produce redundancy gains.

Fifth and last, performance on the condensation (divided attention) task was far worse than on the filtering tasks ($p < .001$), showing further that the dimensions are integral and not configural. This outcome stands in contrast to the results with the Stroop-like stimuli of Experiment 2 and demonstrates that divided attention to relative and common motion was not possible in Experiment 3.

In sum, the results indicate that relative and common motion acted as integral dimensions in Experiment 3, whereas they seemed to act as configural dimensions in Experiment 2.

Conclusions

The results of all three experiments clearly show that the dimensions of relative and common motion are not separable but rather are processed interactively in perception. They also show a nearly perfect parity in their perceptual status: neither relative nor common motion is easier to perceive or harder to ignore than the other.

Whether these two dimensions should be considered as symmetrically integral or as configural is a harder question to answer definitively at present. Experiment 2 showed that divided attention was easier than selective attention (indicating configurality), whereas Experiment 3 showed the opposite (indicating integrality). This discrepancy is probably due to differences between the stimuli in the two experiments. Certain combinations of relative and common motion mix to produce more salient emergent features than other combinations, and that in turn determines whether an integral or a configural result is obtained.

For example in Experiment 2, the stimuli for which the common and relative motions are congruent involve motions whose trajectories fall only along the horizontal and vertical axes, whereas those where common and relative are incongruent involve diagonal trajectories. Thus, the condensation (divided attention) task in Experiment 2 could have been performed by the subject merely noting whether or not any motion (be it common, relative or even absolute) appeared along an oblique path. However, in Experiment 3 there is no obvious emergent feature that distinguishes between the up-and-in and the down-and-out motions as one group, and the up-and-out and the down-and-in motions as a separate group. A unitary explanation that is consistent with all the results is that relative and common motions are processed jointly, as a wholistic percept; that is, attention is always divided across both components of motion. If a suitable emergent feature is available perceptually, it will be used to advantage in the "divided attention" task, but attention will nevertheless be divided between common and relative motion in any condition, including the "selective attention" tasks.

In any case, the data from all three experiments contradict the idea that relative motions (the trees) precede common motion (the forest) in the order of processing. In addition, they appear to contradict Johansson's central notion that we partition the absolute motions of light configurations into phenomenally separate common and relative vectors.

A potential reconciliation between these results and the others that have indicated a precedence for either common or relative motion is that our experiments focus on different and earlier stages of the perceptual process. Whereas the other studies have examined the final perceptual organization of motion events and have used as data what are essentially phenomenal reports, our experiments were intended to zoom in on the microgenesis of the motion percept and use as data RTs and error rates in performance tasks. In this regard, the present experiments are more analytic than the others with respect to the order of processing issue. With this in mind, we conclude that neither the relative nor the common motion component has any detectable temporal or attentional priority in motion perception.

Subjects can clearly perceive these two components as separate entities, as witnessed by their ability to perform the tasks, but they arrive at this separation only after a considerable period of processing during which common and relative motions are anything but separable. In the experiments of others, responses are not made under time pressure, and so subjects can delay responding until the perceptual process has run its full course. It is during this subsequent processing that the relative and common components of motion are segregated. This processing may place different weights on common and relative motion in determining the final perceptual organization. Thus, Cutting and Proffitt (1982) may be correct that the minimization of relative motion receives a higher priority than minimization of common motion. Our present results, however, suggest that the priority of relative motion is not attributable to _faster_ minimization of relative motion vectors.

Footnotes

This research was supported in part by a Biomedical Research Support Grant from the State University of New York at Buffalo. Nelson Toth is now in the Department of Psychology, Rutgers University, New Brunswick, NJ. Requests for reprints should be sent to James R. Pomerantz, Department of Psychology, SUNY at Buffalo, 4230 Ridge Lea Road, Buffalo, New York, 14226.

References

Cutting, J. E. & Proffitt, D. R. The minimum principle and the perception of absolute, common and relative motion. _Cognitive Psychology_, 1982, _14_, 211-246.

Garner, W. R. _The processing of information and structure_. Hillsdale, NJ: Erlbaum, 1974.

Johansson, G. _Configurations in event perception_. Uppsala, Sweden: Almqvist & Wiksell, 1950.

Navon, D. Forest before trees: The precedence of global features in visual perception. _Cognitive Psychology_, 1977, _9_, 353-383.

Pomerantz, J. R. Global and local precedence: Selective attention in form and motion perception. _Journal of Experimental Psychology: General_, 1983, in press.

Pomerantz, J. R. Perceptual organization in information processing. In M. Kubovy & J. R. Pomerantz (Eds.), _Perceptual organization_. Hillsdale, NJ: Erlbaum, 1981.

Pomerantz, J. R. & Sager, L. C. Asymmetric integrality with dimensions of visual pattern. _Perception & Psychophysics_, 1975, _18_, 460-466.

Pomerantz, J. R. & Schwaitzberg, S. D. Grouping by proximity: Selective attention measures. _Perception & Psychophysics_, 1975, _18_, 355-361.

Pomerantz, J. R. & Garner, W. R. Stimulus configuration in selective attention tasks. _Perception & Psychophysics_, 1973, _14_, 565-569.

Table 1

RTs and Error Rates (parentheses) for Experiment I (Dim Ambient Light)
Relevant Dimension by Irrelevant Dimension

Relevant Dimension	Irrelevant Dimension		
	None	Incongruent	Congruent
Common	623 (3.5)	883 (26.3)	653 (3.7)
Relative	629 (3.4)	892 (24.3)	616 (3.0)

Table 2

RTs and Error Rates (parentheses) for Experiment I (Darkness)
Relevant Dimension by Irrelevant Dimension

Relevant Dimension	Irrelevant Dimension		
	None	Incongruent	Congruent
Common	1110 (3.1)	1916 (25.7)	1367 (6.6)
Relative	849 (2.8)	1256 (10.5)	835 (2.9)

Table 3

RTs and Error Rates (parentheses) for Experiment I (Dim Ambient Light)
Relevant Dimension by Interference Level

Relevant Dimension		Interference Level		
	None	Low	Medium	High
Common	623 (3.5)	653 (4.7)	761 (12.8)	892 (27.4)
Relative	629 (3.4)	659 (4.2)	781 (13.9)	821 (22.9)

Table 4

RTs and Error Rates from Experiment 2
Baseline, Selective Attention and Divided Attention Conditions

Condition	RT	Error rate
Baseline: Common	581	2.4
Baseline: Relative	564	4.6
Selective Attn: Common	658	8.2
Selective Attn: Relative	645	10.7
Selective Attn: Mean	652	9.4
Divided Attn (Condensation)	600	7.5

Table 5

RTs and Error Rates from Experiment 3
Relative vs. Common Motion[1]

	Condition								
	CONTROL Common		CONTROL Relative		CORR		FILTERING		CONDEN-SATION
					A	B	Com	Rel	
	1	2	3	4	5	6	7	8	9
Mean RT (msec)	400	400	398	396	384	383	443	446	643
Mean % Error	2.5	2.5	2.6	2.4	2.1	1.9	3.4	3.2	7.2

[1] In this table, "FILTERING Com" refers to filtering conditions in which the Common dimension was relevant and Relative was irrelevant; similarly, "FILTERING Rel" indicates that Relative was relevant and Common was irrelevant. The two different conditions with correlated dimensions are arbitrarily labelled A and B.

Figure 1. A. Johansson's L-shaped motion configuration. B. The absolute motions of A as partitioned into their common and relative components.

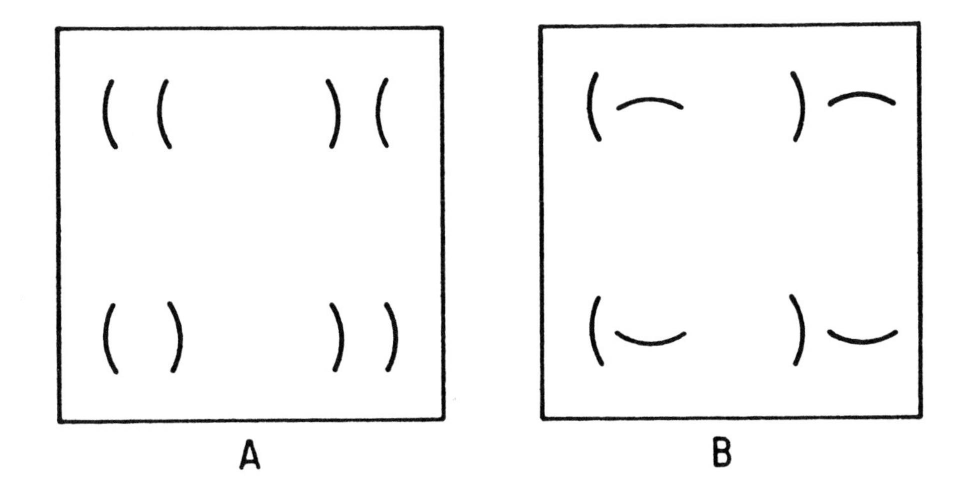

Figure 2. A. An example of configural stimulus dimensions in form perception. B. An example of separable stimulus dimensions.

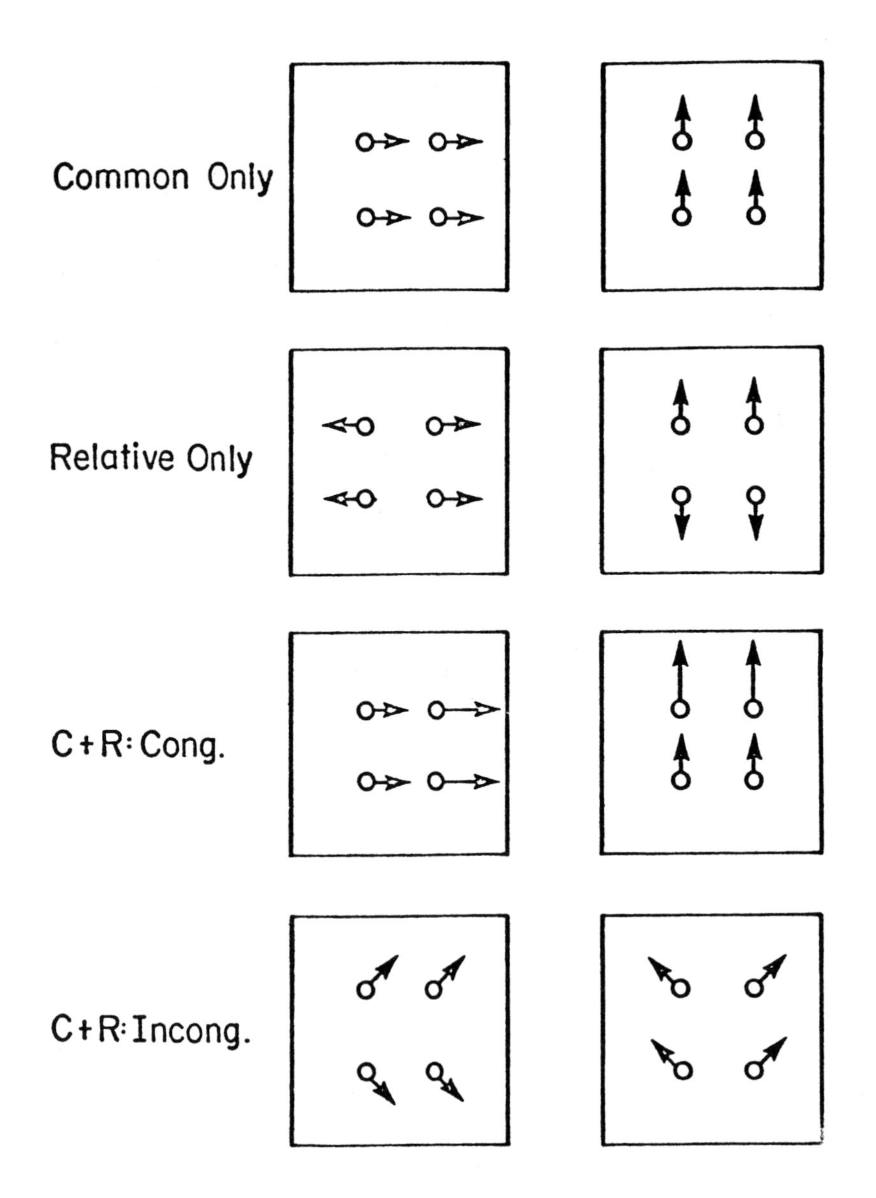

Figure 3. Examples of the motion configurations tested in Experiments 1 and 2. The motions of the four lights are indicated by the length and direction of the arrows.

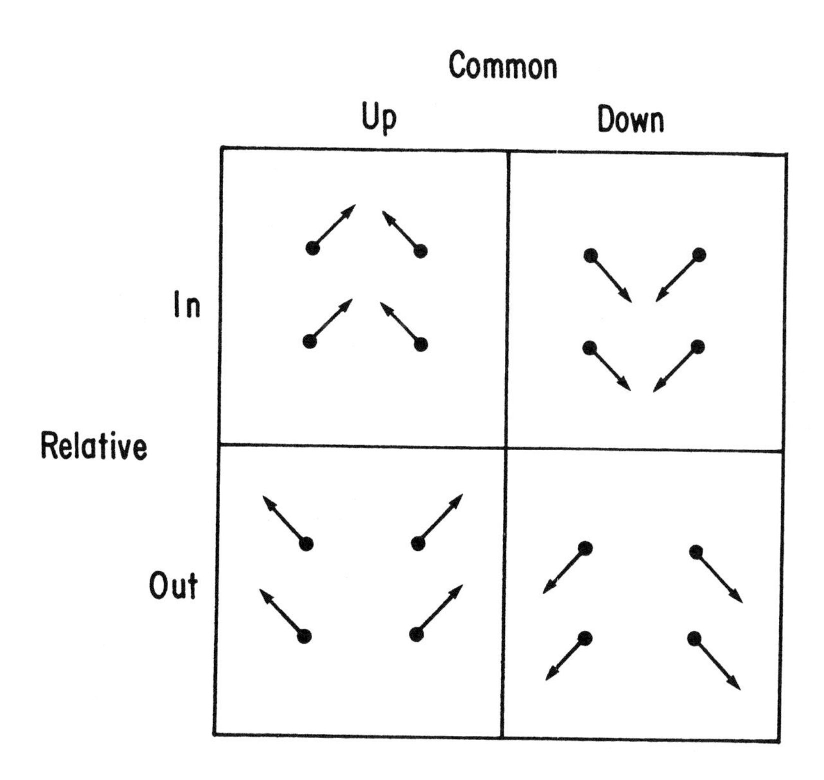

Figure 4. The four motion configurations tested in Experiment 3.

Perceiving and recovering structure from events

James E. Cutting
Department of Psychology, Uris Hall
Cornell University
Ithaca, NY

How do perceivers identify a moving object as seen against a changing background? How do figure and ground separate? Such questions have engaged psychologists for at least seventy years. In particular, the Gestalt psychologists were deeply concerned with the latter, but had only the ill-defined notion of common fate, or uniform density, for dealing with the former. The coherent flow of a moving object is seen, somehow, by extracting those aspects of the whole that segregate it from the ground; the uniform destiny of all parts of the object was thought both to make the whole cohere and to separate the whole from all else. Two pairs of ideas, from two researchers who came out of the Gestalt tradition, helped elucidate the notion of common fate as applied to motion perception.

The first pair of ideas is due to Johansson (1950): The motion of an object can be parsed into common motion and various relative motions. The common motion of elements in a visual display is the vector path shared by all elements; the relative motions of elements are the residuals, moving with respect to the whole. Johansson was quick to see, as others have subsequently (Cutting & Proffitt, 1982), that some decoding principles were needed for a human being or a machine to be able to recover structure from the motion of parts of a visual display. The problem is seen most clearly when we introduce an equation of three terms:

Absolute motion = common motion + relative motion

The new term, absolute motion, is the vector path over time and through space for any given part of a dynamic display without regard to other parts. This motion, of course, is generated by the motion common to the whole and the motion of the particular part in relation to the whole. To be more concrete, let me choose an example used by Johansson (1973) and by Duncker (1929/1938) before him, as shown in Figure 1a. Two lights are mounted on a rolling wheel and all else darkened. One light is mounted at the rim of the wheel and the other at the axle. The absolute motion paths are shown in the top panel of the figure, and two possible interpretations at the bottom. Which of these interpretations is seen depends somewhat on the viewing conditions (Cutting & Proffitt, 1982; Börjesson & von Hofsten, 1975), but the problem is general: The observer could see a smoothly rolling wheel if the motion of the axle-mounted light is seen as the center of the configuration, or she could see a stick with lights mounted on its ends rotating about an unseen center and the stick tumbling through space following the path of a prolate cycloid. The general problem, pointed out by the multiplicity of possible percepts, is one of induction: How does one obtain two component sets of motions when only the absolute motion is given? A priori, there is an indefinitely large set of common motion/relative motion pairs that could, when vectorially combined, yield the same absolute motion. The visual system either needs some rules of thumb to guide choice among possible alternatives in this situation and in others, or it needs some design feature inherent in its construction that constrains possibilities.

The second set of ideas is due to Wallach (1965/1976), and they express essentially the same idea in a different way: The motion of objects can be thought of in terms of two types of displacements, those relative to the object itself and those relative to the observer. It seems most efficacious to discuss these in terms of coordinate systems, and many will immediately recognize these ideas as close to those of Marr (1982). Object-relative displacements are motions or changes of a part of an object with respect to the whole of the object. The origin of this coordinate system is the "center" of the object, and is the origin of Marr's object-centered coordinate system. Of course, the critical issue is to find this center of this coordinate system, and most of this paper is focussed precisely on that point. Observer-relative displacements in Wallach's terms are measured in Marr's viewer-centered coordinate system. In a sketchy way, these ideas can now be applied to the lower two panels of Figure 1. If the observer perceives a rolling wheel from this two-light configuration, the origin of the object-centered system is at Light B. In turn, this coordinate system rotates relative to the observer and follows the translatory vector path of common motion through viewer-centered coordinates. If, on the other hand, the observer perceives the configuration as a tumbling stick then the origin of the system is midway between Lights A and B and the object-coordinate system rotates as it follows the prolate cycloidal path of translation through viewer coordinates.

With this set of interrelated ideas in mind--common and relative motions and observer-relative and object-relative displacements--we can now consider six different types of events where underlying structure is perceived and recovered through motions or displacements of elements in the display.

Motion: Representation and Perception
N.I. Badler and J.K. Tsotsos, Editors
Published by Elsevier Science Publishing Co., Inc.

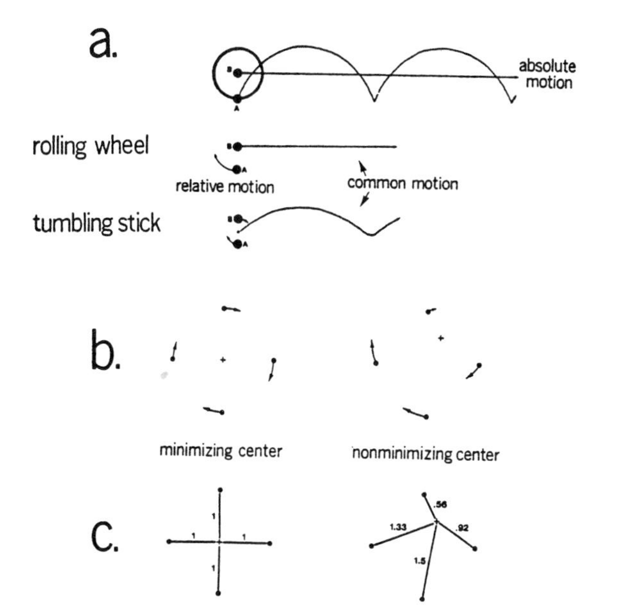

Figure 1. Panel a shows the absolute motion paths of two lights mounted on a rolling wheel, one at the perimeter (Light A) and one at the hub (Light B). Two interpretations of this stimulus are also given, one as a rolling wheel and the other as a tumbling stick. Relative and common motions are indicated for each of these interpretations. What is taken as the center of moment determines perception. Panels b and c show four points in rotation for demonstration of how a minimum principle might apply in extracting relative motions. In Panel b, if these lights are seen to rotate around their centroid, then all momentary vectors sum to zero. If, on the other hand, they are seen to rotate around some other point they will not sum to zero. Instead they will sum to a clockwise rotational vector whose radius is the distance from the centroid. In order to perceive such a configuration, a common vector of the same magnitude but rotating counterclockwise would be needed. The configuration on the left is invariably seen. Panel c demonstrates a second, interrelated minimizng procedure. Rotation around a centroid minimizes the squared lengths of moment arms from the radius to each light. On the left, the summed squares of moment arms is 4.0, and on the right it is 5.18. (After Cutting & Proffitt, 1982).

Event 1: Rolling wheels

The oldest phenomenon of real motion (as opposed to apparent motion) that has come under systematic scrutiny by psychologists is that of rolling motion. What is shown to observers are the movements of a few scattered lights mounted on an unseen wheel. Rubin (1927), following the lead of Galileo and some interests of 17th century mathematicians, noted that the absolute motion paths of such lights (which were always members of the family of cycloids) were rarely seen as such. Instead, viewers had a strong tendency to see objects undergoing rotation (relative motion) and translation (common motion). This finding had strong influence on Gestalt psychologists as demonstrating that perceivers organized noncontiguous elements into coherent wholes. But Rubin and subsequent researchers in the Gestalt tradition (Duncker, 1929/1938; Johansson, 1950, 1973; Wallach, 1965/1976) were always interested in particular stimuli as they elicited particular percepts. In contrast, my colleagues and I have been interested in the whole population of stimuli and in what people generally saw. Such effects cannot be measured through demonstration, but only through experimentation (Proffitt, Cutting & Stier, 1979; Proffitt & Cutting, 1979, 1980; Cutting & Proffitt, 1982).

The results have been remarkably clear-cut and promote a process-orientation as to how the visual system extracts information from these dynamic displays. Stimuli have generally consisted of two to four lights mounted in various places on the wheel--at the rim, within the interior and including the axle, and even exterior to the rim. Although such factors as the number of lights and the symmetry of their arrangement play some role in perception, these are swamped by a separate factor: What is seen by most observers at most times is dictated by the distance of the centroid of the configuration of lights from the wheel's axle. That is, if the center of the configuration is at the axle, then that configuration will look very much like a rolling wheel. If, on the other hand, the center of the configuration is far from the axle (say, three lights mounted closely together near the rim) then it will generally not look like a rolling wheel, but instead will look more like a hobbling or hopping object. Moreover, the degree to which the configuration looks like a rolling wheel is determined by the relative distance between the centroid of the light pattern and the axle. Such a simple notion accounts for about 90% of the variance in all observers' responses across a dozen experiments with over two dozen stimuli.

The process of perceiving these stimuli seems to be as follows: The viewer first minimizes the relative motions of the lights with one another. This minimization process can occur by either of two methods, as shown in Figures 1b and 1c. Imagine these four lights in rotation. What unseen axis will they appear to rotate around? In object-centered terms the rotation will be seen to take place around the centroid because either (1) this is the only point at which the momentary vectors of all points will add to zero, or (2) this is the point at which the length of the moment arms generating the movements are minimal. No other point satisfies these two interlocked minimization procedures. Once the relative motions have been extracted, then the common motion falls out as residual.

Such a procedure--first extract relative motions according to a minimum principle, then observe the residual common motion--is an adequate way to solve the equation given earlier, at least in many situations. It appears that for such wheel-generated motions, this procedure is used most of the time. There are, however, some exceptions. I will consider several later in the discussion of other events, but there is also one here, the rolling-wheel interpretation of the configuration shown in Figure 1. But this is one of the few failures in our object-parsing system, and its perception may be due to an alternative minimization procedure (Cutting & Proffitt, 1982). Moreover, this rolling-wheel percept disappears when external reference frames are removed (Börjession & von Hofsten, 1975).

Let me now introduce a neologism. We have called the point around which these motions appear to take place, the point that also nullifies relative motions in these displays and which minimizes moment arms, the center of moment. Borrowing from Dürer (1528/1972) and capitalizing on the dual meanings of motion and importance, we claim that the center of moment allows the perceiver to recover the structure of the stimulus and to perceive the event. This point is the origin of the object-centered coordinate system. Its own motion through observer-relative coordinates dictates how wheel-like the event will be perceived. If the center of moment (in this case the centroid) moves smoothly and linearly then the observer will see a wheel-like event; if if moves up and down in its translation (following a prolate cycloid) then it will be seen as less wheel-like.

This first event is one in which all elements remain in rigid relation to one another. The second, although it is also basically a rigid stimulus, has many nonrigid relations among its various parts.

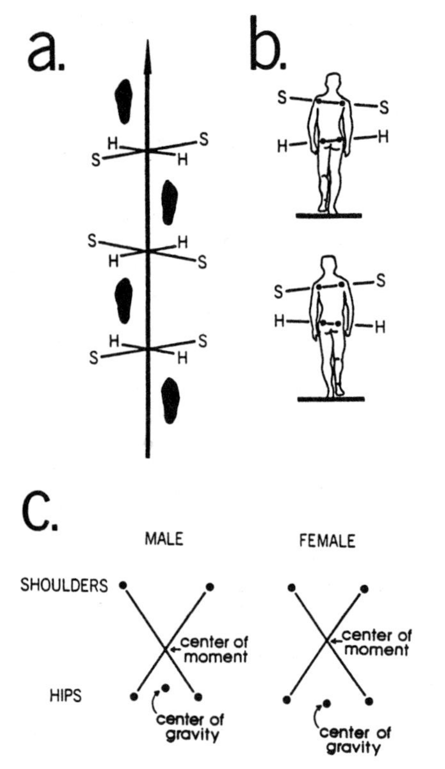

Figure 2. Panel a shows four footfalls of a walker and the relative torsion in the horizontal plane of shoulders against hips in each of the three double support phases. Panel b shows the same torsion in the frontal plane. Panel c draws stress lines across the diagonals of the torso to demonstrate the locales of the center of moment for a male and a female walker. The center of moment is not the center of gravity, also shown. The location of the center of moment in dynamic, computer-generated displays determine the perceived gender of the walker. (After Cutting, Proffitt, & Kozlowski, 1978).

Event 2: Walking People

Johansson (1973, 1976), using a variant of a technique from 19th century photography (Marey, 1895/1972), mounted lights on the joints of an individual, darkened the surround, and asked observers what they saw. Invariably, observers untutored in looking at such displays immediately recognized the presence of a human being. Moreover, they recognized the activities of the individual--walking, running, doing jumping jacks--with as little as 100 msec exposure to the dynamic displays. It is difficult to overestimate the impact and perceptual salience of such configurations; they yield remarkably convincing percepts of people and do so only with an arrangement of a dozen lights or less moving against a dark ground. One is never confused about which lights are dynamically connected to which others: a light at the wrist is seen as connected to the light at the shoulder, and not the light at the hip, even though it is closer to the hip than the shoulder. An adjacency principle (Gogel, 1978) cannot hold in such situations.

My colleagues and I became interested in such displays. We first demonstrated that friends could identify themselves and one another from such information when they walked laterally across the field of view (Cutting & Kozlowski, 1977). We then found that new observers unfamiliar with the particular displays could identify the gender of the walker (Kozlowski & Cutting, 1977; Barclay, Cutting & Kozlowski, 1978). The former result, although probably the most interesting that we have, seemed difficult to pursue: Person identity is wrapped up in too many vaguely known factors about personality and mood. The latter result, then, is the one that engaged us (Cutting, Proffitt & Kozlowski, 1978). The question is: On what informational basis do perceivers make their judgments of gender? After many false starts--finding that such factors as walking speed, step size, and arm swing were all sufficient for gender recognition but not necessary--we began a biomechanical analysis. Ultimately, this is only a little more sophisticated than the idea of "the knee bone is connected to the ankle bone": It is that the human form is a hierarchy of related elements moving through space in an economical way. The economy lies in the pendular motions of the arms and legs and in the dynamically crossed symmetry of arms and legs--left arm moves in synchrony with right leg, right arm with left leg. Everything is either in phase or 180° out of phase. As arms and legs move, so move shoulders and hips. The general arrangement can be seen in Figure 2a.

Since the shoulders move against the hips, the torso, in faithfulness to its etymology, undergoes torsion. This torsion can be likened to that of a flat spring. If stress lines are drawn across the diagonals, as shown in Figure 2c, one has in the intersection a fairly good approximation to the center of moment of a walker. All points in the torso and all points in the arms and legs can be thought of as moving around this point. It is the origin of the object-centered coordinates for the perceiver.

What was interesting to us is that the locus of this point is generally different for males and females. That is, because males have slightly wider shoulders than hips, and because in females these dimensions are roughly the same, male and female torsos differ in their style of torsion. In normal gait, the most ergonomically efficient mode of locomotion, the amounts of torsion and pendular motions of the limbs are constrained (Beckett & Chang, 1969; Murray, 1967). Because of the difference in origin of movements, males and females have many systematic differences in their walks: females swing their arms more, males their shoulders more (see Figure 2b), both in accomplishing the same end of countering the forces generated in the legs and maintaining balance. Because of their generally wider hip girdle, women rotate their hips more and walk more smoothly across a surface. Males, in contrast, tend to bob up and down more when they walk.

What excited us most, however, was the fact that the difference in locus of the centers of movement for males and females accounted for roughly 75% of the variance in gender judgments. That is, as determined by anthropometric procedures adapted to this study (Cutting et al., 1978), males generally had lower centers of moment than females, and the individuals who were systematically misidentified as either female or male had locations of center of moment that more nearly approximated those of the opposite sex than of their same sex.

With such results in hand I set out to computer-synthesize human gait. My purpose was to hold all else constant in the displays except the location of the center of moment and those differences that it generated. The outcome was quite successful (Cutting, 1978a, 1978c). In fact, I was able to generate hypernormal synthetic males and females; they were more often identified as male or female than any of the real people that I had previously videotaped.

Our account of how human observers accomplish the perceptual feat of extracting the information from these dynamic displays is, in its essence, the same as the account for the perception of rolling wheels. Observers appear to extract the relative motions of the lights from one another by some minimum principle that assumes that

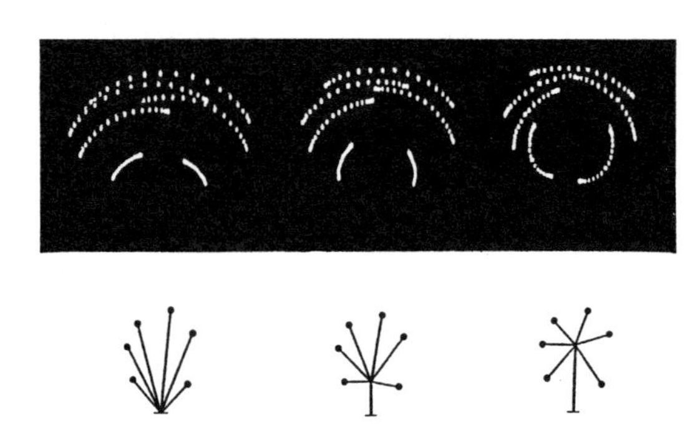

Figure 3. Static representations of three tree- and bush-like stimuli used by Cutting (1982a). At the top are superimpositions of 24 frames in each dynamic stimuli; and at the bottom are stick-figure versions of the same three stimuli. First-order centers of moment are located where the trunks meet the ground; second-order centers are where the limbs meet the trunk. Perceivers can determine the arborization pattern of a bush or tree quite accurately, appearing to use information about centers of moment.

rigid pendular motions are simpler than the various nonrigid comparisons that could be made. Once isolated, the common motion of the whole--an undulating motion with linear translation--falls out as the residual. Evidence for this view is not as solid as in the case of rolling wheels; about all we know is that distorting the common motion of a walker has little effect on perception (Johansson, 1973), but that distorting the relative motions has great effect (Barclay et al., 1978; Cutting, 1981). If extraction of relative motion were logically dependent extraction of common motions, then a different pattern of results ought to have accrued.

Event 1 was a mechanical event (rolling wheels) and Event 2 biomechanical. The former involved rigid and circular relative motions and the latter rigid and nested pendular relative motions, with nonrigid torsion. For purposes of generality I looked at a third mechanical event with somewhat more complex relative motion, one that is completely nonrigid.

Event 3: Swaying trees and bushes

Most of us have seen the following event at a suburban shopping mall. At year's end before the holidays, small lights are placed among the limbs of deciduous trees. When the wind blows, the trees are set in motion, and the pattern of motion of the lights is fairly rich in revealing the limb structure of the tree. Since this so nearly approximated the experimental situations used in the events discussed previously, I decided to extend the technique to study the perception of underlying arborization patterns in the motions of trees and bushes (Cutting, 1982a).

Locations of lights were varied, and the pattern in which the limbs intersected the trunk was also varied. Over a sequence of frames, as before, the displays oscillated back and forth as if the tree or bush were blown by the wind. Here, all motions were not strictly inverse pendular. Instead, all branches and limbs were flexible, and motions slightly phase-staggered. Thus, unlike the displays of the previous two events, there were no rigid relations among the moving lights. Motion depended on the length of the unseen limb (from trunk to light) and on the unseen angle with which the unseen limb intersected the unseen trunk. Static renditions of such displays are shown in Figure 3.

Results of several experiments suggested that viewers are highly attuned to structural information of this kind. For highly stylized trees with six limbs branching from the same locale, observers could not only make systematic comparisons among members of the set of stimuli, but they could also discern the location of the unseen node of all limbs within about a half a degree of visual angle (about 5% of the stimulus height). For less stylized trees, those with six limbs branching from several different locations on the trunk, observers could still make systematic judgments about the relative similarity of the trees or bushes based on the arborization patterns implied by the moving lights on the ends of the limbs.

In this experimental situation there is no common motion; trees and bushes do not uproot themselves and move laterally across the field of view. Thus, no statement can be made for this type of display about priority of relative and common motions or parsing of object and observer coordinate systems. The observer, of course, could move relative to the display (the tree or bush) in the real world. I suspect our same scheme would apply and the motion-segregation problem would be invoked. But the displays here had no such variation. Instead, the displays present differentially nested patterns of relative motions, and the results suggest that observers can parse this structure in order to perceive a coherent whole. Since all motion in these displays is referred to the point at which the trunk meets the ground, but only through the motions of the limbs and the manner and height at which they branch from the trunk, the stimuli represent a complex hierarchy of nonrigid motions. Yet the visual system seems unperturbed by such complexity, and seems to handle such information as well as it handled the complex hierarchy of rigid motions in the walker displays. Any account that assumes that the visual system filters out rigidity in order to perceive objects (e.g. Ullman, 1979) must eventually try to handle such data.

The three types of events considered thus far are all events that occur in real time at a relatively rapid rate. The focus of all of these is a center of moment, the nonarbitrary origin of an object-centered coordinate system around which relative motion takes place for the object as a whole. In the fourth event I will not consider motion at all, but nonrigid change as it takes place over the span of a human lifetime.

Event 4: Aging faces

As we mature our heads and faces change. Orthodontists and surgeons trying to correct cleft palate realize that they must deal with these plastic changes over time: Growth and change must be built into the surgical correction (Todd & Mark, 1981). Much of the change that the human head undergoes can be captured by a relatively simple algorithm--a cardiodial transformation (Todd, Mark, Shaw & Pittenger, 1980). This transformation captures head growth and change, in part, because it mimics the gravitational tugs on tissues.

What interests me about the similarity between growth and this algorithm is that growth changes (Enlow, 1975) and cardiodial transformations (Pittenger & Shaw, 1975) occur around a point. This point is a head-centered origin of a coordinate system that is analogous to the centers of moment found in other events. The experimental question then arose: Can observers judge the goodness of growth changes (represented as cardiodial changes) in profiles of a human head as a function of the origin of the head-centered (cardiodial) origin of those changes?

The procedure and results (Cutting, 1978b) are as follows, as shown in Figure 4. Panel a shows the profile selected, representing an early adolescent, and the matrix of nine points used as cardiodial centers. Panel c shows nine youthened profiles generated from the nine different centers in Panel a, and Panel d shows the nine aged

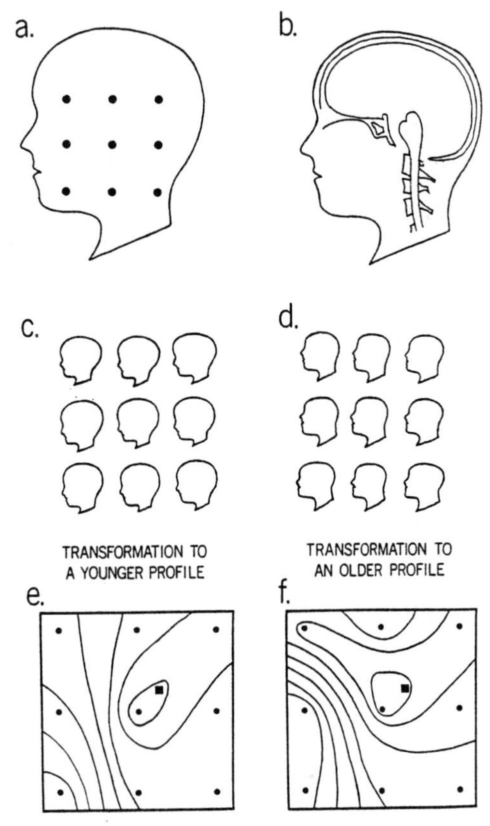

Figure 4. Panel a shows an intermediate-aged (adolescent) profile and the nine centers of moment chosen for cardiodial transformation. Panel b shows the internal anatomy of the same person. Panels c and d show the nine transformed profiles, each one corresponding to the locations given in the 3 x 3 matrix in Panel a. Panel c shows the youth-transformed profiles, and Panel d the aged profiles. Panels e and f show the judgment data of all subjects plotted as contour maps with respect to the 3 x 3 matrix. Squares indicate the maxima of best fitting paraboloids to the data, and should be mapped onto the matrix shown in Panel a and then onto the internal anatomy shown in Panel b. (After Cutting, 1978b).

profiles that correspond to them. Viewers were presented the standard profile (that shown in Panel a without the nine dots) and then the youthened profile and asked to make a judgment on a scale from 1 to 7 as to how well the second profile represented the younger version of the person represented in the first profile. The same procedure was then followed for the standard and the aged profiles. The data for the two tasks are shown in Panels e and f, plotted as contour maps for the averaged judgments of all observers. Because those data approximated elliptic paraboloids, I then calculated the maxima of the best fitting functions, and those are shown as the square dots in the bottom panels. Notice that the locations of these maxima are near the central point in the 3 x 3 matrix, but that they are slightly to the upper right. One can find this locale in the profile of Panel a and then transpose that point to the internal anatomy of a head shown in Panel b. Notice that the maxima of the viewers' judgments correspond to a point very near the foramen magnum, the point where the spinal column disappears into the skull and the brain stem. Interestingly, Enlow (1975) used this point as the origin of the framework for describing the developmental anatomy of the head. What I conclude is that observers are fairly good at discerning proper age and growth changes, and that these are generated by a mathematical transformation that closely mimics growth. In essence, people know the geometry of growth to the extent that they can recognize good and poor exemplars of the process.

As with the blown movements of trees and bushes, age changes in human profiles have no common motion. At least in the manner that the stimuli were generated, only object-centered coordinates and relative change are pertinent. This system and these changes are a structural description which best captures the aging process as it could be used in face perception. As before, a center of moment (here, the origin of the cardiodial transform) proves perceptually useful.

Thus far I have considered relatively fast, rigid (rolling wheels and walking humans) and nonrigid (swaying trees) events, and a relatively slow nonrigid event (aging faces). The fifth event type is a relatively slow rigid event, revealing important information to infrahumans.

Event 5: Rotating night sky

What information do migratory songbirds use to guide their long-distance flights? Among the many sources of information appear the locations of stars in the night sky. In particular, certain birds appear to sit in the upper branches of trees at night, hopping around limb to limb, observing the patterns of stars over time in a relatively cloudless sky. Over the period of several hours, as the earth rotates under the celestial sphere, the vectors paths of the stars begin to make concentric circles around the celestial poles--the North Star (Polaris) in the Northern Hemisphere, as shown in Figure 5. These paths, because they are all referred to the polar direction, could be used by the birds in selection of their migratory direction--south in the Fall and north in the Spring. Emlen (1975) demonstrated that naive birds (those that had neither migrated before nor had a chance to observe the patterns of stars to "memorize" the constellations) appeared to use the information, in Ptolemaic coordinates, about the rotation of the night sky. Placing the birds in small cages in a planetarium, Emlen noticed that they oriented in the direction of Polaris in the Spring and away from it in the Fall. Most convincingly, however, Emlen had the night sky rotate around a completely different point, Betelgeuse, and the birds oriented with respect to it.

Here, the center of moment of the night sky (Polaris for the Northern Hemisphere) is used as information for directing migratory behavior. It is a source that could be invaluable for long flights over dark water or terrain. But the perception of this center of moment must follow a different logic than those generally used in the previous events. Since the night sky is unbounded and is chock full of stars, no location in the sky will nullify relative rotational vectors nor minimize moment arms. All locations should be equally good. All computed relative

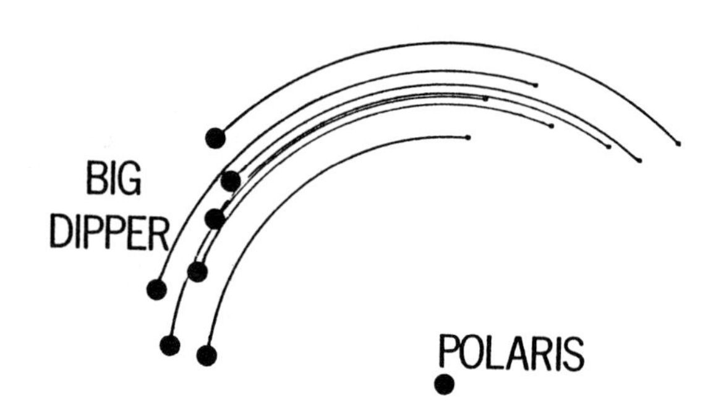

Figure 5. A six-hour rotation of the night sky would reveal circular vector paths around the North star, Polaris. Polaris is the center of moment for the night sky and is used for navigational purposes by migratory songbirds.

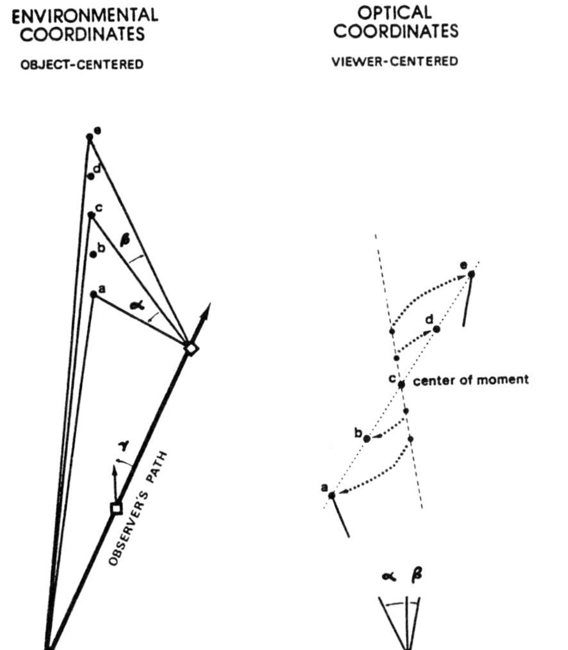

Figure 6. Representations of egomotion through space while observing an object off to the left side. Panel a shows these relations in environmental coordinates, Panel b in viewer-centered coordinates. The latter seems to be used for determining direction of locomotion. Object c is fixated, for example, nullifying common motion in the visual field and the object fixated becomes the center of moment for all rotational (and expansional) vectors. Relative velocities of these vectors in polar projection determine the direction headed: The growth of angle α is greater than that of β for the line of objects, a and e rotating around c. Angle γ, deviation of gaze from direction, is not considered.

motions would be the same. However, for all interpretations of this event other than rotation around the celestial poles, the stars would have a compensatory common counterrotation. But since no such counterrotation is seen, the common motion must be null. Thus, the equation given in the introduction reduces to absolute motion equals relative motion. In this event, the center of moment, or object-centered origin, is determined by superimposing the viewer- and object-centered coordinates, and letting the object coordinates (the night sky) rotate against the viewer coordinates creating a singularity in the sky that is the directional reference, North.

In the five events listed and discussed above, there is a center of moment within them that appears useful to the observer--dictating the wheel-likeness of a rolling object, gender of a human walker, relative arborization of a tree or bush, goodness of growth pattern in a human face, and celestial poles of the night sky. For completeness sake, I must now point out a sixth event that has a global center of moment that observers appear not to be able to use, but which has others that they can use.

Event 6: Expanding flowfields

Gibson (1950) claimed that when moving over a terrain one could tell one direction of motion optically by determining the location of the focus of expansion of visual flow. In my terms, the focus of expansion, the point dead ahead and directly behind one when locomoting, would serve as a center of moment of the event. More recently, however, Koenderink and van Doorn (1976) and Regan and Beverly (1982) have demonstrated that this is not true: The focus of expansion does not unambiguously specify direction because it is confounded with direction of gaze. A focus of expansion will always correspond to where one is looking, provided one is looking at a planar surface within 45° of where one is headed. Thus, this type of center of moment only tells the observer where he or she is looking, not where he or she is going.

One source of salient information for judging direction of motion, however, appears to be motion parallax. If one looks generally, but not directly ahead at an object in the middle distance, that object will lay at the origin of a viewer-centered coordinate system. Consider a concrete example. If one looks 5° off to the left at a tree while walking through a park, the angle between the line of gaze and line of direction will increase. But because registration of ocular rotation may not be very acute, and because researchers in this area are generally interested in optic information rather than kinesthetic information, this increase in gaze direction/motion direction angle is irrelevant: The tree always stays registered on the fovea, and what we are concerned with is the change in relative motion in viewer-centered coordinates of other objects in the field of view. Thus, as one walks through the park staring at the tree, the grass and other objects between the walker and the tree will shear across the line of sight to the left; grass and other objects farther away will shear across the line of sight to the right. In essence, the optic field undergoes rotation around the object of scrutiny, and this object becomes the new center of moment of import. This situation is shown in Figure 6. Direction of gaze, left or right, as it deviates from direction of motion is dictated by the differential rates of shear of textures across the gaze line, or alternatively the rotary motion, around the scrutinized object. Because of polar projection, the more rapidly moving textures always shear in the same direction that gaze deviates from motion. Thus, when looking left the more rapid shears are leftward; when looking right they are rightward. Such information seems adequate for telling gaze/motion deviations as small as a half a degree of visual angle (Cutting, 1982b).

Thus in this event general centers of moment fore and aft, the foci of expansion and contraction, are perceptually useless (Llewellyn, 1971; Johnston, White & Cumming, 1973). What is important are viewer-centered descriptions where the relative motions of objects with respect to the viewer and to an object under scrutiny can be registered. Fixating an object nullifies the common motions in the visual field with respect to the observer, and the perception and recovery of rigid structures can proceed from there. The object under scrutiny serves as the local center of moment for the rotation of the visual world. Motions with respect to that object dictate the direction of motion as it deviates from the object of gaze.

Summary

What I have presented are six types of situations that we might want to call events. In the first pair, rolling wheels and walking humans, I argued that for the viewer to perceive these structures she must generally extract the relative motions of parts from one another, accomplished perhaps through some minimization algorithm, and observe the common motion as the residual. These objects, wheels and walkers, have their own object-centered coordinates and an origin, the center of moment, around which their parts move. As wholes they move through viewer-centered coordinates following the vector of common motion. By definition, the center of moment for configurations on a wheel is the geometric center, or centroid, of that configuration. The center of moment for a walker need not be exactly at the center of the body and is not generally at the center of mass (Figure 2c): The center of moment, instead, is roughly at the intersection of diagonals drawn across the torso from shoulders to opposite hips.

In the second pair of events, swaying trees and aging faces, there is no common motion and thus viewer and object coordinates can be superimposed. Again, structural change is most properly discerned with respect to object-centered coordinates with an origin at the center of moment. The center of moment for a swaying tree is definitely not at the center of the configuration: it is at the point where the trunk meets the ground. The center of moment for a profile is generally near the center, located near the brainstem.

In the third pair of events, rotating night skies and expanding flowfields, the objects to be discerned encompass either most or all of the field of view: In the former event they are the stars in the rigid sky and in the latter they are the objects as laid out in rigid terrain. In both cases, in order to extract perceptually useful information, any common motion of the whole field must be nulled first. In the former case, for celestial navigation, nulled common motion reveals the relative vectors of the stars around the celestial pole. The location of the celestial pole, logically speaking, need not be at the center of the sky, nor does it even have to be visible. It need only be discernible from the vector paths of a few stars. In the latter case, for terrestrial navigation, null common motion of flow in the field of view is accomplished by looking at any object and maintaining fixation on it. That object then becomes the center of moment of the optic array and optic information rotates about that object, clockwise for leftward looking and counterclockwise for rightward viewing. If one happens to be looking directly in the direction one is going, no rotation occurs, only expansion. The center of moment for both events is important in viewer-centered coordinates and directional information available only after that has been determined.

In conclusion, the first four events are about object information. It comes as no surprise, then, that object-centered coordinates are most important and that motions and changes relative to the object's center of moment appear to be central to their perception. The latter two events are about viewer location. That is, the viewer wants to know where she is with respect to the environment. It comes as no surprise, then, that viewer-centered coordinates are most important and that motion changes relative to the viewer, the common motions of the flow, must be dealt with first. It would appear, then, that the task of perception, identifying a moving object or identifying where one is in a moving environment, dictates the use of the various coordinate structures that one might use--object-centered or viewer-centered. It seems likely that this would be true for both human and machine vision.

Acknowledgement

Supported by NIMH Grant MH37467.

References

Barclay, C.B., Cutting, J.E., & Kozlowski, L.T. Temporal and spatial factors in gait perception that influence gender recognition. Perception & Psychophysics, 1978, 23, 145-152.

Beckett, R. & Chang, K. An evaluation of the kinematics of gait by minimum energy. In D. Bootzin & H.C. Muffley (Eds.), Biomechanics. New York: Plenum Press, 1969.

Börjesson, E. and von Hofsten, C. A vector model for perceived object rotation and translation in space. Psychological Research, 1975, 38, 209-230.

Cutting, J.E. Generation of synthetic male and female walkers through manipulation of a biomechanical invariant. Perception, 1978, 7, 393-405. (a)

Cutting, J.E. Perceiving the geometry of age in a human face. Perception & Psychophysics, 1978, 24, 566-568. (b)

Cutting, J.E. A program to generate synthetic walkers as dynamic point-light displays. Behavior Research Methods & Instrumentation, 1978, 10, 91-94. (c)

Cutting, J.E. Coding theory for gait perception. Journal of Experimental Psychology: Human Perception and Performance, 1981, 7, 71-87.

Cutting, J.E. Blowing in the wind: Perceiving structure in trees and bushes. Cognition, 1982, 12, 25-44. (a)

Cutting, J.E. Motion parallax and visual flow: How to determine direction of locomotion. Paper read at the 4th meeting of the International Society for Ecological Psychology, Hartford, CT, October, 1982. (b)

Cutting, J.E. & Kozlowski, L.T. Recognizing friends by their walk: Gait perception without familiarity cues. Bulletin of the Psychonomic Society, 1977, 9, 353-356.

Cutting, J.E. & Proffitt, D.R. The minimum principle and the perception of absolute, common, and relative motions. Cognitive Psychology, 1982, 14, 211-246.

Cutting, J.E., Proffitt, D.R., Kozlowski, L.T. A biomechanical invariant for gait perception. Journal of Experimental Psychology: Human Perception and Performance, 1978, 4, 357-372.

Duncker, K. Induced motion. In W.D. Ellis (Ed.), A sourcebook of Gestalt psychology. London: Routledge & Kegan Paul, 1937.

Dürer, A. The human figure: The Dresden sketchbooks. New York: Dover, 1972. (Originally published in 1528).

Emlen, S.T. The stellar-orientation system of a migratory bird. Scientific American, 1975, 233 (2), 102-111.

Enlow, D. Handbook of facial growth. Philadelphia: W.B. Saunders, 1975.

Gibson, J.J. The perception of the visual world. Boston: Houghton Mifflin, 1950.

Gogel, W. The adjacency principle in visual perception. Scientific American, 1978, 238 (5), 126-139.

Johansson, G. Configurations in event perception. Uppsala, Sweden: Almqvist & Wiksell, 1950.

Johansson, G. Visual perception of biological motion and a model for its analysis. Perception & Psychophysics, 1973, 14, 201-211.

Johansson, G. Spatio-temporal differentiation and integration in visual motion perception. Psychological Research, 1976, 38, 379-393.

Johnston, I.R., White, G.R. & Cumming, R.W. The role of optical expansion patterns in locomotor control. American Journal of Psychology, 1973, 86, 311-324.

Koenderink, J.J. & van Doorn, A.J. Local structure of movement parallax of the plane. Journal of the Optical Society of America, 1976, 66, 717-723.

Kozlowski, L.T. & Cutting, J.E. Recognizing the sex of a walker from a dynamic point-light display. Perception & Psychophysics, 1977, 21, 575-580.

Llewellyn, K.R. Visual guidance of locomotion. Journal of Experimental Psychology, 1971, 91, 245-261.

Marey, E.J. Movement. New York: Arno Press & New York Times, 1972. (Originally published in 1895).

Marr, D. Vision. San Francisco: Freeman, 1982.

Murray, M.P. Gait as a total pattern of movement. American Journal of Physical Medicine, 1967, 46, 290-333.

Pittenger, J.B. & Shaw, R.E. Aging faces as viscal-elastic events: Implications for a theory of nonrigid shape perception. Journal of Experimental Psychology: Human Perception and Performance, 1975, 1, 374-382.

Proffitt, D.R. & Cutting, J.E. Perceiving the centroid of configuraton on a rolling wheel. Perception & Psychophysics, 1979, 25, 389-398.

Proffitt, D.R. & Cutting, J.E. An invariant for wheel-generated motions and the logic of its determination. Perception, 1980, 9, 435-559. (a)

Proffitt, D.R., Cutting, J.E., & Stier, D.M. Perception of wheel-generated motions. Journal of Experimental Psychology: Human Perception and Performance, 1979, 5, 289-302.

Regan, D. & Beverly, K.I. How do we avoid confounding the direction we are looking and the direction we are going. Science, 1982, 215, 194-196.

Rubin, E. Visuell wahrgenommene wirkliche Bewegungen. Zeitshrift für Psychologie, 1927, 103, 384-392.

Todd, J.T. & Mark, L.S. Issues related to the prediction of craniofacial growth. American Journal of Orthodontics, 1981, 79, 63-80.

Todd, J.T., Mark, L.S., Shaw, R.E. & Pittenger, J.B. The perception of human growth. Scientific American, 1980, 242 (2), 132-144.

Ullman, S. The interpretation of visual motion. Cambridge, MA: MIT Press, 1979.

Wallach, H. Visual perception of motion. In G. Kepes (Ed.) The nature and the art of motion. New York: George Braziller, 1965. Revised in H. Wallach, On perception. New York: Quadrangle/New York Times, 1976.

MOTION ANALYSIS OF GRAMMATICAL PROCESSES IN A VISUAL-GESTURAL LANGUAGE

Howard Poizner
The Salk Institute for Biological Studies
La Jolla, California

Edward S. Klima
University of California, San Diego
The Salk Institute for Biological Studies
La Jolla, California

Ursula Bellugi
The Salk Institute for Biological Studies

and

Robert B. Livingston
University of California, San Diego

Abstract

Movement of the hands and arms through space is an essential element both in the lexical structure of American Sign Language (ASL), and, most strikingly, in the grammatical structure of ASL: it is in patterned changes of the movement of signs that many grammatical attributes are represented. These grammatical attributes occur as an isolable superimposed layer of structure, as demonstrated by the accurate identification by deaf signers of these attributes presented only as dynamic point-light displays. Three-dimensional computer graphic analyses were applied in two domains, to quantify the nature of the 'phonological' (formational) distinctions underlying the structure of grammatical processes in ASL. In the first, we show that for one 'phonological' oppostion, evenness/unevenness of movement, a ratio of maximum velocities throughout the movement perfectly captures the linguistic classification of forms along this dimension. In the second, we map out a two-dimensional visual-articulatory space that captures in terms of signal properties, relevant relationships among movement forms that were independently posited as linguistically relevant. The fact that we are finding direct correspondences between properties of the signal and properties of the 'phonological' system in sign language, may arise in part because in sign languages, unlike in spoken languages, the movements of the articulators themselves are directly observable, and, also in part, because of the predominantly layered 'phonological' organization of sign language.

INTRODUCTION

Modality and Language

Current research shows that American Sign Language (ASL) has developed as a fully autonomous language with complex organizational properties, (Baker and Cokely, 1980; Bellugi and Studdert-Kennedy, 1980; Klima and Bellugi, 1979; Lane and Grosjean, 1980; Siple, 1978;

Motion: Representation and Perception
N.I. Badler and J.K. Tsotsos, Editors
Published by Elsevier Science Publishing Co., Inc.

Wilbur, 1979). ASL exhibits formal structuring at the same two levels as spoken language (the internal structure of the lexical units and the grammatical scaffolding underlying sentences); it also reveals similar kinds of organizational principles (constrained systems of features, rules based on underlying forms, recursive grammatical processes). While ASL shares underlying principles of organization with spoken languages, the instantiation of those principles occurs in formal devices arising out of the very different possibilities of a visual-gestural mode.

Despite the commonalities in principles of organization between signed and spoken languages, there are aspects of linguistic form in signed languages that stand out as most strongly resulting from the difference in modality. A fundamental difference between ASL and spoken language is in surface organization: signed languages display a marked preference for co-occurring layered (as opposed to linear) organization. The inflectional and derivational devices of ASL, for example, make structured use of space and movement, nesting the basic sign stem in spatial patterns and complex dynamic contours of movement. In the lexical items, the morphological processes, the syntax and even the discourse structure in ASL, the multi-layering of linguistic elements is a pervasive structural characteristic (Bellugi, 1980).

Signs of American Sign Language are related by virtue of a wide variety of inflectional and derivational processes. Different lexical items have families of associated forms, all interrelated by formal patterning, based on modifications of the movement of signs in space. Thus a single root form, for example that glossed as ASK or QUESTION, has a wide variety of manifestations, as shown in Figure 1. These different forms mark grammatical categories such as Person, Number, Reciprocity, Temporal Aspect, and Distributional Aspect as well as form the basis for derivationally related noun-verb pairs and a host of other derivational processes. The variety of inflectional and derivational devices indicate that ASL is an inflective language, more like Hebrew, Latin, and certain African languages than like English or Chinese (Bellugi and Klima, 1980; Klima and Bellugi, 1979).

In the *kinds* of distinctions that are morphologically marked, ASL is like many spoken languages; in the *degree* to which morphological marking is a favored form of patterning in the language, ASL is again similar to some spoken languages. However, in the *form* by which its lexical items are systematically modified, ASL has aspects that are unique. Grammatical processes (inflectional and derivational) are conveyed by changes in movement and spatial contouring, and occur as isolable superimposed layers of structure. One of the most distinguishing and pervasive characteristics of ASL is the multilayered nature of its structural organization. In individual forms the lexical and grammatical components occur in distinct co-occurring layers, and most significantly, this layering is largely maintained in the surface structure of these forms. There is a separation in the dimensions of contrast used to build morphemes at each layer. Take for example, the inflected form GIVE['to her'][Imperative],[1] representing the lexical morpheme meaning 'to give,' and two inflectional morphemes, one a pronoun-index morpheme for third person and, the other, the imperative morpheme. The lexical morpheme like other lexical morphemes in the language is represented

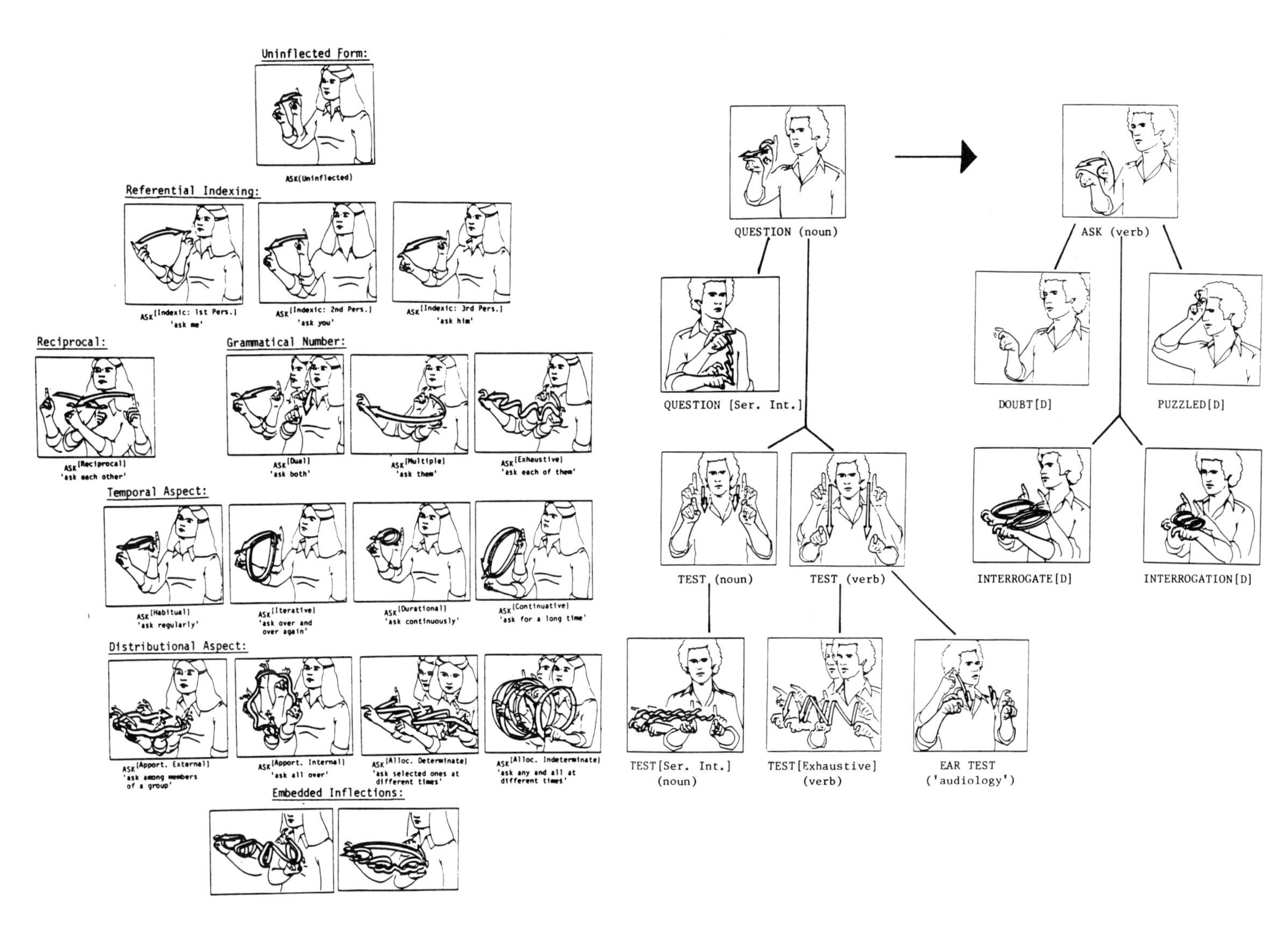

Figure 1. A variety of morphological processes in ASL layered on a single root.

by a specific hand configuration (in this case 'O'), a specific location (in this case the space in front of the body), and specific path shape of movement (in this case linear). These lexical components themselves are present in layered form (i.e., separate and co-occurring) in the surface. The third person pronominal index has as its form a particular diagonal direction of movement that is superimposed in layered fashion over all the other components of the form. The imperative morpheme itself is realized by tense quality of movement, again, simultaneously spread over the entire form. It is in this sense that sign language is predominantly multilayered with respect to form--with differences in levels of grammatical function mirrored by differences in layers of form. Thus, within the general domain of movement, path shape, direction, and quality each has its independent morphological value within this form.

In some morphologically complex forms, sequences of contrasting values occur within one or more layers, but this does not change the picture of the predominantly layered organization. The layering in addition to the sequencing within layers is the basis for a very rich morphological system.[2]

We have identified some fifty different derivational and inflectional processes in ASL. Linguistic analysis has revealed that the derivational and inflectional forms do not differ holistically, but rather differ along a limited number of spatial and temporal dimensions. Spatial dimensions such as geometric array (circle, line, arc), planar locus (vertical, horizontal), and direction of movement (upwards, downwards, sideways) involve primarily the manipulation of forms in space and figure significantly in the structure of inflections for Person, Reciprocity, Grammatical Number, and Distributional Aspect. Movement qualities such as end-manner (continuous, hold); tension (tense, lax); and rate (fast, slow) figure significantly in the structure of inflections for Temporal Aspect, Focus, Manner, and Degree. Two dimensions--cyclicity (singular, reduplicated) and hand use (one hand, two hands)--interact with the others in the formation of inflections in several grammatical categories. Thus the large number of morphological processes in ASL are combinations of a limited number of formal dimensions of space and movement.

LINGUISTIC MOVEMENT AS DYNAMIC POINT-LIGHT DISPLAYS

Extracting Movement from Sign Form

In our research linguistic analysis and experimental studies are linked together. We have the following objectives in carrying out these experimental investigations of such movement in ASL. First, these investigations may help answer general questions about the role of dynamic events in visual perception. Second, we want to put the multi-layered linguistic model of ASL to the strong test of isomorphism between posited elements of the linguistic structure and properties of the signal. Third, we investigate whether the perception of movement is modified by the special linguistic experience involved in the acquisition of a visual-gestural language. In order to experimentally investigate linguistically significant movement in ASL directly, we needed a way to isolate movement of the hands and arms, that is, to extract movement from sign forms. We used small light-emitting diodes placed at selected spots on the body, adapting a technique introduced by Johansson for studying the perception of biological motion (Johansson, 1975).

With respect to our first objective, evidence from perception of dynamic events has already offered clues to such longstanding questions as the nature of perceptual constancies, and the organizing principles for the visual system. Johansson (1973, 1975), Cutting and Kozlowski (1977), Ullman (1979) and others have demonstrated the importance of studying perception of dynamic point-light displays, displays in which form cues are drastically reduced. In our research, we extended this technique of studying biological motion to the complicated movement patterns within the linguistic system of ASL. We reduced the sign image to nine points of light, placing these lights at points on the body corresponding to the major joints of the arms and hands (shoulders, elbows, wrists, index fingertips). We record signing in a darkened room so that on the videotape only the pattern of moving points of light appear against a black background.

The Information in Point-Light Displays

Using a standard placement of nine point lights, we found that signers could accurately *match* lexical and inflectional movement presented only in dynamic point-light displays to those presented in normal form on videotape. Signers, furthermore, could accurately *identify* ASL inflections presented in point-light displays and signs of a constant hand configuration. Indeed, the signs were identified almost as well when presented as point-light displays in two dimensions as when presented in three, reflecting in part the information that moving dots carry about depth. Since dynamic point-light displays can convey sign forms, we investigated the information-carrying components within these point-light displays. We found that the movement of the fingertips, but not of any other pair of points, is necessary for sign identification, and that in general the more distal the joint, the more information its movement carried for sign identification (Poizner, Bellugi and Lutes-Driscoll, 1981).

Thus, even with such greatly reduced information, deaf signers were highly accurate at recognizing and identifying morphological processes of ASL presented in these point-light displays, demonstrating that these patterns of dynamic contours of movement form a distinct isolable (but co-occurring) layer of structure in ASL. These results broaden the data base on the perception of dynamic point-light displays to include perception of dynamic forms of a formal linguistic system. In this domain, as in others, moving points of light have a strong coherence, allowing perception of complex events. These displays, furthermore, provide a foundation for the development of techniques that will allow the rigorous analysis and control over important stimulus parameters of ASL for psycholinguistic investigations. We have shown that point-light displays accurately convey morphological processes in ASL and the movement of lexical signs. In order to quantify such movement trajectories, we have developed procedures for the computer graphic analysis of these displays and present some first results of these analyses below. We first turn briefly to studies of the perception of ASL movement dimensions.

The Interplay Between Perceptual and Linguistic Processes

One of the uses to which we have put the point-light displays is in the study of the interplay of basic perceptual processes and higher order linguistic ones. To pursue this, we have contrasted the psychological representation of movement by native deaf signers with that of hearing nonsigners. Triads of basic and of inflected ASL signs were presented as point-light displays for judgements of movement similarity. Multidimensional scaling and hierarchical clustering of judgements for both groups of subjects revealed, in the first place, that lexical and inflectional movements were perceived in terms of a limited number of underlying dimensions. Secondly, that the perceptual dimensions for the lexical level in general differ from those of the inflectional level (Poizner, in press). This result supports, with perceptual data, our previous linguistic analysis; namely, that the linguistic fabric of the two levels of structure in ASL is woven from different formational materials. Furthermore, the psychological representation of movement types within each level differs for deaf and hearing subjects, with perception of movement form tied to linguistically relevant dimensions for deaf, but not for hearing subjects. Thus, the data suggest that acquisition of a visual-gestural language can modify the natural perceptual categories into which these movement forms fall (Poizner, 1981, in press). These experiments extend previous studies of the perception of other formational categories of ASL, i.e., configuration of the hands (Lane, Boyes-Braem, and Bellugi, 1976; Stungis, 1981) and of location of the hands (Poizner and Lane, 1978). However, in these previous studies, the patterns of results for deaf signers and hearing nonsigners were the same; no modification of perception due to linguistic experience was found for static sign attributes. Perception of ASL movement (and perhaps movement in general, as a category) may in fact differ crucially in nature from perception of static parameters such as handshape and location.

THREE-DIMENSIONAL COMPUTER GRAPHICS AND LINGUISTIC ANALYSIS

Figure 2 illustrates aspects of the computer graphic analysis of ASL's elaborate system of movement contrasts. Up to now there has been little precise quantification of the movement signal itself. We have developed methods that allow both accurate and rapid three-dimensional movement measurements and the three-dimensional reconstruction of movement so that interactive control over the movement signal is attained (Loomis, Poizner, Bellugi, Blakemore, and Hollerbach, Note 1). In our early studies, we digitized trajectory information frame by frame from the output of a single camera (Bellman, Poizner, and Bellugi, in press). We now use a three-dimensional movement monitoring apparatus, the Selspot System, designed to permit rapid high-resolution digitization of hand and arm movement. Two opto-electronic cameras track the positions of light-emitting diodes attached to the hands and arms, and provide a digital output directly to a computer which calculates three-dimensional trajectories. From the position measurements, the movements are reconstructed in three-dimensions on an Evans and Sutherland Picture System. This system allows display and interactive control over the three-dimensional movement trajectory, so that various trajectory and dynamic characteristics can be calculated for any portion of the movement (see Figure 3). Figure 3 presents illustrations of four ASL inflections together with the reconstructed movement of the fingertip. Velocity and acceleration of the hand along the movement path are shown for each inflection. A rectangular solid is drawn around the movement path to delimit the maximum deviation of the movement along each of the three Cartesian dimensions. A three-dimensionsal calibration index, 50mm on a side, is rotated with the data to calibrate absolute displacement congruent with the perspective of each rotation.

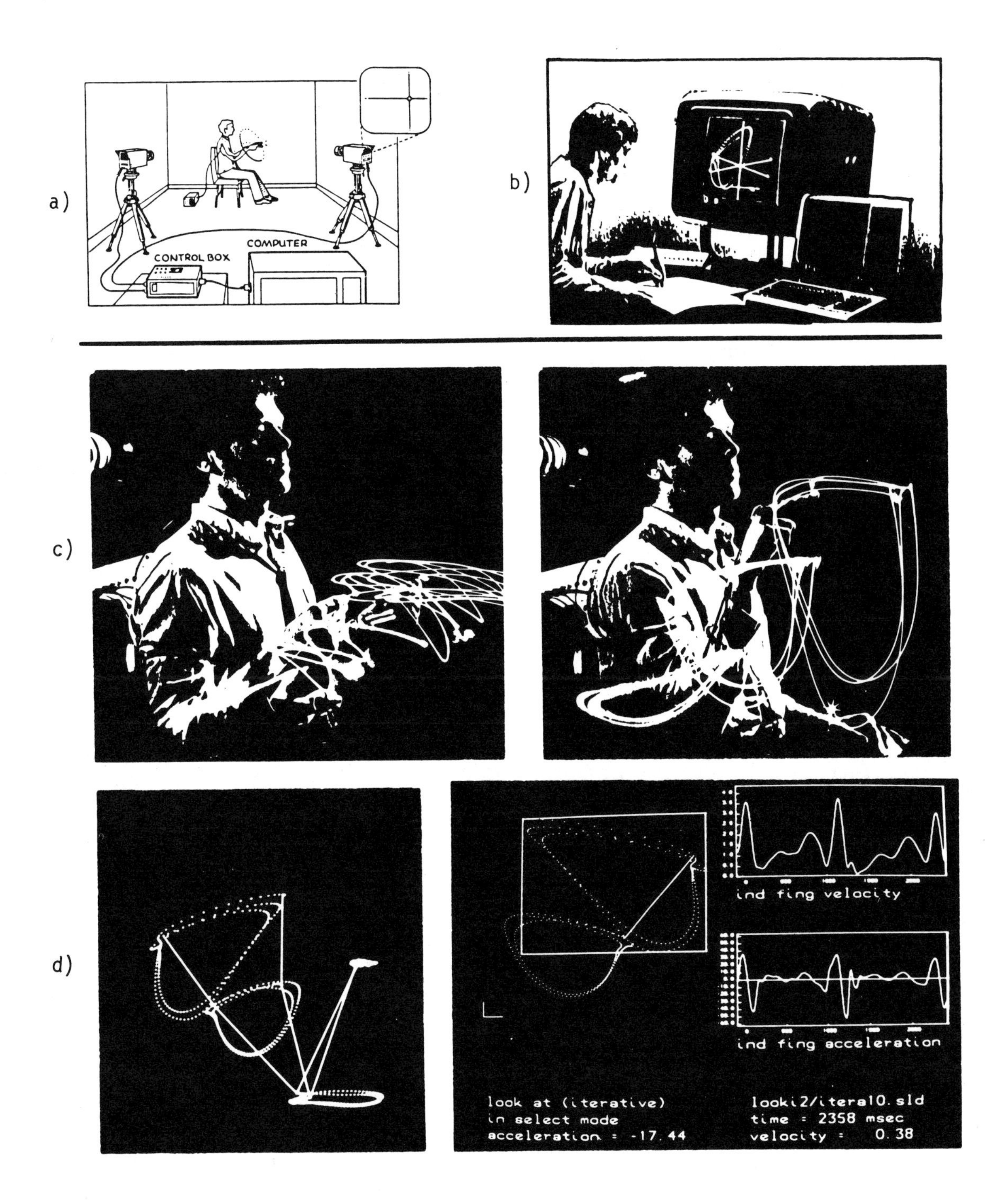

Figure 2. Computer graphic analysis of ASL movement. a) Movement digitization by Selspot cameras. b) Evans and Sutherland Picture System. c) Dynamic point-light displays. d) Three-dimensional computer graphic reconstructions.

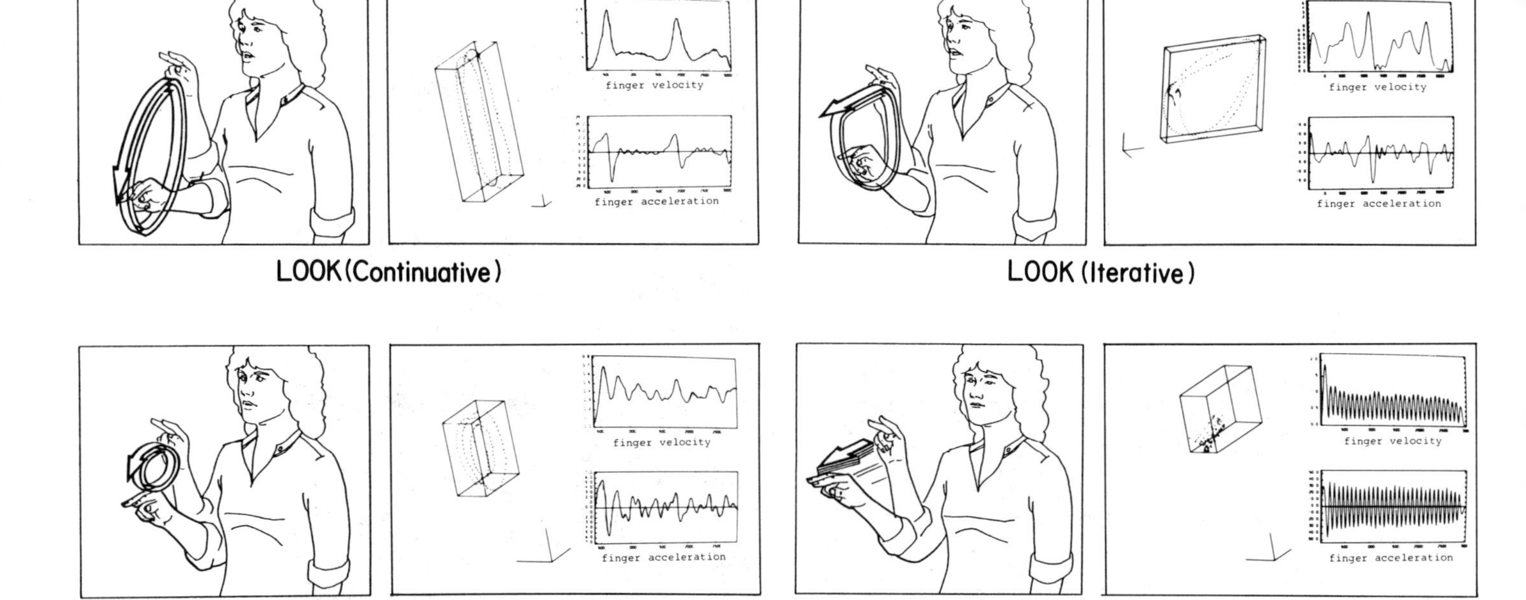

Figure 3. Three-dimensional computer graphic reconstructions of four ASL inflections. Velocities and accelerations are along the path, given in meters/sec and meters/sec^2, respectively. The calibration index is 50mm on a leg.

These procedures provide potent methods for extracting the movement from sign forms and for applying three-dimensional analyses to these movements. With these procedures coupled to our linguistic analyses, we can address some basic issues in the organization and structuring of language. We link these two modes of analyses below in two domains. In one, the analysis of physical characteristics from three-dimensional reconstructions of movement helps reveal the physical correlates of linguistic dimensions and in the process, helps refine our understanding of the linguistic dimensions. In the second, we present a study designed to investigate the biological boundary conditions that may help determine the structure that sign languages take in general. From three-dimensional movement analyses we derive a visual-articulatory space in which some ASL movements are embedded. From the metric that a series of these spaces will define, we will be able to make predictions about contrasts that might likely appear in other sign languages, about the order of acquisition of movement forms by deaf children, and about the structure underlying perception of ASL movement forms.

Computer Graphics and Linguistic Dimensions

A discrete, if not binary, system of oppositions characterizes the 'phonological' structure of ASL movement (Bellugi, 1980). The physical structure of ASL movement, however, is on the whole continuous in nature. It is therefore of interest to investigate the nature of the physical correlates of the 'phonological' distinctions.

Similar investigations of the physical correlates of phonological segments in speech reveal a lack of correspondence between segments in the signal and phonological segments (Liberman, Cooper, Shankweiler, and Studdert-Kennedy, 1967). It is possible that the situation may differ in sign for the following reasons. In the first place, the phonological structure of ASL retains its layered nature at the surface (e.g., the handshape retains its form throughout the movement), the elements of the separate layers differ dramatically in formational composition (e.g., handshape vs. location vs. movement shape vs. movement quality). Furthermore, these layers are independently controllable and to a large extent do not interact. Finally, in sign, unlike in speech, movements of the articulators themselves are directly observable rather than their secondary effects. We expect that this will result in less of a discrepancy between an articulatory and a visual description of the movement event. For one 'phonological' opposition in ASL, <u>evenness</u>, we have already found a relatively direct mapping between the structure of the signal and the phonological distinction.

We have posited eleven spatial and temporal linguistic dimensions along which inflectional-derivational processes vary and differ from one another. (For a fuller discussion see Klima and Bellugi, 1979.) In our first linguistic analysis, we described some inflectional forms as <u>even</u> (e.g., Durational, Habitual, Nominalization) and some as <u>uneven</u> (e.g., Continuative, Iterative, Resultative). That is, under some inflections movement seems constant in rate (<u>even</u>); under others there is a decided acceleration (<u>uneven</u>). The assignment was in accordance with the judgements of native informants and researchers.

In order to investigate the physical manifestation of this linguistic dimension, we used the interactive three-dimensional segmentation capability that the Evans and Sutherland Picture System provides, whereby movement cycles can be segmented into portions roughly corresponding to movement of the verb stem and return movement to the beginning position for the next cycle. In the case of LOOK[Habitual] (see Figure 3), the division can be made at the end of the outward movement from the signer's body; for LOOK[Continuative] and LOOK[Durational], segmentation is performed along the major axis of the ellipse. In general, then, repetition cycles can be thought of as consisting of (at least) two parts, as can some single cycle movements.

In order to capture that aspect of the movement signal conveying <u>evenness</u> or <u>unevenness</u> of movement, a variety of forms were first segmented into two major parts for each repetition cycle, and peak velocity measured for each segment. Figure 4a presents peak velocities across half cycles for LOOK[Continuative], LOOK[Habitual], and LOOK[Durational]. The solid line represents peak velocity of the first half cycle, and the dashed line, peak velocity of the second half. Figure 4a shows that for LOOK[Habitual] the hand reached about the same maximum velocity moving away from the body as when returning toward the body for the last 15 of 16 cycles. Similarly, maximum velocity of the first half cycle of each cycle of LOOK[Durational] was about the same as of the second half. However, for LOOK[Continuative], the hand reached a maximum velocity in the first half of a cycle nearly three times that of the second half. The Continuative inflection, has been considered to have an uneven quality of movement, whereas Habitual and Durational have an even movement quality.

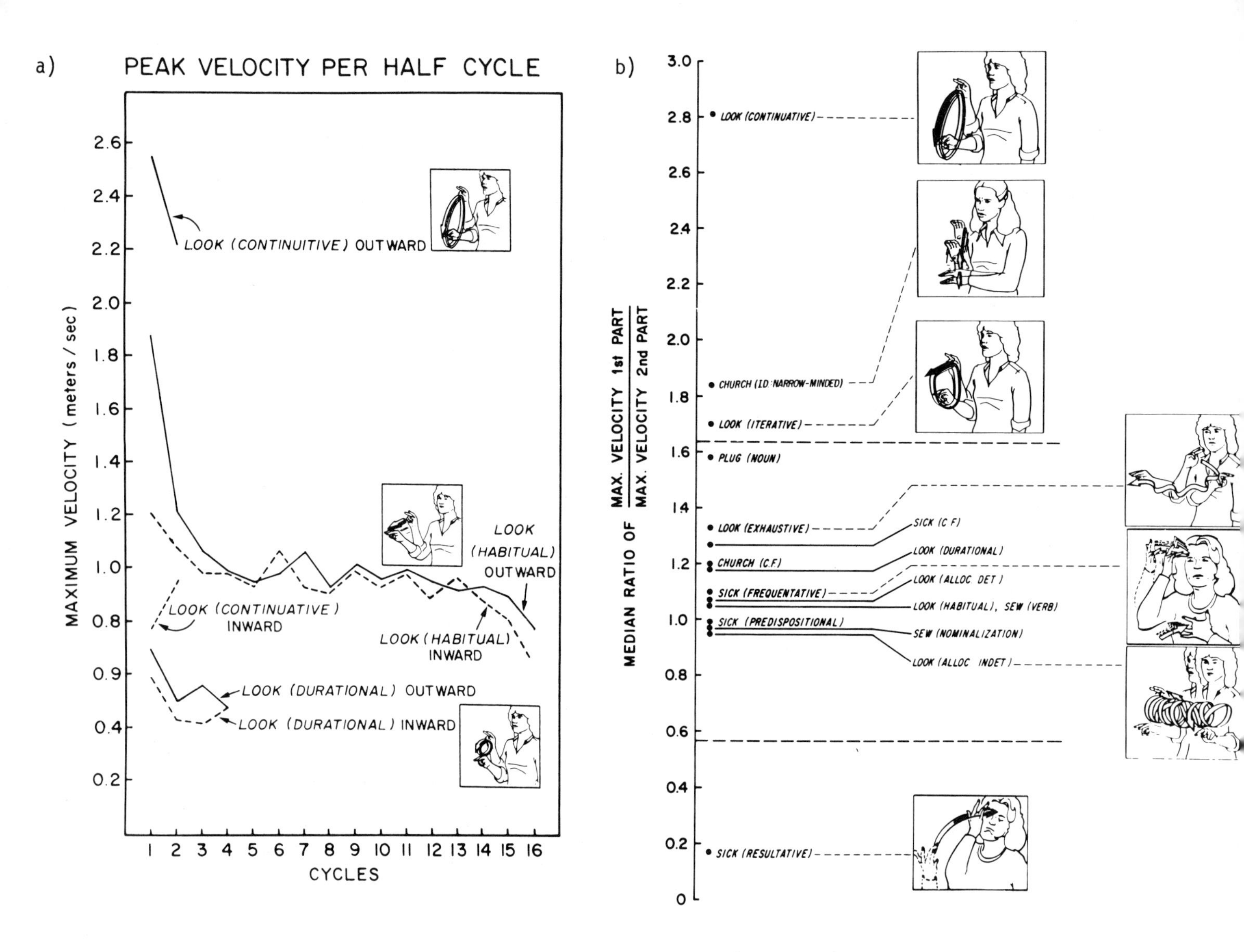

Figure 4. Physical correlate of a phonological opposition (evenness).

To evaluate the possibility that differential peak velocities across half cycles may indeed index this aspect of ASL movement, ratios of maximum velocity in the first half of the movement cycle to that of the second half were calculated for 15 multipart ASL movements. The median ratio across repetition cycles was used to index the evenness of the movement. Figure 4b presents these median ratios. When the hand moves faster in the first half cycle than in the second, the ratio is greater than 1.0; when the movement is faster in the second half, the ratio is less than 1.0. The Figure shows that the ratios vary from about .2 to 2.8, with a cluster of points at 1.0. The two dashed lines in the figure form a boundary around movements that have been described linguistically as being <u>even</u>. The three movements above this group and the one movement below it have all been classified as <u>uneven</u>. Thus, as Figure 4b indicates, the distribution of velocity ratios corresponds without exception to the linguistic classification of the opposition <u>even/uneven</u> and therefore provides a quantitative index of this movement attribute. Thus in the case of <u>evenness</u> we have found a correspondence between signal properties and 'phonological' components. We have made substantial progress in uncovering the physical correlates of other posited featural distinctions, including those of tension and restrained manner. We are particularly pleased about the possibility of capturing the physical correlates of these latter two movement attributes, since their descriptions have been couched purely in articulatory terms.

The preceding analysis is by no means comprehensive but is intended to illustrate a general approach to developing a visual-'phonetics' of sign language movement.

Biological Boundary Conditions On Language

The second domain in which we link computer graphic analysis and linguistic analysis is the search for biological bounday conditions on the form of language. If specific requirements of perception and motor control inherent in characteristics of the transmission system help determine 'phonological' structure, then the 'phonological' structures of signed and spoken languages should differ correspondingly. This issues is all the more interesting in light of the fact that in humans, specialized structures and functions have evolved for spoken communication: vocal tract morphology, lip, jaw and tongue innervation, mechanisms of breath control all have undergone evolutionary change from nonhuman primates to man that have shaped the vocal apparatus into a more efficient transmission system for the production of a variety of sounds (Liberman, 1970, 1974; Lieberman, 1975; Lenneberg, 1967).

In many ways the transmission system of sign language (visual-gestural) is radically different from that of speech and offers remarkably different possibilities and constraints. The auditory system, for example, is particularly adept at temporal discrimination (Julesz and Hirsch, 1972) and, indeed, speech exhibits a strong temporal patterning. The visual system is extremely well-suited for spatial discrimination and for simultaneous processing of visual parameters. Common to both modes, however, is the pressure for rapid and efficient communication. Our previous studies have suggested that the rate of transmitting propositional information in ASL is the same as that in English, despite radical differences in the size of the articulators and the time required to produce basic lexical items in two modes (Klima and Bellugi, 1979). The pressure towards maximizing output rate, and the differing processing capacities of the transmission modalities could very well help shape the form that languages take. Liberman (1982), in fact, has suggested that phonetic segments of speech are complexly encoded in the acoustic signal in order to bypass certain sensory limits of the auditory system (the resolving power of the ear). If, as Julesz and Hirsch (1972) suggest, the resolving power of the eye for temporal differentiations is far less than that of the ear, then this boundary condition may well have an effect on shaping the surface form of language when it has been so transplanted into another channel.

Mapping a visual-articulatory space. This paper presents one visual-articulatory space into which a subset of the movements of one sign language (ASL) fall. The sign forms used for this mapping all involve repetition of movement. These forms were mostly ones inflected for temporal aspect and degree, inflections that convey duration or recurrence over time and, in the case of "degree," the relative intensity of the state or action. Table 1 presents four morphologically distinct sets of ASL movements including the ones we focus on here, as well as some dimensional contrasts distinguishing morphological processes.

The nature of the dimensions of the spaces we will derive and the positioning of movement forms in these spaces will reveal the ways in which ASL has used the possibilities of the visual-gestural modality to create its movement contrasts.

Table 1

Four Morphologically Distinct Sets of Movements for Which Visual-Articulatory Spaces Will be Derived

Sample Movement Forms for Each Set

Lexical Movement Types	Derivational Forms
Stokoe Notation	SEW[D: Nominalization]
Vertical Action:	COMPARE[D: 'comparison']
Sideways Action:	CHURCH[D:'pious']
Horizontal Action:	DIGRESS[iD:'instead']
Rotary Action:	QUIET[iD:'acquiesce']
Hand Internal Action:	LOOK[iD:'anticipate']
Bending, Opening,	WRONG[iD:'unexpectedly']
Nodding, Closing,	CHURCH[iD:'narrow-minded']
Wiggling	
Circular:	

Inflections for Temporal Aspect and Degree	Inflections for Distributional Aspect
Durational	Multiple
Habitual	Exhaustive
Continuative	Seriated External
Iterative	Seriated Internal
Predispositional	Allocative Determinate
Frequentative	Allocative Indeterminate
Incessant	Apportionative External
Approximative	Apportionative Internal

Dimensional Contrasts Distinguishing Morphological Processes

End Manner

-End-Marked	+End-Marked
Continuative	Frequentative
Predispositional	Iterative
Durative	Resultative
Facilitative	Punctual

Tension

-Tense	+Tense
Predispositional	Frequentative
Approximative	Intensive
Habitual	Resultative
Durational	Predicative

Geometric Contour

-Linear	+Linear
Multiple	Seriated External
Exhaustive	Seriated Internal
Apportionative External	Augmentative
Allocative Determinate	Diminuative

Evenness

-Uneven	+Uneven
Exhaustive	Continuative
Habitual	Iterative
Intensive	Resultative
Frequentative	Idiomatic Derivative

Quantification. Each sign form from which the movement was to be extracted was made in canonical form, with the movement digitized by the Selspot Movement Monitoring System, and reconstructed in three dimensions on the Evans and Sutherland Picture System. We quantified the movements in a number of ways, including measuring such kinematic parameters as duration, amplitude, velocity, acceleration along the path and Fourier analysis. From these measurements, we constructed low dimensional spaces in which the movements are located. We motivated our selection of particular spaces by their ability to reveal relevant linguistic relationships.

The rhythmic properties of many ASL morphological processes are part of their essential defining characteristics. For example, the inflections for Continuative and Durational aspect are distinguished essentially by their rhythmic dynamic properties throughout cycles of movement. We present one method of quantification, Fourier analysis, particularly relevant to cyclic movements. Fourier analysis allows the description of rhythmic (and spatial) structure of events by summing suitable sinusoidal components. We attempt to describe relevant relationships among ASL movements in terms of their physical characteristics.

Fourier analysis. We have applied Fourier analysis to cyclic ASL movements. Figure 5 presents the Fourier spectra of movement components in the sagittal and vertical directions for the LOOK[Continuative] and the LOOK[Durational] inflections. The sign forms used in this study have their primary movement components in these directions. Line drawings of these two inflected signs and three-dimensional reconstructions of the movements are also presented. The spectra for the Continuative inflection exhibit a first major peak at .80hz., reflecting a basic repetition cycle rate of .8 cycles per second. Higher harmonics occur at integral multiples of this fundamental frequency, with their amplitude (in dB) decreasing in a roughly linear fashion. The Durational inflection, however, shows a very different pattern of Fourier components. First of all, its fundamental repetition rate is 1.68 hz., approximately twice as fast as that of the Continuative. Unlike the relatively high energy of the harmonics of the Continuative, the Durational shows essentially one major energy component (the fundamental) with very reduced harmonics. Spectra of movements with sinusoidal components in the sagittal and vertical directions are precisely what we would expect for the smooth, circular, repeated movement of the Durational inflection. Since the Continuative inflection has an uneven rhythm, however, its spectra show substantial energy components at the higher harmonic capturing this difference in dynamics. The relative amplitudes of the harmonics with respect to the fundamental, and the frequencies at which they occur provide a concise index that captures relevant relationships among ASL movement forms.

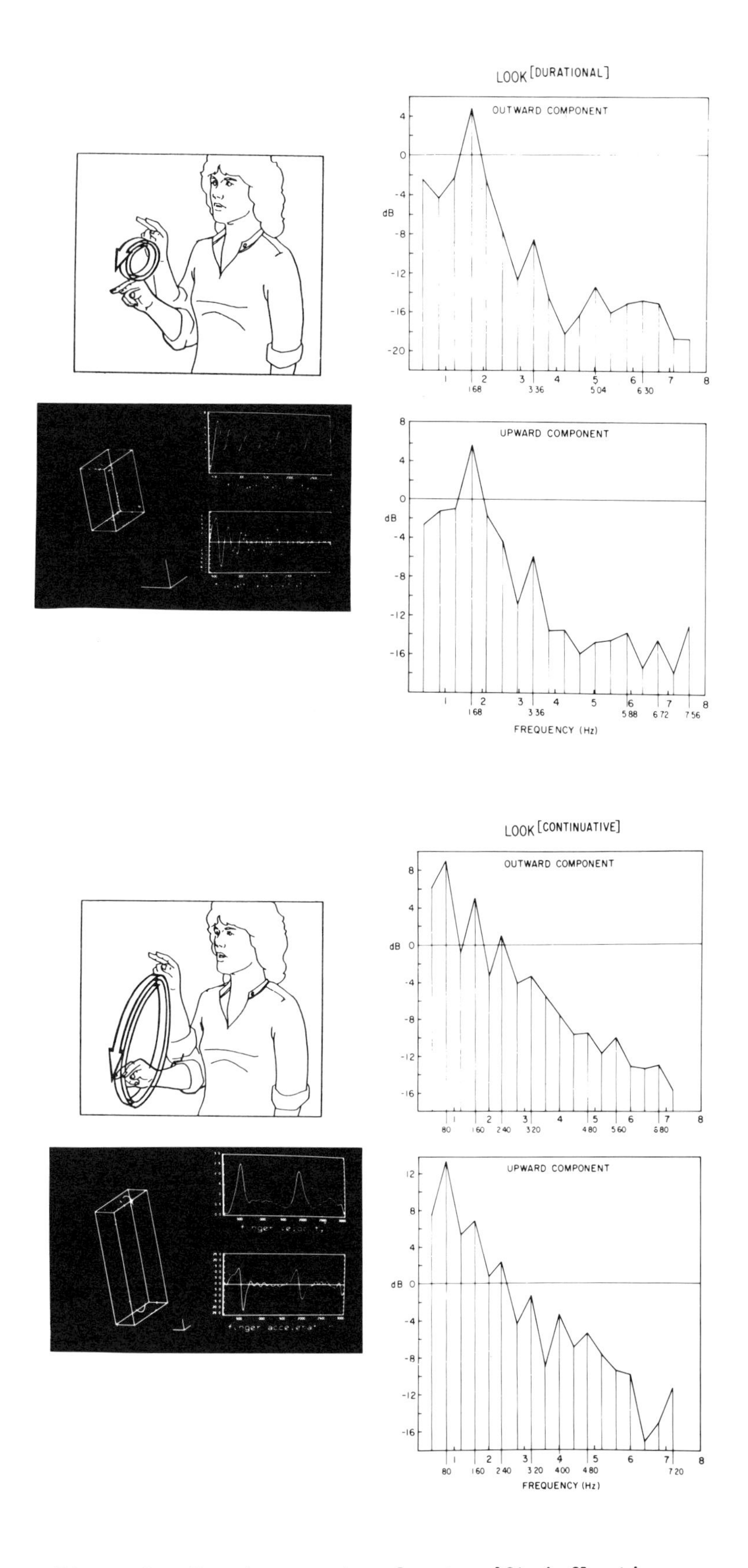

Figure 5. Fourier spectra for two ASL inflections.

Deriving the space. Figure 6 illustrates one visual-articulatory space for 11 cyclic ASL movements. This figure presents energy of the 3rd harmonic relative to that of the fundamental versus energy of the 2nd harmonic relative to that of the fundamental for the movement component in the sagittal axis (toward and away from the signer's body). The figure shows several clusters of movement forms. _Even_ movements without any endmarking (points A,B,C) form a tight cluster in the upper right corner of the figure (all have energy at essentially only the fundamental). Diametrically opposite to these movements, lie movements that are uneven in quality (points G and I are forms having high energy at the 2nd and 3rd harmonics as well as at the fundamental). Points D, E, and F lie intermediate along this major axis of the figure, and are _even_, but _endmarked_. The final two points in the figure (J and K) are grouped together, away from the other movements, and exhibit amplitudes of the 2nd harmonic equal to or even greater than that of the fundamental. The Allocative Determinate and Allocative Indeterminate are the only two movement forms of the figure that are displaced from the sagittal axis, a movement characteristic important to the linguistic structure of the morphological processes and to the perception of these processes (Klima and Bellugi, 1979; Poizner, in press). Fourier components reflect this pattern of displacement. Thus this preliminary construction of a physical space in which these movements lie, already captures in terms of signal properties, relevant relationships among movement forms that were independently posited as linguistically relevant. This approach should illuminate the physical basis for these linguistic relationships.

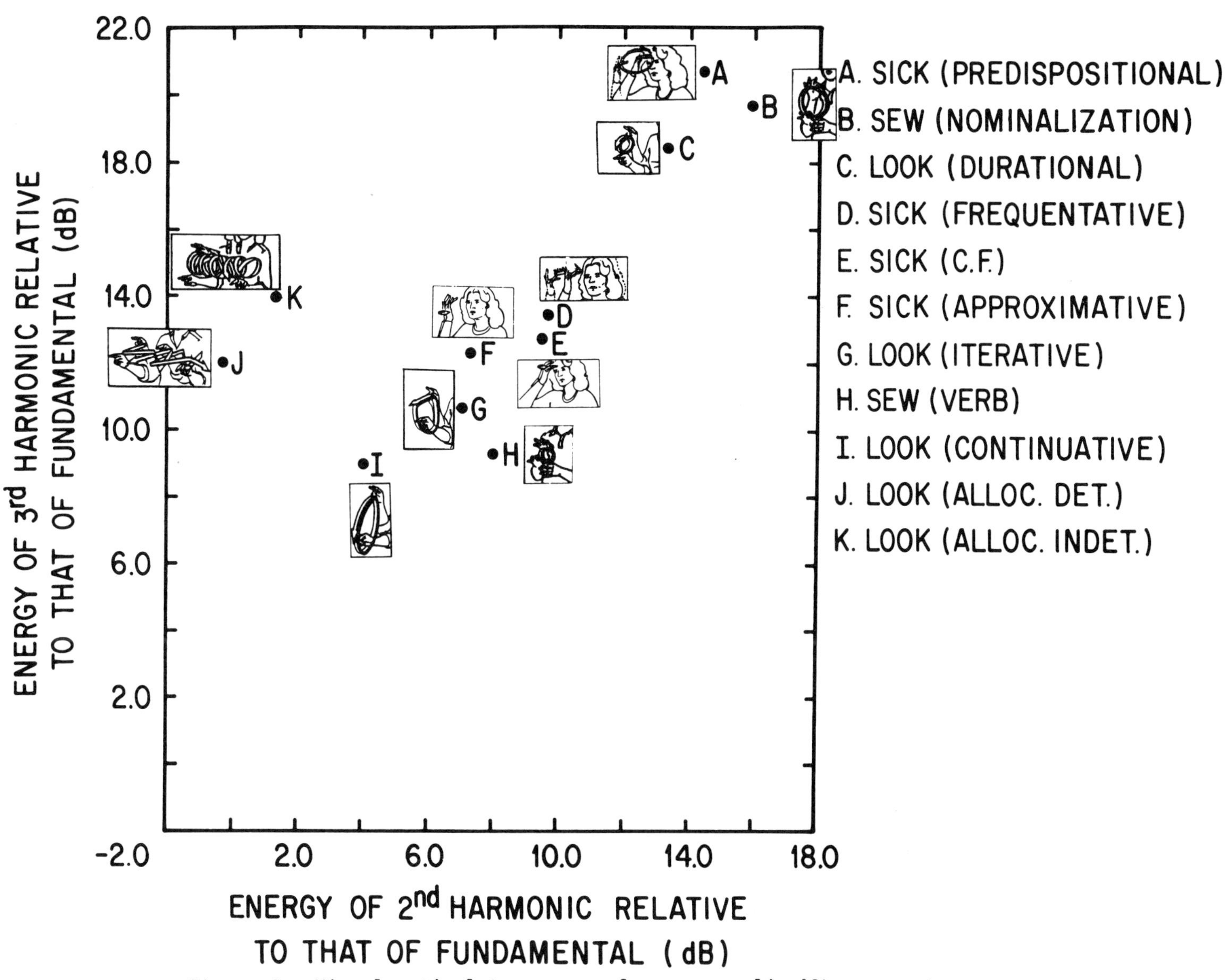

Figure 6. Visual-articulatory space for some cyclic ASL movements.

From the metrics that such spaces define, we will be in a position to make predictions about the following three areas: 1) the general form that sign languages take; 2) the order of emergence of featural contrasts as the child acquires the languages; and 3) structure underlying perception of movement. We predict that those constrasts that are minimally separated in the space are less likely to appear universally in other sign languages, will emerge late in the child, and are more easily confused in perception and we predict the reverse for contrasts that are maximally separated in the space. For example, from the space presented in Figure 6, the contrasts even/uneven and displaced/nondisplaced are salient contrasts of the movements' properties. We predict that the linguistic featural oppositions underlying these contrasts will likely appear in other sign languages and be easily recognized in the perceptual process.

Modality issues. Derivational of the physical spaces underlying the sounds of speech has also been attempted. There is a growing realization that the science of phonetics can produce insights for the understanding of phonological structure (Ladefoged, 1980; Lindblom, 1978; Ohala, 1974; MacNeilage and Ladefoged, 1976). Lindblom and his colleagues have tried to map out how the phonological structure of vowels is shaped by requirements of speech perception and production (Lindblom, 1978; Lindblom, 1979; Liljencrantz and Lindblom, 1972). Lindblom derives a possible acoustic-auditory space based on a model of vocal-tract articulation. Within this possible acoustic-auditory space, he calculates, by algorithm, the maximal separation for a vowel system with a given number of vowels. He then compares these calculated vowel spaces with those observed in the world and finds a strikingly good match. Maximizing perceptual contrast thus seems a strong determinant of the phonological structure of vowel systems.

While the work of Lindblom and his colleagues is very intriguing, they have as yet only attempted to derive spaces for steady-state vowels, sounds that are somewhat artificial in speech (Strange, Jenkins, and Edman, 1977; Strange, Verbrugge, Shankweiler, and Edman, 1976). The severe acoustic-phonetic problem that dynamic speech sounds show (i.e., the lack of correspondence between acoustic and phonetic segments), may be a limiting factor in this domain. The study of sign language has advantages in this exploration of determinants of phonological structure and may prove an especially promising domain. We are encouraged by our initial studies that we will not face the same kind of acoustic-phonetic problem inherent in speech. In conclusion, since the positioning of the movements in the visual-articulatory space presented above captured important 'phonological' patterns, we thus are finding in this domain, also, a direct correspondence between properties of the signal and properties of the 'phonological' system. Perhaps this situation exists in part because in sign languages, unlike in spoken languages, the movements of the articulators themselves, rather than their secondary effects, are directly observable, and also in part because of the predominantly layered 'phonological' organization of sign language.

Acknowledgements

This work was supported in part by National Science Foundation Grant #BNS81-811479 and by National Institutes of Health Grant #HD13249 to the Salk Institute for Biological Studies. We are indebted to Dr. Emilio Bizzi for use of the Selspot Movement Monitoring System. We thank Dr. Harlan Lane for the suggestion to apply Fourier analysis to quantify sign movement.

Footnotes

1. Lexical bases of signs are denoted by English glosses in full capitals (e.g., GIVE). Bracketed labels, as in GIVE['to her'] refer to specific inflectional processes signs have undergone. Multiple bracketing, as in GIVE['to her'][Imperative] refer to hierarchically nested inflectional processes.

2. Some spoken languages also have a multilayered organization. In Hebrew for example, the word is composed of a tri-consonantal lexical root at one layer, and vowel patterning conveying inflectional and derivational morphemes as a second layer. The surface form of thee words however is nonetheless linear, with consonant and vowels following one another. Thus, unlike spoken languages which show a pronounced layered structure, ASL forms retain their layered structure in the surface structure--components of the lexical root, derivational, and inflectional morphemes co-occur in a multilayered structure.

Reference Note

Note 1. Loomis, J., Poizner, H., Bellugi, U., Blakemore, A., and Hollerbach, J. "Computer Graphic Modeling of American Sign Language." Ms., The Salk Institute for Biological Studies, La Jolla, Ca.

References

Baker, C. and Cokely, D. American Sign Language: A Teacher's Resource Text on Grammar and Culture. Silver Spring, Md.: National Association of the Deaf, 1980.

Bellman, K., Poizner, H. and Bellugi, U. "Invariant Characteristics of Some Morphological Processes in American Sign Language." Discourse Processes, in press.

Bellugi, U. "The Structuring of Language: Clues from the Similarities Between Signed and Spoken Language." In U. Bellugi and M. Studdert-Kennedy (Eds.), Signed and Spoken Language: Biological Constraints on Linguistic Form. Dahlem Konferenzen. Weinheim/Deerfield Beach, Fla.: Verlag Chemie, 1980, 115-140.

Bellugi, U. and Klima, E. "Morphological Processes in a Language in a Different Mode." In W. F. Hans, C. Hofbauer, and P. R. Clyne (Eds.), The Elements: Linguistic Units and Levels. Chicago, Ill: Chicago Linguistic Society, 1980, 21-42.

Bellugi, U. and Studdert-Kennedy, M. (Eds.). Signed and Spoken Language: Biological Constraints on Linguistic Form. Dahlem Konferenzen. Weinheim/Deerfield Beach, Fla.: Verlag Chemie, 1980.

Cutting, J. E. and Kozlowski, L. T. "Recognizing Friends by their Walk: Gait Perception Without Familiarity Clues." Bulletin of the Psychonomic Society, 1977, 9, 353-356.

Johansson, G. "Visual Perception of Biological Motion and a Model for Its Analysis." Perception and Psychophysics, 1973, 14, 201-211.

Johansson, G. "Visual Motion Perception." Scientific American, 1975, 232, 76-89.

Julesz, B. and Hirsch, I. J. "Visual and Auditory Perception: An Essay of Comparison." In E. E. Davis and P. B. Denes (Eds.), Human communication: A Unified View. New York: McGraw-Hill, 1972, 283-340.

Klima, E. S. and Bellugi, U. The Signs of Language. Cambridge, Mass.: Harvard University Press, 1979.

Ladefoged, P. "What are Linguistic Sounds Made of? Language, 1980, 56, 485-502.

Lane, H., Boyes-Braem, P., and Bellugi, U. "Preliminaries to a Distinctive Feature Analysis of Handshape in American Sign Language." Cognitive Psychology, 1976, 8, 263-289.

Lane, H. and Grosjean, F. (Eds.). Recent Perspectives on American Sign Language. Hillsdale, N.J.: Erlbaum, 1980.

Lenneberg, E. Biological Foundations of Language. New York: Wiley, 1967.

Liberman, A. M. "On Finding that Speech is Special." American Psychologist, 1982, 37, 148-167.

Liberman, A. M. "The Grammars of Speech and Language." Cognitive Psychology, 1970, 1, 301-323.

Liberman, A. M., Cooper, F. S., Shankweiler, D. P., and Studdert-Kennedy, M. "Perception of the Speech Code." Psychological Review, 1967, 74, 431-461.

Liberman, A. M. "The Specialization of the Language Hemisphere." In F. O. Schmitt and F. G. Worden (Eds.), The Neurosciences: Third Study Program. Cambridge, Mass.: MIT Press, 1974, 43-56.

Lieberman, P. On The Origins of Language. New York: Macmillan Publishing Co., Inc., 1975.

Liljencrantz, J. and Lindblom, B. "Numerical Simulation of Vowel Quality Systems: The Role of Perceptual Contrast." Language, 1972, 48, 839-862.

Lindblom, B. "Phonetic Aspects of Linguistic Explanation." Studio Linguistics, 1978, 23, 137-153.

Lindblom, B. "Experiments in Sound Structure." Revue de Phonetique Appliquee. Belgique: Universite de l'Etat Mons, 1979, 51, 155-189.

MacNeilage, P. and Ladefoged, P. "The Production of Speech and Language." In E. Carthette and M. P. Friedman (Eds.), Handbook of Perception VII. New York: Academic Press, 1976, 75-120.

Ohala, J. J. "Experimental Historical Phonology." In J. M. Anderson and C. Jones (Eds.), Historical Linguistics II. Theory and Description in Phonology. (Proceedings of the First International Conference on Historical Linguistics. Edinburgh, 2-7 September 1973.) Amsterdam: North Holland Publishing Co., 1974, 353-389.

Poizner, H. "Perception of Movement in American Sign Language: Effects of Linguistic Structure and Linguistic Experience." Perception and Psychophysics, in press.

Poizner, H. "Visual and 'Phonetic' Coding of Movement: Evidence from

American Sign Language." _Science_, 1981, _212_, 691-693.

Poizner, H. and Lane, H. "Discrimination of Location in American Sign Language." In P. Siple (Ed.), _Understanding Language Through Sign Language Research_. New York: Academic Press, 1978, 271-287.

Poizner, H., Bellugi, U. and Lutes-Driscoll, V. "Perception of American Sign Language in Dynamic Point-Light Displays." _Journal of Experimental Psychology: Human Perception and Performance_, 1981, _7_, 430-440.

Siple, P. (Ed). _Understanding Language Through Sign Language Research_. New York: Academic Press, 1978.

Strange, W., Jenkins, J. J., and Edman, T. R. "Identification of Vowels in 'Vowel-less' Syllables." _Journal of the Acoustical Society of America_, 1977, _61_, S39(A).

Strange, W., Verbrugge, R. R., Shankweiler, D. P., and Edman, T. R. "Consonant Environment Specifies Vowel Identity." _Journal of the Acoustical Society of America_, 1976, _60_, 213-224.

Stungis, J. "Identification and Discrimination of Handshape in American Sign Language." _Perception and Psychophysics_, 1981, _29_, 261-276.

Ullman, S. _The Interpretation of Visual Motion_. Cambridge, Mass.: MIT Press, 1979.

Wilbur, R. _American Sign Language and Sign Systems: Research and Applications_. Baltimore, Md.: University Park Press, 1979.

Graphics, Control, and Description

Motion Graphics, Description and Control

Norman I. Badler

Computer and Information Science

Moore School D2

University of Pennsylvania

Philadelphia, PA

Perhaps the single most important reason to investigate methods of motion graphics, description and control is that these modalities _verify_ that a representation is _adequate_ to describe motion phenomena. Whether designed for human or computer vision, a motion representation often only serves as a target for information extracted to accomplish a particular task, such as pattern recognition, classification, change detection, or shape description. Often this motion representation is selected by convenience and need only distinguish between phenomena of interest, possibly incorporating rather rudimentary notions of direction, velocity, and shape.

Such representations may exhibit notable failings when they are asked to reconstruct (draw, animate, describe, control) what they purport to represent. Research in computer graphics, robotics, and expert systems is now confronting these representation questions directly by requiring the _understanding_ and _implementation_ of the _semantics_ underlying a motion representation system. Our representations can no longer be just instruments of convenience;

Motion: Representation and Perception
N.I. Badler and J.K. Tsotsos, Editors
Published by Elsevier Science Publishing Co., Inc.

rather they must be rich enough to simulate a reasonable slice of reality.

The papers in this session address motion representations in a variety of ways. I shall try to bind them together within a framework of motion understanding issues:

1. What representations are needed for complex, jointed object motion?

2. Should the representation be high level or low level?

3. How is the motion information within the representation visualized?

4. How many "uses" can a representation support?

To examine these issues more readily, I shall first very briefly summarize each paper in the session, then return to characterize the issues in terms of the papers.

In the first paper, Ginsberg and Maxwell of the MIT Architecture Machine Group discuss a low cost data acquisition system for human motion, and demonstrate a graphical cloud-like human figure model to "play back" those acquired motions. While this method gives accurate and voluminous motion data, it is somewhat more difficult to manipulate (to vary a motion within desired ranges or constraints) and generalize.

Where the motion information arises from manual or procedural generation techniques or key-frames, an interactive animation system editor becomes necessary. Fortin, Lamy, and Thalmann of the University of Montreal describe such an editor which permits graphical manipulation of an animated "score." Here units of action, their timings, and mutual relationships may be varied at a symbolic level to achieve an overall animation goal. The motion semantics are implicitly supplied by the animator's own juxtaposition of events and the "built-in" (graphical or mathematical) interpolation between key-frames.

Zeltzer from Ohio State University proposes a much more knowledge-driven approach to human movement (in particular). His human figure, for example, is driven by computational analogues of the human motor system, including local (low level) motor programs, (intermediate level) skills, and overall (high level) goals. The representation expects to support moving objects which interact with, and respond to, their environment. The visualization model is a highly detailed skeleton which is able to demonstrate walking and jumping as well as simpler jointed motions.

Murthy and Raibert of Carnegie-Mellon University concentrate on the physical verification of a motion representation for locomotion in a one-legged hopping "robot." The real physical world (with gravity, friction, inertia, and obstacles) provides the verification environment for the dynamic equations which control the hopper. Both simulated and actual motions are obtained.

Finally, Simmons and Davis from the MIT Artificial Intelligence Laboratory describe an expert system for representing the causes and results of slow physical (geological) changes, such as erosion, deposition, and intrusion. Graphical generation is used to visually validate the model, but the heart of the system is the reasoning component which understands the mechanisms of change and can reason about and with change processes.

Let us return now to our issues.

1) What representations are needed for complex jointed object motion?

Clearly some of the papers specifically address the problem of human motion: Ginsberg and Maxwell, Zeltzer, and indirectly, Murthy and Raibert. Ginsberg and Maxwell use the joint position traces over time as their representation; Zeltzer uses "hardwired" motor control procedures modeled as a hierarchy of finite-state automata; and Murthy and Raibert use dynamics equations. The latter two demonstrate the tendency towards more (bio-) mechanical or robotics models of motion, a realization by them (and others) that "shortcut" graphical solutions to motion control (such as relying totally on animator-generated key-frames) may be eventually displaced by programs which understand joints, linkages, and physics. (The <u>animator</u>, by the way, is not being displaced; the issue is how much effort he or she has to expend in creating necessary but "mundane" motions such as reaching or walking. While artistically valid works may indeed be created by drawing individual frames (with or without computer assistance), this approach must be viewed as an extreme where the only "representation" is the image itself. It is not a readily modifiable <u>symbolic</u> view of the motion.) The range of representations, therefore, may be seen to stretch from complete frame-by-frame specifications to more economical symbolic (and often linguistic) descriptions of what ought to be happening.

Fortin, Lamy, and Thalmann present a graphical tool for effectively dealing with the temporal aspects of a motion representation. Since temporal properties cannot be completely orthogonal from motion parameters, their scheme may have significant impact on the user view of an animation system. The complexity of jointed motion is partially due to the parallel nature of the individual part motions, and any motion representation which expects to deal with several interacting objects must confront this issue directly. Perhaps there are useful lessons to be learned from experiences in parallel programming languages. It is certainly significant that recent animation systems (such as Reynolds' ASAS) have begun to take on the appearance of partial or complete simulation systems with objects, actors, clocks, and message-passing mechanisms.

This issue therefore easily leads to the next one:

2) Should the representation be high level or low level?

At the lowest level, the trace of joint positions used by Ginsberg and Maxwell is certainly an adequate representation for reproducing motion on the graphics model. Very little interpretation is needed in the program since the semantics are so direct. Murthy and Raibert's hopper may also be considered low level in the sense that the equations of motion, once solved, result in directly usable control signals. Besides the high level goals embodied in the equations (balance, directional motion, jumping height, etc.), there is no explicit planning or reasoning level at which the hopper can presently operate. Of course, such a layer may be readily added to such a system, as the motion equations provide the necessary low level semantics to carry out high level goals. A clear example of this is described by Zeltzer, as his simulated human skeleton can (or will) navigate terrain, respond to obstacles, and yet walk with a certain "style" dictated by embedded (but communicating) motion control programs. Simmons and Davis demonstrate the potential of high level representations by showing that "why" questions are answerable with a high-level representation: demonstrating a representation's knowledge structure not only of <u>how</u> to move (or change), but <u>why</u>. The Fortin, Lamy, and Thalmann paper falls into an intermediate level where temporal relationships can be handled with graphical relationships and motions themselves may be of any sort supported by the remainder of the animation system.

Since a principal function of a motion representation is verification that that representation is indeed adequate for a task, we must ask:

3)How is the motion information within the representation visualized?

The obvious answer is, of course, with computer-generated graphics. The less obvious methods are with "natural language" descriptions, physical (robot) control, and question-answering. Significantly, most of these methods are exemplified by these papers. Zeltzer, Ginsberg and Maxwell, and Murthy and Raibert all use graphical models; the latter also controls the real thing. Simmons and Davis use both graphics (diagrams and "movies") and reasoning processes (text) to demonstrate that their system knows "more" than just the input data.

Among the specific graphical issues are the display methods used to simultaneously portray motion and shape, graphics hardware limits, and any human perceptual limits. There are interesting connections here between the graphical synthesis process and the human perceptual process which observes and checks its validity. For example, Ginsberg and Maxwell use a cloud-like figure because it is efficient to display on existing point and vector drawing hardware, yet relies on three-dimensional moving dot perception to integrate it as a connected jointed unit. Zeltzer uses a shaded skeleton figure, sacrificing real-time generation to solid shaded detailed imagery animated only by playback. Murthy and Raibert's hopper must in fact, actually perform. Single snapshots are sufficient to convince us that the Simmons and Davis system can reason about geological change. But for very fast or subtle motions, sophisticated graphical tools such as motion blur and aliasing effects must be taken into account.

The last issue addresses the generality of a motion representation:

4) How many "uses" can a representation support?

Since motion, and especially human motion, is expressible in so many different forms, can a representation be adapted or integrated to permit alternative modes of visualization or expression? For example, Murthy and Raibert's dynamic model supports graphics and control, but not reasoning. Ginsberg and Maxwell's representation is a natural target of a motion acquisition system (for example, by computer visual imaging), and supports graphics, but not (directly) motion description. Fortin, Lamy, and Thalmann could use their system to temporally organize any of the others, although it would lack reasoning processes.

Zeltzer's system should demonstrate maximum utility for animation and robotics control. Murthy and Raibert's control system could be used as a low level (but goal-directed) component of Zeltzer's hierarchic system. The ideas in the Simmons and Davis paper could form the basis of a planning and description system built on Zeltzer's model. And so on. While this happens to be of particular interest to me, I find that the other authors are all aware of the multiple possibilities of their systems. The sheer enormity of the task of integrating all these modalities of motion expression leaves considerable challenging work for the future.

With this short introduction, I hope to have shown that motion generation, description, and control are perhaps more central to the motion understanding problem than might previously have been imagined. It is certainly important to understand the workings of the human visual system and the structure of effective computer vision systems for processing moving objects, but the representations that are chosen to elucidate some mechanism cannot be viewed in isolation. Rather, their semantics should be implementable and transformable into a more readily visualized form amenable to either visual inspection or analytical verification. Design and implementation of motion semantics is an exciting and challenging research field stimulating our representation structuring skills while serving scientific and artistic applications.

"GRAPHICAL MARIONETTE"

Carol M. Ginsberg
Delle Maxwell

Architecture Machine Group
Massachusetts Institute of Technology

INTRODUCTION

Many person-modelling 3-D animation systems are currently being developed, but often suffer from confusing and elaborate user interfaces. Given over 200 degrees of freedom, the human form is capable of such intricate motion that its specification and display presents considerable difficulty to both animators and animation systems designers. Given such difficulties with single figures, the orchestration of several in parallel remains a major challenge. In pursuit of understanding thoroughly this complex motion of human beings, while faced with the difficulty of modelling the human form using conventional computer graphics techniques, the actual physical mien of the graphic figure is sometimes relegated a secondary status. A primitive and unconvincing appearance is the usual result. Our research in this field addresses these issues of user interface, motion, and expressive appearance within an animation application.

The perception of motion is instrumental in our understanding of scenes with missing or partial information [23]. Detail that is sacrificed in articulating the geometry of a life-like character can be compensated for by realistically depicting the character's motion. It is theorized that the most natural and effective means of accomplishing motion representation is to have a human "scriptor" directly act out the motions that will eventually be interpreted graphically. This method is termed "scripting-by-enactment" [4]. The animated output may be referred to as a "graphical marionette". A graphical marionette that is brought to life through scripting-by-enactment also has the advantage of having motion represented quantitatively. Once the motion is parametrized, the animating process can then depart from the realism as desired to allow for added flexibility, i.e., position coordinates representing the instantaneous location of body parts can be exaggerated for effect. For example, the graphical marionette can be made to "leap tall buildings in a single bound", or gain weight in a matter of seconds.

Running parallel with the drive to incorporate true-to-life human forms into computer graphics is the drive to incorporate true-to-life, i.e. natural, human actions into the use of computers and other machines. Where we began by communicating non-interactively with computers using an alpha-numeric format, we have progressed to highly interactive styles and expanded the I/O repertoire to include such means as touch-sensitive displays and voice recognition/synthesis. The next logical step to enhancing the human-machine interface is to incorporate kinetic input, or input by gesture. This technique furthers the realization of the "fully responsive interface" [16], facilitating conversation between the animator and the computer, its various peripherals, and its graphic inhabitants.

Section 1 will provide some background on the history and issues of person-modelling systems. Current research at other facilities investigating these issues are mentioned. The main elements of the "Graphical Marionette" project itself are delineated in Section 2. Body tracking is discussed in Section 3. Section 4 covers figure representation, and Section 5 considers animation scripting in terms of the topics in Sections 2, 3, and 4.

1 BACKGROUND

1.1 Body-Modelling

The concept of person-modelling is

Motion: Representation and Perception
N.I. Badler and J.K. Tsotsos, Editors
Published by Elsevier Science Publishing Co., Inc.

by no means new or novel; one can refer back to the myths of Pygmalion and Galatea, or to the "animated statues" of Daedalus. "Man ... has used all possible ingenuity to cause inanimate matter to perform the functions of living beings: whether it be human or animal, playing musical instruments, eating, or whatever else was stirring people's imagination..." [20].

As early as the 15th century, a more rigorous approach to the accurate and expressive depiction of human proportion and motion is embodied in the vast oeuvre of Leonardo da Vinci, and in the numerous canons of proportion devised by Quattrocento artists. At this time, extraction of an arithmetic norm from natural phenomena provided a basis for dimensioning the human form. Leonardo's studies of kinetics and anatomical investigations of structure, from bones to muscle to flesh, exemplify this desire to both understand and powerfully portray human appearance, motion, and character [22]. Current mathematical modelling techniques for motion and human form comprise a signifcant advance over Renaissance methodologies, yet represent a continuation of this mathematical abstraction of nature.

Recent developments in related computer fields such as artificial intelligence, graphics, robotics, and animation enable us to more closely approximate human movement and gesture than ever before. Numerous movement representation methods have been devised, both "notation systems designed for recording movement, and animation systems designed for the display of movement" [1]. The aim of these systems ranges from entertainment, characterized by a degree of artistic license in the animation, to more detailed research applications; simulations with the goal of modelling "exactly...a process, theory or system" [19], drawing upon and augmenting the knowledge of other related discplines such as biomechanics and kinesiology. Simulation systems generally can be classified as kinematic or dynamic - that is, motion without reference to cause or mass as compared to motion influenced by forces within and external to the body.

Unfortunately, because of the computational complexity of human modelling, and particularly simulation, actual image quality has yet to approach the artistic standards set five hundred years ago by Leonardo. Nevertheless, a common goal, realistic portrayal of human form and motion, constitutes a focal point for diverse research. Ultimately, emerging solutions to some fundamental computational problems pave the way towards creating compelling yet accurate synthetic, graphic counterparts.

1.2 Issues and Current Research

Regardless of application, the fundamental problems of representing human motion remain the same: specifying the movements, and creating a graphical image based on a physical model. The motion may be directed by descriptive or analytic inputs, as exemplified by analog position sensors or symbolic inputs (as in the Labanotation dance notation system [13]). Processing the input data, at the simplest level, requires translation to movement primitives which in turn are used to animate the database model.

The database, or description of the figure, provides, at least, names and connection specifications for joints and segments of the body. Stick figures provide a basis from which to work, but more realism is achieved through wire frame or solid models. A more natural-looking, but more complex figure may model surface "skin" over the linear segments using a large number of polygons, or surface patches. Neither polygons nor patches are entirely satisfactory, as problems such as deformations of the surfaces near moving joints occur. Computational cost is high in trying to rectify these problems. Other approaches, such as volume models (cylinders, ellipsoids, or spheres) are also being explored as alternatives to these methods. At the other end of the scale of "realism" are NYIT's whirling geometric solids suggestive of a dancing figure, exemplifying the creative use of such animation systems. Brief descriptions of current systems under development are presented as examples of some of the techniques mentioned above.

The University of Pennsylvania group, Badler, Smoliar, O'Rourke, and Korein [1,2,3] are conducting human movement research based primarily in these areas: computer graphics used for motion synthesis, computer vision, Labanotation [13] for movement notation, language analysis, and robotics [1]. One of the applications of this system is the simulation of the "activities of several people in a workstation environment" [1]. Directions, facings, revolutions, contacts, and shapes, concepts culled from the Labanotation system, are used as positional and directional movement primitives. In order to represent the dynamic qualities of movement, i.e, how one moves (slow, quick, forceful), other notation systems are being investigated which can be incorporated with the concepts of Labanotation. Describing these dynamic qualities using natural language would require using motion verbs and their modifiers developed within a special purpose task specification language. This method extends this research into ancillary fields such as language analysis, thereby increasing the scope and complexity of the research. The graphical representation is modelled using spheres, which are mapped onto the display as circles and ellipses, making for a somewhat lumpy looking, albeit versatile character known as the Bubble Person. The emphasis in this system is more on the representation and synthesis of movement, and the development of task specification languages, rather than on the physiognomy of the figure.

At Ohio State University, Zeltzer and the Computer Graphics Research Group are modelling human and animal forms as skeletons, simulating the bone and joint structures before tackling the problem of muscle and flesh [24,25,26,27]. Articulated, realistically modelled skeletal figures combined with motor control programs have been used to create convincing walking sequences over flat terrain.

This system is not based on a movement description notation such as the Labanotation [13] used by Badler, et al [1,2,3], as they doubt its extendability to other (non-human or imaginary) figures. Their method is likened to the GRAMPS [18] system, with its "facilities for defining articulated objects" [27]. Joint movement primitives, "bends", are called in two ways. The first is by keyboard commands, entering rotation amounts for each joint. Alternatively, an animator may choose to describe "tasks" to a task manager, which calls motor control programs, which then in turn activate local motor programs. These local motor programs, or LMPs, act upon the skeleton database by changing joint rotation values, and thus represent bend primitives. An eventual goal here, as well as at the University of Pennsylvania, is further development of systems which accept natural language scripting as input [27].

Yet another approach to human animation, under development by Calvert, Chapman, and Patla at Simon Fraser University applies both symbolic (Labanotation-based) and analog (electrogonimeter) inputs to motion control [8,9]. In this case, the graphics display can be driven by both simultaneously. The notation pattern can be built to produce an animation duplicating that obtained from the analog data. Modifications can be made after observing the actual movement to check for subtleties which may have been overlooked by the electrogoniometer's input. Three potential applications are being investigated. The system can first function as a tool assisting in dance notation and its visualization. Second are its clinical usages, primarily in studying motion abnormalities. Lastly, the macrolanguage developed for the human animations is also is used in robot manipulator control [9]. Previous difficulties in the synchronization of the two inputs [8] appear to have been resolved in later versions [9]. Its implementation as a kinematic simulation limits the types of motion that can be performed, but this is viewed as a reasonable compromise in terms of practicality.

A goal of many simulation systems is the use of natural language for specifying dynamic motion simply at the top level. Giving a command such as "reach out and grasp the handle firmly" is certainly more natural and straightforward than relating trajectory descriptions composed of joint angles, velocities and accelerations to joint torques, which eventually are used to graphically realize that trajectory. However, this kind of simplicity at the animator's level presupposes a complex "hidden" system drawing upon knowlege gained from artificial intelligence and robotics research as well as from language analysis. The manipulation of a graphic entity appearing to move autonomously and smoothly, capturing the subtleties of human gesture, remains a significant issue and challenge.

2 THE "GRAPHICAL MARIONETTE" PROJECT

In response to this challenge, we opt to circumvent the complexity of a true dynamic simulation in favor of a more manageable yet unique set of animation programming tools to animate our marionettes. This project focuses on the themes of scripting-by-enactment of the bodily motions of the marionette, designing the figures, the elaboration of facial expressions from multiple inputs, and the refinement of natural motion [4].

Capturing the subtle dynamics of human motion via scripting-by-enactment is accomplished through body tracking. The system being developed for this project (described in Section 3) will parameterize human motion in real time and make this information available to the other system components. This has been tested on a smaller scale in a software simulation of a Unimation Puma robot arm. The graphic robot can be "taught" to move by a human trainer using a Polhemus cube system in our Media Room [15]. A movement processing module accepts input data used to update current locations of the figure. The figure is represented as a data structure composed of joints and segments. Different display programs utilize the final screen coordinate output to produce each frame of animation on a Ramtek raster display. Frames recorded on a write-once videodisk for immediate playback are also controlled through the animation programs.

Future work will investigate the specification of emotion and expression for the face. Providing direct and effective interfaces for image manipulation will be a primary concern. In close-up views we wish to control eye position, head attitude, and expression change. Eye movement can be controlled in several ways; eye tracking, finger touch on a touch-sensitive display, or by joystick. Scriptor inputs from manipulation of a joystick along two axes or from other means can adjust the degree of distortion in the image. The caricature generator of Susan Brennan is a precursor to this work [6]. These are still long-range goals: at present we are considering line drawings as a suitable starting point.

3 BODY TRACKING

3.1 Background

Machine analysis of human motion has been studied in several ways. Pattern recognition of video or film is potentially the most powerful in terms of the richness of the data available and the ease in acquiring it. However, precisely because of the abundance of information and also because of limits on present technology, it is an extremely complicated problem in machine vision which is not presently suited to tracking of unconstrained body motion.

Moving Light Displays (MLDs) are a simplified representation of motion for pattern recognition and have traditionally been used for studies in motion perception [23]. The object or objects whose motion is to be studied is represented as a set of points or lights distributed on its (their) surface(s). These lights then constitute a mapping from object space to image space and can be analyzed for motion as the image varies over time. In the context of scripting-by-enactment, the human subject, or scriptor, would wear a set of continuously illuminated lights located at strategic points. This amounts to parallel acquisition of data which is presented as successive sets of dots, each set representing a "snapshot" or frame of the object in question. The problem in this case would be to assign meaning to each dot within each frame. Inotherwords, correspondance must be established between each dot of the captured image and points on the surface of the object. This is a complicated yet tractable endeavor but when two separate perspectives and triangulation for depth information are required then problems with speed, occlusion of lights and the added correspondance factor all become severe [21]. It then becomes unreasonable to track in three-dimensions gross body movements with a flexible, array-of-lights setup in real time at low cost.

Several products that illustrate other approaches are available commercially, including the Polhemus, the Coda-3, and the Selspot. The Polhemus system [5] consists of a transmitting device, a sensing device, and support software. The sensing device, attached to a cord and worn by the subject, is a plastic cube into which three mutually orthogonal coils are epoxied. The transmitter radiates a nutating dipole field by which the sensing cube is able to yeild six degrees-of-freedom positional information. Cubes like this, with their cords, are unwieldy for total body tracking, not to mention the fact that on the order of thirty transmitting/sensing sequences would have to be multiplexed.

The Coda-3 is a device, not yet available in the U.S., that provides remote and non-contacting means of monitoring the movement of up to eight so-called landmarks. It uses optical scanning to sweep three fan-shaped beams of light across the field of view and then senses the reflections from the landmarks. It is a good, if costly, solution to the problem but the limit of eight position indicators is too restrictive for body-tracking as an animation tool.

In the Selspot system [10,14], the subject wears a set of LEDs emitting in the infrared range which are pulsed in sequence. Typically, thirty LEDs may be worn. A camera which is sensitive primarily to these LEDs and synchronized to them senses their projected position by means of a two-axis lateral effect diode. Depth is determined by triangulation from the output of two cameras. This method of movement monitoring corresponds to serial acquisition of the data and also poses the most likely solution to body-tracking for driving a graphical marionette. The Selspot is, however, too restrictive in terms of the placement of the LEDs and too expensive.

The application is tracking of gross body movements for driving a body model. The Selspot approach is a good place to start but is inappropriate for several reasons. It is typically used in situations, such as determining muscle forces, where great precision is required. Lower resolution and lower accuracy are tolerable in the present context and even necessary to reduce costs and simplify processing requirements. On the other hand, greater flexibility is required in LED placement. The Selspot uses arrays of LEDs which are strapped to the body segment. The scriptor of the graphical marionette should be able to don a comfortable, ready-to-wear garment. The point is to enhance the human-machine interface, not complicate it. Many more LEDs will have to be used to compensate for the non-rigidity of the garment. This fact can be used to advantage since many LEDs in the vicinity of one body point may pulsed simultaneously, thus constituting a "virtual point" and providing a brighter light source. Unlike Selspot, domain specific knowledge, i.e. body geometry and dynamics, may also be used for error checking and interpolation.

The body-tracker should be automated, i.e. it should require no operator intervention. It should produce a set of position points (x,y,z) in real time and provide visual feedback for the scriptor. It should also be as unobtrusive as possible to the scriptor and aim for portability in the sense of being adaptable for use on a personal computer.

3.2 Procedure

The heart of the body tracking system is a two-axis lateral effect diode made by United Detector Technologies (UDT). This optical sensor detects the position of a spot of light on its surface. Two-dimensional position information is then obtained via four electrode connections at the edges of the detector. The signals from these connections are fed to an interface module, called the Op-Eye (UDT), where they are amplified and digitized to 12 bits. The output is compatible with an Apple II which then sends the data to a Perkin Elmer 3230 after a certain amount of low-level processing. The camera system uses ordinary optics and does not require precise focus since the detector senses the "centroid" of the light spot on its surface. Motion is constrained to the intersecting fields of view of two cameras. Both the tracked LEDs and the photo-detector operate in the infrared range so the system is relatively unaffected by ambient light. (See Fig. 1)

The prototype garment being developed is like a lab coat. A suit will be devised or pants will be added to enable tracking of the entire body. LEDs will be distributed across the surface of the prototype garment and fastened by piercing their leads through the fabric

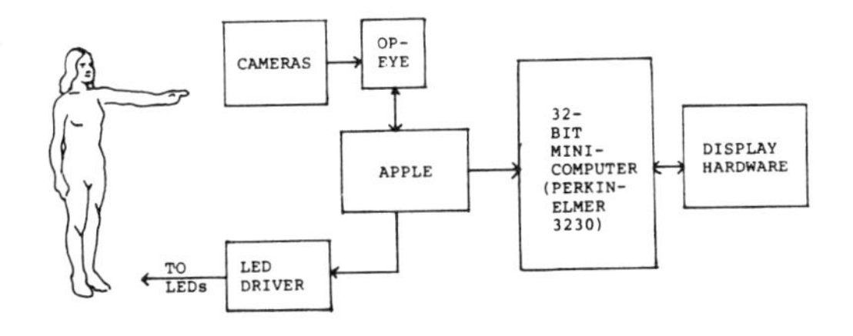

Fig. 1: Body-tracking block diagram

and wire-wrapping the leads on the inside of the garment. Wiring for the LEDs will thus be on the inside and will be protected by a lining. Placement of the LEDs will be on points corresponding to joints and other bony prominences.

Control circuitry must be as portable as possible so as not to encumber the wearer. No umbilical cord would be necessary given an initialization sequence for each "body-cycle". One is planned for now, however, so that the sequencing of the LEDs can be done in a flexible asynchronous manner, so that problems with loss of synch can be avoided, and so that a standard power supply can be used. Since each LED will be pulsed individually, the LED wiring inside the garment is forced to be in parallel. To minimize this wiring, a decoding scheme can be employed. In this case, the LEDs are divided into subgroups with the LEDs belonging to each subgroup connected in parallel. Each subgroup corresponds to a major subdivision of the human body, like a leg-hip section, for instance. The master controller is implemented with the Apple and thus is a separate unit located away from the scriptor.

The center frequency of the LEDs is such that it matches the peak response of the cameras and such that ambient light does not interfere. If necessary, optical filters may be used to eliminate incandescent sources. The diodes must operate with large included angle between half-intensity points in order for the camera to be able to detect them when they are at various orientations in space. For a radiation angle of 120 degrees, at least three LEDs must be used to represent a joint such as an elbow. More LEDs are actually needed since extra information helps offset the difficulties encountered by using a loose-fitting garment. More LEDs also yield a higher signal to noise ratio at the detector. In this context, the spatial averaging response of the detector is a feature which is exploited. Since all the LEDs associated with one joint are pulsed simultaneously, they do not, by themselves, represent increased bandwidth. The Op-Eye is capable of sampling at up to 5 KHz but the speed of the overall system is much greatly reduced by the present throughput limitations of the host (960 char/sec).

The diodes must radiate with adequate intensity to be detected by the cameras. They will be operated with a very short duty factor, enabling them to handle a large amount of instantaneous current. The Op-Eye's sensing photodiode has a minimum detectable intensity at its surface of 0.1 microwatts per square centimeter which translates to a required minimum LED power in the vicinity of 4 mW [11], considering optical parameters and scripting space dimensions. The diodes used can radiate on the order of 100 mW in derated situations.

Upon transfer to the host, the data should first be used as entries into a lookup table to correct for non-linearities due to each camera. A calibration routine using a fixed grid of light sources is required to initialize the lookup table. The raw data are then low-pass filtered. The resulting x-y coordinates from each camera and corresponding to a particular LED are to be used along with focal length information to reconstruct by triangulation the position of the LED in object coordinates. Occluded points may be inserted by interpolation or neglected when feasible. All (x,y,z) coordinates may be checked for errors on the basis of past history and anatomical constraints [12]. The final points are conditioned for the animation program and then passed to it. The goal is to process an entire body's worth of LEDs every thirtieth of a second, although the real bottleneck is likely to be the animation program.

The "scripting space" is defined by the fields of view of the cameras and will be located in the MIT Architecture Machine Group's "Media Room" (see Fig. 2). This room is about sixteen feet long, eleven feet wide and eight feet high. The far wall is a back-projection screen illuminated by a projector situated in an adjoining room. For visual feedback, the scriptor may make use of any combination of a CRT terminal, a graphics monitor, the projection screen, a "head-mounted display" [7], and a 3-D display [17]. The latter two devices were created at the Architecture Machine Group.

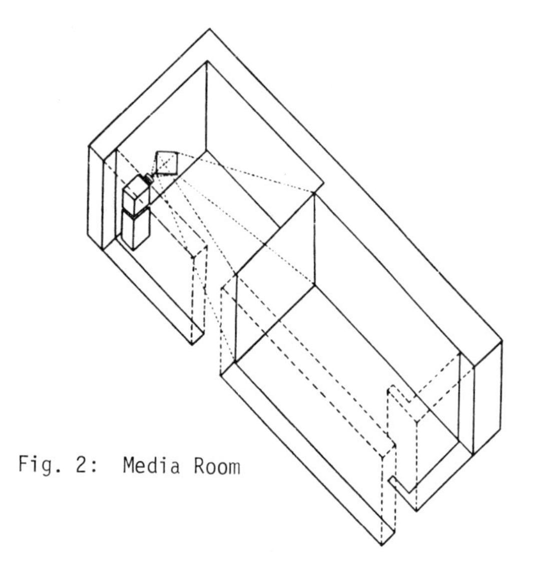

Fig. 2: Media Room

4 BODY REPRESENTATION

The human body, for modelling purposes, can be regarded as a sequence of segments connected by joints, whose movement results from rotatory motion about those joints. Albeit an oversimplification of human structure, this format provides a connected "skeleton" on which to build. This stick figure models the primary segments and joints in the body: pairs of feet, lower legs, thighs, hips, shoulders, upper arms, lower arms, and hands, and singly, the trunk, neck, and head. This level of representation is enough to approximate motion, yet not sufficent to show such subtleties as hand gesture employing the fingers. Separate "close-ups" are necessary to show such detail. However, the scope of the joints and limbs can be enlarged as the project progresses.

The initial body models are two: a stick figure, and a "cloud" figure based on ellipsoids (see Fig. 3), to be viewed in perspective. Both of these provide simple, quick visual feedback to the animator. Subsequent "fleshed-out" models can be constructed from this initial database after the animation has been scripted and previewed.

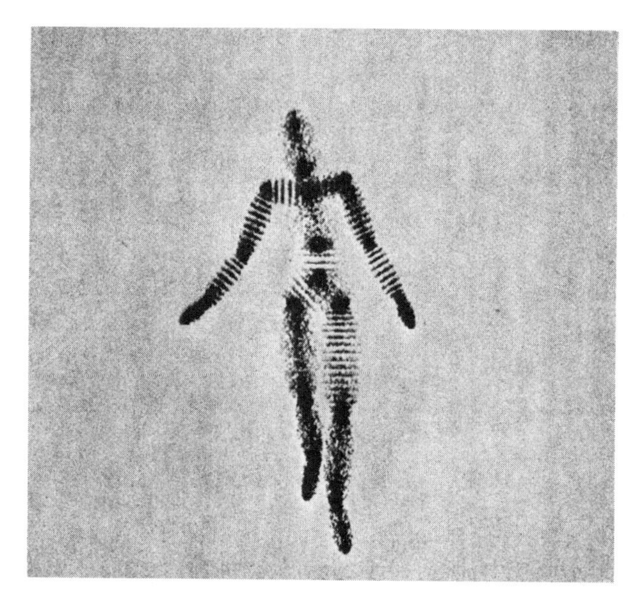

Fig. 3: "Cloud Person", based on ellipsoids

The sequence of rigid segments connected by joints can be treated as a tree structure to describe the body's connectivity. A joint is seen as a 4x4 matrix containing rotational and translational data. Positional data from the body tracker can be used to map motions in the global scripting space into the separate coordinate systems in which each segment and its corresponding joint are embedded, by traversing the tree from the center hip node to all leaf nodes of the data structure.

Limits on degrees of freedom and rotational amounts, and segment lengths are kept in this database. While tracking motion on a limited number of segments initially, there is a need to determine the position and orientation of the remaining segments. The fully implemented version of the body tracker will obviate the need to perform this extra computation. It is also at this point that segment length data can be obtained directly from the scriptor while wearing the LEDs. The body description of the marionette can be further altered to characterize different body types or to exaggerate certain features. The simplicity of this database enables the construction of other jointed figures, not necessarily human.

5 ANIMATION AND SCRIPTING IN CONTEXT

For each frame, the display processing software is passed positional data for each segment, as well as viewing data, the body database, and commands for specifying rendering style and control of peripheral devices such as the write-once videodisk. Mapping of the position and orientation data from the coordinate space of the tracking device, to the body coordinate systems, to the display device coordinates is done by the graphics routines. Appropriate rendering programs produce the desired graphic output (see Fig. 4).

During the scripting process, the stick figure will serve as feedback to the animator. Once satisfied with a motion sequence, the animator may save this script for a later playback with a marionette "fleshed-out" as a solid figure. An extension of the scripting-by-enactment technique will be its application to scenes requiring more than one marionette performing in a sequence, or perhaps within a certain virtual environment - that is, scripting in context [4].

In this scenario there could be two figures walking about and conversing in an imagined space. The scriptor performs the actions of the first marionette while wearing 3-D head-mounted optics [7] which enables him/her to view this previously created and stored "virtual space", as well as the actual environment. This motion sequence, now stored, may be played back and viewed again with the head-mounted display. The scriptor similarly takes the part of the second or nth actor, in synchrony with the prior figure in that "space". This allows the scriptor to move about within the same space as the animated figures, so that the scripting occurs in both these 3-space contexts.

We envision the graphical versions of the scriptor to be capable of dancing, sauntering, falling, and a wide variety of movements according to the disposition of its' human counterpart.

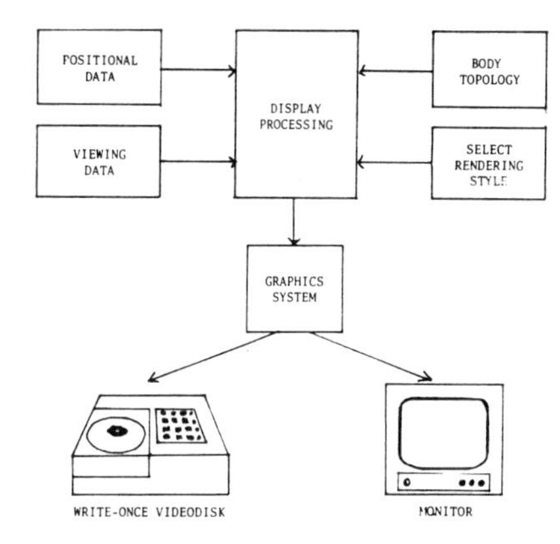

Fig. 4: Software block diagram

SUMMARY

In attempting to accurately simulate human movement in relation to our physical environment, complex modelling systems have been devised. For the purpose of animation, such completeness is perhaps less important than the model's characterization and appearance, and ease in animation. Our focus here is on those concerns. Since the marionette figure uses a generalized jointed figure database, its application is not restricted to human animation. One could conceivably envision a scenario with dragons or robots or insects, employing similar type of database for their definition. The technique of scripting-by-enactment is being pursued as a more natural method for animation scripting. It will also aid in constructing parallel motions, one of the unwieldy aspects of computer animation. Actions in a virtual 3-D world can be the consequences of motion, speech, or touch performed in the user's 3-D world. Having the graphic output able to "do as I do, and not as I type", will circumvent some of the awkwardness at the top level of traditional animation systems.

REFERENCES

1. Badler, N., "Design of a Human Representation Incorporating Dynamics and Task Simulation", Siggraph '82 Tutorial, July 1982.

2. Badler, N., O'Rourke, J., and Kaufman, B., "Special Problems in Human Movement Simulation", Computer Graphics, Vol. 14, No. 3, 1980, pp. 189-197.

3. Badler, N., and Smoliar, S., "Digital Representations of Human Movement", Computing Surveys, Vol. 11, No. 1, March 1979, pp. 19-38.

4. Bolt, R. A., Funding proposal for the development of a Graphical Marionette, Architecture Machine Group, MIT, 1981.

5. Bolt, R. A., "'Put-That-There': voice and gesture at the graphics interface", Proc. Siggraph, Vol. 14, pp. 262-270.

6. Brennan, S., Caricature Generator, Master's Thesis, Architecture Machine Group, MIT, 1982.

7. Callahan, Mark, Three Dimensional Display Headset for Personalized Computing, unpublished Master's Thesis, Architecture Machine Group, MIT, 1983.

8. Calvert, T.W., Chapman, J., and Patla, A., "The Integration of Subjective and Objective Data in the Animation of Human Movement", Computer Graphics, Vol. 14, No. 3, 1980, pp. 198-203.

9. Calvert, T.W., and Chapman, J., "Aspects of the Kinematic Simulation of Human Movement", Computer Graphics and Applications, Vol. 2, No. 9, November 1982.

10. Conati, F.C., Real-Time Measurement of Three-Dimensional Multiple Rigid Body Motion, Master's Thesis, MIT, 1977.

11. Gage, S., ed. Optoelectronics Applications Manual, Hewlett-Packard Company, McGraw-Hill, N.Y., 1981.

12. Hoffman, D.D., and Flinchbaugh, B.E., "The Interpretation of Biological Motion", A.I. Memo No. 608, Artificial Intelligence Laboratory, MIT, December, 1980.

13. Hutchinson, A., Labanotation, Theatre Arts Books, N.Y., 1970.

14. Mann, R.W., Rowell, D., Dalrymple, F.C., Conati, F.C., Tetewsky, A., Ottenheimer, D., and Antonsson, E., "Precise, rapid, automatic 3-D position and orientation tracking of multiple moving bodies", Memo of the Dept. of Mechanical Engineering, MIT, 1981.

15. Miller, Lynn, unpublished Bachelor's Thesis, Dept. of Computer Science and Electrical Engineering, MIT, 1982.

16. Negroponte, Nicholas, "Media Room", Proceedings of the SID, Vol. 22/2, 1981.

17. Nisselson, Jane, A Model Kit: a system for constructing three-dimensional interactive graphic models, Master's Thesis, Architecture Machine Group, MIT, 1983.

18. O'Donnell, T.J., and Olsen, A.J., "GRAMPS - A Graphics Language Interpreter for Real-Time, Interactive, Three-Dimensional Picture Editing and Animation", Computer Graphics, Vol. 15, No. 3, August 1981, pp. 133-142.

19. Parke, Frederic I., "Parameterized Models for Facial Animation", Computer Graphics and Applications, Vol. 2, No. 9, November 1982.

20. Reichardt, Jasia; Robots: Fact, Fiction, and Prediction; Penguin, 1978.

21. Rashid, R.F., Lights: A System for the Interpretation of Moving Light Displays, Ph.D. Thesis, University of Rochester, 1980.

22. Richter, Jean Paul, The Literary Works of Leonardo da Vinci, Vol. 1, University of California Press, 1977.

23. Ullman, S., "The Interpretation of Structure from Motion", A.I. Memo 476, Artificial Intelligence Laboratory, MIT, October, 1976.

24. Zeltzer, D., "3D Movement Simulation", Siggraph '82 Tutorial, July 1982.

25. Zeltzer, D., "Motor Control Techniques for Figure Animation", Sigggraph '82 Tutorial, July 1982.

26. Zeltzer, D., "Motion Planning Task Manager for a Skeleton Animation System", Siggraph '82 Tutorial, July 1982.

27. Zeltzer, D., "Representation of Complex Animated Figures", Proceedings, Graphics Interface '82, May 1982.

A MULTIPLE TRACK ANIMATOR SYSTEM FOR MOTION SYNCHRONIZATION

D. Fortin, J.F. Lamy and D. Thalmann

Département d'informatique et
de recherche opérationnelle
Université de Montréal
Montréal, Canada

Abstract

MUTAN (MUltiple Track ANimator) is an interactive system for independently animating three-dimensional graphical objects. MUTAN can synchronize different motions; it is also a good tool for synchronizing motion with sound, music, light or smell. To indicate moments in time, marks are associated with appropriate frame numbers. MUTAN enables the marks to be manipulated. An animator can also adjust one motion without modifying the others. To make this possible, MUTAN handles several tracks at a time (as in sound reproduction). All animation constraints for a graphical object are recorded on each track. Some simple but powerful commands allow the animator to manipulate marks, tracks and frames. MUTAN is part of a complete 3D shaded animation system including the CINEMIRA computer animation language based on actor and camera data types, the 3D HORIZON graphics editor and a 3D digitizing program.

1. INTRODUCTION

The production of three-dimensional computer animated films using a graphical programming language is time-consuming. For example, it took 14 months to produce the 13 minute film Dream Flight [1]. This was the prize-winning film at the Computer Graphics '82 Animation film festival in London. Like other graphical languages (e.g. ASAS [2,3]) the MIRA-3D [4] language used to produce Dream Flight offers more possibilities than are currently available with interactive systems. For example, the synchronization of movements of different objects is generally difficult to handle interactively, although there exist several excellent computer animation systems [5,6]. The system presented in this paper provides the animator with a good way of synchronizing the movements of different graphical objects in a scene.

Generally, in 3D animation systems, parallel commands allow the user to animate multiple objects. Instructions are individually scheduled to be active over a range of time during the animation sequence. This approach does not allow the animator to easily handle the synchronization of movements.

The design of our system, called MUTAN, was influenced by the conventional animation process and by research in the use of the computer in cartoon animation [7,8,9]. Although MUTAN is a 3D animation system, it is mainly a key frame animation system and certain features resemble the "exposure sheet" system used for animation in the cartoon industry. An exposure sheet is basically a visual synopsis that includes frame-by-frame information pertinent to the scene. It is the basis of the final production step in a conventional animation studio where all the pieces of artwork, background and cels are placed and photographed. There are several columns in the exposure sheet: the background column, the cel columns and the columns for technical specifications such as cross-dissolves, fades and zooms. The cels columns indicate the cels that the animation cameraman must photograph in the proper sequence. The cel columns can be considered as tracks and this is this concept that we have introduced into MUTAN. In summary, MUTAN has been designed as a tool:

- for synchronizing the motion of multiple objects
- for controlling the flow of drawings
- for producing inbetweens

2. THE MUTAN SYSTEM

MUTAN (MUltiple Track ANimator) is an interactive system for independently animating three-dimensional graphical objects

Copyright 1986 by ACM
Motion: Representation and Perception
N.I. Badler and J.K. Tsotsos, Editors
Published by Elsevier Science Publishing Co., Inc.

(or parts of graphical objects), as in a scene. The main purpose of the system is to define by key positions the best motions for a graphical object. For example, MUTAN can synchronize motion with sound, light or smell. Suppose, for example, that an animator would like to produce a film sequence in which two little girls are skating with music. There are two problems: the synchronization of the motion with the music and the synchronization of the motion of the two girls who must dance at the same time.

To obtain synchronization of the motion with the music, marks are associated with appropriate frame numbers. These marks indicate that at this moment in time, the person is performing exactly a specific motion. MUTAN is very useful because it enables the marks to be manipulated. As the intermediate frames are produced by MUTAN, it is not necessary to start again if the synchronization is found inadequate. The marks merely have to be moved.

An animator might also wish to adjust the motion of one individual without modifying the motion of the others. To solve this problem, we introduced the concept of a TRACK (as in sound reproduction). All animation constraints of a graphical object are recorded on each track.

The name MUTAN stands for "MUltiple Track ANimator" which means that the system can handle several tracks. It also provides a visual image that allows the animator to see exactly what is being done.

3. THE SCREEN

During an interactive session with MUTAN, the screen is divided into three areas as shown in Fig. 1.

1. The CHRONOGRAM/VISUAL area

   This area can display either a schema of the tracks with the marks (chronogram mode) or a specific frame (visual mode) depending on the command.

2. The animator-computer dialog area

   This area allows the user to enter commands and receive error messages.

3. The information area

   This area may contain information about the tracks or about the current frame when it is displayed.

4. THE MUTAN COMMANDS

There are presently 19 commands in the MUTAN system as shown in Fig. 2. Almost all commands have similar parameters:

- a track or a set of tracks
- a mark or a set of marks
- a frame number

CREATE <track> allows the animator to create a new track. This track of course has no marks. The tracks can be deleted by using the command DELETE <track>.

If a film consists of a little girl MELANIE playing with a dog, the animator will use the commands CREATE MELANIE and CREATE DOG. These commands will initialize the tracks. To display them, the command TRACKS MELANIE DOG must be entered. This command has the following effect: two red lines are displayed in the chronogram area. Each line identifies a track.

Now the marks must be placed on the tracks to define animation constraints. The command MARK <frame number> <track> allows the animator to create or modify a mark on a track and to define the frame number where the mark is placed. For example, suppose the dog hears a whistle, stops running and then starts again. Two marks would be used.

```
MARK 15  DOG
MARK 18  DOG
```

For each MARK command, the animator receives the following message in the dialog area: "ENTER THE NAME OF THE FRAME"; he/she then types respectively "DOG STOPS", "DOG STARTS".

The MARK command can also be used without parameters. In this case the animator enters the mark by a graphical input on the red line representing the track. The principle is valid for other commands.

When the MARK command is used, the animator has to enter in the dialog area the name of a key-position which identifies an image. The key positions can be created by a graphics editor, by using the graphical programming language MIRA-3D or by digitizing. The different key positions can be grouped into indices. The command MENU <index> allows the animator to see which key-positions are available on the file indicated by the index.

A mark can be erased by using the command ERASE, while the command GET memorises a mark before erasing it and the command PUT replaces it on the track.

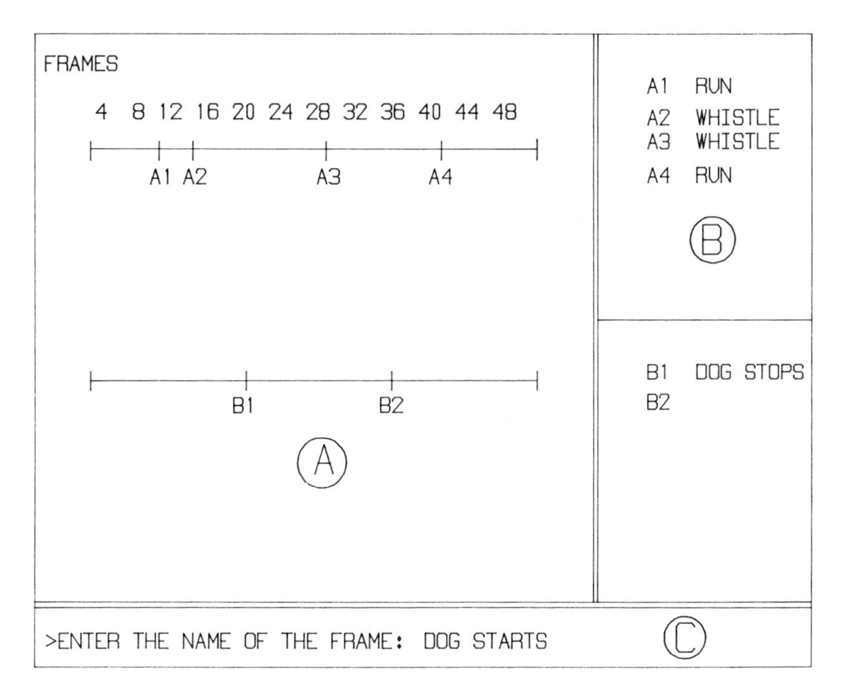

Fig. 1 Screen area

A: chronogram area
B: information area
C: dialog area

ADD	-	adds a certain number of frames
BACKWARD	-	moves backward along the film
CREATE	-	creates a track
CHRONOGRAM	-	displays a part of film in the chronogram area
CUT	-	cuts frames on one or several tracks
DELETE	-	deletes a track
END	-	ends the session
ERASE	-	erases a mark
FILM	-	produces the frames on a file
FORWARD	-	moves forward along the film
GET	-	memorises a mark
IDENTIFY	-	selects invisible marks
LOOK	-	looks at any frame in the visual area
MARK	-	creates or modifies a mark on a track
MENU	-	displays a file indicated by an index
PUT	-	replaces a mark on a track
READ	-	reads mark information
TRACKS	-	displays tracks in the chronogram area
WRITE	-	writes mark information

Fig. 2 The MUTAN set of commands

The animator can move a window along the film by using the command CHRONOGRAM. This command has different forms. For example, CHRONOGRAM 50 75 MELANIE DOG, displays the frames 50 to 75 for the tracks MELANIE and DOG. At the same time, information about the different marks on this part of the film are displayed in the information area. If there are too many marks to display, only some of them are visible and others can be seen by using the command IDENTIFY.

As the work of an animator is essentially sequential, we have added two commands to move forwards and backwards along the film. These commands FORWARD or BACKWARD perform a shift to the right or the left by the same number of frames as that presently displayed in the chronogram area.

The animator can cut one or several frames on one or several tracks by using the command CUT <set of frames> <set of tracks>. All marks will be shifted to the left.

Conversely, the command ADD <frame number> <set of tracks> adds a certain number of new frames to the right of the frame specified and on the chosen tracks. The number of frames added has to be entered by the animator in the dialog area.

The LOOK command allows the animator to look at any frame of the film in the visual area. For the frames where no marks exist, an interpolation is performed. It is also possible to select only one or several tracks. In this case, only a part of the frame is displayed in the visual area.

The command FILM produces the frames on a file to be recorded on a videorecorder or to be displayed for a camera.

There are also commands to read and write MARK information on a file.

5. IMPLEMENTATION

MUTAN has been implemented on the CDC Cyber of the University of Montreal. It consists of about 4200 lines of MIRA source program. It can be used for two-dimensional animation and three-dimensional animation with and without hidden lines.

The tracks are implemented as double-link lists of marks. This data structure has been chosen because the work of an animator is generally sequential. Commands correspond to insertion, deletion and movements forwards and backwards.

Fig. 3 shows the definitions of the types used for defining the data structures representing the tracks. Each mark in a track includes pointers for retrieving the definition of a key position. As the user specifies a key position by an index and a name, two pointers are stored, one to the index and the other to the associated key position. Acceleration when leaving position and deceleration when arriving at the position also have to be given. An acceleration or a deceleration can be linear, cosinusoidal, arithmetic or geometric.

MUTAN commands can manipulate one track, all tracks or a subset of tracks. This is done using the PASCAL declaration "SET OF INTEGER". This means that a set of tracks is represented by a set of of track numbers. To retrieve track number N, two arrays called TRACKNAME and TRACKPTR are used. TRACKNAMES[N] contains the name of the track number N and TRACKPTR[N] a pointer to the track descriptor.

When the commands LOOK and FILM are used, inbetweens have to be computed except when the required frames correspond exactly to marks or key positions.

6. FUTURE DEVELOPMENTS

MUTAN is part of a general computer animation system as shown in Fig. 4. Basic figures can be created by using the HORIZON [10] graphics editor, by using a 3D digitizing program that we have developed or by using the MIRA-3D programming language. MUTAN can then produce complete film sequences without programming. This approach is quite satisfactory, but may limit animation creativity. In particular, synthetic camera movements are not supported and real complex physical movements can only be simulated by inbetween computation. For this reason, we are now implementing the CINEMIRA language [11,12]. This language is based on animated basic types, script subprograms, actor types and camera types. It also includes all graphical facilities defined in MIRA-3D. A program in CINEMIRA will be able to read mark files produced by MUTAN and also write such files. This means that it will be possible to produce films completely without programming, using MUTAN to synchronize motions. This is difficult to do with programming.

ACKNOWLEDGEMENT

The authors are grateful to the referees for their helpful comments. They also would like to express their gratitude to Ann Laporte who revised the English text. The research is supported in part by the Natural Sciences and Engineering Council of Canada and the Department of Education in Quebec.

```
TUserName = packed array [1..MaxLgUserName] of char;
TFileName = packed array [1..MaxLgFileName] of char;
BigAlfa = packed array [1..MaxLgBigAlfa] of char;
TLabel = packed array [1..MaxLabelChars] of char;
TVariation = (Linear, Cosinusoidal, Arithmetic, Geometric);

(* Frame description associated to a mark *)
TPtInstance = ^ TInstance;
TInstance =
  record
    Username: TUsername;
    Filename: TFileName;
    SegmentNo: integer;
    NextInstance: TPtInstance;
  end;

(* Index description *)
TPtIndex = ^TIndex;
TIndex =
  record
    Name: TFileName; (* Index name *)
    FirstInstance: TPtInstance;
    NextIndex: TPtIndex;
  end;

(* Mark description *)
TPtMark = ^ TMark;
TMark =
  record
    FrameNo: integer;
    Index: TPtIndex; (* to retrieve the index name *)
    Instance: TPtInstance; (* associated key position *)
    Amplitude: real; (*size factor *)
    TypeAcc,TypeDec: TVariation; (* type of accel./decel. *)
    Acceleration, (* when leaving position *)
    Deceleration: (* when arriving to position *)
      real;
    Identified: boolean; (* true if the mark will be described *)
    NextMark, PrevMark: TPtMark; (* Left and right links *)
  end;

TViewport =
  record
    LowerLeft, UpperRight: vector; (* corners *)
  end;

(* track definition *)
TPtTrack = ^ TTrack;
TTrack =
  record
    Name: Bigalfa; (*associated name of the track *)
    FirstMark: TPtMark;
    CurrentMark: TPtMark;
    Empty: boolean; (* true if the track is empty *)
    Shown: boolean; (* true if the track must be displayed *)
    Explained: boolean; (* true if the marks have been identified*)
    TrViewport, (* where to display the track *)
    LabelViewport, (* part of track for labels *)
    DescViewport (* where to display associated informations*)
          : TViewport;
    NbId: integer; (* number of identified marks *)
    FirstLabel: TLabel;
    NextTrack: TPtTrack;
  end;
```

Fig. 3 Data structure

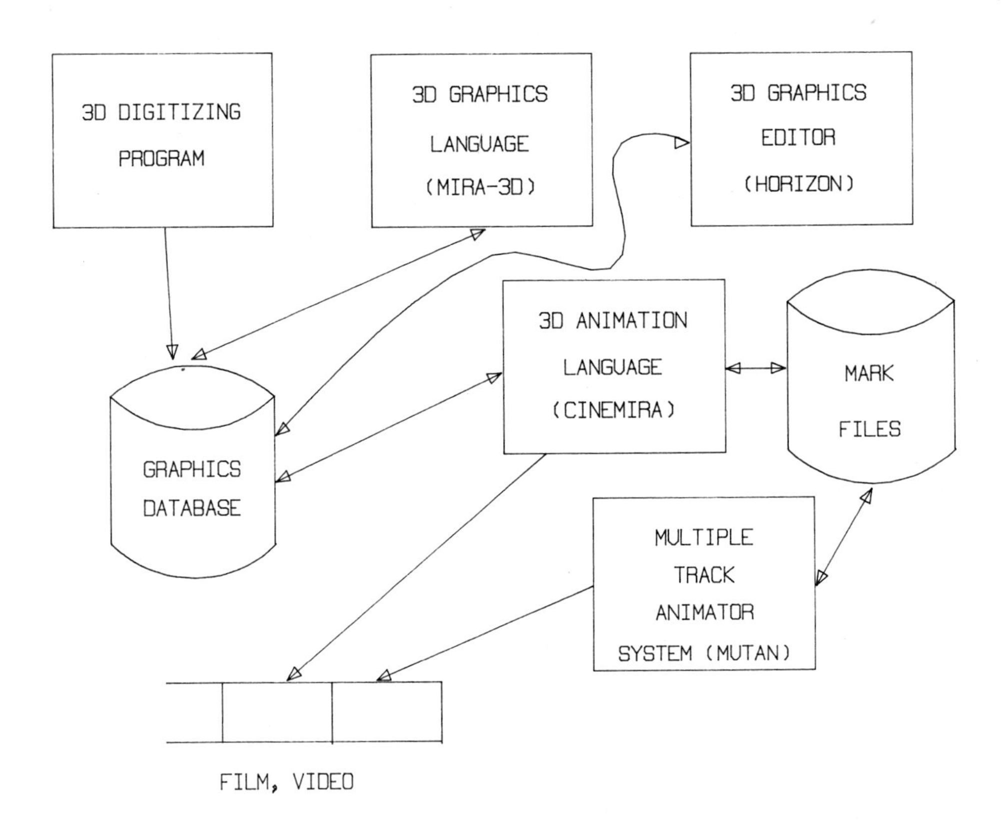

Fig. 4 A complete 3D animation system

REFERENCES

[1] Thalmann, D.; Magnenat-Thalmann, N. and Bergeron, P. "Dream Flight: A Fictional Film Produced by 3D Computer Animation", Proc Computer Graphics '82, Online Conf. Ltd. 1982, pp. 353-368.

[2] Reynolds, C.W. "Computer Animation with Scripts and Actors", Proc. SIGGRAPH '82, pp. 289-296.

[3] Reynolds, C.W. "Computer Animation in the World of Actors and Scripts", SM Thesis, MIT, 1978.

[4] Magnenat-Thalmann, N. and Thalmann, D. "MIRA-3D: A Three-Dimensional Graphical Extension of PASCAL", Software-Practice and Experience, John Wiley, Vol. 13, 1983.

[5] Csuri, C. et al. "Towards an Interactive High Visual Complexity Animation System", Proc. SIGGRAPH '79, Computer Graphics, 1979, Vol. 13, No. 3, pp. 289-299.

[6] Hackatorn, R.J. "ANIMA II: a 3D Color Animation System", Proc. SIGGRAPH '77, Computer Graphics, 1977, Vol. 11, No. 2, pp. 54-64.

[7] Catmull, E. "The Problems of Computer-assisted Animation", Proc. SIGGRAPH '78, pp. 348-353.

[8] Levoy, M. "Computer-assisted Cartoon Animation", M.S. Thesis, Cornell University, 1978.

[9] Wallace, B.A. "Merging and Transformation of Raster Images for Cartoon Animation", Proc. SIGGRAPH '81, Computer Graphics, Vol. 15, No. 3, 1981, pp. 253-262.

[10] Magnenat-Thalmann, N.; Larouche, A. and Thalmann, D. "An Interactive and User-oriented Three-dimensional Graphics Editor", Proc. Graphics Interface '83, Edmonton.

[11] Magnenat-Thalmann, N. and Thalmann, D. "CINEMIRA: A 3D Computer Animation Language Based on Actor and Camera Data Types", Publ. Dép. Inform. et Rech. Opér., Université de Montréal.

[12] Thalmann, D. "Actor and Camera Data Types in Computer Animation", Proc. Graphics Interface '83, Edmonton (invited paper).

Knowledge-based Animation

David Zeltzer
Computer Graphics Research Group
The Ohio State University

ABSTRACT

In constructing a goal-directed system for automatic motion synthesis for computer animation, the essential problem is to account for the extraordinary flexibility and adaptability exhibited by moving creatures. The selective potentiation and depotentiation of elements of a hierarchy of motor control programs is a key to the generation of adaptive motor control. The constraints on motion sequences are analyzed, and mechanisms for achieving continuity of movements are discussed. The organization of two data bases containing knowledge about the simulated environment and about available movements is described. An example showing the interaction between the motion control system and the data bases is presented.

1. Introduction

While the output of computer graphics systems in general has become increasingly sophisticated and realistic, the objects which appear on computer displays are for the most part monolithic entities or objects with only a few moving parts. The reason is that articulated motion is difficult to represent and control. The human skeleton, with over 200 degrees of freedom, is capable of motion so complex that we are still learning how to measure and define it [11,12,15].

The Skeleton Animation system (SA) is a prototype system for controlling the movements of complex figures. It is capable of generating animated sequences showing three-dimensional, perspective views of articulated objects. An objective of this research is to construct a goal-directed system for automatic motion synthesis in which the burden of specifying the details of figure motion is left to the motion-control software rather than the animator.

Biological movement systems are inherently goal-directed, an observation made as early as the time of Leonardo da Vinci [22]. Such systems seem to be based on distributed control of a hierarchy of motor programs. The key notion is that systems with potentially many degrees of freedom can be controlled by a set of low level programs, each invoked with only a few degrees of freedom. The basis for distributed, hierarchical motor control has been established by a number of investigators [1,5,6,10,23].

In section 2.0 I discuss the kinds of problem-solving employed by motor control systems, and the principle of selective potentiation and depotentiation of functional units. Section 3.0 analyzes the constraints on motion sequences and mechanisms for achieving continuity of movements. In sections 4.0 and 5.0 I discuss the organization and structure of the scene and movement data bases. Section 6.0 gives an example showing the interaction between the task manager and the data bases. For a review of recent efforts involving the representation and control of the human figure for computer animation, the reader is referred to the November, 1982 special issue of IEEE Computer Graphics and Applications.

2. The Organism as Expert

All living vertebrates continually solve problems involving the coordination and control of complex skeletal motion. Naturally most creatures are unable to tell us how they accomplish this, and indeed, humans have difficulty describing their physical coordination strategies ("How did I do that?"). Nevertheless, enough has been learned from neurophysiology and psychology to provide a basis for constructing a first approximation model of such motor control systems.

The essential problem is to account for the extraordinary flexibility and adaptability exhibited by moving creatures. Specifically, a given movement may be executed in many different ways depending on the circumstances at hand. For example, if I stand up, the precise motion sequence will depend on whether I am sitting on a hard desk chair, an overstuffed easy chair, or lying on the lawn. My motor control system is capable of problem-solving to the extent that a given goal must be achieved under a variety of unforeseen circumstances, including physical damage to the organism. There are many conceivable sequences of

Motion: Representation and Perception
N.I. Badler and J.K. Tsotsos, Editors
Published by Elsevier Science Publishing Co., Inc.

movements available at a given time, but only a few of them can be assembled to achieve the target motion or posture. How is the correct sequence of movements constructed -- rapidly, and without any apparent conscious intervention?

What has emerged from the study of natural and artificial motion control systems is that the notion of expectation is fundamental, echoing an hypothesis advanced by Simon, suggesting that complex behavior is a reflection of a complex environment [20]. Units of animal behavior evolve as solutions specific to the problems confronted in a given ecological niche. Evolution or learning can structure these behavioral units into sets of potential responses to anticipated conditions. In a process called selective potentiation and depotentiation, the activation of one behavioral unit may enable -- but not activate -- other units [5,21]. Depending on the circumstances, these potentiated units may or may not become active. If a unit is activated, it may in turn potentiate a further set of behavioral units. At the same time that some units are enabled, others may be disabled, i.e., prevented from becoming active. The pattern of potentiation and depotentiation of an organism's available motor and behavioral control programs determines its repertoire of behavior, and its flexibility in adapting to changing circumstances.

The use of rules to select from a set of expected alternatives and possibilities is a basic mechanism for controlling cognitive activity in a number of expert systems (see, e.g., [8,9]). Several important theories from artificial intelligence, including Minsky's theory of frames [14], and Schank's scripts [19], embody the notion of selective potentiation and depotentiation. A frame, for example, may contain pointers to procedures which can be invoked whenever certain conditions arise. That is, these attached procedures are potentiated by the given frame, to be activated if and when appropriate. Results from physiological investigations suggest that this same mechanism is central to the control of movement and behavior.

Gallistel [5] and Tinbergen [21] both give numerous examples, ranging from the most complex behaviors to simple components of locomotor programs, in which selective potentiation plays a key role. For example, consider the reflex mechanisms that prevent a cat from tripping should its foot encounter an obstacle as it walks. Tapping the paw of a cat while the leg is in swing phase will elicit a flexion reflex as the cat tries to lift its foot over the obstacle. Tapping the paw in the same way while the leg is in support phase will elicit a different reflex, extension, as the cat tries to quickly complete support phase so that it can lift its leg over the obstacle. What has happened is that the flexion reflex is potentiated during swing phase, while the extension reflex is depotentiated. During support phase, the opposite relationship holds. Note that neither reflex may be triggered at all during gait -- they are simply in a state of readiness for activation [5].

In the next sections, I will show how a hierarchy of movement frames and attached rules can establish patterns of potentiation in an adaptable motor control system.

3. Movement Continuity

The task manager is the top node of the hierarchical control structure of the Skeleton Animation system. Input to the task manager is in the form of a description of a specific movement sequence called a task description, e.g., "Go to the door and open it", "Run over to the soccer ball and kick it", "Sit down". The task manager must extract from each task description a set of movement functions, called skills. Each skill represents some class of motions the figure can perform, e.g., walking, running, grasping, and so on. Skills are implemented by procedures called motor programs. Finally, the motor programs themselves each invoke a set of primitive procedures called local motor programs (LMPs). Thus the control structure of SA is a hierarchy in which the task manager is at the top, the branch nodes represent skills (and associated motor programs), and the leaf nodes are the local motor programs. A more detailed description of motor programs, LMPs, and the organization of SA can be found in [25].

Given a list of skills extracted from a task description, we wish to generate a continuous animation sequence which satisfies the following criteria: Every skill has certain preconditions which must be satisfied before it can be executed, e.g., both feet must be on the ground before the figure can walk. Preconditions might require that a certain skill be ongoing with or executed immediately before a second skill; or preconditions may specify a given body posture and/or various relations between the figure and the environment, e.g., the figure must be facing the direction in which it is to start walking. If all of the preconditions for a given skill are satisfied, we say that skill is feasible. Transitions between skills must be smooth, i.e., transitions must conform to kinematic data, if it exists for a given movement, or conform to everyday observations of moving figures. Smoothness clearly requires that a movement be executed in some sense correctly. For example, a smooth transition between running and kicking a soccer ball requires, among other things, that the figure initiate the kick at the proper distance from the ball, and this requires that the stridelength be adjusted as the figure approaches the target. A planning mechanism is needed to coordinate the activities of such motor programs in order to ensure smoothness.

3.1. Feasibility

The movements of figures in the real world are constrained by the dynamics and kinematics of motion. Animals do not move from one arbitrary posture to another. On the contrary, biological motor systems behave as if motion were being optimized with respect to a variety of criteria, e.g., energy expenditure [6]. That is, a figure arrives at some posture or configuration through an orderly sequence of motions. The converse is also true: starting from a given posture, there are some movements a figure can execute, and some movements that the figure cannot. As we have seen, these constraints (and others) are expressed by patterns of potentiation and depotentiation.

How can we characterize movements in a general way? McGhee [13], Muybridge [16], and others have characterized gait by describing support patterns which list, for a given time interval, which limbs are supporting a figure and which are free-swinging. Such a description is called a support list. The support list can be generalized to show other kinds of relationships, e.g., one limb supporting another, as when a figure is sitting in a chair with legs crossed.

For movement in general, a little more information than just a support list is required. Each figure has a set of end-effectors, e.g., hands and feet for a human, which are used to manipulate objects. An end-effector list maintains the status of each end-effector, e.g., free, supporting, in contact with some object, or blocked. In addition, the balance of the figure can be characterized by calculating the projection of the center of gravity with respect to the support polygon, i.e, the convex hull of the support points.

Necessary preconditions for movement can be stated in terms of information described by the balance, support list, and end-effector list for a figure at a given time. (It may also be necessary to include information on the current dynamics of the figure, as well as the kinematic configuration. This is an open question, and will not be discussed further here). Preconditions for human walking might specify that the support list contain the left and right feet, that the hands be free (or at least not grasping a fixed object), and that the figure be balanced. Preconditions may also include statements about the immediate environment, e.g., there must be no obstacles within a given distance in front of a figure in order for walking to begin. (I will discuss interaction with objects in the environment in section 5.0). Furthermore, for each skill there may be a set of postconditions that can also be stated in these terms.

Feasibility criteria need be checked once only at the onset of movement. A table of facts about the figure can be updated as movement progresses, with facts being added and deleted as necessary. Motor programs can communicate with each other, and preconditions can be checked, by consulting this table, which we call the blackboard, after Erman and Lesser [4].

The purpose of the blackboard is to provide a communication channel to coordinate the execution of asynchronous motor programs and LMPs. The blackboard is divided into several areas, corresponding to the support list, the end-effector list, balance, the names of the pending motor programs and the names of the motor programs active in the last animation frame. Preconditions are checked by examining the appropriate area of the blackboard. Motor programs and LMPs access the blackboard as they execute. The LMPs for walking, for example, can check for heel-strike and toeoff by noting when, say, the left foot enters the support list, and when it is removed from the support list. Thus programs can signal events by adding and deleting data from various areas of the blackboard.

3.2. Smoothness

While feasibility criteria address the problem of selecting the correct skills to invoke, smoothness criteria specify that movements should appear to blend in a natural way. Consider the problem of directing the figure to "walk to the chair and sit down". Assume no other objects or obstacles are in the environment. First, we need to compute the direction to walk. Now, what does "to the chair" mean? Assuming that the object has some specified center, we can calculate direction and distance from the center of the figure to the center of the object. However, if the figure is to sit on the chair, the figure must somehow arrive at a position in front of the chair facing away from the chair and within the correct distance. The task manager must have available the following information: The orientation of the chair Some reference point on the chair from which to compute distance, i.e., the figure needs to arrive at a certain distance from the front edge of the chair How close to the chair the figure must be in order to sit Speed and stride-length such that the start-walk, walk, and stop-walk motor programs position the figure at the desired distance from the chair

These requirements concern knowledge about the environment rather than the figure itself. Thus the task manager must have access to a store of facts about the environment in which the figure has been placed -- the scene data base. The design of algorithms that compute paths and velocities, as in the soccer ball example, is a separate issue which I will not deal with here. In the context of generating smooth sequences, I will confine my discussion to the representation of environmental knowledge.

4. Skills and Frames: The Movement Data Base

Whenever a skill is selected for execution, we need to ascertain whether all its associated preconditions are satisfied. We can think of each skill as an expert at its particular motion that "knows" why it was invoked and what preconditions, if any, prevent its execution. Thus it is appropriate to attach knowledge about precondition satisfaction to the skills.

Each precondition has attached triggers that invoke the appropriate skills if the precondition is not true. In effect, each skill selectively potentiates a set of other skills that may be useful in satisfying its preconditions. (In a multi-processor environment, where many motor programs may be active, e.g., humans and other animals, the depotentiation of inappropriate skills is also necessary). In living systems these patterns of potentiation are subject to change through evolution, and in more complex organisms, learning. For the Skeleton Animation system, however, the set of potentiated skills is fixed when a particular skill is coded.

In general, for each skill we need to know the following: What are the movement preconditions? If a precondition is not satisfied, what skills should be invoked? If the movement is feasible, what motor programs should be invoked? What facts about the figure will no longer be true after execution begins? What new facts about the figure will become true after execution begins? What are the nominal or default parameters for this skill?

This collection of information about a stereotypical entity can be represented by a frame data-structure called a movement frame. The attached preconditions (rules) determine whether a skill, once potentiated, can be invoked. Once invoked, the attached rules are tried in order. Each rule that is triggered invokes one or more skills that must complete execution before the current skill can execute.

Once all preconditions are satisfied, the pending movement can begin. However, the actual form of a movement may depend on the current configuration of the figure, or aspects of the current scene. A set of rules, called execution rules, may be attached to a movement frame whenever the execution of a motion depends on the information contained in the blackboard or the scene data base. The attached execution rules define further patterns of potentiation in the same way as the preconditions. They select other skills, or define parameters for the current skill, appropriate to a given situation.

5. The Scene Data Base

In constructing a scene the user must have tools for defining objects in terms of their three-dimensional coordinates or other data used by the various display algorithms. In addition, software must be available which allows the user to describe other attributes of these same objects. The problem of defining and displaying objects is fairly well-understood [17]. The creation, manipulation, and display of higher-order aggregates has hardly been attempted. That is, we would like to be able to talk about a room full of objects, each of which possesses attributes in addition to the usual "graphical" properties which include information on how to color the object, and how it is to be scaled, translated, and rotated in 3-space. Such a scene might consist of a room with doors, windows, and furniture; in which some objects have impenetrable surfaces while others have flexible components; in which some objects are fixed, and some may be moved; and so on.

In most current animation systems, access to object data is restricted to the display routines (see, for example, [2,3]). If an articulated figure is to move through such an environment, then at least the locations of all the objects must be accessible in a controlled way to other modules of the animation system. More than just the position of the center of the object is needed; the coordinates of the surface of the object must be accessible as well. Environmental data must be organized carefully, however, to avoid triggering exhaustive searches through the data base in every frame in which a figure moves. One technique for accomplishing this using the idea of a hierarchically organized "environmental working set" is discussed in [24].

In order for a figure to interact with its environment, rather than simply navigate through it, the figure controller must have access to other kinds of information about the environment and the objects in it. Consider the command "Go to the door and open it". The task manager would be unable to construct a continuous animation sequence unless it knew something about doors and how they usually work. One could conceive of a skill for opening doors, with appropriate motor programs for grasping and turning door knobs, and indeed, such procedures are certainly necessary. The problem is that not all doors have door knobs. The skill for opening doors must be generic and adaptable to a variety of circumstances, but knowledge about particular environments must be separate from the movement data base. This way the figure can interact with the environment at the appropriate level of abstraction. Otherwise the figure controller would become rigid and restricted to particular environments, and the movement data base would become cluttered with unnecessary detail.

The scene data base is a long-term structure, built up over time. Here the frame is the appropriate data-structure since we are concerned with stereotypical information about common objects. An FRL-like implementation 8,7 is well suited to such an application. This allows us to instantiate a room, say, making use of all the applicable generic knowledge in the scene data base, yet particularizing the details of the surroundings as necessary. The task manager and skills can access the scene data base, examine object instances, and update them if necessary.

Common-sense knowledge about objects can be embedded in the scene data base as well. For example, suppose an animation script instructs the figure to "walk out of the room" it is currently in. The task manager cannot fill in a definite location for the walk controller, so the frame for the "current room" in the scene data base must be checked. The required information is not found there, so the type link is followed to the generic room frame where the information "exit rooms through doors" is found. Returning to the "current room" frame, the slot for "door" is found, and the location of the door is passed to the walk controller. (If this particular room had no doors, or several, processing would halt and the human animator would be consulted).

In general, interaction with objects could proceed in the following way. Given some skill S that involves object o. The movement frame for S contains precondition and execution rules for interacting with the class of objects O of which o is an instance. S must ascertain the details of instance o that are important to S. If any preconditions are not satisfied, other skills will be triggered. When control returns to S, and all preconditions are satisfied, the execution rules will be tried until the appropriate motor programs for dealing with o are invoked. (If no applicable rule is found, an error has occurred, meaning the figure has insufficient skills. However, at this point it is known which object is causing difficulty, and which skill is being attempted). The postconditions for S are checked, and any changes to o are saved.

6. An Example

Let us suppose that skills are defined and have the frame representation described in Figure 1. Each frame includes a slot with with a pointer to the motor program to be invoked just in case all preconditions are satisfied.

Suppose the figure is currently sitting in a chair with its legs crossed. Suppose further the task manager is passed the single skill "walk". Since no movement specifications are present, the skill for walking is called and default ("normal")

```
Frame: WALK
Speed: "normal" (default)
Direction: "forward" (default)
Preconditions: --standing
               --feet on ground
Triggers:
        if not standing, call STAND_UP.


Frame: STAND_UP
Speed: "normal" (default)
Direction: "ahead" (default)
Preconditions: --feet on ground near center of gravity
Triggers:
        if current posture is sitting, S1
        if current posture is lying, S2
        if current posture is kneeling, S3


Frame: S1 (prepare to stand from a sitting position)
Speed: "normal" (default)
Preconditions: none
Procedure: move_legs
```

FIGURE 1. Example Frames.

values for speed and direction are used. Now preconditions are checked. The blackboard is consulted to see if the skeleton is in a "standing" posture. Since this is not the case, the first trigger is invoked, i.e., the stand_up skill is invoked.

Since the skeleton is currently in a sitting position with feet NOT on the ground, skill S1 is invoked. S1 is specifically constructed to generate a transition sequence between sitting with legs up and standing. (No transition is needed between sitting with legs on the floor and standing; standing may be invoked directly). It can determine whether one or both legs need moving, which leg needs to move first if both must be moved, and it can compute desired goal positions for whichever leg needs to be moved. S1 passes these start configurations and goal configurations to the motor program move_legs which finally updates the skeleton data base such that the figure on the display appears to move.

When both legs have been adjusted (if necessary) S1 signals completion to STAND_UP, which can now invoke the necessary motor programs for causing the skeleton figure to stand. When complete, STAND_UP returns control to WALK which can now execute.

7. Conclusion

I have outlined the design of the knowledge-based aspects of a system for automatic motion synthesis for the animation of articulated figures. The notion of expectation is central, as reflected in the patterns of selective potentiation of elements of the hierarchy of implemented skills. These patterns are established by the interconnections among movement frames. The motor control hierarchy interacts with the scene data base in two ways:

Skills interact with object frames to obtain (and possibly update) physical attributes of objects (e.g., current position, size, weight, etc.).

Skills interact with object frames to determine functional attributes (e.g., "How can I pick it up?", "What part of a chair is for sitting on?", "How far in front of a door do I have to stand in order to open it?").

Many difficult problems remain. I have not addressed the problem of planning movements, as in the soccer ball example, where perceptual input plays a strong role in controlling motor output. Servomechanistic systems are promising models for this application [5,18]. It is also not clear what is the best way to decompose the motor system. Should skills be constructed strictly along functional lines, with a skill or set of skills designed around a specific task model? Or should task-oriented skills ultimately be constructed from a set of movements based on the structure of the figure, e.g., a set of primitive skills for moving the arms and legs? Modification of the motor hierarchy, i.e., "training" the figure controller, is another critical issue, especially for a practical animation system. An animator must be able to create new movements and modify existing ones. How can a new skill be integrated into the existing potentiation network in the movement data base?

As of this writing, too few skills have been implemented to provide a demonstration of movement continuity. Nevertheless, interesting sequences of animation have been generated. As development of the Skeleton Animation system progresses and the range of movements of the animated figures increases, so too will the complexity of interaction between movements, and between figures and the environment. In the near future this will provide a practical testbed for the proposed control mechanisms.

References

1. Bernstein, N., The Coordination and Regulation of Movements, Pergamon Press, Oxford (1967).

2. Crow, F. C., "A More Flexible Image Generation Environment," Proc. ACM Siggraph 82, pp. 9-18 (July 1982).

3. Csuri, C., Hackathorn, R., Parent, R., Carlson, W., and Howard, M., "Towards an Interactive High Visual Complexity Animation System," Proc. Siggraph 1979, (August 1979).

4. Erman, L. D. and Lesser, V. R., "A Multi-Level Organization for Problem-Solving Using Many Diverse Cooperating Sources of Knowledge," Proc. 4th International Joint Conf. on Artificial Intelligence, pp. 483-490 (September 1975).

5. Gallistel, C. R., The Organization of Action: A New Synthesis, Lawrence Erlbaum Associates, Hillsdale, New Jersey (1980).

6. Gelfand, I.M., Gurfinkel, V.S, Tsetlin, M.L., and Shik, M.L., Models of the Structural-Functional Organization of Certain Biological Systems, MIT Press, Cambridge (1971).

7. Goldstein, I.P. and Roberts, B., "The FRL Manual," AI Memo 409, MIT, Cambridge, MA (September 1977).

8. Goldstein, I.P. and Roberts, B., "Using Frames in Scheduling," pp. 256-284 in Artificial Intelligence: An MIT Perspective, ed. R.H. Brown, MIT Press, Cambridge, MA (1979).

9. Gomez, F. and Chandrasekaran, B., "Knowledge Organization and Distribution for Medical Diagnosis," IEEE Transactions on Systems, Man, and Cybernetics, Vol. SMC-11, (1) pp. 34-42 (January 1981).

10. Greene, Peter H., "Problems of Organization of Motor Systems," Progress in Theoretical Biology, Vol. 2 , pp. 303-338 Academic Press, (1972).

11. Herman, R., Wirta, R., Bampton, S., and Finley, F.R., "Human Solutions for Locomotion: Single Limb Analysis," pp. 13-49 in Neural Control of Locomotion, ed. R. Herman, Plenum Press, New York (1976).

12. Lamoreux, L. W., "Kinematic Measurements in the Study of Human Walking," Bulletin of Prosthetics Research, Vol. BPR 10 , (15)(Spring 1971).

13. McGhee, R.B., "Robot Locomotion," pp. 237-264 in Neural Control of Locomotion , ed. R. Herman, Plenum Press, New York (1976).

14. Minsky, M., "A Framework for Representing Knowledge," in The Psychology of Computer Vision, ed. P. Winston, McGraw-Hill, New York (1975).

15. Murray, M. P., Drought, A. B., and Kory, R. C., "Walking Patterns of Normal Men," Journal of Bone and Joint Surgery, Vol. 46A , (2) pp. 335-360 (1964).

16. Muybridge, E., Animals in Motion, Dover, New York (1957).

17. Newman, W. and Sproull, R., Principles of Interactive Computer Graphics, 2nd Edition, McGraw-Hill, New York (1979).

18. Powers, W. T., Behavior: The Control of Perception, Aldine Publishing Co., Chicago (1973).

19. Schank, R. C. and Abelson, R. P., "Scripts, Plans, and Knowledge," Adv. Papers of the 4th International Joint Conf. on Artificial Intelligence, (September 1975).

20. Simon, H. A., The Sciences of the Artificial, MIT Press, Cambridge, MA (1979).

21. Tinbergen, N., The Study of Instinct, Oxford University Press, London (1969). Chapt. 5.

22. Vinci, Leonardo da, The Notebooks of Leonardo da Vinci, Random House, New York (1957). pp. 206.

23. Weiss, P., "Self-Differentiation of the Basic Patterns of Coordination," pp. 217-274 in The Organization of Action: A New Synthesis, ed. C. R. Gallistel, Lawrence Erlbaum Associates, Hillsdale, New Jersey (1980).

24. Zeltzer, D. and Csuri, C., "Goal-Directed Movement Simulation," Proc. Conf. Canadian Society for Man-Machine Interaction, pp. 271-279 (June 1981).

25. Zeltzer, D., "Motor Control Techniques for Figure Animation," IEEE Computer Graphics and Applications, Vol. 2, (9) pp. 53-59 (November 1982).

3D Balance in Legged Locomotion:

Modeling and Simulation for the One-Legged Case

Seshashayee S. Murthy
Department of Electrical Engineering

Marc H. Raibert
Department of Computer Science

Robotics Institute
Carnegie-Mellon University
Pittsburgh, PA.

Abstract

This paper explores the notion that the motion of dynamically stable 3D legged systems can be decomposed into a planar part that accounts for large leg and body motions that provide locomotion, and an extra-planar part that accounts for subtle corrective motions that maintain planarity. The large planar motions raise and lower the legs to achieve stepping, and they propel the system forward. The extra-planar motions ensure that the legged system remains in the plane. A solution of this form is simple because 3D dynamics do not play an important role.

We develop a model of a 3D one legged hopping machine that incorporates a springy leg of non-zero mass and a two axis hip. The hopping machine is modeled as an open loop linkage that has different configurations in flight and in stance. Behavior at transitions between phases is calculated by invoking conservation of momentum. We have decomposed control of the model into four parts that control hopping height, forward velocity, body attitude, and spin. Hopping height is controlled by regulating vertical energy. Velocity is controlled by placing the foot fore or aft during flight. Body attitude is controlled by torquing the hip during stance. Spin is controlled by placing of the foot outside the plane of motion. Simulation data are presented which show that these control algorithms result in good control of velocity, body attitude and spin, while traveling along a straight desired path.

I. Introduction

The locomotion of legged systems is a form of motion that has gained the attention of biologists seeking to understand the behavior they observe in natural organisms, and of engineers who attempt to build useful legged vehicles. Animators and film makers have also shown interest, but mostly in simulating the appearance of systems that use legs to locomote. Our interest is not so much in the appearance or description of locomotion, as it is in the underlying mechanisms that are responsible for production and control of such motion. In particular we have focused on the problem of controlling the motion of systems that balance as they run.

Dynamic stability is a key ingredient in the mobility exhibited by legged systems. Systems that balance can move on a narrow base of support, permitting travel where obstacles are closely spaced or where the support path is narrow. A dynamically stabilized system need not be supported at all times and may therefore use support points that are widely separated or erratically placed. These characteristics relax the constraints on the type of terrain a legged system can negotiate.

Casual observation of a running animal, say a cat, a horse, or a kangaroo, might lead one to conclude that running in a straight line is a 2D activity. The legs swing fore and aft through large angles while the body bobs up and down. The body may also undergo pitching motions that are quite pronounced. These large motions of the legs and body propel the animal upward so that the feet may be picked up and placed on a new spot, they allow the animal to balance itself so that it does not tip either forward or backward, and they propel the animal forward so that transportation takes place. However, these large planar motions do not tell the whole story. Natural legged locomotion takes place in a 3-space where motion with six degrees of freedom is possible. Roll and spin have to be controlled and the direction of motion must sometimes be altered.

To study dynamically stable locomotion in 3D we have modeled a system with just one leg and a very small foot. This simple one-legged model allows us to address the dynamic stability problem squarely, while totally avoiding the coupling problem that complicates the analysis of systems with many legs. Our goal is to test the idea that control for legged systems running in 3-space need not explicitly deal with the complications of 3D dynamics. Rather it may be feasible to decompose the problem into a planar part that controls locomotion using the large motions described above, and an extra-planar part that uses only very subtle motions to restrict behavior to the plane. Decomposition results in a very simple solution that seems to be in concert with what we observe in natural systems.

Background

Previous studies of balance in 3D legged systems have been carried out by a number of workers, most notably in Europe and Japan. Vukobratovic and his co-workers [1, 12, 13] have developed the notion of *zero moment point, ZMP*, control. They have shown in simulation how a 3D multi-linked walking biped can be balanced by manipulating the relationship between the projection of the center of gravity and the support areas provided by the feet. Kato et. al. [2] have studied quasi-dynamic walking in the biped. In their studies a physical biped with 10 hydraulically driven degrees of freedom used a preplanned sequence of quasi-static motions to dynamically transfer support from one leg with a

Notion: Representation and Perception
N.I. Badler and J.K. Tsotsos, Editors
Published by Elsevier Science Publishing Co., Inc.

large foot to the other. Miura and his students [6] have built a number of small electrically powered walking bipeds that balance using tabular control schemes. Their most advanced model demonstrates dynamic balance without large feet. It has three actuated degrees of freedom that permit each leg to move fore and aft, to move sideways, and to lift slightly off the floor. This machine balances with a shuffling gait that reminds one of Charlie Chaplin's stiff-kneed walk.

Hopping has also been studied. About fifteen years ago Seifert [11] explored the idea of using a large pogostick for transportation on the moon where low gravity would permit very long hops. Matsuoka [4] analyzed 2D hopping in humans with a one-legged model, assuming that the leg could be massless, and that the stance period could be of very short duration. He derived a time-optimal state feedback controller that stabilized his system. Matsuoka [5] also implemented a very simple one-legged hopping machine that lay on a table inclined 10° from the horizontal, and controlled it using this controller.

Raibert and his co-workers modeled and built a physical one-legged hopping machine that was constrained to move in a plane [8, 9, 10]. They found that the control of locomotion in this device could be subdivided into three largely independent parts; regulation of hopping height, control of forward velocity, and control of attitude. The parts of their control system are summarized here.

- Height: The control system regulated hopping height by manipulating hopping energy. The machine had a springy leg, so hopping was a bouncing motion that was generated by an actuator that excited the leg spring. Hopping energy was conserved by the leg spring from hop to hop. The height to which the machine hopped was determined by the energy recovered from the previous hop, and by the losses in the hopping cycle. Since all energy in the system takes the form of potential energy at the peak of a hop, hopping height could be regulated by injecting an appropriate amount of energy during each step.
- Velocity: The control system manipulated forward velocity by moving the leg during the flight part of each hop to properly position the foot with respect to the *CG-print*. The CG-print is the locus of points on the ground over which the center of gravity of the system will pass during the next stance period. If the foot is placed in the center of the CG-print, the device will tip neither forward nor backward, but will continue its forward motion at about the same rate as before. If the foot is placed rearward of the center of CG-print, then the device will tip forward, increasing its forward velocity. If the foot is placed forward of the center CG-print, then the device will tip backward, decreasing its velocity. The control system calculated the length of the CG-print from the measured forward velocity of the device and the estimated duration of stance. The control system then used the error in forward velocity to position the foot to control and correct the forward speed of locomotion.
- Attitude: The control system maintained an erect body posture during running, by generating hip torques during stance that servoed the body angle. During stance friction between the foot and ground permitted large torques to be applied to the body without causing large accelerations of the leg. These torques were used to implement a simple linear servo that returned the body angle to an erect posture once each step.

In this paper we extend this approach to 3D locomotion for the case of straight line running. To do so we have modelled and simulated a 3D one-legged hopping machine that moves freely on a flat open floor. Hopping height is controlled as before in the 2D case. In order to extend the 2D control system to 3D, we must modify it to handle three new degrees of freedom: lateral translation, roll orientation, and yaw orientation.

II. Modeling and Simulation

The model, shown in Fig. 1, has two primary parts; a body and a leg. The body is represented by a rigid mass with substantial moment of inertia about its three primary axes. (See Appendix for parameters.) The leg is a long slender linkage that is springy in its axial dimension, with a small foot at one end. The leg is connected to the body by a universal hip joint that provides two degrees of freedom.

All three joints in the machine are actuated. The hip is driven by a pair of torque actuators that can be used to orient the leg with respect to the body, or to change the attitude of the body when the foot is in contact with the ground. The leg is driven by a third actuator that operates in series with a passive spring. Changes in the length of this actuator are used to excite the leg spring and to make the machine hop.

In addition to the model's three actuated joints, called *internal* degrees of freedom, the model has *external* degrees of freedom that permit it to move with respect to its surroundings. During stance, when the foot touches the ground, there are three external degrees of freedom that give the model's orientation in space. During flight when the foot leaves the ground, there are three additional external degrees of freedom, a total of six, that specify Cartesian position in space. Unlike the three internal joints that are driven directly by actuators, and therefore easily controlled, the external degrees of freedom are indirectly driven by dynamic interactions among the model, gravity, and the ground. At the heart of the legged locomotion problem lies the need to find ways to use direct control of internal degrees of freedom to achieve indirect control of external degrees of freedom.

Basic operation of the model is to bounce on the springy leg in a hopping motion that alternates between periods of support and periods of flight. Four events characterize this alternating cycle; *lift-off* - the moment in the hopping cycle when the foot looses contact with the ground, *top* - the moment the body achieves peak altitude, *touch-down* - the moment the foot first touches the ground, and *bottom* - the moment the body has minimum altitude and the leg is fully compressed. These four events help to synchronize the control algorithms to the behavior of the model.

Equations of motion

The 3D hopper is modeled as an open loop dynamic linkage. This allows simulation using existing computer programs originally intended for robot arms [14]. In order to model the hopper as an open loop linkage the external degrees of freedom were modeled as joints with zero mass and zero moment of inertia. The kinematics of these joints are described in the Appendix.

The state of the model is described by $\mathbf{Q}$ and $\mathbf{Q}'$, where $\mathbf{Q}$ is the position vector describing the position of each of the links and $\mathbf{Q}'$ is the velocity vector describing the velocity of each of the links. To obtain the equations of motion we must find the acceleration vector $\mathbf{Q}''$ given the state of the model and the forces acting at each joint:

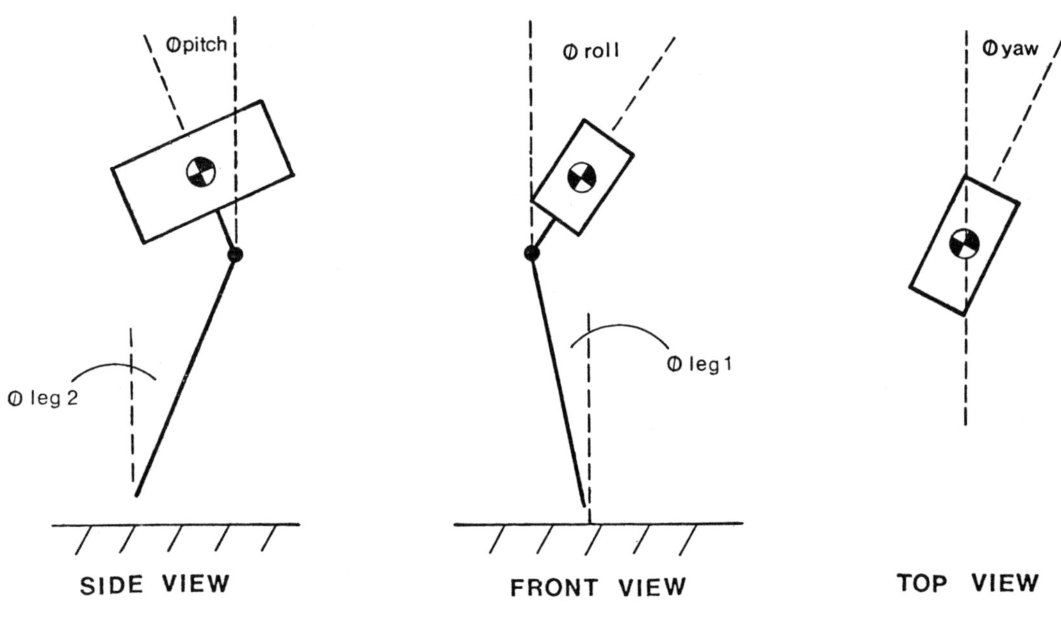

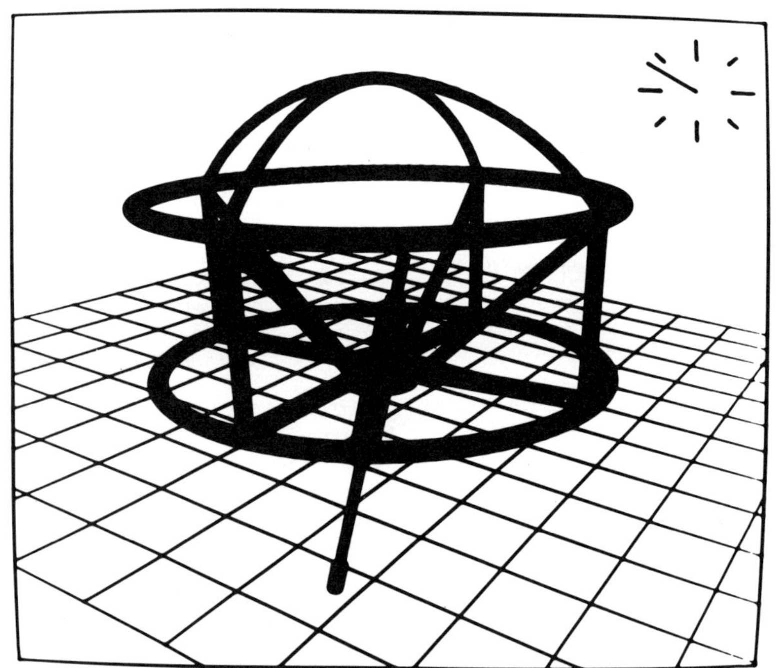

Figure 1: The 3D one-legged system modeled in this paper. It has a body and a leg, connected by a hip. The body is a rigid structure with mass and moment of inertia. The leg has mass, moment of inertia, and an actuated spring along its major axis. The body and leg are connected by a hip with two orthogonal axes of rotation, each driven by a torque source. Top: Kinematic diagram. Bottom: graphical model used for simulations. Model parameters are given in the Appendix.

$$Q'' = f(Q, Q', \tau, g) \qquad (1)$$

where

τ is the vector of torques that acts on the joints of the mechanism, and

g is the gravity vector.

The derivation of the equations of motion were described by Luh, Walker, and Paul [3], and efficient methods for solving them numerically were given by Walker and Orin [14]. We used Walker and Orin's third method to determine $\mathbf{Q''}$.

The analysis is divided into a set of equations that describes the system when it is on the ground, and another set that describes the system when it is in flight. It is also necessary to determine what happens at transitions between these two phases. During stance the machine is an inverted pendulum that can tip in two directions, in addition to rotating about the major axis of the leg. Ground forces resulting from impact, internal forces and torques, and gravity affect the angular and linear momentum of the hopper during stance. The hopper is modelled as a 6 joint open loop linkage. The first three links represent three rotational degrees of freedom at the foot, and the next three represent the internal joints of the hopper. In flight overall motion of the system is ballistic, affected only by gravitational forces. The horizontal component of the linear momentum and the angular momentum about the center of gravity remain unaltered. The hopper is modelled as a 9 joint open loop linkage. The first three links represent translational degrees of freedom at the base of the foot. The next three represent rotational degrees of freedom and the last three represent the internal joints of the hopper. Solution of Eq. 1 gives the trajectory of the system as a function of time, during either stance or flight.

The simulation has discontinuities at lift-off, the transition between stance and flight, and at touch-down, the transition between flight and stance. At these transitions the laws of conservation of momentum are invoked in order to determine changes in state. If we assume the ground to be rigid with no compliance, then at touch-down an impulse force of duration ε acts on the foot. As a result the velocities $\mathbf{Q}'$ are changed while the position $\mathbf{Q}$ remains unaltered. Simulation of the hopper in the ground phase requires calculation of the change in $\mathbf{Q}'$. The following assumptions are made about the nature of the impact.

- An impulse of magnitude $\delta = \int_{t-\varepsilon/2}^{t+\varepsilon/2} F\,dt$ acts on the foot at the time of impact. No torques act on the foot.
- Internal forces and torques acting along the z axis of each joint are finite in magnitude.
- The duration of transition between phases, ε, is very small.

From these assumptions it follows that

$$\Delta L = \int_{t-\varepsilon/2}^{t+\varepsilon/2} \tau \, dt = \tau \, \varepsilon = 0 \qquad (2)$$

i.e. $L_{t-\varepsilon/2} = L_{t+\varepsilon/2}$

where

τ is a vector of internal forces and torques that acts along the z axis of each joint, and

L is the vector of linear or angular momenta about the z-axis of each joint.

If joint i is rotational then L_i is the total angular momentum of links i through 9, about the z-axis of joint i. If joint i is translational then $\mathbf{L}_i$ is the total linear momentum, along the z-axis of joint i, of links i through 9. From Eq. 2 it follows that, during impact, angular momentum is conserved about the z-axis of a rotary joint and linear momentum is conserved along the z-axis of a translational joint. By applying conservation of momentum about the z-axis of the three external rotational joints and the three internal joints we get six simultaneous equations:

$$\sum_{j=i}^{9} a_{ij}\, Q'_{j+} = L_i \text{ for i = 4 through 9} \qquad (3)$$

where

Q'_{j+} is the velocity at joint j at time $t + \varepsilon/2$, and

a_{ij} is a linear constant that depends on $\mathbf{Q}_i$ to $\mathbf{Q}_9$.

L is calculated from the state, $\mathbf{Q}$ and $\mathbf{Q}'$, of the hopper at $t - \varepsilon/2$. Eq. 3 is then solved for the velocity vector after touch-down. The state at $t+\varepsilon/2$ is thus computed without knowledge of the impulse forces acting on the foot.

At lift-off the leg and body assume the same velocity, which is the velocity of the center of gravity. An inelastic collision is assumed. The effect of this impact is calculated in a similar fashion. Conserving momentum about the remaining eight joints provides the velocity vector at lift-off. This procedure permits modelling of the transitions from flight to stance and from stance to flight with very little computation.

III. Control

The strategy employed here to control locomotion of the 3D model is to decompose its motion into a planar part and an extra-planar part. There is a plane that contains the gravity vector, the center of gravity, and the forward velocity vector. We call this the *plane of motion*. We call the line, where this plane intersects the ground, the line of motion. If the control system were to always place the foot on the line of motion, and there were no roll or yaw motion, then all forces acting on the model would lie in the plane of motion. In that case the machine would never leave the plane of motion and the planar control system mentioned earlier would be adequate to regulate hopping, attitude, and forward travel.

The extra-planar control part corrects three types of errors introduced by external disturbances and noisy control. These errors are roll rotation, yaw rotation, and lateral translation. Corrections for roll error are made by the attitude control algorithm, much as pitch corrections are made. Yaw errors are corrected by placing the foot outside the plane of motion, and applying suitable hip torque during stance. Lateral translations are not actually corrected in the present scheme, but they are taken into account when the plane of motion is redefined on each step. The planar control part operates properly in 3D only when the extra-planar part successfully limits each of these error motions to small magnitude.

By augmenting the planar 3-part controller with additional extra-planar controls, we arrive at a 3D control system with 4 separate control algorithms. They control hopping height, forward velocity, body attitude, and spin.

Height control

Control of hopping height for the 2D case has been explored elsewhere in simulation and through physical experiments [7, 8, 9]. Simple control of hopping height is a one dimensional problem that is substantially the same for locomotion in two and three dimensions. Therefore we have simplified the present model to generate accurate vertical motion without representing all the necessary details. For instance the simulation of the hopper does not incorporate the various losses that occur due to friction in the actual machine. The only losses represented are those due to impact at touch-down and lift-off. Therefore, very little has to be done to maintain correct hopping height once it is attained.

Velocity control

The primary mechanism used for controlling the velocity of the model is proper placement of the foot at touch-down. During flight the control system orients the leg so as to position the foot with respect to the center of the CG-print. The algorithm is described below.

At touch-down:

$$L_0 = m\,\mathbf{r} \times \mathbf{v} + L_{cg} \qquad (4)$$

$$= m\,v\,r \sin(\phi_{gv} - \theta_{tdn}) + L_{cg}$$

where

- L_0 is the angular momentum of the model about the point of touch-down,
- L_{cg} is the angular momentum of the model about its own center of gravity,
- r(t) is the vector from the point of touch-down to the center of gravity of the model,
- v(t) is the velocity of the center of gravity,
- m is the combined mass of the body and leg,
- ϕ_{gv} is the angle formed by the gravity vector and **v**,
- θ(t) is the angle formed by the gravity vector and **r**, and
- θ_{tdn} is θ(touch-down).

See Fig. 2.

We make the assumption that **r**, **v**, and the gravity vector lie in the same plane. The first term in Eq. 4 is dependent on the placement of the foot. It determines whether the system tips forward or backward after touch-down. During the normal hopping cycle, the horizontal velocity at touch-down is large and the angular momentum about the center of gravity is small. The major component of the angular momentum is the first term.

During stance the change in angular momentum is:

$$\Delta L = \int_{touch-down}^{lift-off} \mathbf{r} \times m\mathbf{g}\, dt = \int_{touch-down}^{lift-off} m\,g\,r \sin(\theta) dt \qquad (5)$$

$$= \int_{touch-down}^{lift-off} m\,g\,r \sin(\theta) d\theta / \theta'$$

Prior to touch-down, the velocity control algorithm calculates the angle θ_{tdn} necessarry to achieve the angle the change ΔL. The foot is then oriented at this angle in the plane of motion. The change ΔL is a nonlinear function of ϕ_{gv}, θ_{tdn}, v(t), r(t) and θ(t). Its exact evaluation would require solution of the equations of motion for the stance phase. We have no closed form solution. Data generated by a systematic set of simulations for a range of initial conditions are shown in Fig. 3. They show that the relationships between lift-off velocity, on the one hand, and touch-down velocity, vertical velocity, and leg angle, on the other hand, are all nearly linear over a wide range of values. We have used the linear approximation to calculate θ_{tdn} with good success.

This control algorithm was used to control the model in simulation. Figure 4 shows the trajectory of the center of gravity as a function of time when this control was used. A constant desired velocity was specified until the model had translated 2 m, at which point the desired velocity was set to zero. Average velocity was controlled with good precision. The temporary deviations from the average velocity visible in the plot were caused by the attitude control servo, which begins to erect the body right after touch-down.

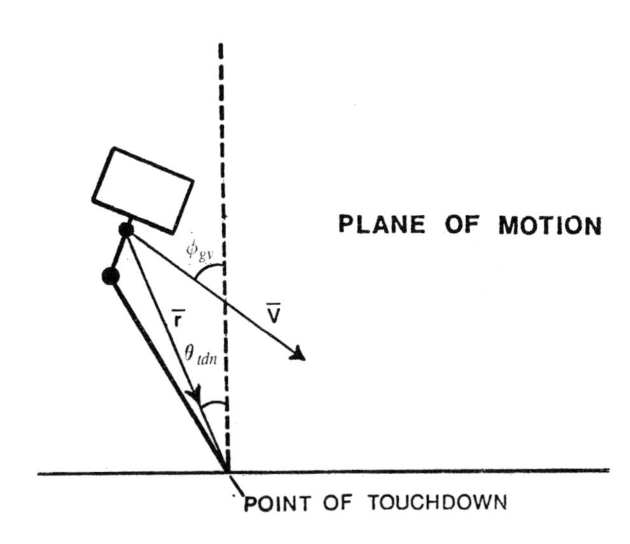

Figure 2: Configuration of model in plane of motion.

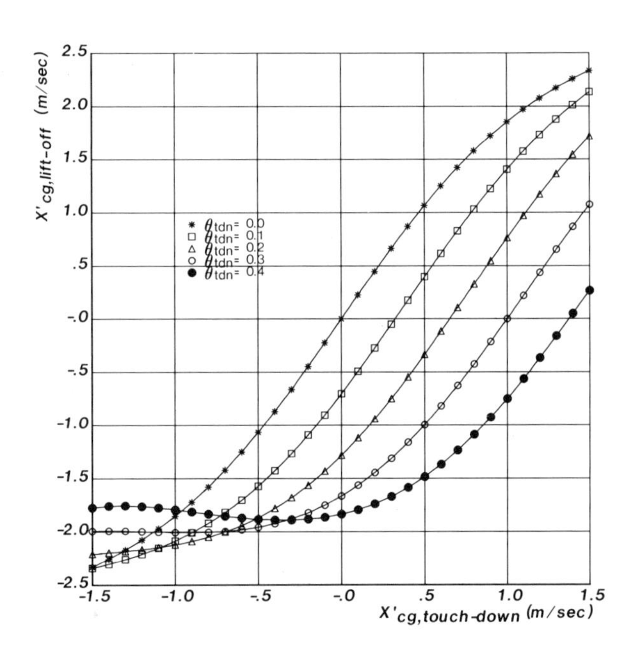

Figure 3: The effect of foot placement on forward velocity was determined empirically by simulating stance for a set of initial conditions and foot placements. Once forward velocity is known, it is possible to use these data to select a foot placement that will change the forward velocity to a desired value. The data in this figure are for $Z'_{cg,touch-down} = 1$ m/sec.

Attitude control

The attitude controller must correct errors in both pitch and roll. Pitch errors are caused by the reaction of the body to the swinging motion of the leg made when the foot is swung forward in preparation for the next step. Roll errors are caused by disturbances. Roll errors will generally be small while pitch errors are large. During stance it is possible to take advantage of friction between the foot and the ground to generate hip torques that will erect the body.

The attitude controller is a linear servo that must neither be too stiff nor too soft. It must be stiff enough to have settled by the end of stance. If this is not so the servo will have disturbed the body's attitude by lift-off, rather than corrected it. The servo must not be so stiff that it causes the foot to slip when it generates hip torques. The weight of the system and the coefficient of friction of the foot are the limiting factors. Independent servo controllers are used about the pitch and roll axes, each with $\zeta = 0.707$.

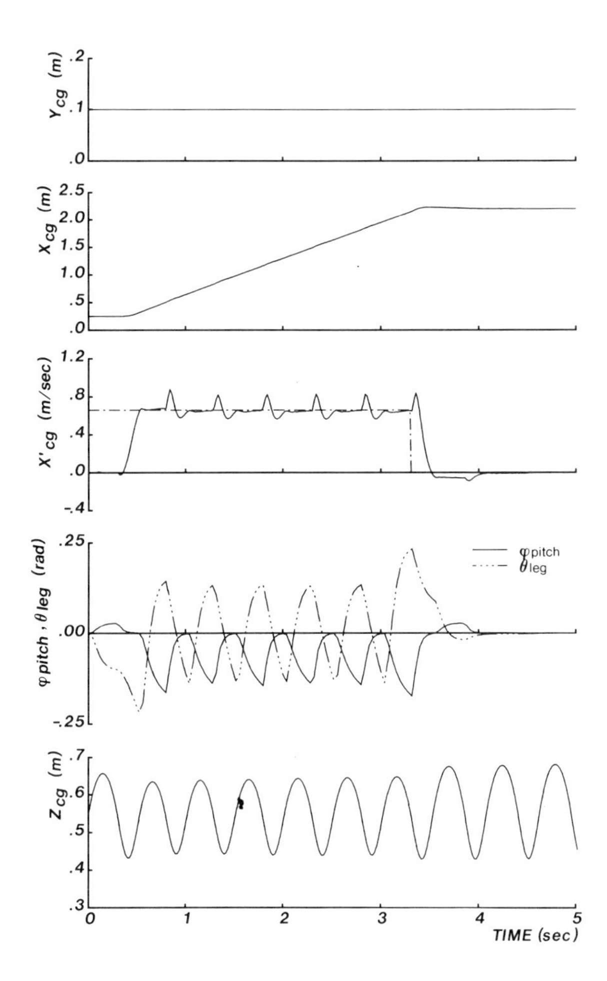

Figure 4: The 3D one-legged model traveling in a straight line for 2 m. Plots show simulation data of the trajectory of the center of gravity of the model, velocity of the center of gravity in the plane of motion, body and leg angle in the plane of motion, and the vertical position of the center of gravity. (Dashed line in the velocity plot represents the desired value.)

$$\tau_{pitch} = - k_p(\varphi_{pitch} - \varphi_{pitch,d}) - k_d \varphi'_{pitch} \quad (6)$$

$$\tau_{roll} = - k_p(\varphi_{roll} - \varphi_{roll,d}) - k_d \varphi'_{roll} \quad (7)$$

where

τ_{pitch}, τ_{roll} are the torques applied about the pitch and roll axes,

φ_{pitch}, φ_{roll} are the pitch and roll angles of the body,

$\varphi_{pitch,d}$, $\varphi_{roll,d}$ are the desired pitch and roll angles of the body at lift-off, and

k_p, k_d are the proportional and derivative feedback gains.

The fourth curve in Figure 4 is a plot of body pitch angle and the leg angle as a function of time during constant velocity running. The body tilts forward in the flight phase as the leg swings forward. At touch-down the body angle reaches a maximum. During stance the controller forces the body angle toward zero. Roll motions are similarly corrected. ($\varphi_{pitch,d} = \varphi_{roll,d} = 0$)

Spin control

The control system suppresses spin by placing the foot outside of the plane of motion. Fore and aft forces on the foot, generated both by the foot's impact during touch-down and by hip torque during stance, produce a torque about the yaw axis when the foot is placed outside of the plane of motion. Let the distance of the foot from the plane of motion be $d_\perp$. On touch-down an impact torque causes a change in spin momentum:

$$\Delta L_{yaw} = d_\perp m \Delta v_{horz} \quad (8)$$

where L_{yaw} is the angular momentum about the yaw axis.

In this equation mv_{horz} is the linear momentum of the hopper in the direction of the line of motion, and $d_\perp$ is the horizontal distance between the center of gravity and the point of touch-down in a direction perpendicular to the plane of motion.

When a hip torque is applied to the body about the axis perpendicular to the plane of motion, it causes a ground reaction force at the base of the foot in the direction of the line of motion.

$$F_{horz} = \tau_{hip} \; r \; \cos\theta \quad (9)$$

This force causes a torque to act about a vertical axis through the center of gravity. The total change in the spin angular momentum during stance is:

$$\Delta L_{yaw} = \int_{touch-down}^{lift-off} F_{horz} d_\perp \, dt \quad (10)$$

Thus, if the foot is placed asymettrically, outside the plane of motion, the spin angular momentum is altered due to impact forces at touch-down and hip torques during stance. The spin of the hopper can thus be controlled. During flight the distance $d_\perp$ is calculated. The foot is then oriented at an angle θ_{tdn} from the vertical, in the plane of motion, and a distance $d_\perp$ from the plane of motion. After touch-down a torque τ_{hip} is applied at the hip to reduce spin.

$$\tau_{hip} = - k_d \varphi'_{spin}$$

Figure 5 (top) plots spin about the yaw axis as a function of time for four different initial spin rates when this controller was used to suppress spin. For initial rates of 0.2 rad/sec or less, spin is quickly controlled. We are not sure how the present technique will work when the initial spin rate exceeds 0.2 rad/sec. Figure 5 (bottom) shows the path followed by the hopper during the spin suppression maneuver. Placement of the foot outside the plane of motion to correct spin, has the side-effect of changing the direction of travel.

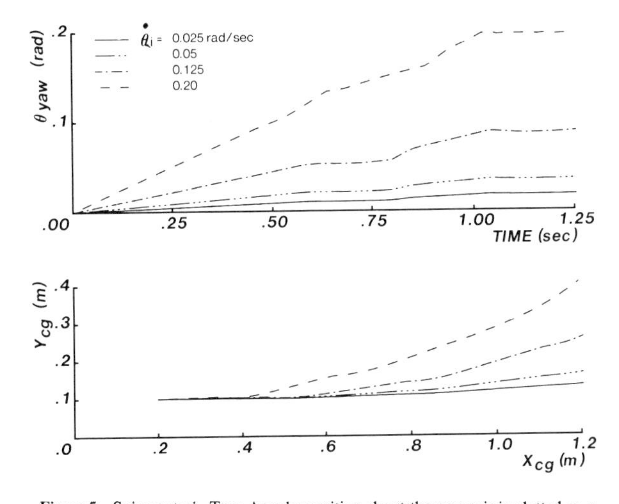

Figure 5: Spin control. Top: Angular position about the yaw axis is plotted as a function of time for various initial spin rates. The spin velocity is suppressed in a short time for initial spin rates of up to .2 rad/sec. Bottom: Path followed by center of gravity while spin is suppressed. Spin suppression places the foot outside the plane of motion, with the side-effect of altering the direction of travel.

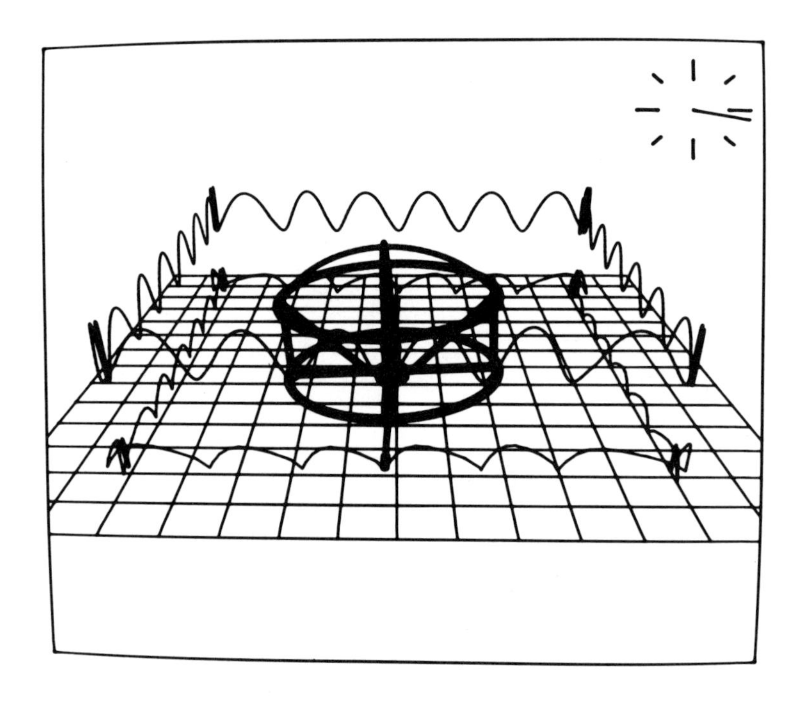

Figure 6: Path control. The 3D model was made to follow a square path. It started at (1,-1), lower right, and progressed clockwise through (-1,-1), (-1,1), and (1,1), finally returning to the starting point. (Grid spacing is 0.2 m. Center is (0,0).) The upper trace marks the path of the center of gravity. The lower trace marks the path of the foot. Total time around the square was 24 sec.

The various controllers may require conflicting actions to achieve their goals. There are two cases where this is apparent. First, spin control may require the foot to be placed at a large distance from the plane of motion. However if $d_\perp$ is large the trajectory in the plane of motion is affected. Care should be taken to see that $\sin^{-1}(d_\perp/|r|) << \theta_{tdn}$. Second, spin control may require certain body torques to be applied according to Eq. 10. However correction of body pitch errors may require that other torques be applied. Priority is given to control of spin. Correction of the body angle can often be deferred to a later time.

Path control

The control algorithms just described were used to implement a simple form of path control. If a desired path is decomposed into a set of straight line segments, then the path can be followed by stopping the machine at each vertex and changing its direction of travel. The plane of motion is not uniquely defined when there is no forward travel, so the control system is free to choose the plane that includes the next straight path segment. Figure 6 shows a cartoon of the one-legged machine, and the trajectory it took in traversing a square path. The settling time required at each vertex made progress quite slow, about 24 sec for the circuit, but the accuracy of the path was reasonably good.

Even for the simple case of a one-legged system, the algorithms presented here are not yet complete solutions to the 3D locomotion control problem. First, we do not yet know how to change heading while running. In order to change the direction of travel, the existing control system brings the machine to a halt, selects a new direction, and then accelerates. It would be very tedious to follow a winding contour using this approach. Second, choosing the plane of motion to incorporate lateral velocity errors permits the system to balance using the planar algorithms, but errors in heading cannot be corrected once travel starts. Yaw, roll, and lateral velocity errors will all contribute to heading drift. Third, these algorithms depend on a system that can travel equally well in all directions. Although it is possible to change heading, the algorithms provide no way to change the machine's facing direction.

V. Conclusion

Legged locomotion is a largely planar activity that takes place in 3D space. We argue that such behavior can be accomplished by providing one set of control algorithms that balance and generate travel within a plane, and a second set of control algorithms the eliminate motions that deviate from that plane. Control within the plane of motion can be further decomposed. The entire system consists of four control algorithms:

- Height: The springy leg is driven to cause hopping oscillations of the machine. Hopping height is regulated by using a measurement of the system's vertical energy to determine the correct amount of thrust to be delivered by the leg.
- Forward velocity: The CG-print is the locus of points over which the machine's center of gravity will travel during the next stance period. The control system regulates forward velocity by manipulating placement of the foot relative to the center of the CG-print.
- Attitude: During stance when friction holds the foot in place, hip torque is used to erect the body. Pitch and roll angles are both corrected with a linear servo.
- Spin: The foot is placed outside of the plane of motion and

a torque is generated at the hip. This produces a torque about the yaw axis that retards spin motion.

Simulation data showing effective forward velocity control, spin suppression, and straight segment path control encourage us to further test the feasibility of these ideas with a set of physical experiments.

Acknowledgments

This research was sponsored by the Systems Science Office of the Defense Advanced Research Projects Agency under contract MDA903-81-C-0130, and by the System Development Foundation. Kirk Botula and Mike Chepponis did the computer graphics on an E & S PS300 donated by Evans & Sutherland Computer Corp. We thank Ben Brown for his many useful ideas and suggestions.

Appendix

Dimensions of the one-legged model:

Link	leg	cylinder	ring	body	Units
Description	Cylinder	Cylinder	Ring	Ring	
Mass	.8626	0.5902	0.1	14.755	Kg
Length	0.8	0.15	0.01		Meters
Radius	0.01	0.02	0.03	0.3	Meters
J_{xx}	0.0474	0.00674	0.00009	1.152	Kg-m^2
J_{yy}	0.0474	0.0004	0.000046	0.807	Kg-m^2
J_{zz}	0.0001	0.00674	0.000046	1.152	Kg-m^2
J_{xy}	0	0	0	0	Kg-m^2
J_{yz}	0	0	0	0	Kg-m^2
J_{xz}	0	0	0	0	Kg-m^2

Denavit-Hartenberg description of the model:

#	θ	d	a	α	Description
1	$\pi/2$	Q_1	0	$\pi/2$	Translational degree of freedom.
2	$\pi/2$	Q_2	0	$\pi/2$	Translational degree of freedom.
3	$\pi/2$	Q_3	0	$\pi/2$	Translational degree of freedom.
4	Q_4	0	0	$\pi/2$	Rotational degree of freedom.
5	Q_5	0	0	$\pi/2$	Rotational degree of freedom.
6	Q_6	0	0	0	Rotational degree of freedom.
7	$\pi/2$	Q_7	0	$\pi/2$	1st body translational joint.
8	Q_8	0	0	$\pi/2$	1st body rotational joint.
9	Q_9	0	0	$\pi/2$	2nd body rotational joint.

References

1. Juricic,D., Vukobratovic,M. "Mathematical modelling of bipedal walking system." *ASME Publication 72-WA BHF-13* (1972).

2. Kato, T., Takanishi, A., Jishikawa, H., Kato, I. The realization of the quasi-dynamic walking by the biped walking machine. 4th Symposium on Theory and Practice of Robots and Manipulators, IFTMoM, 1981.

3. Luh, J.Y.S., Walker, M.W., Paul, R.P. "On-line computational scheme for mechanical manipulators." *Dynamic Systems, Measurement, and Control 102* (June 1980), 69-77.

4. Matsuoka,K. A model of repetitive hopping movements in man. Proc. of Fifth World Congress on Theory of Machines and Mechanisms, IFIP, 1979.

5. Matsuoka,K. "A mechanical model of repetitive hopping movements." *Biomechanisms 5* (1980), 251-258. In Japanese

6. Miura, H., Shimayama, I. "Computer control of an unstable mechanism." *J. Fac. Eng.*, 17 (1980), 12-13. In Japanese

7. Raibert, M.H., Brown, H.B.,Jr., Chepponis, M., Hastings, E., Shreve, S.T., Wimberly, F.C. Dynamically Stable Legged Locomotion. Tech. Rept. CMU-RI-81-9, Robotics Institute, Carnegie-Mellon University, 1981.

8. Raibert, M.H. "Hopping in legged systems -- Modelling and simulation for the 2D one-legged case." *IEEE Tran. Systems, Man, and Cybernetics*, Submitted (1983).

9. Raibert, M.H., Brown, H.B.,Jr. "Experiments in balance with a 2D one-legged hopping machine." *ASME J. Dynamic Systems, Measurement, and Control*, Submitted (1983).

10. Raibert, M.H., Sutherland, I.E. "Machines That Walk." *Scientific American 248*, 1 (1983), 44-53.

11. Seifert,H.S. "The lunar pogo stick." *J. Spacecraft and Rockets 4*, 7 (1967).

12. Vukobratovic,M., Stepaneko,Y. "Mathematical models of general anthropomorphic systems." *Mathematical Biosciences 17* (1973), 191-242.

13. Vukobratovic,M., Okhotsimiskii,D.E. "Control of legged locomotion robots." *Proc. International Federation of Automatic Control Planary Session* (1975).

14. Walker, M.W., Orin, D.E. "Efficient Dynamic Computer Simulation of Robotic Mechanisms." *ASME J. Dynamic Systems, Measurement, and Control 104*, 3 (1982), 205-211.

REPRESENTING AND REASONING ABOUT CHANGE

Reid G. Simmons & Randall Davis

The Artificial Intelligence Laboratory
Massachusetts Institute of Technology
Cambridge, MA

1. INTRODUCTION

A recent trend in artificial intelligence research is the construction of expert systems capable of reasoning from a detailed model of the objects in their domain and the processes that affect those objects [4]. We describe a system being built in this fashion, designed to solve a class of problems known as geologic interpretation [23]: given a cross-section of the Earth's crust (showing formations, faults, intrusions, etc.), hypothesize a sequence of geologic events whose occurrence could have formed that region. Solving the geologic interpretation problem requires reasoning about change, in particular, spatial change. The shape of a formation, for example, can be altered by the process of erosion. Doing this reasoning, in turn, requires representing objects, which show the effects of change, and processes, which are the causes of those changes.

The main focus of this research is to explore the machinery needed to represent and reason about both mutable objects and the processes that induce changes in them. To do this, we have developed two representations of objects, one involving *histories* and the other involving *diagrams*. We have also developed two corresponding representations of physical processes, each suited to reasoning about one of the object representations. We have been careful to keep the two representations well separated, limiting their interaction to a relatively small and clearly defined interface (Figure 1).

Fig. 1. Information Flow Among Representations

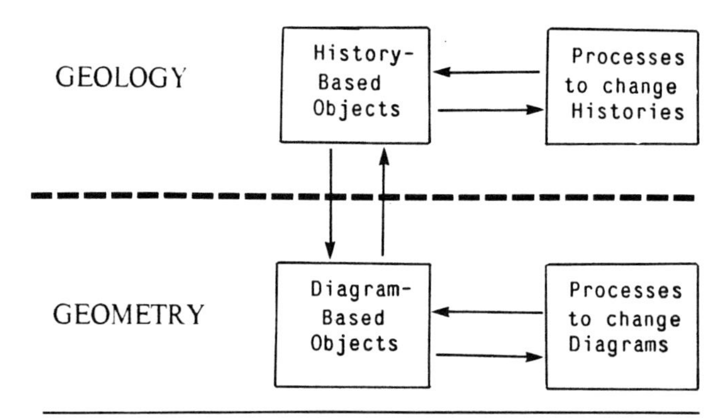

We have used these representations to model a subset of geology large enough to allow us to solve most geologic interpretation problems. In particular, the model allows us to *imagine* a sequence of geologic events. *Imagining* is a new form of qualitative simulation similar to *envisioning* [5].

Section 2 provides an outline of the major foci of interest in the paper. In Section 3 we describe the basic task of geologic interpretation, present a simple problem and demonstrate its solution. Section 4 describes the two representations of objects, while Section 5 presents the corresponding representations of physical processes. In Section 6, we show how our representations facilitate the *imagining* of a sequence of events. Section 7 presents a comparison with related work.

2. OVERVIEW

Our concerns in this paper focus around three main issues, reviewed briefly here and explored in more detail in the remainder of the paper.

2.1 The Representation of Mutable Objects and Processes

In order to solve the geologic interpretation problem, we need to reason about two basic forms of changes to objects. First, objects have a life-span, that is, they exist for a certain period of time and can be created or destroyed. Second, an object has various attributes whose values can change over time. In our current domain of geology, for example, a rock can be created by deposition or destroyed by erosion. The attributes of a rock include its composition, thickness, and location in space, all of which are subject to change over time.

Since all changes to objects are caused by physical processes, we are also concerned with representing processes. The process representation must facilitate reasoning about both types of change, that is, reasoning about which objects were created or destroyed and reasoning about how the attributes of an object may have changed.

Motion: Representation and Perception
N.I. Badler and J.K. Tsotsos, Editors
Published by Elsevier Science Publishing Co., Inc.

2.2 The Organization of Representations to Facilitate Reasoning About Time and Space

One part of solving the geologic interpretation problem requires reasoning about the specific change to an object between two instances in time. Since most geologic changes are spatial in nature (e.g. a change in shape due to erosion), we have developed a specialized representation for reasoning about the spatial characteristics of objects at instances of time. Another part of the problem requires reasoning about the cumulative effects of changes over time (e.g. the overall effect on the location of a rock due to a sequence of uplifts and tilts). We have developed a second specialized representation specifically suited to reasoning about such changes. In addition, we have developed corresponding representations for processes, one suited to reasoning about spatial change, and the other suited to reasoning about temporal change.

Spatial reasoning is done using diagrams, represented as collections of *vertices*, *edges*, and *faces*. The character and organization of diagrams facilitates inferences about changes in shape, location, orientation, etc. Temporal reasoning is done using a *history-based* representation. This representation is frame-like, but the value of an attribute is a *history*, which is a sequence of values over time, rather than a single value. This history of values facilitates reasoning about the sequence of changes to an object.

2.3 The Use of Simulation in Problem Solving

Our overall approach to the problem of geologic interpretation has much in common with generate and test. One part of the system generates a candidate solution while another tests it against the given cross-section (see Section 3).

Since a candidate solution is a sequence of processes, it is tested, in a process we call *imagining*, by simulating the effects of each process in turn and comparing the final result against the given cross-section. Unlike traditional generate and test, however, the test is not simply a binary predicate, and failing the test does not necessarily disqualify the candidate. A discrepancy between the result of the simulation and the cross-section can provide important information for augmenting the solution, information that may be impossible to infer otherwise.

This interaction between candidate generation and simulation illustrates a useful approach to the integration of local and global information in problem solving. By "local", we mean the kind of information that can be found by examining a single rock or single boundary in the diagram. By "global" we mean the overall consistency of the proposed solution. Although each individual process may be plausible, we need to determine the plausibility of the entire sequence, that is, does it produce the desired result? As we will see in the next section, the hypotheses are pieced together from local information; the *imagining* then provides an important check on the global consistency of the solution.

3. GEOLOGIC INTERPRETATION

3.1 An Example

In the problem of geologic interpretation, we are given a diagram that represents a vertical cross-section of a region, along with a legend identifying each kind of rock formation (Figure 2a). The task of geologic interpretation is to construct a sequence of geologic events that plausibly could have formed that region.

Typically, a geologist approaches this problem by looking at boundaries between rocks and making a collection of simple inferences in an attempt to build up a sequence of events. In this case, for example, he might note that, since the **mafic-igneous** crosses the **schist**, it intruded through (i.e. forced its way through) the **schist** and hence is younger (Figure 2b, step 1; the

Fig. 2. Simple Geologic Interpretation Problem

2a. Geologic Cross-Section and Legend

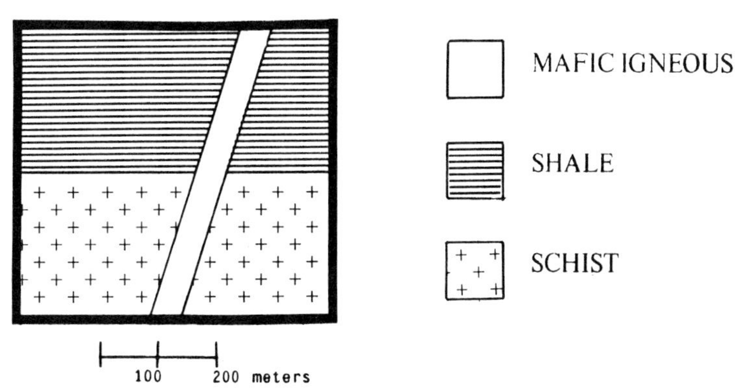

2b. Partial Orders at Each Stage of the Problem

1. mafic-igneous ⇒ schist
2. mafic-igneous ⇒ schist
   ⇒ shale
3. mafic-igneous ⇒ schist ⇒ shale
4. mafic-igneous ⇒ uplift/erosion ⇒ schist ⇒ shale

2c. Solution of Geologic Interpretation Problem

1. Metamorphose schist
2. Uplift and erode to uncover the schist
3. Deposit **shale** on schist
4. Intrude mafic-igneous through schist and shale

collection of partial orders shows the geologist's solution at each stage of development). The same reasoning would indicate that the **mafic-igneous** also intruded through the **shale** (Figure 2b, step 2). Thus the **shale** and the **schist** were both in place before the **mafic-igneous** intruded through them. To determine in what order the **schist** and the **shale** appeared, the geologist would infer that, since sedimentary rocks are deposited from above onto the surface of the Earth, the **shale** (a sedimentary rock) must have been deposited on top of the **schist**, and hence is younger than the **schist** (Figure 2b, step 3). The geologist knows that the **schist** was created from existing rock by the process of metamorphism. However, metamorphism occurs to rocks buried deep in the Earth and deposition occurs on the surface, so somehow the **schist** must have gotten from the depths to the surface, in order for the shale to have been deposited upon it. The geologist might infer that a combination of the processes of *uplift* and *erosion*, neither of whose effects are reflected in the diagram of Figure 2a, would suffice to bring the **schist** to the surface (Figure 2b, step 4). The final inferred sequence of events is shown in Figure 2c.

3.2 Problem Solving Technique

The problem solving technique used in the example above consists of two basic phases [23]. In the first phase, we use a technique we call *scenario matching* to generate a sequence of geologic events that might explain how the cross-section came into existence. In the second phase, we use a technique we call *imagining* to test if the hypothesized sequence is correct. In addition, if the hypothesis is not correct, we debug the hypothesis using a technique we call *gap filling*.

3.2.1 Scenario Matching

Scenario matching is a way of inferring a sequence of geologic events by reasoning backwards in simple, one-step inferences from the effects of processes to their causes. A scenario is a pair consisting of a diagrammatic *pattern* and a sequence, called an *interpretation*, that could have caused that pattern. For example, in solving the example in Figure 2 we used the following scenario twice:

pattern	*interpretation*
<rock> | igneous | <rock>	**igneous** intruded through the **<rock>**

A *pattern* represents the local effects of a geologic process and typically involves the boundaries between two or three formations. An *interpretation* is a sequence of events that is a possible causal explanation for the pattern's occurrence. Each pattern may have several plausible interpretations.

By matching scenario patterns throughout the diagram and combining the local interpretations obtained from the matches, we generate sequences that purport to explain how the region was formed. However, these sequences might not be completely valid for two reasons. First, local consistency does not imply global consistency. Second, the evidence for the occurrence of some physical processes might no longer exist in the geologic record (as reflected in the diagram). For instance, there is no evidence in Figure 2a for the occurrence of the processes of uplift and erosion of the schist, because the erosion has removed whatever once covered the schist. To detect both types of inconsistencies, some form of global reasoning is needed.

3.2.2 Imagining

We are developing a new technique called *imagining* to detect inconsistent hypotheses. An outgrowth of the notion of *envisioning* [5], imagining takes as input an initial environment, a goal state that is the final environment (in this case, a diagram), and a sequence of processes. The imaginer simulates each of the processes in turn, producing a final environment that is compared with the goal environment to test for a qualitative match. In effect we are "imagining" what would happen if those processes had actually occurred, in order to check whether they could have produced the desired result.

Aside from the final qualitative match, the imaginer has three basic tasks to perform for each process in the sequence.

1. It determines whether a process is applicable in the current environment.

2. It determines quantitative values for the process parameters.

3. It simulates the process by redrawing the diagram to reflect the geologic changes induced by the process.

In the rest of this section we discuss these tasks.

For each process, the imaginer must determine if it can be applied to the environment produced by the simulation up to that point. For example, one process might indicate "erode shale to sea-level", but clearly this would be inapplicable if the top of the shale was currently below sea-level. If the imaginer cannot continue, it returns an explanation of the problem encountered. This explanation consists of the process that the imaginer could not simulate and the difference between the current state and the state that would be needed in order to simulate that process. In the above example, the difference reported would be that the shale is below sea-level, but should be above sea-level in order for the erosion to occur.

The sequence hypothesized by the scenario matcher does not indicate values for the parameters of the processes, such as the thickness of a deposition or the angle of an intrusion. In order to make tractable the problem of qualitatively matching the goal environment and the final environment produced by the simulation, the parameters used in the simulation of a process must closely match those parameters used in the actual geologic process. For example, in order to simulate "deposit shale on schist" the imaginer must have some indication of the thickness of the shale deposit. Thus, to do the imagining the system must be able to infer values for the parameters of the geologic processes being simulated.

The system uses measurements taken from the diagram, along with knowledge of geologic processes, to determine these parameters. Since each parameter represents some real-world quantity, we begin by determining the value of that quantity at some point. Usually this is accomplished by measuring the quantity in the diagram corresponding to that time. Then we need to compensate for any changes that occurred to the quantity between the time when it was measured and the time when the process occurred.

A simple example will illustrate this parameter determination process. Suppose we wish to find the thickness of the **schist** when it was originally deposited. In Figure 2a, which represents the final point in the sequence, we can measure the thickness of the **schist** deposit at that time (which turns out to be 300 meters). Since we also know that part of the original **schist** deposit had been eroded away earlier (in step 2, Figure 2c), we infer that the original thickness of the **schist** must have been greater than the measured thickness in the diagram. Since we cannot infer the exact amount of the erosion, the best we can do is to say that the original thickness was "greater than 300 meters". Reasoning in this fashion, we can establish ranges of values for the parameters of all the processes involved in creating the final environment, and thus can approximate in our simulation the effects of the actual geologic processes.

The actual simulation phase of the imaginer is accomplished by producing a sequence of diagrams, one for each process in the hypothesized sequence, to reflect the effects of our model of geologic processes. The use of diagrams is not crucial to the concept of imagining, but is useful in this case for two reasons. First, most geologic effects are spatial in nature, hence their changes are easier to represent in a diagram, which is a spatially organized representation. Second, an important check on the validity of the hypothesized sequence of events is to match the input diagram against the final diagram produced by the simulation. Diagrams are thus useful for describing the effects of the changes and for validating the hypothesized sequence of processes.

3.2.3 Gap Filling

If the imaginer detects a "gap" between the state needed for some process to occur and the actual state of the environment (as would have occurred if we had not inferred the presence of the uplift and erosion in Figure 2), we need to hypothesize some sequence of events to fill the gap. As described in Section 3.2.2, the imaginer indicates why it could not continue in terms of the difference between two states, and, from that, one can reason about which process or sequence of processes would have the effect of minimizing or eliminating that difference. This is means-end analysis [20] used in a restricted context.

4. REPRESENTING CHANGE IN OBJECTS

The remainder of this paper concentrates on the representations and reasoning necessary to do the imagining phase of solving the geologic interpretation problem. In the previous section, we saw that in order to *imagine* a sequence of events, we need to determine values for the parameters of the processes in the hypothesized sequence. Given these values, we can approximate the effects of the actual geologic processes, by reasoning about how objects change over time due to the effects of the processes. This, in turn, requires reasoning about the relationships between the value of an attribute before and after the change occurs.

We have developed a representation for physical objects that facilitates such reasoning. The objects are represented as frame-like structures ([3],[24]), organized into a type hierarchy. Each type of object has certain attributes associated with it, and possibly some associated constraints. For example, a **sedimentary deposit** has a **thickness** that is constrained to be positive.

In fact, there are two types of objects - *temporal* and *abstract*. Temporal objects correspond to real-world physical objects. They have an associated life-span and can be created or destroyed. Trees, people, and sedimentary deposits are temporal objects. Abstract objects are non-physical objects that always exist. Geometric planes, vertices, and numbers are abstract objects.

4.1 Histories

Since we want to represent the situation in which the attributes of objects can change over time, the value of an attribute is represented as a *history* (the term is adopted from [12]), rather than as a single value. A history is simply a totally ordered sequence of values over time. Each value in the history represents the value of the attribute over a particular temporal interval. For instance, the *thickness* of a sedimentary deposit consists of the sequence of all thickness values of that deposit over time.

Each distinct point in the history represents, by definition, an interval during which some change occurred to the object. Since we assume a "causal model" of the universe, that is, only physical processes can cause changes, each distinct point in the history is also associated with a process that caused the change. For example, if the thickness of a deposit decreases between time **t0** and **t1**, this change is represented in the "thickness" history by an interval which is associated with a geologic process (such as erosion) occurring from **t0** to **t1**.

4.1.1 The @ Operator

Since the attribute of an object is not a single value, but is a history (that is, a sequence of values), we need a way to select the value of a history at a particular point in time. We have defined the @ operator for this purpose.

To illustrate the use of the @ operator, suppose that S represents a sedimentary deposit. **S.thickness** is a history of values of the thickness of the deposit, and the referent of the expression **S.thickness@t0** is the thickness of S at time t0. If later S were partially eroded, then the thickness of S would change, and **S.thickness@t1** would not equal **S.thickness@t0** (assuming **t1** postdates the erosion process).

We have developed a formal notation that enables us to refer to the attributes of objects at a point in time. The BNF grammar for this notation is:

<temporal expr> :: = = <historical expr>@<time>

<historical expr> :: = = <object> | <historical expr>.<attribute>

<object> :: = = <temporal object> | <abstract object> | (<temporal expr>)

This notation is especially useful in dealing with more complex temporal expressions. For example, **S.top** is the history of the highest points of the sedimentary deposit S (Figure 3). **S.top@t0** refers to the highest point of the formation at time t0 (Figure 3a), and **S.top.height@t0** refers to the height of that point at time **t0**. If more deposition occurred between **t0** and **t1**, then the point referred to by the expression **S.top@t0** would not be the same point as the one referred to by the expression **S.top@t1** (Figure 3b). Note, however, that **S.top@t0** refers to a point that is still part of the deposit S at time **t1**, although it is no longer the top.

Fig. 3. Top of S Before and After Deposition

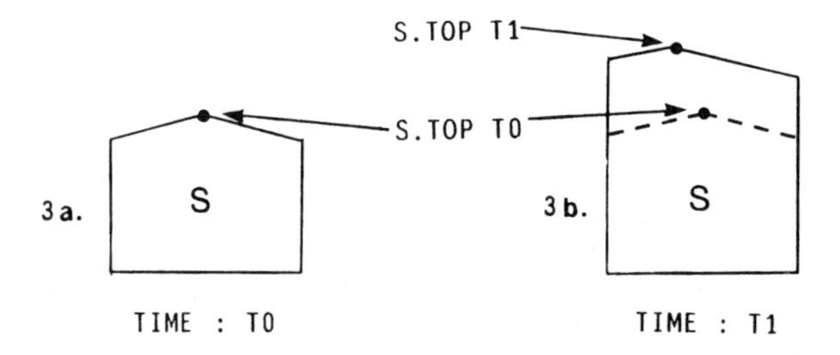

Thus it makes sense to talk about (**S.top@t0).height@t1**, that is, the height at time **t1** of the point that was the top of S at time t0. This could be different from **S.top.height@t0** if, for instance, uplift occurred between **t0** and **t1**.

Since temporal objects can be created and destroyed, it is useful to define the @ operator over temporal objects as well as over histories.[1] If **A** is a temporal object, we define the value of **A@t** to be **A** if **A** exists at time **t**, otherwise the value is ⊥. ⊥ (bottom) is a "null" object, and it is *strict* (that is, any function or operator applied to ⊥ returns ⊥).

In light of this, let us re-examine the interpretation of the expression **S.thickness@t0**. Since the referent of S might be ⊥ at t0, we need to "distribute" the @ operator through the expression to determine the value of the expression. The expression **S.thickness@t0** is in fact shorthand for **(S@t0).thickness@t0**. This is interpreted as follows: if **S** exists at **t0** then the value of the expression is the same as before; if **S** does not exist (e.g. it was "destroyed" by erosion or not yet deposited), then the referent of **S@t0** is ⊥ and the value of the whole expression is ⊥.

The general rule for expanding temporal expressions is to recursively replace occurrences of the form

<historical expr>.<attribute>@<time>

by the form

(<historical expr>@<time>).<attribute>@<time>

Thus the expression **S.top.height@t0** is shorthand for **((S@t0).top@t0).height@t0**, and **(S.top@t0).height@t1** is shorthand for **(((S@t0).top@t0)@t1).height@t1**.

4.1.2 Implementation of Histories

Histories are implemented as a time-line divided into two types of intervals - quiescent and dynamic. A quiescent interval indicates that nothing happened to the attribute during the interval, hence the value within the interval is constant. A dynamic interval indicates that some process induced a change during that interval to the attribute represented by the history. For reasons discussed in section 5.1, the value within a dynamic interval is defined to be *unknown*.

To determine the value of an attribute at a particular time, the @ operator searches the time-line of a history to find the interval that contains that time point and returns the value found there. If the time point falls outside of the extent of the history time-line, then the value ⊥ is returned.

1. For any abstract object **B**, **B@t** equals **B**.

4.2 Diagrams

Histories are useful for dealing with certain types of changes, essentially characterized as one-dimensional. For example, the fact that the height of a point in a formation will increase if the formation undergoes uplift is well described using histories. However, many of the effects of geologic processes are two- or three-dimensional in nature, such as the change in shape of a formation caused by erosion, or the change in which point is the "top of the formation" that occurs because the formation undergoes tilting. To facilitate reasoning about these types of changes, we have developed methods for representing, reasoning about, and manipulating diagrams.

In our system, a diagram represents a cross-section of the geologic environment, or more precisely, a 2-dimensional spatial abstraction of the environment at a particular point in time. By "spatial abstraction" we mean that diagrams represent only the geometric aspects, such as the size, shape and location of objects, and spatial relationships, such as *above* and *below*. In particular, there is no reference in the diagram to geology. In general, we have been careful to distinguish and separate the geologic representation from the geometric representation. They interact only through a small, simple, and clearly defined interface (section 4.3). This separation allows us to independently develop and reason about the two representations.

4.2.1 Diagrammatic Representation

A diagram consists of a collection of *vertices, edges*, and *faces*. Part relations, such as all the edges surrounding a face, or the end-points of an edge, are explicitly represented. Spatial relations, such as adjacency, above, or below, are easily determined using the diagram. In addition, many metric properties are easily measured, including the length of an edge, the location of a vertex, or the maximum width of a face.

To illustrate the use of diagrams, we present a typical cross-section in Figure 4. The shale *rock-unit*[2] is represented by the diagram faces S1, S2, S3 and S4. The granite rock-unit is represented by the faces G1 and G2. In addition, the fault *boundary*[3] is represented by the diagram edges b1, b2, b3, b4 and b5. We can easily determine which rock-units are adjacent to the

Fig. 4. Simple Diagram Cross-section

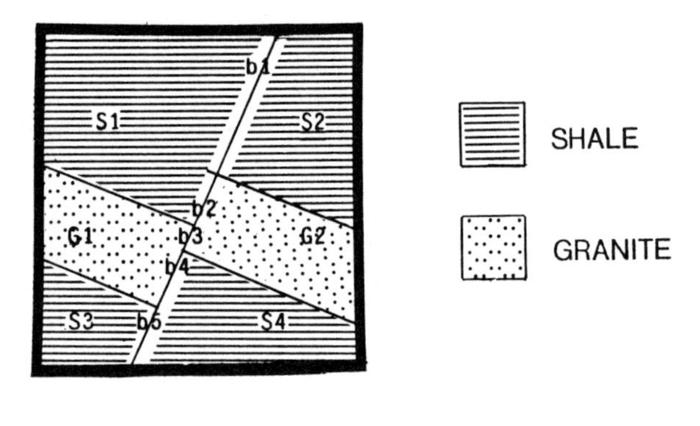

2. A *rock-unit* is a collection of one or more *primitive rock-units*. A *primitive rock-unit* is a single contiguous piece of homogeneous rock (for example, the rock represented by face S2 in Figure 4).

3. A *boundary* is the surface formed at the interface of two or more rock-units.

fault boundary by finding the faces adjacent to the edges b1 - b5 (that is, the faces S1, S2, S3, S4, G1 and G2) and determining which rock-units those faces represent (that is, the shale and granite). We can determine the orientation of the fault by averaging the angles of all the edges that represent the fault boundary.

Another use of diagrams is in representing the effects of changes. For the simulation phase of the imaginer, we need to represent how objects change over time. Since diagrams are a spatial abstraction of geologic objects, we can represent how those objects change spatially by manipulating the diagrams in accordance with our model of geologic processes. For example, deposition can be simulated by drawing the new formation in the diagram. This is illustrated in section 5.2.2, which describes the representation of processes used in conjunction with the diagram representation.

4.2.2 Implementation of Diagrams

Our implementation of diagrams is based on the wing-edge structure of [2], adapted to 2-dimensional diagrams.

The primitive objects in this representation are *vertices, edges,* and *faces*. A vertex is represented by (X,Y) coordinates and has a pointer to one of the edges surrounding it. A face has a pointer to one of the edges of its perimeter. An edge is represented as shown in Figure 5. Each edge has pointers to exactly two faces, two vertices, and four "wings" (that is, its connections to other edges). From these connections, we can easily compute such things as the perimeter of a face, the length of an edge, or the spatial relationship between two faces.

The wing-edge structure is well suited to our needs for three reasons. First, the primitive objects used in the representation -- vertices, edges, and faces -- have a natural correspondence with the primitive objects used in the geologic representation -- geologic points, boundaries, and rock-units. Second, the representation allows easy access to the spatial relationships (such as "above") and metric properties (such as "angle of slope") that we need to solve the geologic interpretation problem. Third, the wing-edge representation was designed to facilitate manipulation of the geometric structures, which makes it easy to do the diagrammatic simulation of geologic processes. In particular, local changes to a diagram (such as adding or deleting edges or faces) can be accomplished with only local changes to the wing-edge structures.

Fig. 5. The wing-edge representation of an edge

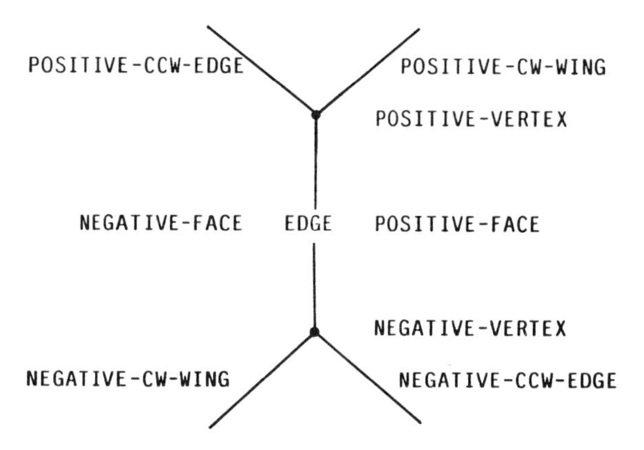

There are only a few types manipulations that we need to perform on diagrams in order to achieve a simulation of the set of geologic processes we currently handle.[4] These manipulations are adding and deleting edges, faces and points; rotating and translating the entire diagram; splitting one diagram into two diagrams; and joining two diagrams into one. The relatively small number of primitive operations needed to simulate a large class of geologic processes suggest that diagrams are an appropriate form of representation, and that our vocabulary of primitive operations is reasonably well chosen.

4.3 Geology-Geometry Interface

As mentioned, the interface between geologic and diagrammatic representations is relatively simple. Basically, it consists of a one-to-one mapping between primitive elements in each domain. A diagram corresponds to the world at a particular instant of geologic time. Each edge in the diagram corresponds to a single geologic boundary. Each face corresponds to a single rock-unit. Each vertex corresponds to a geologic point, such as the top of a rock-unit. Similarly, a collection of primitive rock-units (that is, a rock-unit) maps into a collection of faces. So, for example, the collection of faces S2, G2, and S4 in Figure 4 corresponds to the rock-unit which is known as the *up-thrown block* of the fault.

In addition, there are several functions mapping the spatial and metric relations in the diagram to the corresponding relations in the geologic world. For instance, we can determine if one primitive rock-unit is above another by seeing if the corresponding faces in the diagram are above one another. Similarly, we can determine the orientation of a boundary by measuring the angle of slope of the corresponding edge.[5]

5. PROCESSES

In this paper, our chief interest in physical objects is reasoning about how they change. Since processes are the cause of change, our representation of processes is focused on describing them in terms of the changes they produce.

In the previous section, we discussed the two representations for geologic objects we developed to facilitate reasoning about different types of changes. Similarly, we have developed two corresponding representations for processes, one suited to dealing with the history-based representation and the other suited to the diagram-based representation.

5.1 Level of Representation

Both types of process representation make use of an "end-point" model of geologic processes. This model assumes that we know the values of the affected attributes only at the beginning and end of a process, and nothing can be assumed about the intermediate values. For example, the composition of a rock-unit is known before and after metamorphism, but the exact composition during the process is unknown. Using an end-point model means that, in

4. They are deposition, erosion, uplift, subsidence, intrusion, faulting and metamorphism.

5. The definitions easily generalize for geologic objects corresponding to collections of faces or edges.

general, we cannot deal with simultaneous interacting processes, that is, processes that simultaneously affect the same attribute of the same object.[6]

Since most occurrences of geologic processes are non-interacting (although they may be simultaneous), the use of the end-point model has proven sufficient in solving most geologic interpretation problems. In addition to being sufficient, the end-point model is also appropriate for two reasons. First, there are many cases where we do not know what occurs during a complex geologic process, as in metamorphism where the composition of a rock-unit during the process is not well understood. Hence, in many cases the end-point model is the best that we can do. Second, even in those cases where we have a fairly accurate model of a process (as in uplift), representing it in more detail (see, for example, the representation of processes in [9]) would lead to a computationally infeasible solution in our case.

5.2 Representation of Processes

In this section, we examine our two representations of process. One representation is designed for use with the history-based representation of objects, the other is designed for use with the diagram-based representation.

5.2.1 Process Representation for Changing Histories

Figure 6 shows the deposition process represented in a form useful for reasoning about changes in the history-based representation of objects.

1. The INTERVAL field describes the temporal interval during which the process is active. A temporal interval I is simply an interval of time represented by its end points I_{start}) and I_{end}.

2. PRECONDITIONS is a set of conditions which must be true in order for the process to occur.

3. PARAMETERS is a list of parameters that indicate the magnitude of the effects of the process. The imaginer must determine values for these quantities in order to simulate the process.

4. AFFECTED is a list of the objects that exist at the time the process began and which are changed in some way by the process.

5. CREATED is a list of objects that are created by the process.

6. The EFFECTS field is a set of statements that describe how the process changes the various attributes of the affected and created objects.

7. RELATIONS is a set of assertions that are constrained to hold as a result of the occurrence of the process.[7]

Fig. 6. Description of the Deposition Process

```
DEPOSITION
  INTERVAL        I : temporal-interval
  PRECONDITIONS   {(< SURFACE.bottom.height@i_start SEA-LEVEL)}
  PARAMETERS      DLEVEL : positive-real, DCOMPOSITION : sedimentary-rock
  AFFECTED        SURFACE
  CREATED         A : sedimentary, BA : boundary
  EFFECTS         {(change = A.thickness DLEVEL I DEPOSITION)
                   (change = A.bedding-plane.y-angle 0.0 I DEPOSITION)
                   (change = BA.side-1 {a} I DEPOSITION)
                   (change = BA.side-2 C I DEPOSITION)
                   (change = A.composition DCOMPOSITION I DEPOSITION)
                   (change = A.top (dfn DLEVEL SURFACE@i_start) I DEPOSITION)
                   (change = A.bottom SURFACE.bottom@i_start I DEPOSITION)
                   (change = SURFACE.bottom (dfn DLEVEL SURFACE@i_start) I DEPOSITION)}
  RELATIONS       {(= SURFACE.bottom.height@i_end (+ DLEVEL SURFACE.bottom.height@i_start))
                   (equiv A.bedding-plane.y-angle A.orientation i_end)
                   (< A.top.height@i_end SEA-LEVEL)
                   (= C {r : rock-unit | (exists? r@i_start) ->
                                            (and (on-surface r i_start)
                                                 (< r.bottom.height@i_start
                                                    (+ DLEVEL SURFACE.bottom.height@i_start)))))})
                    (equiv A.orientation BA.orientation i_end)}
```

6. However, we can deal with simultaneous, non-interacting processes.

7. (EQUIV H1 H2 T) means that after time T, histories H1 and H2 are equivalent, that is, their values at all points in time are identical.

For purposes of reasoning about change, the field of primary interest here is the list of EFFECTS. The general form is (CHANGE <type> <attr> <change> <interval> <cause>). ATTR is an expression describing the attribute changed by the process, INTERVAL is when the change occurred, and CAUSE is the process that causes the change. TYPE and CHANGE jointly describe how the old and new values of the attribute are related. TYPE can be " = ", in which case the new value after the process equals CHANGE. For example, the form

(CHANGE = A·thickness DLEVEL I DEPOSITION)

represents the fact that after the deposition process, the thickness of the created sedimentary deposit equals the value of the parameter DLEVEL. TYPE can also be an arithmetic operator (+, -, *, /), in which case the new value is found by applying the operator to the value of the attribute at the start of the process and the CHANGE. For example, one effect of the uplift process is represented by

(CHANGE + A·height UPLIFT-AMOUNT I UPLIFT),

which indicates that the height of a rock-unit after the uplift process equals its height before the uplift plus the amount of the uplift. Finally TYPE can be "function" in which case the CHANGE is any function that is applied to the old value.[8]

5.2.2 Process Representation for Changing Diagrams

The representations of processes used with the diagram-based representation are simply end-point style algorithms for manipulating the diagrams. The simulation of deposition, for example, is described as shown in Figure 7. Figure 8 shows the effects of running that algorithm.

Fig. 7. An Algorithm for Simulating Deposition in a Diagram

1. Find the lowest end-point of all the edges that represent the surface of the Earth.

2. Draw a horizontal line "DLEVEL" above that.

3. Erase all parts of the line that cut across a face corresponding to a rock-unit.

4. All other newly created faces below the line are part of the newly created sedimentary rock unit.

Note that although the diagram-based representation itself makes no reference to geology, the diagram manipulation algorithms need to make reference not only to properties of geometric objects in the diagram, but also to correspondences between geometry and geology. For example, in Figure 7 a property of geometric objects is "the lowest point of all the edges", and a correspondence is "all the edges that represent the surface of the Earth".

One of the main differences between the simulations produced by the two process representations is that the history-based processes produce a qualitative simulation and the diagram-based processes produce a metric simulation. That is, for the diagram-based simulation, all the process parameters must be assigned exact values. This is due to the metric nature of diagrams. For example, a point in a diagram must be placed in a specific coordinate location -- it cannot have a "fuzzy" position in the diagram. Thus, the system can do the history-based simulation when given the sequence of events, but it needs to determine the process parameters before it can do the diagram-based simulation.

5.2.3 Implementation of Processes

We have implemented a program that instantiates a process at a particular point in time by making changes to the history-based representations. The input to the program is an assertional representation of a process, of the sort shown in Figure 6, along with some additional information which specifies the values of some of the expressions in the process description. Examples of this additional information are "DLEVEL = 10 meters", and "BA·side-2@I_{end} = {BEDROCK}" (that is, "bedrock" lies on one side of the newly created depositional boundary).

To instantiate a process, the system carries out three steps. First, it creates a representation for each member of the list of "created" objects. Second, it updates the histories of the affected and created objects, according to the "change" statements in the EFFECTS field, by inserting a dynamic interval into the appropriate place on the history time-line. It does this by splitting a quiescent interval into two pieces and inserting the dynamic interval in between. Third, the program asserts that all of the statements in the RELATIONS field hold.

Fig. 8. (Diagram numbers correspond to the steps in Figure 7)

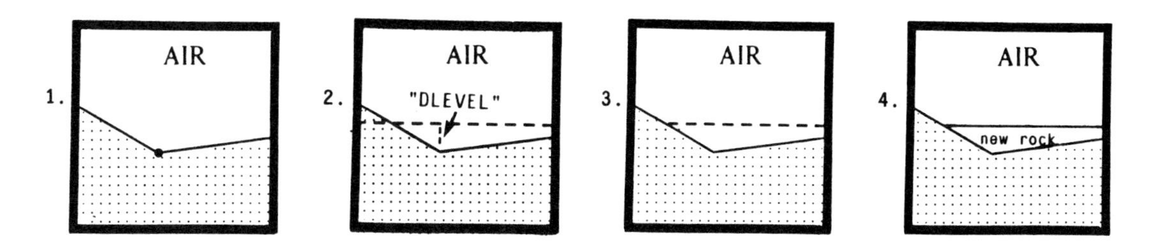

8. The type "function" is the most basic type; all other types can be defined in terms of it. For example, the "+" type with change Q is equivalent to the "function" type with change (LAMBDA (X) (+ X Q)).

For example, to instantiate the deposition process in Figure 6, the system carries out the following:

1. It creates the new rock-unit A (the sedimentary deposit), and the new boundary BA (the boundary between A and whatever it was deposited upon).

2. It updates the appropriate histories for all the CHANGE statements. For example, it updates the (newly-created) history of the thickness of A, inserting the dynamic interval from $I_{start)}$ to I_{end}. Prior to time $I_{start)}$, the thickness is 0, between $I_{start)}$ and I_{end}, the thickness is defined to be *unknown*, and after I_{end} the thickness is "DLEVEL".

3. It asserts that all of the RELATIONS shown in Figure 6 now hold.

Since the representation of processes used with diagrams is algorithmic rather than assertional, the diagram-based representation of a process is implemented simply as a LISP function. These functions access the diagrams directly, and indirectly access the history-based representation through the geologic-geometric interface (see section 4.3). The simulation produces a sequence of diagrams by copying the current diagram, associating it with the end point of the process interval, and modifying the copy to account for the effects of the geologic processes.

6. IMAGINING -- AN EXAMPLE

In this section, we present an example of the imagining process, showing how the representations we have developed enable us to do the imagining. The input to the imaginer is shown in Figure 9 -- a cross-section representing the current geologic environment and a sequence of events (produced by the scenario matcher) hypothesized to have caused that environment.

The first step in doing the imagining is to instantiate each of the processes in the sequence, using the history-based representation. This is accomplished as described in section 5.2.3, simulating each process by creating objects and inserting dynamic intervals into their histories to represent the changes. This step produces sequences of changes to the attributes that allows us to reason about the cumulative effects of the changes on the attributes. For example, after the instantiation step the thickness history of the sandstone would contain dynamic intervals due to the initial deposition (step 1, Figure 9b), the intrusion of granite (step 2), and erosion (step 4). However, at this stage the actual value of the thickness at any point in time is not known, beyond the fact that it is positive.

In order to do the next step, the diagram-based simulation, we need to determine the parameters used in each process. As an example, we consider how to determine the parameter DLEVEL, the amount of deposition (see Figure 6), for the deposition of the sandstone in step 1.

Parameter determination requires two steps. First, we measure the current value in the diagram; second, we take into account the changes that have occurred to the parameter. For example, the system knows that the thickness of the sandstone (a sedimentary formation) corresponds to the maximum width of the corresponding diagram faces, measured perpendicular to the "y-angle" of the bedding-plane. From the instantiation of the deposition process, the system knows that at the time of deposition the y-angle was 0° (see Figure 6). However, by examining the history **sandstone.bedding-plane.y-angle** the system knows that there was a change in the y-angle of 13°, due to the tilt in step 5. Thus, the system measures the maximum width, perpendicular to 13°, of the sandstone faces (Figure 9a) and determines that the thickness of the sandstone in the current environment is 500 meters.

Fig. 9. A Geologic Interpretation Problem and Hypothesized Solution

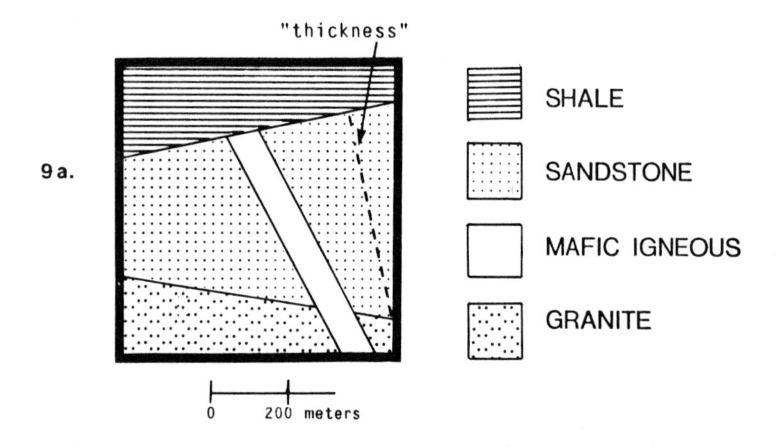

9b.

1. Deposit Sandstone on Bedrock
2. Intrude Granite into Sandstone
3. Intrude Mafic-Igneous through Granite and Sandstone
4. Erode Sandstone and Mafic-Igneous
5. Tilt by 13°
6. Deposit Shale on Sandstone and Mafic-Igneous

Next, the system examines the thickness history and determines that the changes due to the granite intrusion (step 2) and erosion (step 4) must be accounted for. The model of geology we are using states that the thickness of a formation being intruded into is decreased by the amount of the thickness of the intruding formation. Thus, to account for the change in thickness due to step 2, the system needs to determine the thickness of the granite at the time of intrusion. It does this by measuring the width of the faces corresponding to the granite formation. Using the same reasoning as above, the system determines that it also must measure this width perpendicular to 13°. The measured thickness is "greater than 200 meters" ("greater than" because some part of the granite formation continues outside the boundary of the diagram), so the current estimate for the thickness of the sandstone is "greater than 700 meters". Finally, the system knows that the thickness was decreased by the amount of erosion in step 4. The system tries to determine an exact value for the amount of erosion, but this information is not determinable from the diagram. The best the system can do is to determine that the amount of erosion was greater than zero. Thus, the estimate of the initial amount of deposition is "greater than 700 meters".

Since the value of the other parameter of the deposition process, DCOMPOSITION, is given explicitly to be sandstone (step 1, Figure 9b), the imaginer can now simulate the deposition process in the diagram. The initial diagram the imaginer starts with is blank, which represents that just "bedrock" exists at this point. Next, the imaginer chooses an exact value for DLEVEL within the allowable range of "greater than 700 meters" (we have chosen 800 meters), and uses the algorithm from Figure 7 to create the new diagram shown in Figure 10, diagram 1.

An exact value is needed because the simulation is done using diagrams, which are metric in nature. For example, it is impossible to actually draw a horizontal line in the range "somewhat greater than 700 meters" because a line drawn in the diagram defines an *exact* equation for that line. So, to draw a line, exact parameters must be chosen.

The question remains -- which value do we choose from within the range? Recall that the purpose of the parameter determination is to choose values which approximate the actual geologic parameters used, in order to make tractable the qualitative matching of the input diagram and the final result of the simulation. Is the matching process affected by our choice of value from within the allowable range?

The answer is -- no, it does not matter; choosing any arbitrary value within the range will eventually lead to the same final diagram. This can be seen by recalling that the reason a

Fig. 10. Diagrammatic Simulation of Hypothesized Sequence

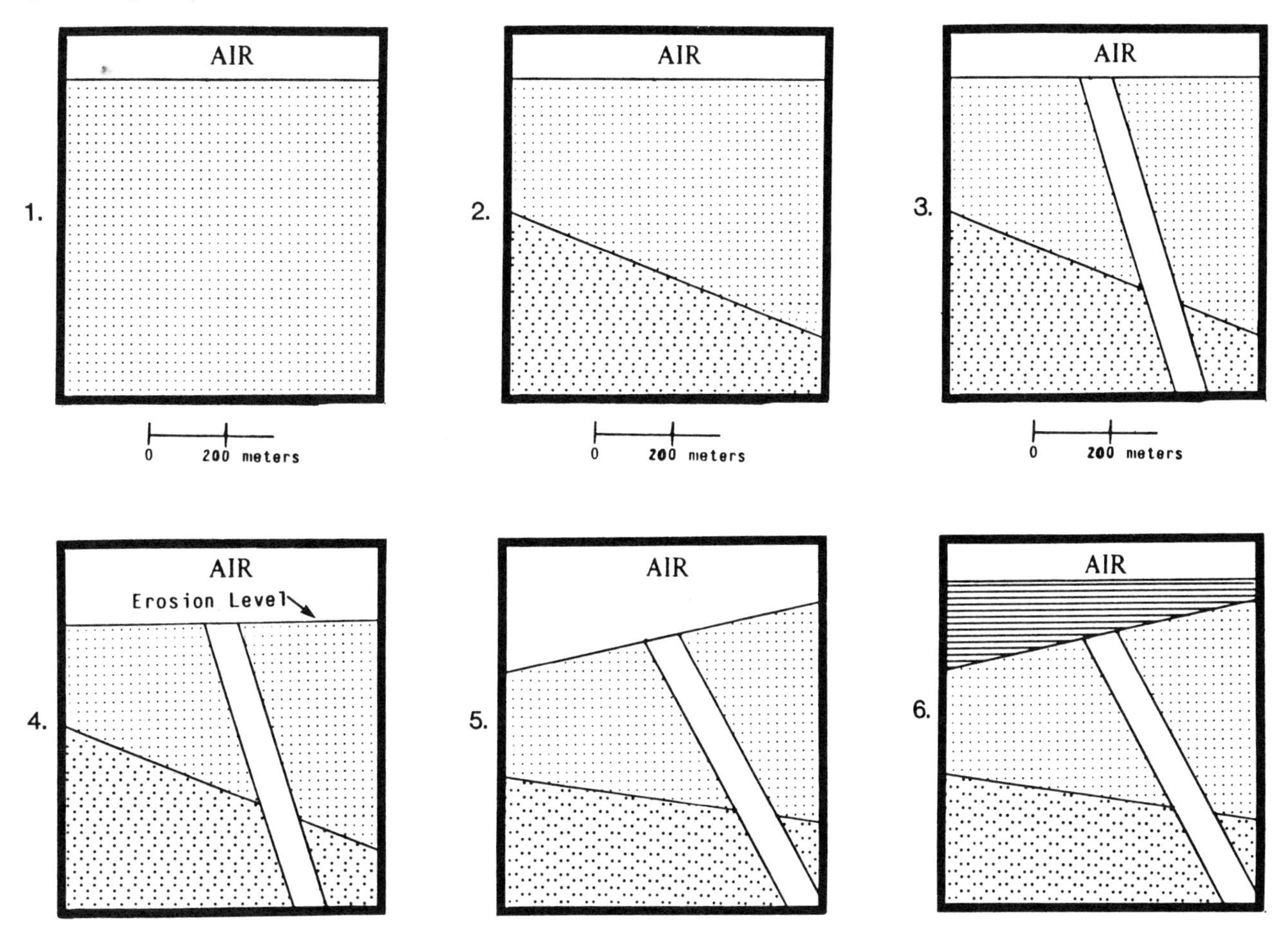

measurable attribute (such as the thickness of a rock-unit) is known only within a range is that, although the measurement from the input diagram is exact, the magnitude of some subsequent change to that attribute is known only within a range. By choosing an exact value for the parameter, in order to do the simulation of one step, we also determine an exact value for the magnitude of the subsequent change. When the process which caused that change is later simulated, the change value will already be determined exactly.

For example, by choosing DLEVEL to be 800 meters, we constrain the change due to the granite intrusion and the erosion to be exactly 300 meters (since the measured thickness in the input diagram was 500 meters). When the intrusion of granite is simulated (step 2, Figure 9b), the amount of granite is constrained to be between 200 and 300 meters (non-inclusively). If we (arbitrarily) choose the amount of granite to be 250 meters, we automatically constrain the amount of erosion (step 4) to be 50 meters. After the erosion is simulated, the thickness of the sandstone in the final diagram will be 500 meters, the same as in the input diagram. Thus, the cumulative effect on a measurable attribute due to the diagram-based simulations will be the same as the value measured in the input diagram, no matter what value is actually chosen.

This technique of determining the process parameters and then simulating the process to produce a new diagram continues for each step in the sequence. The results of those simulations are shown in Figure 10, diagrams 1-6. Finally, the end result of the simulation (Figure 10, diagram 6) is compared with the input diagram (Figure 9a) to check that they do, in fact, qualitatively match. The system would conclude that the sequence in Figure **9b.** is a valid hypothesis for describing how the geologic environment arose.

The problem of matching the diagrams has not yet been adequately explored. However, the basic algorithm has two steps. First, we check existence: each rock-unit or boundary in the input diagram should have a corresponding entity in the simulated diagram. Second, we check adjacency: the rock-units adjacent to each rock-unit or boundary in the input diagram should correspond to the rock-units adjacent to the corresponding entity in the simulated diagram.

7. RELATED WORK

Several of our ideas on representing physical objects have been influenced by the work on naive physics of Hayes ([12][13]). Specifically, Hayes suggested the use of *histories* [13] as a technique for reasoning about temporal knowledge. Also, we have followed Hayes' recommendation [12] to develop separate *clusters* of related pieces of knowledge, and to keep the interaction among clusters to a minimum. We have developed the beginnings of two clusters -- a geologic and a geometric cluster -- and have kept the interface between them relatively small and well defined.

The @ operator, designed as a temporal selector to be used in conjunction with histories, is adopted from work on tense logic (see, for example, [18]). Other researchers concerned with representing change (such as [9],[19], and [22]) have developed operators similar in function to the @ operator. Our @ operator differs from their selector functions in being defined over temporal objects as well as over sequences of values.

Several researchers have investigated using diagrams to aid in problem solving (e.g. [8][10][11]), and many different diagrammatic representations have been developed (e.g. [2][8][10][11][16]). Our approach to representing diagrams is similar to [8] and [11] in that the primitive objects in the diagram vocabulary closely correspond to the primitive objects in the domain being represented. This correspondence, along with the spatial organization of the representation, is a source of the power in reasoning with diagrams.

Much previous and current work is focused on representing physical processes (e.g. [7][9][14][19][21]). Our use of an end-point model is similar to the approach of [7] and [21]. However, unlike those models, we maintain a record of all changes (by using histories) instead of deleting old values from a data-base, and we maintain relationships between the values at the start and end of a change. This facilitates reasoning about the cumulative effect of changes to an attribute, which we saw in Section 6 was necessary to determine the value of an attribute.

The representations of [9],[14], and [19] can handle interacting processes, although the work in [14] has limitations in that it cannot automatically determine where the interactions occur -- the interactions must be indicated within a process definition. We believe that reasoning about interacting processes will be needed to solve some geologic interpretation problems, and this will no doubt require models more detailed than ours are currently. However, a model of intermediate detail can be built using the current process representation. This might take the form of knowledge about how values change. For example, we might specify that the height of a rock-unit during uplift increases monotonically. This seems to be the approach adopted by [19], which makes the assumption that all changes are monotonic and continuous.

One of the capabilities present in our system, but lacking in some other representations that can handle interacting processes (e.g. [9][14]), is our ability to deal with change to non-numeric attributes (the representation described in [19] can also deal with such changes). We can, for example, reason about the change to the composition of a rock-unit, as well as about changes to its height.

Recently, there has been some discussion over whether points or intervals are the primitive units of time [1][19][25]. We take a middle ground between the two arguments (a similar position is advocated by [1]) -- time points are really intervals that, at higher levels of abstraction, have no duration and thus *behave* as points. From lower levels of abstraction they appear as intervals with a real duration. Thus, for geologic events it makes sense to talk about **S-thickness**@1982, where 1982 is viewed as a single point in time with respect to geologic time, even though 1982 is clearly an interval with respect to common human time scales.

8. CONCLUSIONS

The research presented in this paper has been motivated by a desire to use the technique of *imagining* as part of solving the geologic interpretation problem. Imagining simulates a sequence of events by manipulating both histories and diagrams. We have found the explicit representation of changes to objects useful in doing imagining. We reason about these changes in determining parameter values from an input diagram in order to do the diagrammatic simulation.

We have developed two representations of objects to facilitate reasoning about such changes. The first representation is *history* based. This representation is frame-like with histories as the value of attributes. We designed the @ operator, which ranges over temporal objects and histories, and which returns the value of the object or history at a particular point in time. We also have implemented an efficient representation for histories to facilitate using the @ operator and adding changes to an object. The second representation is *diagram* based. It incorporates a diagram system that facilitates spatial reasoning, both in accessing and changing spatial attributes of geologic objects.

Finally, we have presented two representations of processes. Each representation is geared to one of the two object representations. These process representations facilitate changing the history- and diagram-based representations in order to simulate and reason about the effects of geologic processes on the real world.

ACKNOWLEDGMENTS

We would like to thank Ken Forbus and Chuck Rich for their valuable suggestions and comments.

REFERENCES

[1] Allen, James F - An Interval-Based Representation of Temporal Knowledge, IJCAI 7, August 1981, p. 221.

[2] Baumgart, Bruce - "Geometric Modelling for Computer Vision," Stanford AIM 249, October 1974.

[3] Bobrow, D. G; Winograd, T - An Overview of KRL, a Knowledge Representation Language, Cognitive Science, vol. 1, no. 3, 1977.

[4] Davis, Randall - "Expert Systems: where are we and where do we go from here," **AAAI Magazine**, Summer 1982.

[5] DeKleer, Johan - "Qualitative and Quantitative Knowledge in Classical Mechanics," MIT AI-TR-352, 1975.

[6] Erman, Lee D; Hayes-Roth, Frederick; Lesser, Victor R; Reddy, Raj D - The Hearsay-II Speech-Understanding System: Integrating Knowledge to Resolve Uncertainty, Computing Surveys, Vol. 12, No. 2, June 1980.

[7] Fikes, Richard E; Nilsson, Nils J - STRIPS: A New Approach to the Application of Theorem Proving to Problem Solving, Artificial Intelligence, Vol. 2, pp. 189-208, 1971.

[8] Forbus, Kenneth D - A Study of Qualitative and Geometric Knowledge in Reasoning about Motion, MIT-AI-TR-615, February 1981.

[9] Forbus, Kenneth D - Qualitative Process Theory, MIT-AIM-664, February 1982.

[10] Funt, Brian V - Problem-Solving with Diagrammatic Representations, Artificial Intelligence 13, 1980.

[11] Gelernter, H - Realization of a Geometry Theorem Proving Machine, in *Computers and Thought*, eds. Feigenbaum and Feldman, 1963.

[12] Hayes, Patrick J. - The Naive Physics Manifesto, in *Expert Systems in the Micro-Electronic Age*, ed. D. Michie, Edinburgh University Press, May 1979.

[13] Hayes, Patrick J. - Naive Physics I: Ontology for Liquids, University of Essex, August 1978.

[14] Hendrix, Gary - Modeling Simultaneous Actions and Continuous Processes, Artificial Intelligence 4, 1973, p. 145.

[15] Hillis, William Daniel - The Connection Machine (Computer Architecture for the New Wave), MIT-AIM-646, September 1981.

[16] Hunter, Gregory M.; Steiglitz, Kenneth - Operations on Images Using Quad Trees, IEEE Transactions on Pattern Analysis and Machine Intelligence, Vol. PAMI-1, No. 2, April 1979.

[17] Kahn, Kenneth M - Mechanization of Temporal Knowledge, MAC-TR-155, September 1975.

[18] McArthur, Robert P - Tense Logic, D.Reidel Publishing Co, 1976.

[19] McDermott, Drew - A Temporal Logic for Reasoning About Processes and Plans, Cognitive Science 6, pp. 101-155, 1982.

[20] Newell, Allen and Simon, H.A. - "GPS, A Program that Simulates Human Thought," in *Computers and Thought*, eds. Feigenbaum and Feldman, 1963.

[21] Rieger, Chuck; Grinberg, Milt - The Declarative Representation and Procedural Simulation of Causality in Physical Mechanisms, IJCAI 6, 1979, p. 250.

[22] Shapiro, Daniel G - A Proposal for Sniffer: a System that Understands Bugs, MIT-WP-202, July 1980.

[23] Simmons, Reid G - Spatial and Temporal Reasoning in Geologic Map Interpretation, Proceedings of AAAI-82, August 1982, Pittsburgh, PA.

[24] Stefik, Mark - Planning with Constraints (MOLGEN: Part 1), Artificial Intelligence, Vol. 16, pp. 111-140, 1981.

[25] Vilain, Marc - A System for Reasoning About Time, AAAI 1982, Pittsburgh PA, August 1982.

AUTHOR INDEX

T5-BZR-478

INTEGRITY, CIVILITY, INGENUITY:
A REFLECTION OF GEORGE WASHINGTON

The Making of the
FORD ORIENTATION CENTER
and the
DONALD W. REYNOLDS MUSEUM AND EDUCATION CENTER
at
MOUNT VERNON

CREO PRESS
BALTIMORE, MARYLAND

For Kids
Audio Tours

FORD MOTOR
THE ANNENBERG FOUNDATION

To Keep Him First
DONALD W. REYNOLDS FOUNDATION • ROBERT H. AND CLARICE SMITH
THE RICHARD AND HELEN DEVOS FOUNDATION • J. HAP AND GEREN FAUTH • GAY HART GAINES
THE RICHARD AND HELEN DEVOS FOUNDATION • THE MARY HILLMAN JENNINGS FOUNDATION
VIRA I. HEINZ ENDOWMENT • ELIZABETH AND STANLEY DEFOREST SCOTT
FRANK PETERS MOORE • ROBERTSON FOUNDATION • A. ALFRED TAUBMAN
THE GORDON V. AND HELEN C. SMITH FOUNDATION • THE CHARLOTTE AND WALTER KOHLER CHARITABLE TRUST
WILLIAM R. KENAN, JR. CHARITABLE TRUST • THE MARY MORTON PARSONS FOUNDATION
MARS FOUNDATION • RICHARD KING MELLON FOUNDATION • DR. AND MRS. ROBERT C. SEAMANS, JR.
SARAH SCAIFE FOUNDATION INCORPORATED
SKAGGS AND MARY C. SKAGGS FOUNDATION

Title page:
Copper detail on the Ford Orientation Center entry canopy.

Following title page:
Ford Orientation Center lobby.

Overleaf:
Donald W. Reynolds Museum and Education Center lobby.

Published by Creo Press
Baltimore, Maryland

Library of Congress Control Number: 2006935056

ISBN: 0-9789779-3-9

ISBN-13: 978-0-9789779-3-1

First printing: November 2007

Designed by Kyle Isfalt, Baltimore, Maryland

Manuscript edited by Phil Freshman, St. Louis Park, Minnesota

Printed by Friesens, Canada

CONTENTS

Foreword

Roger K. Lewis, FAIA

You are reading a book, replete with narrative, photographs, drawings, diagrams, and detailed captions, which easily could have been another slick architectural volume, one to reside comfortably on the most elegant of coffee tables. Happily, that is not the case. Instead it is a documentary in print tracing the arduous, stimulating process by which new architecture on historic grounds is envisioned and realized, about what it takes to bring such buildings to life and meaningfully reshape a fragile, significant landscape. But rather than being a history book about George Washington, this is an account of how his tangible legacy, Mount Vernon, has been reconsidered and constructively transformed through creative design.

As the subtitle clearly states, this book is about "making." It focuses on ideas and inspiration, people and organizations, opportunities and constraints, and means and methods for generating and implementing an ambitious design. Finally, it presents and explains the tangible culmination of this process: the new Ford Orientation Center and Donald W. Reynolds Museum and Education Center at the first American president's home and property overlooking the Potomac River, eight miles south of Alexandria, Virginia.

For me this book strikes an especially resonant chord. Its holistic recounting of the history of a particular design, as opposed simply to presenting a design outcome, recalls an exhibition I organized and installed in 1985 at the University of Maryland. The goal of the exhibition, whose title was *IDEAS—IDEAL—DEAL—REAL*, was to show how architectural projects evolve in actual practice, to reveal the complex, sometimes messy, always challenging design process. The title was a poetic shorthand for that process: incipient "ideas" lead to "idealized" concepts; functional, technical, regulatory, and economic forces then exert pressure on concepts until a single design emerges and is embraced by all, consummating the architectural "deal"; and finally, the design becomes "real" when built. For the exhibition, six accomplished architects submitted design-study documents—program data, napkin doodles, sketchbook diagrams, preliminary models and renderings, and even working drawings—of one of their projects, including a few images of the finished building. Thus the graphically displayed history of each design's evolution allowed visitors, especially students, to better see, understand, and appreciate why and how architecture is made.

Mount Vernon's design history is covered here in considerably greater breadth, depth, and detail, since far more of the story can be presented in a book than can be displayed on a gallery wall. Consequently, readers first see what it took to save Mount Vernon, an effort begun in the early 1850s. They then learn how the Mount Vernon Ladies' Association undertook heroic initiatives not only to gain control of, preserve, protect, and maintain the estate but also to document fully the exceptional life and character of George Washington, helping sustain his legacy through research, interpretation, and public education. As the authors justifiably observe in the Introduction, "Mount Vernon stands as a model of dynamic stewardship."

In treating aspects of the latter-day Mount Vernon story directly related to designing 71,000 square feet of new

construction, the book covers matters as diverse as site archaeology, financing and capital campaigns, functional programming, and master planning. It also addresses the need of the design team at GWWO, Inc./Architects to grasp what it termed the "spirit of George Washington" as well as to discover "a single, unifying concept that would link the master plan with the history of the Mount Vernon Estate." This required gaining a profound understanding of the story and form of the entire 500-acre landscape: topography, vegetation, microclimate, landscape details and materials, views, approaches and pathways, existing structures and their various architectural vocabularies. It meant discovering how to reconcile 18th-century traditions and techniques with 21st-century requirements and aspirations. And it meant choosing guiding aesthetic principles and an architectural language that balanced and synthesized the competing forces of client demands and expectations, functional needs, state-of-the-art technology, site problems and opportunities, cultural context, and budget limitations.

The outcome of that lengthy process comprises this book's second half, which details how the two new buildings occupy their site, what they look like, and what happens inside and around them. A topographically complex design approach—placing the buildings partly underground—yielded an unconventional but didactically logical journey for visitors through hallways, galleries, and theaters. On exhibit are sculptures, paintings and portraits, models, photographs, films, documents, various artifacts, succinct explanatory descriptions, and pithy quotations describing Washington's life, times, and accomplishments. Visitors are offered a well-organized, easily understood history lesson about 18th-century America and its "indispensable" founding father. This inventive design strategy not only fulfills the objectives of the orientation and education program, but it also responds to the Mount Vernon Ladies' Association's strong desire to ensure preservation of the historic, pastoral setting of the Mansion.

However, history-minded readers of this book may be surprised to discover that the architectural language of

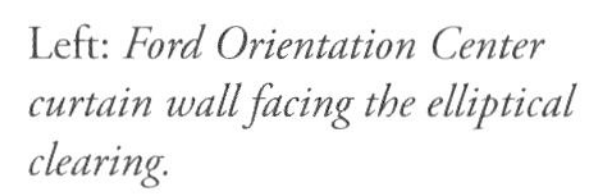

Left: *Ford Orientation Center curtain wall facing the elliptical clearing.*

the Ford Orientation Center and Donald W. Reynolds Museum and Education Center is decidedly non-historicist. The new structures are devoid of 18th-century design motifs, neoclassical riffs, and allusions to colonial architecture. There are no Doric, Ionic, or Corinthian columns, no Greek or Roman entablatures and pediments, no Palladian windows, no dentils, cornices, arches, or keystones. A mix of traditional and contemporary materials, assembled with consistently modern detailing, fleshes out the reinforced concrete and steel structural skeletons. Building facades are made of glass, metal, and red brick, and interiors are finished with Spanish limestone, plaster, hardwoods, monochromatic ceiling and wall fabrics, carpeting, and tile—all deployed appropriately. This is not in-your-face modernism, not a modernism of bombast, exotic geometry, or architectural hyperbole that so often competes with and detracts from historic-site renovations. The Mount Vernon Ladies' Association and their design team wisely decided that the new buildings should be elegant and memorable but also aesthetically comfortable for both exhibits and people.

For more than 200 years, the Mount Vernon Estate has taught an architectural lesson: how a building, simple in form and function, can take possession of and dominate its site while commanding spectacular views extending far beyond the site. In contrast to George Washington's Mansion, the new architecture defers to rather than dominates its site, merging with the land while still achieving compelling aesthetic character and presence. This book adds a new and very important lesson: It convincingly demonstrates that visually rich, well-functioning, contemporary architecture can be contextually harmonious without depending on the importation of trendy aesthetic formulas or ideologies, nor on the replication or mimicry of historic architectural styles.

Roger K. Lewis, FAIA, is a practicing architect, professor emeritus of architecture at the University of Maryland, and *Washington Post* columnist. His "Shaping the City" essays and cartoons have appeared in the newspaper on Saturdays since 1984.

The Ford Orientation Center from the elliptical clearing.

Ladies, the Home of Washington is in your charge. See to it that you keep it the Home of Washington! Let no irreverent hand change it; no vandal hands desecrate it with the fingers of—progress! Those who go to the Home in which he lived and died, wish to see in what he lived and died! Let one spot in this grand country of ours be saved from "change!" Upon you rests this duty.

— Ann Pamela Cunningham, 1874

A Challenge

This was the parting challenge that Ann Pamela Cunningham presented to the Mount Vernon Ladies' Association when she retired as Regent in 1874. Little did she know the sweeping effect those words would have on all historic preservation in America. Nor could she have anticipated the needs imposed by modern visitation to historic sites. Today Mount Vernon, preserved as it was in George Washington's time, stands as a model of dynamic stewardship. Welcoming more than one million visitors annually, it is the nation's most popular historic house.

The Mount Vernon Ladies' Association of the Union was founded in 1853 for the express purpose of saving Washington's home from deterioration and opening his estate as a national landmark embodying his legacy. In 1986 the Association expanded its mission: to educate Americans about Washington's remarkable life and myriad accomplishments. It pursued that goal by conducting an ambitious capital campaign, culminating in the opening of the Ford Orientation Center and the Donald W. Reynolds Museum and Education Center in October 2006. These buildings now provide an enriching educational experience for estate visitors and anchor an outreach program that touches people around the world.

The challenge of designing 71,000 square feet of space for these two new facilities—almost 10 times the area of the Mansion—was especially challenging when the design team reflected upon Ann Pamela Cunningham's enduring words. Any new edifices had to complement the historic site, effectively and efficiently performing their important functions without detracting from the experience of visiting the home of George Washington as it was in his time.

Left: *The roof of the Donald W. Reynolds Museum and Education Center entrance is elevated by a ribbon of clerestory windows.*
Right: *Detail of the glass-and-copper canopy at the entrance to the Ford Orientation Center.*

1853–1986 History: Saving Mount Vernon

George Washington is certainly best known for his singular career of public service, most notably his role as commander in chief of the Continental Army during the Revolutionary War and as first president of the United States. Less known are his other talents and accomplishments. Washington was an enthusiastic and innovative farmer and estate manager; an entrepreneur who, among other things, operated one of the largest whiskey distilleries in 18th-century America; and a skilled architect who continued to design and redesign the buildings and landscape of his beloved home throughout his years there. His training and experience as a land surveyor equipped him with technical skills that served him well in these endeavors. Similarly, his years as a military officer honed his administrative and leadership qualities, preparing him to undertake complex building projects on his estate.[1]

Mount Vernon became Washington's personal passion. He twice enlarged the house inherited from his half-brother Lawrence and continually remade the surrounding gardens and grounds to complement the evolving architectural assemblage. He revamped his agricultural operations, exchanging tobacco for wheat as his cash crop and experimenting with such advanced techniques as crop rotation and using fertilizer to enrich his soils. Washington even invested in a revolutionary automated milling system so he could produce enough flour to meet demand as far away as the West Indies and Europe.

Rather than simply adopt current architectural designs imported from England and elsewhere, Washington selected styles and features that pleased his discerning eye and combined them to create a striking yet functional integration of buildings and landscape. The two-story piazza running the entire 94-foot length of the Mansion's east front and the open-sided colonnades linking the house to flanking outbuildings are two examples of his inventive design. When completed in 1787, the Mansion was surrounded by a carefully orchestrated landscape that took full advantage of the site's natural beauty and provided a perfect context for the house.

Guests at Mount Vernon frequently commented on Washington's mastery in blending the manmade and natural environments to form a homogenous scene that capitalized on the site's inherent beauty. For example, a Polish nobleman who visited the year before Washington died noted that "the garden, the plantations, the house, the whole upkeep, proves that a man born with natural taste can divine the beautiful without having seen the model. The G[enera]l has never left America. After seeing his house and his gardens one would say that he had seen the most beautiful examples in England of this style."[2]

Above are examples of George Washington's inventive design.
Left: *Two-story piazza on the east side of the Mansion.*
Right: *Curving arcade flanking the Mansion to the north.*

Chronology of the Mount Vernon Estate

In 1674 Lord Culpeper granted Nicholas Spencer and John Washington 5,000 acres of land as payment for bringing a group of settlers into the Virginia Colony; this acreage ultimately included the Mount Vernon Estate. By 1690, the tract was divided among the heirs of the original grantees, with John Washington's share going to his son Lawrence. When Lawrence died, he left the estate to his daughter, Mildred, and her husband, Roger Gregory. In 1726 Mildred sold the land to her brother Augustine Washington. George Washington was born on February 22, 1732, to Augustine and Mary Ball Washington at Popes Creek Plantation along the Potomac River in Westmoreland County, Virginia. By 1735, Augustine moved his family to Little Hunting Creek Plantation (land that would later become part of the Mount Vernon Estate), presumably living in a modest house on the property. The family moved again, four years later, to Ferry Farm near Fredericksburg, Virginia, along the Rappahannock River.

Augustine died in 1743, when George was 11 years old. In 1740 he had deeded the land to his son Lawrence, who was George's older half-brother and became his surrogate father after Augustine's death. Lawrence either built or expanded a house on the estate, added some dependencies, and changed its name from Little Hunting Creek Plantation to Mount Vernon—in honor of Admiral Edward Vernon, under whom he had served in the British Navy in the Caribbean.

Lawrence died in 1752, bequeathing lifetime tenure at Mount Vernon to his widow, Ann Fairfax. Two years later, she leased the house and 2,126 acres of farmland to her brother-in-law George, the residual heir of the estate. During the French and Indian War, George Washington served as an aide to the British General Edward Braddock and as an officer in the Virginia militia. Despite his extended absences from Mount Vernon, he began expanding the house there in 1757. After he married the wealthy widow Martha Dandridge Custis, in 1759, the pace of expansion quickened. Over the next 40 years, Washington would continue to alter his estate, often experimenting with innovative building techniques, beginning new businesses, and expanding his landholdings.

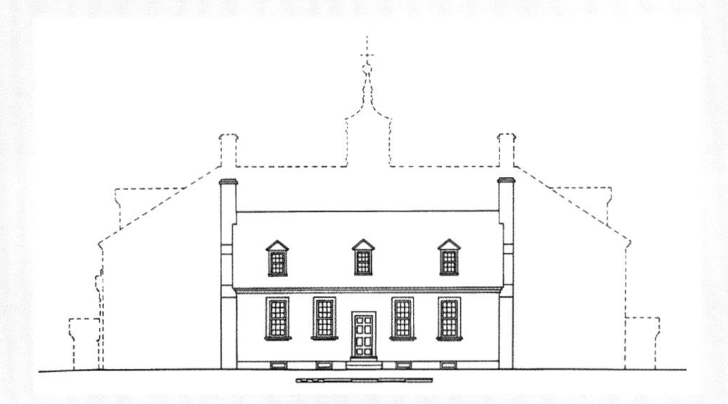

Top: *Drawing representing the development of Washington's residence. It grew from a one-and-a-half-story home to a two-and-a-half-story mansion. The Mansion reached its present size in 1787.*
Middle: *Peter Waddell,* George Washington: Architect, *1998.*
Bottom: *Attributed to Edward Savage,* West Facade of Mount Vernon, *c. 1792.*

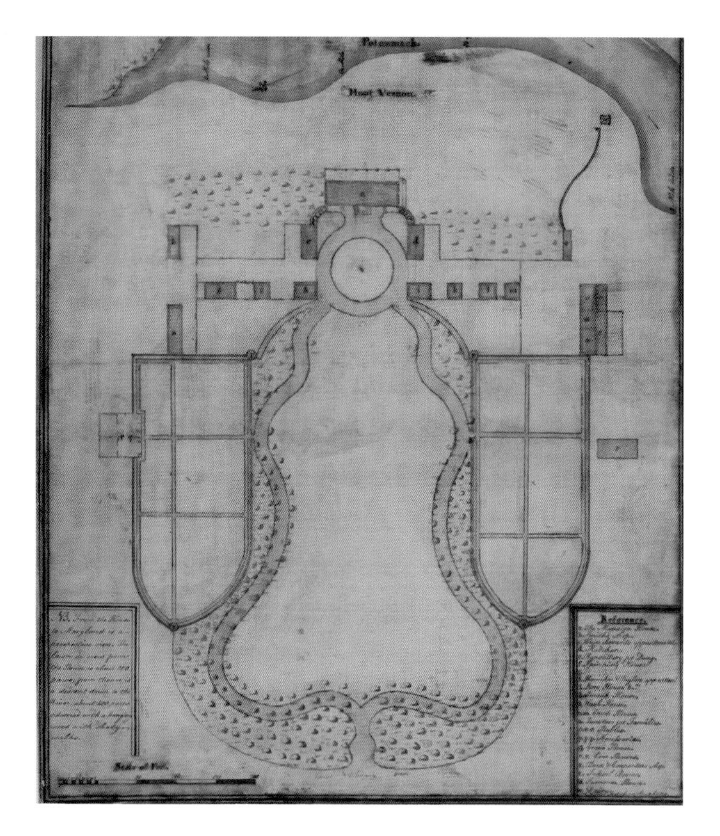

Samuel Vaughan, Plan of Washington's house and gardens drawn (detail), 1787.

The Mount Vernon Mansion has become an icon of American architecture, and the site itself claims a preeminent place in the history of this country's landscape design. The distinctive two-story piazza is one of the most copied features in American architecture, lending distinction to ordinary homes and palatial mansions alike. So completely have Mount Vernon's distinctive elements been incorporated into the canon of American architecture that it seems Washington's congenial balance of style and practicality continues to resonate within the national consciousness.

In the five decades following George Washington's death in 1799, ownership of Mount Vernon changed hands among five family members. In 1850 it finally passed to the great-great nephew of the former president, John Augustine Washington III, who soon found himself saddled with two burdens: decreasing profits from depleted fields and the arrival of hordes of pilgrims eager to see what was rapidly becoming a national shrine. As his funds dwindled and the imposition of visitors increased, John Washington could not properly maintain the Mansion and its surrounds. Even so, when he received an offer of $300,000 for the property, he would not consent to selling Mount Vernon without absolute assurance that it would be protected for future generations.[3]

Washington first approached Congress and then the Virginia General Assembly, suggesting they purchase the estate. But his timing was unfortunate. Factions from the North and South were immersed in the tumultuous debates about land and slavery that within a few years would result in the Civil War, so little attention was paid to his proposal.

Ultimate salvation would arrive through an unexpected and timely source. On a moonlit evening in 1853, a southern matron named Louisa Dalton Bird Cunningham was aboard a steamship gliding down the Potomac River and awoke to the traditional bell-ringing salute as the craft passed Washington's home. Gazing out from the deck,

John Augustine Washington III owned Mount Vernon from 1850 to 1858, when he sold it to the Mount Vernon Ladies' Association.

Mrs. Cunningham was appalled to see the first president's home in near ruin. She immediately wrote to her invalid daughter, Ann Pamela, describing the Mansion's shocking condition. Ann Pamela Cunningham was, in turn, moved to action. If men have neglected the home and burial place of America's most respected hero, she reasoned, perhaps women should band together to save it.

Despite her physical infirmities, Miss Cunningham vigorously campaigned to raise funds to purchase Mount Vernon. She began by organizing the Mount Vernon Ladies' Association of the Union, establishing herself as Regent and enlisting 22 prominent women to serve as Vice Regents. Several men outside the realm of government stepped in to help, the best-known being Edward Everett of Boston. A respected orator as well as a former U.S. senator and president of Harvard College, Everett frequently delivered a lecture titled "Oration on the Character of Washington." Miss Cunningham accepted his offer to support her cause by donating his speaking and writing fees to the Association's campaign chest. Between 1856 and 1861 he gave his Washington lecture 129 times, and as a result, earned—and donated—$69,551.06, more than one-third of the ultimate purchase price.

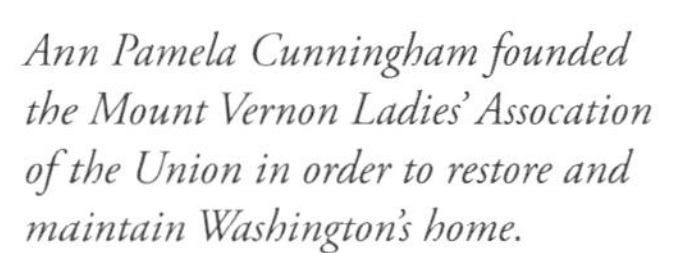

Ann Pamela Cunningham founded the Mount Vernon Ladies' Association of the Union in order to restore and maintain Washington's home.

In the spring of 1858, John Washington agreed to sell the Mansion and 200 acres of adjoining land to the Association for $200,000. On February 22, 1860, the anniversary of George Washington's birth, John and his family relinquished the Mansion keys to Ann Pamela Cunningham and her secretary, Sarah Tracy, who moved in. The house was virtually empty, with a few priceless exceptions—the key to the Bastille, presented as a gift to General Washington by the Marquis de Lafayette; Washington's terrestrial globe tracing the travels of Captain James Cook; and the original terra-cotta bust of Washington made by the French sculptor Jean-Antoine Houdon at Mount Vernon in 1785.

The Mansion was in a state of disrepair when the Association bought it in 1858.

Washington-owned artifacts that remained in the Mansion when the Association purchased the estate in 1858.
Left: *Key to the Bastille.*
Right: *Terrestrial globe tracing the travels of Captain James Cook.*

Now, Miss Cunningham began the challenging task of restoration. But where to begin?

Perhaps Ann Pamela Cunningham's most important talent was for selecting Vice Regents. Although they represented upper-crust families and brought with them a considerable share of influence both within the boundaries of their states and beyond, the original Vice Regents also comprised a colorful, almost exotic lot. With her very first choice, Anna Cora Ogden Ritchie of Richmond, Virginia, Miss Cunningham demonstrated that she was looking for women who were action-oriented, who had already established reputations beyond the confines of polite society. A successful actress, Ritchie was known on two continents by her stage name, Anna Mowatt. She had also written poetry, plays, and novels. Her status as a descendant of Declaration of Independence signer Francis Lewis and the granddaughter of an Episcopal priest tendered questionable her decision to appear on the professional stage—a role considered socially unacceptable for a woman of such a pedigree. Yet by all accounts, Ritchie was attractive, charming, and persuasive. It also helped that her husband was an influential member of Virginia's General Assembly and editor of the *Richmond Enquirer*.

To serve as the first Vice Regent for Massachusetts, Miss Cunningham selected Louisa Ingersoll Greenough, widow of the famous sculptor Horatio Greenough, whose immense, classically inspired marble statue of a seated Washington was as controversial as it was popular. In New York, she chose Mary Morris Hamilton, granddaughter of Alexander Hamilton, and in Wisconsin, Martha Reed Mitchell, wife of the prominent Milwaukee railroad executive and politician Alexander Mitchell. The first Vice Regent for Florida, Catherine Daingerfield Willis Grey Murat, was not only a great-grandniece of George Washington himself but was also a bona fide princess by virtue of her marriage to Achille Murat, a nephew of Napoleon Bonaparte.

Such was the group of distinctive women who banded together, determined to save Mount Vernon. Equipped with a negligible bank balance, Miss Cunningham and her cohorts faced a decidedly uphill struggle—and there were no government agencies waiting in the wings to help.

Time was also a factor. The condition of the Mansion was growing more tenuous by the day, and the nation was enduring the ravages of civil war. How could Americans focus on protecting George Washington's home when the very survival of their country was in question? Fortunately, Miss Cunningham, who possessed the sensibilities of a true preservationist, continued to dismiss all proposals

Artifacts returned to the Mansion through the Association's efforts.
Left: *The Washingtons' bed.*
Right: *Nelly Custis's harpsichord.*

to tear down Washington's outbuildings or to remold his landscape. What visitors see today is indeed the most authentic surviving 18th-century plantation in America. This is not because the Mount Vernon Ladies' Association purchased the Mansion and then made time stand still; rather, exhaustive research and a continuing quest for original furnishings have enabled the Association, meticulously and respectfully, to turn back the clock.

Over time, the interior of the Mansion has been extensively transformed. In the 19th century, some members of the Association assumed responsibility for decorating individual rooms, and as a result, the décor grew decidedly eclectic. Many original items, including a harpsichord that Washington bought for his step-granddaughter, Nelly Custis, and the Washingtons' bed, were returned to the Mansion by family descendants. As other original pieces were acquired, the idea of restoring the Mansion to its appearance the year Washington died became a more feasible goal, and the detailed list of objects in the house that was made at the time of his death served as a guide for furnishing the house.

Thomas Edison in front of Washington's tomb, c. 1916.

The large dining room, undergoing restoration to the colors it was painted when Washington lived there.

One of the most perplexing issues regarding historic preservation is whether or not to install modern utility systems. Such was the case when Association members assembled at Mount Vernon in 1916 to discuss a proposal by Thomas Edison's firm to "electrify" the Mansion for the first time. The kerosene lamps and candles then in use presented a serious fire hazard. Edison offered to set up a system powered by generator-fed storage batteries, at a cost of $3,325, with a "guarantee of absolute safety." Although Association members were skeptical, fearing a loss of authenticity, the concern for safety prevailed. Edison's system was approved and installed.

In the early 1980s, the Association embarked on a study to document the colors of every painted surface in the house. In all, some 2,500 samples were gathered from the walls, ceilings, and woodwork in all three stories of the building, revealing up to 26 layers of paint. The chemical compounds of the samples were analyzed, and the pigments were microscopically examined. This was one of the most definitive paint studies ever conducted on an 18th-century house, and when it was finished, a complete paint history was established for each room.[4]

As the Mansion was slowly returned to its 1799 appearance, the surrounding gardens and grounds and the various outbuildings also were restored. As early as 1859, Upton Herbert, the estate's first superintendent, took steps to repair deteriorating garden walls, walkways, and other site amenities. In the 1930s, outbuildings were thoroughly researched and restored. In 1951 the Association undertook its most ambitious reconstruction effort since purchasing the property: it rebuilt Washington's brick greenhouse and slave quarters, which had burned in 1835. Fortunately, the foundations of the original buildings remained, and matching bricks left over from President Harry Truman's concurrent renovation of the White House were available to rebuild the structures.

As the Association approached the 100th anniversary of acquiring Mount Vernon, a new danger to the estate's 18th-century atmosphere began to materialize. The post–World War II economy stimulated real estate development farther and farther from the core of the nation's capital. Washington's remarkable view across the Potomac River to the green forests and fields of Maryland was on the verge of succumbing to what Ann Pamela Cunningham had characterized in 1874 as the dangerous "fingers of progress." In the spring of 1955, speculators looking for a site to build an oil refinery targeted 500 acres of land directly across the Potomac from Mount Vernon. Their efforts were thwarted by one of the Association's Vice Regents, Frances Payne Bolton, who also happened to be a U.S. congresswoman from Ohio. The Bolton family possessed considerable wealth, and when the threat became clear, Mrs. Bolton purchased the property in question for $330,000 so it could be preserved in perpetuity.

Other threats surfaced in subsequent years, but Mrs. Bolton had a number of effective allies—the National Park Service, the Accokeek Foundation, the Moyaone Association, and others. The campaign to save the view garnered national attention as tens of thousands signed petitions, called their congressmen, and made their

Left: *Washington's greenhouse was reconstructed in 1951.* Right: *Attributed to Edward Savage,* View of Mount Vernon from the Northeast (detail)*, c. 1792. George Washington's slave quarters were located near the Mansion.*

voices heard. Finally, in 1974, President Gerald R. Ford signed into law bills that provided for the establishment of Piscataway Park and permanently safeguarded the view directly across from Mount Vernon. Today, the park encompasses some 5,000 acres, and active efforts to expand the park into still-unprotected areas, which span 80 square miles, continue in earnest, with computer-aided systems monitoring potential encroachments.

By 1986, the Mount Vernon Ladies' Association had achieved its laudable mission of preserving the Mansion and its surrounding gardens, enabling more than a million visitors a year to see where the Father of Our Country lived and died. The members had even saved from intrusion the vistas from his Mansion to the river and much of the landscape beyond. With this singular achievement behind them, and with a system of continual upkeep well established, the Association began broadening its vision of advancing the appreciation and understanding of George Washington. In addition to preserving his buildings and grounds, its members would begin to preserve his rich legacy to the nation. The goal was a lofty one—to educate the world about Washington's extraordinary life, achievements, and character.

The view from the Mansion across the Potomac River was preserved from development through the generosity of Frances Payne Bolton, Vice Regent for Ohio and U.S. congresswoman, among others.

1986–1995: Broadening the Mission to Tell the Whole George Washington Story

While the Mount Vernon Ladies' Association continued to follow the principal philosophies and methods of the historic preservation movement, by the mid-1980s, its members acknowledged the need to reconsider seriously the messages Mount Vernon conveyed as well as the way they were presented. In 1987 the Association created an ad hoc committee charged with assessing the group's "strengths and weaknesses, studying . . . all [its] . . . resources and capabilities and [determining] their adequacy to meet future needs." In particular, the committee was directed to "define the essence of George Washington" and to develop "interpretive themes that should be communicated as part of the Association's ongoing program." The decision to move the personality and character of George Washington to the forefront of the interpretive programs was not based on intuition. Rather, the need was clearly identified by Mount Vernon's most important constituency—the visitors themselves. A survey concluded that even after they had toured the estate, visitors felt they had not learned as much about Washington as they wanted. Furthermore, the survey revealed that visitors perceived him to be a "one-dimensional and somewhat uninteresting" man. The Regent and Vice Regents knew that nothing could be further from the truth: Washington was not just the most important and powerful American leader of the period but also one of the most interesting and multifaceted men of his day.

Commemorative postcards of George Washington and Mount Vernon, c. 1910–15.

The Association was equally disturbed to learn how American history was being taught in many schools across the nation. Specifically, its members were discouraged by current educational trends that affected Americans' knowledge of and appreciation for George Washington. Among their most startling findings were the following:

- A survey of 55 of the nation's top universities revealed that most students lacked an adequate understanding of U.S. history. Barely one-third of those surveyed correctly answered that Washington was the victorious general in the Battle of Yorktown, the final engagement of the Revolutionary War.

- In a survey of fourth graders, 7 of 10 students thought that Illinois, California, or Texas were among the 13 original colonies.

- George Washington had almost vanished from the nation's classrooms. One survey revealed that his coverage in history textbooks had declined dramatically to less than 10 percent of what it had been 40 years earlier. His likeness—for so long displayed in every classroom—had virtually disappeared.

With a serious decline in the attention paid to George Washington in schools and other evidence of his flagging significance in American life—such as the combination of his birthday with that of Abraham Lincoln to create the Presidents' Day holiday—it became clear that an increasing number of visitors to Mount Vernon arrived with little knowledge about the real man's life and legacy. And once they arrived at the estate, they were exposed to only part of the first president's remarkable story, since the principal focus remained on the Mansion, its furnishings, Washington's domestic life on the plantation, and on his tomb. His experiments as an innovative farmer, astute businessman, soldier, statesman, and architectural visionary responsible for the design of all the buildings and the surrounding landscape were largely ignored.

George Washington, *c. 1804, one of many portraits of Washington by Gilbert Stuart.*

Members of the Mount Vernon Ladies' Association determined that nothing short of an all-out campaign was required to revive interest in George Washington and that their organization was best equipped to lead it. Their initial plan was twofold: to enhance on-site interpretation of Washington's home by incorporating stories that previously had been ignored or incompletely told; and to expand the educational mission so it would include his entire life story. Almost immediately, the Association took steps to improve the existing interpretive efforts. Meanwhile, a planning committee began exploring options for erecting a new museum and educational facility, where a well-rounded portrayal of Washington's life could be presented.

With education as its focus, the Association also created the Advisory Council of George Washington Scholars—25 authors and professors, each possessing extensive knowledge about at least one key aspect of Washington's life. This veritable Who's Who of scholars provided critical advice in the planning of exhibits and programming.

In order to emphasize Washington's diverse accomplishments, numerous interpretive themes were identified that would help reveal his fascinating character. These included refocusing the Mansion interpretation to reveal how the furnishings reflected his personal sense of style, and how the activities in the rooms tell us about his personality. New exhibits were added to give educators appropriate settings to address little-discussed topics, such as Washington's farming and surveying careers, his entrepreneurial efforts as a miller and distiller, and his forward-thinking attitudes about conservation.

Actors portraying historical characters were placed at key locations, greatly improving the visitor's experience. Several rooms in the Mansion were refurbished as a result of new research, and objects that had been in the house when Washington was alive were acquired. As Mount Vernon craftsmen learned more about the original appearance of the interior spaces, they were able to restore elements such as the ornamented plaster ceilings and woodwork with greater accuracy. Scientific analysis made it possible to approximate the late-18th-century paint colors with even greater precision.

As the authenticity of the Mansion was being enhanced, the appropriateness of collections housed in the outbuildings was reassessed, and numerous changes were made. Three of the original 18th-century outbuildings were extensively restored so they could be interpreted according to their original functions. After years of serving as administrative offices, the servants' hall now looks as it did when it housed servants and slaves of visitors to the estate. Likewise, the spinning house now resembles the structure that contained Washington's extensive cloth-making operation. And finally, the gardener's house, where

George Washington as portayed by William Sommerfield.

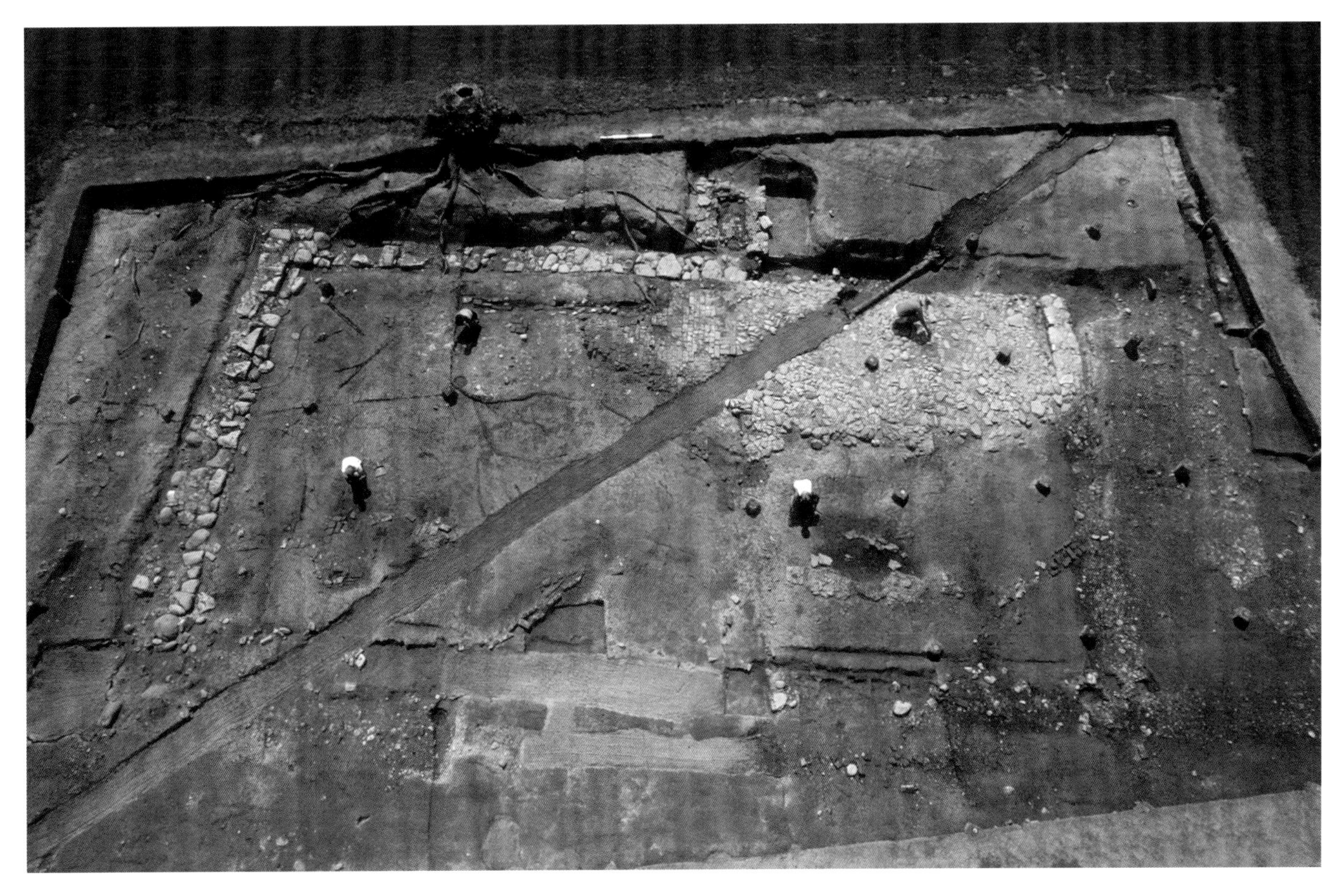

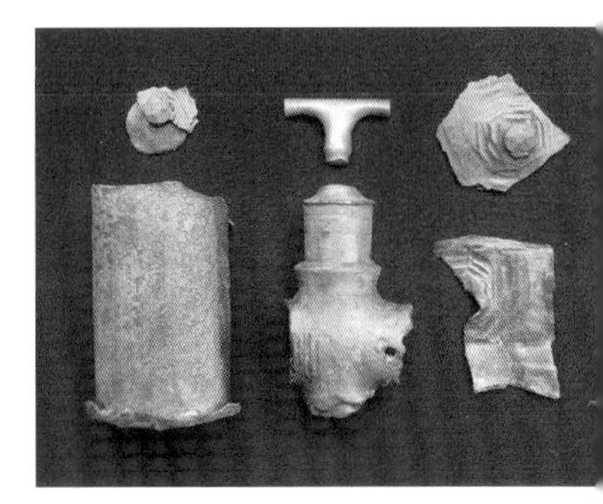

Left: *Archaeological dig at Washington's distillery.*
Above: *Fragments of distillery-related objects found at the site.*

Mount Vernon's security-control room was located from 1893 until 2005, was restored to resemble the residence it once had been for Washington's gardener; it opened to the public in 2007.

In 1987 the Mount Vernon Ladies' Association established a permanent program of archaeological research, charged with studying various activities once pursued on the plantation. Over the years, participating archaeologists have excavated numerous sites on the property, including a trash-filled cellar situated beneath the main slaves' quarters, the blacksmith shop, a trash dump associated with the Washington household, the site of the fruit garden and nursery, Washington's whiskey distillery, and even the site of the dung repository, where organic waste was composted for fertilizer. Their findings made possible expanded educational programs focusing on the daily lives of the enslaved workers; provided essential information to guide the re-creation of the dung repository, the fruit garden, and the distillery; and informed plans for reconstructing the blacksmith shop.

The combination of archaeological evidence and ongoing research using Washington's voluminous writings provided

Right: *The only surviving photograph of Washington's innovative 16-sided treading barn, believed to have been demolished after 1870.*

"Thanks to a generous grant from the W. K. Kellogg Foundation, Washington's remarkable 16-sided treading house *has been reconstructed at Mount Vernon on a four-acre site near the Potomac River. The barn exactly follows Washington's plans and drawings from the 1790s, and the adjoining stables and corn houses are part of Washington's design. Every building was made of hand-shaped bricks, lumber hewed and pit-sawed to final dimension and nails hammered at an open-fire forge. In season, horses and mules tread wheat on the upper floor of the treading barn, just as happened under Washington's direction."*

—Mount Vernon Official Guidebook, 2002

Below: *Replica of the 16-sided treading barn.*

the basis for organizing several tours devoted to specialized topics. Visitors now can join a guided tour focused on the Mount Vernon plantation community, for example, or one that concentrates on the estate's 18th-century landscape and gardens. An audio tour describes various plantation service buildings and the activities carried out in them, providing additional insight into the daily lives of the slaves.

When the Association determined that Washington's accomplishments as an innovative farmer would be suitable subjects for engaging new educational programs, it transformed a vacant four-acre tract on the Potomac River into a living-history experience called George Washington—Pioneer Farmer. The centerpiece of the exhibit is a meticulously researched and crafted replica of the unique 16-sided treading barn that Washington himself designed and had built at his outlying Dogue Run Farm in 1794 and 1795. Since 2002, a fully functional replica of Washington's state-of-the-art automated gristmill has operated on its original site. Additionally, an authentic

George Washington's Gristmill was restored through a partnership with the State of Virginia in 2002, and continues to operate as it did in his time.

Left: *A replica of George Washington's Distillery demonstrates how whiskey was produced in 18th-century America.*
Right: *The Distillery was reconstructed and opened to the public in 2007.*

Claude Regnier, Life of George Washington: The Farmer, *1853, engraving after a painting by Junius Brutus Stearns.*

reconstruction of his whiskey distillery, one of the largest in America at the time, has been completed.

Developing lively new programs to maximize resources on the Mount Vernon Estate was only part of the overall plan to return George Washington to his rightful prominence in 21st-century America. For no matter how successful that effort might be, the story of the estate alone would still neglect vast portions of its owner's life—in particular, the stunning accomplishments of his public career. After years of detailed study and with the assistance of numerous consultants, the Association concluded that the estate needed at least one major new building where the full story of Washington's life could be told. To address this need, the members began to explore the implications of funding such an undertaking and to consider who might help support its realization.

Mount Vernon Ladies' Association Mission Statement, 1874

Ladies, the Home of Washington is in your charge. See to it that you keep it the Home of Washington! Let no irreverent hand change it; no vandal hands desecrate it with the fingers of—progress! Those who go to the Home in which he lived and died, wish to see in what he lived and died! Let one spot in this grand country of ours be saved from "change!" Upon you rests this duty.

— *Ann Pamela Cunningham's farewell address to the Vice Regents in 1874 established a concise and eloquent mission statement.*

Early members of the Mount Vernon Ladies' Association, including Ann Pamela Cunningham (seated left of the famous Houdon bust of George Washington), 1870.

Mount Vernon Ladies' Association Mission Statement, 1999–2007

The mission of the Mount Vernon Ladies' Association is to preserve, restore and manage the estate of George Washington to the highest standards and to educate visitors and people throughout the world about the life and legacies of George Washington, so that his example of character and leadership will continue to inform and inspire future generations.

— *Crystallized in this statement are the ideas that had been evolving since the Association began broadening its aims and purposes in 1986.*

The Regent and Vice Regents of the Mount Vernon Ladies' Association pose on the bowling green during their spring 2006 Grand Council.

Seated, left to right: Mrs. William H. Borthwick, California; Mrs. Lloyd A. Semple, Michigan; Mrs. Charles B. Mayer, Louisiana; Mrs. Frank X. Henke III, Oklahoma; Mrs. James Evan Allison, Washington; Mrs. Stanley N. Gaines, Regent, Florida; Mrs. John F. Bookout III, Texas; Mrs. J. Hap Fauth, Minnesota; Mrs. Stewart Gammill III, Mississippi; Mrs. Charles G. Lane, South Carolina; Mrs. Jared I. Edwards, Connecticut; Mrs. Joseph W. Henderson III, District of Columbia.

Standing, left to right: Mrs. James M. Walton, Pennsylvania; Mrs. P. Coleman Townsend, Delaware; Mrs. Randolph H. Guthrie, New York; Mrs. Robert W. Lawson III, West Virginia; Mrs. John F. Mars, Wyoming; Mrs. Melody Sawyer Richardson, Ohio; Mrs. James F. Crumpacker, Oregon; Mrs. P. William Moore, Jr., Virginia; Mrs. Robert E. Lee IV, Maryland; Mrs. J. Schley Rutherford, Alabama; Mrs. Richard Simplot, Idaho; Mrs. Sam Buchanan, Arkansas; Mrs. Shepard B. Ansley, Georgia; Mrs. Everette C. Sherrill, North Carolina.

To Keep Him First: Mount Vernon's Capital Campaign

Recognizing—and disheartened by—the dimming appreciation and knowledge of George Washington, the Mount Vernon Ladies' Association accepted its greatest challenge since rescuing the property in the 1850s. Not since Ann Pamela Cunningham's campaign to raise the then seemingly unattainable $200,000 needed to purchase the estate had such a bold and ambitious effort been undertaken. Determined to rescue and restore Washington's name and rightful place in American history, the Association launched a capital campaign that eventually would generate tremendous public support.

In 1995 Mrs. H. Taylor Morrissette of Alabama, then serving as Regent, announced the Association's commitment to the construction of a new facility that would provide visitor orientation and education spaces. Plans also called for an expansion of the restaurants and gift shop. To achieve these ambitious goals, the Association would need to raise what she identified as "tens of millions of dollars to construct and endow" the new facilities. Designated *To Keep Him First*, the campaign established a goal of $60 million to transform the Mount Vernon experience by effectively introducing the personality, achievements, and character of Washington and by showcasing the Association's many rare and fragile objects. The campaign also would facilitate an aggressive education program to reach students, teachers, and people throughout the world. Finally, the Association sought a substantial increase in Mount Vernon's endowment to cover the additional costs of maintaining and operating these new facilities.

Right: *Cover of capital campaign brochure, 2002.*

As fundraising began in earnest, the Association's 1998 *Annual Report* noted, for the first time, the year-end status of the campaign—a less-than-impressive $784,000. By 1999 the campaign had gained some traction, with total gifts and pledges amounting to $14 million. One year later, in the fall of 2000, a new Regent, Mrs. James M. Walton of Pennsylvania, reported a total of $34.5 million. As the campaign passed the halfway mark toward reaching its goal, she declared 2000 "the most productive year ever."

The main cause for celebration was the Ford Motor Company Fund's announcement of a $7 million commitment, at that time the largest gift ever received by the 147-year-old Mount Vernon Ladies' Association. Announcing the pledge at a black-tie gala attended by some of Mount Vernon's most loyal supporters, Ford executives challenged others to follow their lead. They also noted that founder Henry Ford had made the company's initial contribution to Mount Vernon in 1923: the estate's first fire engine. During this critical period of the campaign, the Association also received significant commitments from the Vira I. Heinz Endowment, the Mars Family and Foundation, the William Randolph Hearst Foundation, the Mary Hillman Jennings Foundation, and Robert H. and Clarice Smith. Despite the nation's vulnerable economy and its somber mood in the wake of the terrorist attacks of September 11, 2001, *To Keep Him First* persevered, and Americans responded to the Mount Vernon project with extraordinary generosity. The difficult times seemed to renew patriotic sentiment and heightened America's appreciation of public symbols and historic treasures.

On April 27, 2002, some 300 guests gathered for another elegant black-tie affair, this one held on the Mansion's piazza, to support the first official fundraising event for the *To Keep Him First* campaign. The historic evening featured dinner and dancing under a Tiffany Blue tent inspired by the one First Lady Jacqueline Kennedy had

ordered for a presidential gala at Mount Vernon in 1961, the only state dinner ever held outside the White House. Celebrants expected to be impressed by the four-course menu of French-inspired favorites—matching the one Mrs. Kennedy had selected—and the music of renowned dance-orchestra conductor Peter Duchin. But they were hardly prepared for the climactic announcement delivered by officials from the Donald W. Reynolds Foundation: a $15 million pledge would support a new education center named for the late Mr. Reynolds, founder of the Donrey Media Group.

With gifts and pledges of more than $53 million now safely in hand, including commitments of a million dollars or more apiece from 12 donors, some thought the path ahead would be smooth. But with the construction industry now regaining momentum, it became clear that the planned facilities, as well as the endowment needed to fund their operation, would be more expensive than originally predicted: the Regent and Vice Regents voted to increase the campaign's official goal to $85 million.

By now, Association members were certain that George Washington was a figure around whom Americans could

"An Auction, by George!" in October 2003 raised more than $550,000 for the capital campaign.

rally. "As we face the clear and present danger of a world in great chaos," Mrs. Walton said, "he can be a true touchstone for the American leaders who will make critical decisions that have lasting impact on the entire globe."

The momentum of the campaign continued in October 2003 with "An Auction, by George!," which attracted 675 guests who bid on 365 items that included jewelry, antiques, artwork, silver, and luxury excursions around the world. The auction raised more than $550,000 for the capital campaign, making it the single most successful benefit in Mount Vernon Ladies' Association history.

Meanwhile, the Association was stepping up its direct-mail campaign, hoping to involve thousands of Americans from all 50 states. One package, which featured a 20-minute videotape narrated by former U.S. Senator Fred Thompson, resulted in new commitments totaling more that $1.2 million. Several hundred donors answered the call to contribute $25,000 or more; they would have their names inscribed in stone on a wall of honor within the complex. In addition, more than a dozen donors created named endowments, sponsored galleries, or underwrote theaters. Exceptional leadership gifts were received from the F. M. Kirby Foundation, the Annenberg Foundation,

ANNOUNCING THE MOST SIGNIFICANT PROJECT AT
Mount Vernon
IN 150 YEARS
THE CONSTRUCTION OF OUR NEW, WORLD-CLASS EDUCATION CENTER
Most of the Center is underground so that the greenery above still appears as it did in George Washington's day with sheep peacefully grazing in the pasture.
The serene, unobstructed views of the house, the grounds, and the Potomac River that Washington so loved will stay as they were in his day.

Left: *Direct mail campaign during the later phases of the 2003 fundraising year.*
Right: *Fundraising event for the capital campaign held at the Library of Congress on October 2, 1999.*

Reynolds Foundation board members and family participated in groundbreaking-ceremony events on June 17, 2004.

Donald and Nancy de Laski, the Richard and Helen DeVos Foundation, the Gilder Foundation, J. Hap and Geren Fauth, and Gay Hart Gaines.

Then, in the spring of 2004, the Donald W. Reynolds Foundation significantly increased its commitment, from $15 million to $24 million. "We are convinced more than ever that George Washington should be a role model for future generations," said foundation Chairman Fred W. Smith, "and no place can tell his story in a more complete and compelling way than Mount Vernon."

Ground was finally broken for the new facilities on June 17, 2004, with U.S. Supreme Court Associate Justice Sandra Day O'Connor delivering the keynote address. Before that day was out, the campaign received still another significant boost when longtime Mount Vernon friends Robert H. and Clarice Smith of Arlington, Virginia, increased their total commitment to $10 million.

Writing in her first *Annual Report* as Regent, in 2004, Gay Hart Gaines of Florida acknowledged that there is "something both exhilarating and frightening about breaking ground on a pair of buildings and exhibits costing $50 million." By the end of that year, the Mount Vernon Ladies' Association had raised $83 million, an impressive sum but by no means enough to impart a sense of security. There were a number of factors adding to the facilities' projected costs, including new multimedia programs, skyrocketing expenses for the film that would be continuously screened in the orientation center, and the upkeep of the new technologies that would be employed in the buildings and exhibits. As Mrs. Gaines informed Mount Vernon's most loyal supporters, "We have miles to go before we sleep."

Several major donors, including Donald and Nancy de Laski, Robert H. and Clarice Smith, John and Adrienne Mars, Richard M. Scaife, and the Ford Motor Company Fund, made additional gifts of over a million dollars. Similar leadership gifts were received from the Robertson Foundation, Elizabeth and Stanley DeForest Scott, Douglas and Eleanor Seaman, the Gordon V. and Helen C. Smith Foundation, and A. Alfred Taubman. In addition, a new direct mail appeal gave already-generous donors a chance to sponsor individual trees; hundreds responded promptly. When construction of the facilities was well under way, Regent Gaines organized a campaign "to fill the Wall of Honor with the names of true patriots." The response was enthusiastic. By the time the campaign officially came to a close, in June 2007, Mount Vernon had raised an astonishing $116 million from some 5,000 donors—nearly twice the original $60-million goal.

Left: *George Washington, as portrayed by Dean Malissa, toasts guests gathered for lunch following 2004 groundbreaking ceremonies for the new facilities.* Below: *Recognition certificates placed on the structure of the new buildings acknowledge project donors.*

Overleaf: *The names of all donors of $25,000 or more to the* To Keep Him First *capital campaign are engraved on a wall of honor in the lobby of the Donald W. Reynolds Museum and Education Center.*

A CAPITAL CAMPAIGN

FORD MOTOR COMPA

THE ANNENBERG FOUNDATION • DONA

THE GILDER FOUNDATION, INC. • W

THE F. M. KIRBY FOUNDATION • JOHN

UGLAS AND ELEANOR SEAMAN • ALBERT A

ANNE AND JAMES F. CRUMPACKER FAMIL

GERRY AND MARGUERITE LENFEST • CH

THE RICHARD S. REYNOLDS FOUNDATION

FOUNDATION • MARGARET AND JAMES ALLISON • BRUCE AND JOY AMM

URDICK • CASTLE ROCK FOUNDATION • THE CHISHOLM FOUNDATIO

ERS, WASHINGTON COMMITTEE FOR HISTORIC MOUNT VERN

A HENDERSON • BONNIE AND FRANK X. HENKE III • HI

RMAN • FRAYDA AND GEORGE L. LINDEMANN • A.

ND MRS. P. WILLIAM MOORE, JR. • MRS. H. T.

S. NELSON T. SHIELDS III • T. EUGENE

MR. AND MRS. W. TEMPLE W

To Keep Him First
DONALD W. REYNOLDS FOUNDATION
THE RICHARD AND HELEN DEVOS FOUNDATION
FRANK PETERS MOORE • ROBERTSON FOUNDATION
THE GORDON V. AND HELEN C. SMITH FOUNDATION
THE JOHNSTON-LEMON GROUP • WILLIAM R. KENAN, JR. CHARITABLE TRUST
MARS FOUNDATION • RICHARD KING MELLON FOUNDATION
MELODY SAWYER RICHARDSON • RICHARD M. SCAIFE • SARAH SCAIFE FOUNDATION INCORPORATED
RICHARD AND ADELIA SIMPLOT • THE L. J. SKAGGS AND MARY C. SKAGGS FOUNDATION
JEAN VAN HORNE BABER • BACARDI USA, INC. • ROYCE AND KATHRYN MCCORMICK BAKER
THEODORE AND ST. CLAIR CRAVER • DIAGEO • DISTILLED SPIRITS COUNCIL OF THE UNITED STATES
FUTURE BRANDS LLC • LYNN AND STEWART GAMMILL • DOUGLAS AND CAROLYN GOMEZ
THE HISTORY CHANNEL • MICHAEL AND LINDY KEISER
MR. AND MRS. BENJAMIN F. LUCAS II • THE LYNCH FOUNDATION
ELEANOR AND PHILLIP NEWMAN • GEORGE AND CAROL OVEREND
MR. AND MRS. FRANZ T. STONE • JACK AND SUSAN TAYLOR
THE JOE WEIDER FOUNDATION • CHUCK AND LUCY WELLER • ANN AND STEPHEN WEST

Robert H. and Clarice Smith

Repeatedly over the course of the capital campaign, a singular Virginia couple stepped forward to play a major role in the design and construction of Mount Vernon's new facilities. Robert H. and Clarice Smith initiated their support in 2000 by sponsoring the creation of an auditorium in the expanded Mount Vernon Inn complex. To help bring this new space to life, they endowed a three-part lecture series in honor of their daughter Michelle.

The Smiths also played a significant role in the building of both the Ford Orientation Center and the Donald W. Reynolds Museum and Education Center, underwriting the larger theater in the first facility and the multimedia presentation on the Revolutionary War in the second—the latter in honor of their son David and his wife, Elizabeth. They also endowed the position of senior curator, created a book-publishing fund, underwrote the cost of numerous brick walkways connecting the new facilities to the historic area, funded a redesign of the Mount Vernon website, and sponsored the sculpting of the four life-sized bronze statues of George and Martha Washington and their two grandchildren, posed as if greeting visitors entering the Ford Orientation Center. A nationally recognized collector of Renaissance bronzes, Mr. Smith was consulted in the creation of these statues from beginning to end.

In addition, Mr. Smith volunteered to create and fund what is now known as the "big tree program." Sixty-six mature trees, some of them 40 feet tall, have given the landscape around the new facilities the kind of presence only nature can bestow.

Robert H. Smith's family-owned company is best known for developing the Crystal City complex in northern Virginia. He serves as chairman of Vornado/Charles E. Smith Commercial Realty, a division of Vornado Realty Trust, and as chairman of Charles E. Smith Residential, a division of Archstone-Smith. Clarice Smith is a painter whose subjects include portraits, florals, landscapes, still lifes, and horses. She has had numerous solo exhibitions in New York and abroad. For a number of years, she taught art on the faculty of George Washington University.

At the groundbreaking ceremony on June 17, 2004, a 30-foot red maple tree was planted to commemorate the occasion. Honored guests added soil imported from 23 different sites related to George Washington's life to help nurture the tree.

The Smiths' philanthropy has touched numerous local, national, and international organizations, and in every case, their support has also included lending energy and expertise. Their generosity has had a lasting impact at several other historic sites (including Monticello, Montpelier, and President Lincoln's Cottage) as well as at the University of Maryland, the National Gallery of Art, the Wilmer Eye Clinic at Johns Hopkins University, the Mayo Foundation, and Hebrew University in Jerusalem.

Robert H. and Clarice Smith

Regents of the Mount Vernon Ladies' Association: 1995–2007

Mrs. H. Taylor Morrissette
(Vaughan)

Regent 1993–96

Mrs. Robert E. Lee IV
(Carew)

Regent 1996–99

Mrs. James M. Walton
(Ellen)

Regent 1999–2004

Mrs. Stanley N. Gaines
(Gay)

Regent 2004–07

Building Committees: 1995–2007

In October 1992, Mrs. H. Taylor Morrissette of Alabama was elected Regent of the Mount Vernon Ladies' Association. She established the Planning and Resources Ad Hoc Committee, chaired by Mrs. Donald J. Nalty of Louisiana, and including Vice Regents Mrs. Robert E. Lee IV of Maryland, Mrs. Stewart Gammill III of Mississippi, Mrs. Phillip Newman of Kentucky, Mrs. Douglas Seaman of Wisconsin, Mrs. Shepard B. Ansley of Georgia, Mrs. James M. Walton of Pennsylvania, and Mount Vernon Executive Director James C. Rees.

In October 1996, Mrs. Robert E. Lee IV of Maryland was elected Regent of the Mount Vernon Ladies' Association. She established the Construction Oversight Committee for the Mount Vernon Inn, a subcommittee of the Mount Vernon Inn (standing) Committee. Chaired by Mrs. Richard W. Call, Vice Regent for California, its members included Mrs. Lee, Mrs. P. William Moore, Jr., of Virginia, Mrs. Nelson T. Shields III of Delaware, and Mrs. Stephen K. West of New Jersey. Assisting the committee were four advisors with construction-industry experience—W. O. (Bill) Jones III of Richmond, Robert Fitton of Alexandria, Mark Lowham of McLean, and Preston Caruthers of Arlington—as well as Executive Director James C. Rees.

In October 1999, Mrs. James M. Walton of Pennsylvania was elected Regent of the Mount Vernon Ladies' Association. A New Facilities Oversight Subcommittee was established from the Planning and Resources Ad Hoc Committee to oversee design and construction of the Ford Orientation Center and the Donald W. Reynolds Museum and Education Center. Chaired by Mrs. Shepard B. Ansley of Georgia, its members included Mrs. Walton, Mrs. Stanley N. Gaines of Florida, Mrs. Joseph W. Henderson III of the District of Columbia, Mrs. Jared I. Edwards of Connecticut, Mrs. Robert W. Lawson III of West Virginia, and Executive Director James C. Rees. Sean Regan served as owner's representative and committee consultant.

In October 2004, Mrs. Stanley N. Gaines of Florida was elected Regent of the Mount Vernon Ladies' Association. She kept the New Facilities Oversight Subcommittee in place and established the Film Ad Hoc Committee to oversee development and production of the orientation movie, *We Fight to Be Free*. Chaired by Mrs. Melody Sawyer Richardson, Vice Regent for Ohio, the committee also included Mrs. Gaines, Mrs. James Evan Allison of Washington, Mrs. John F. Bookout III of Texas, Mrs. Sam Buchanan of Arkansas, Executive Director James C. Rees and Mount Vernon Advisory Committee Chairman Roger H. Mudd. The new buildings were completed and opened during the first two years of Mrs. Gaines's tenure.

E
W

4 Master Planning

The expanded mission of the Mount Vernon Ladies' Association clearly redefined the educational needs to be met by the Mount Vernon Estate. A more gracious and informative introduction to the estate, a more efficient organization for the circulation of visitors, and improved infrastructure to accommodate the million-plus visitors each year were other goals to be defined before beginning the design process.

In the winter and spring of 1995, the Association worked with programming consultants to develop a preliminary description of the proposed new facilities. In May, this effort produced a summary document, titled *A Case Statement*, which anticipated one building that would house an orientation center of approximately 20,000 square feet, an education center of 25,000 square feet, and a 6,000-square-foot expansion of the existing restaurants and gift shop associated with the Mount Vernon Inn.

The Association devised a three-phase process to select the design team and asked about two dozen architects to submit their qualifications. The request suggested that "most, if not all, of the new construction will be underground" and that the architects should have a proven ability "to integrate new construction with an existing historical architectural vocabulary and aesthetic." The architects were instructed to submit credentials that highlighted their similar design experience, explained their design philosophy, and identified the key individuals and consultant firms to be involved.

Once these new facilities and experiences are introduced, virtually no question about Washington will go unanswered, and no aspect of his remarkable life will be ignored. Visitors will depart from the estate with a full understanding of Washington's achievements. Students will once again understand the difference between a celebrity and a genuine hero. And perhaps most importantly, they will comprehend the qualities of leadership and character that have made Washington the most celebrated leader in history.

— Mount Vernon Ladies' Association case statement for the new facilites, May 1995

Phase one of the selection process required submission of written materials by June 2, 1995. By the following Wednesday, four firms—GWWO, Inc./Architects of Baltimore; Keys Condon Florance of Washington, D.C.; Hartman-Cox Architects, also of Washington; and Shepley Bulfinch Richardson & Abbott of Boston—were chosen to participate in the second and third phases. Each firm had a solid history in the design of visitor-related projects for "high-visibility" institutional clients as well as considerable experience working on historically sensitive sites. Phase two of the process included a site walk-through at the estate, a

Aerial photograph of George Washington's Mount Vernon Estate and Gardens in 1995, before master planning began. North orientation is to the right.

Site plan indicating designated boundaries for the project, as delineated in the 1995 project guidelines.

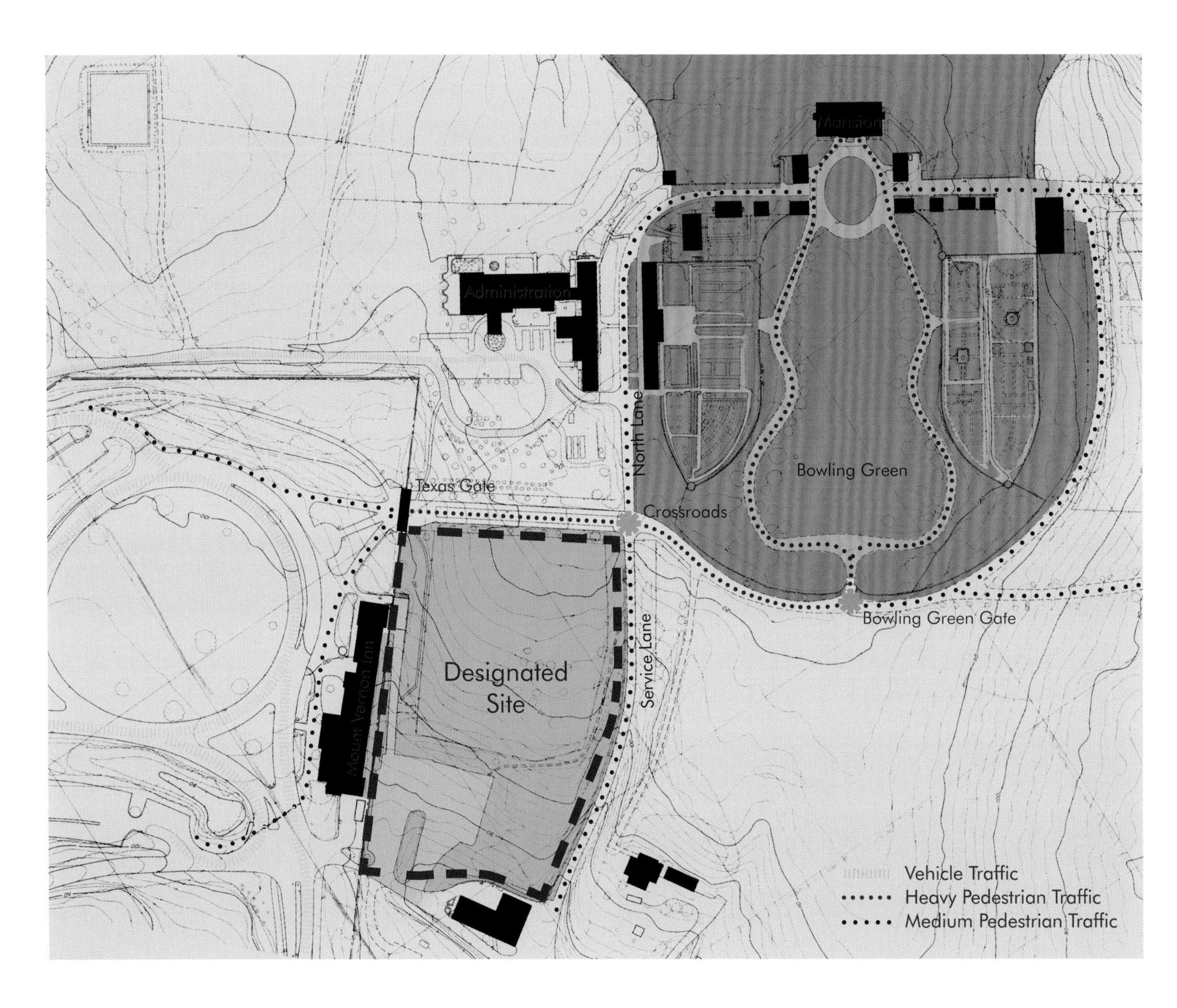

Called North. Image not to scale.

second written submission solidifying the team members and consultants, and visits by the Mount Vernon Ladies' Association building committee to each of the firms' offices. For phase three, each short-listed firm presented ideas for executing its design to the building committee and key estate staff members.

On June 19, Mount Vernon Executive Director James C. Rees led the walk-through of the grounds. After summarizing the Association's expanded mission and reviewing the scope of the program and its requirements, he itemized and discussed the physical boundaries of the project and construction site. At the time, the proposed site was an open field of approximately four acres, used for a temporary orientation tent and bordered by the Texas Gate and the existing brick boundary wall. Beyond this wall were the Mount Vernon Inn, built around 1931, and a small post office. Everyone recognized the importance of having the site set apart from the historic core of the estate, and it was a given that any new construction should be invisible from the bowling green gate. Tests conducted on the designated construction site revealed that it had no archeological significance.

When ownership of the Inn, which is listed on the National Register of Historic Places, was transferred to the Association in 1992, the National Park Service added protection of its north facade, which fronts the George Washington Memorial Parkway, to the agreement. Therefore, any modifications to the building's appearance had to be in keeping with its historic character. The site of the Texas Gate and the structure itself, built in 1899, could be integrated into the overall organization of the new facilities, but it, too, had been deemed a historic feature of the estate.

Following the walk-through, the contending design firms had three weeks to develop an overall concept plan that would satisfy the program within the defined boundaries of the site, enhance the visitor's experience, and most important of all, honor the legacy of George Washington. GWWO enthusiastically embraced this challenge by forming a cohesive project team, which began to consider a variety of design approaches.

Program Analysis

Members of the GWWO team identified several key concepts that would prove to be crucial to the design they eventually submitted. Although it was originally assumed that most, if not all, of the new construction would be underground in order to minimize its visual impact upon the historic estate, the design team had concerns about the effect of such a solution upon visitors. Wouldn't it feel rather uninviting to enter the estate and immediately descend 12 to 15 feet into an underground area for orientation? Since the site rose about 15 feet between the Texas Gate and the bowling green gate, an underground approach would work against the topography, placing visitors 25 feet below the level of their ultimate destination—the estate's historic core and Washington's Mansion.

As the team continued to consider the program, they identified other concerns. For example, combining all the

Tented facilities housing estate orientation film and information, 1995.

necessary spaces into a single complex could create crowd-control problems, especially during busy tourist seasons. Visitors just starting their tour—leaving the orientation area on their way to the estate and Mansion—would likely encounter people returning after the tour to the education center, gift shop, and restaurants; the resulting pedestrian "traffic jam" would be disorienting and inhospitable. Designating portions of the facility as "free zones," areas where no tickets are required, was also an important matter. The restaurants and gift shop, for instance, had to be separated from the rest of the ticketed venues.

Finally, the design team considered the realities of funding the construction. The possibility of having to phase the overall construction into smaller components was a real one, contingent upon the success of a major capital campaign that could take several years to be conducted. A design with one large, interconnected underground building might be difficult to build in a phased manner. Because the Mount Vernon Ladies' Association wanted to keep the estate fully open and accessible throughout construction, any phasing increments would have to be carefully gauged both to satisfy capital campaign goals and to minimize visitor inconvenience.

Design Response

Knowing that the main reason people visit Mount Vernon is to see the Mansion, the GWWO team realized that a sound design would enable them to do so as soon after arriving at the estate as possible. It was also important, however, for visitors to access the orientation facilities, including the ticketing operation, immediately upon arrival. Based on these factors, the design team chose to separate the program components physically—orientation center, education center, and the Mount Vernon Inn—in order to achieve ideal visitor-flow sequence: orientation, Mansion and estate, education center, and finally, the Mount Vernon Inn restaurants and gift shop.

Recognizing that the Texas Gate had for a century served as the main public entry to the estate, the designers felt it merited respect and thus should be preserved. After passing through the gate, visitors would enter a bucolic environment that would evoke a sense of 18th-century agrarian life. The orientation center would be an aboveground building located to the left of an entry plaza; the structure actually would be slightly outside the prescribed site but well within the nonhistoric areas,

Site section from the Texas Gate to the crossroads, showing conditions that existed prior to the project. Image by GWWO, Inc./Architects.

tucked behind brick garden walls that would shield its volume. Besides buying tickets there, visitors could watch a short film about George Washington to be shown in two theaters. After leaving the theaters, visitors would ascend a path to the traditional crossroads and proceed to the Mansion and the rest of the estate.

After completing their estate tour, visitors would take a new path from the crossroads heading northwest toward the education center, along a walk that sloped down by the far edge of the entry pasture. According to the design, the gently rising pasture that would be visible upon entering the Texas Gate actually would cover nearly all of the education facility! Visitors would experience the new interactive exhibits in a building that would be buried on all sides except its northwest face. The restaurant and gift shop facilities would be expanded until they connected seamlessly with the education center. At the end of their tour, visitors could gaze at the lower pasture while enjoying a bite to eat. By allowing visitors to then exit through the gift shop, the circulation problem that had existed since 1899—because the Texas Gate served as both the estate's entrance and exit—would be solved.

Although GWWO's design was astute, it ignored a dictate of the Mount Vernon Ladies' Association building committee by separating the program into several pieces and locating almost half of the buildings above ground. It exceeded the designated construction limits. And finally, it proposed a visitor flow that the client had not anticipated.

Top: *The Texas Gate, built 1899; restored 1989.*
Bottom: *Upon entering the estate, visitors first see a bucolic sheep pasture. Rendering by EDAW, Inc.*

Design Presentation

On July 10, 1995, the GWWO design team presented its concept plans to the Association's building committee. The architects described their vision for providing a visitor flow that would be welcoming and uncongested. They stressed that the natural beauty of the site would greet arriving visitors while concealing more than 50 percent of the required project square footage. Within a week, Mount Vernon Executive Director James C. Rees informed

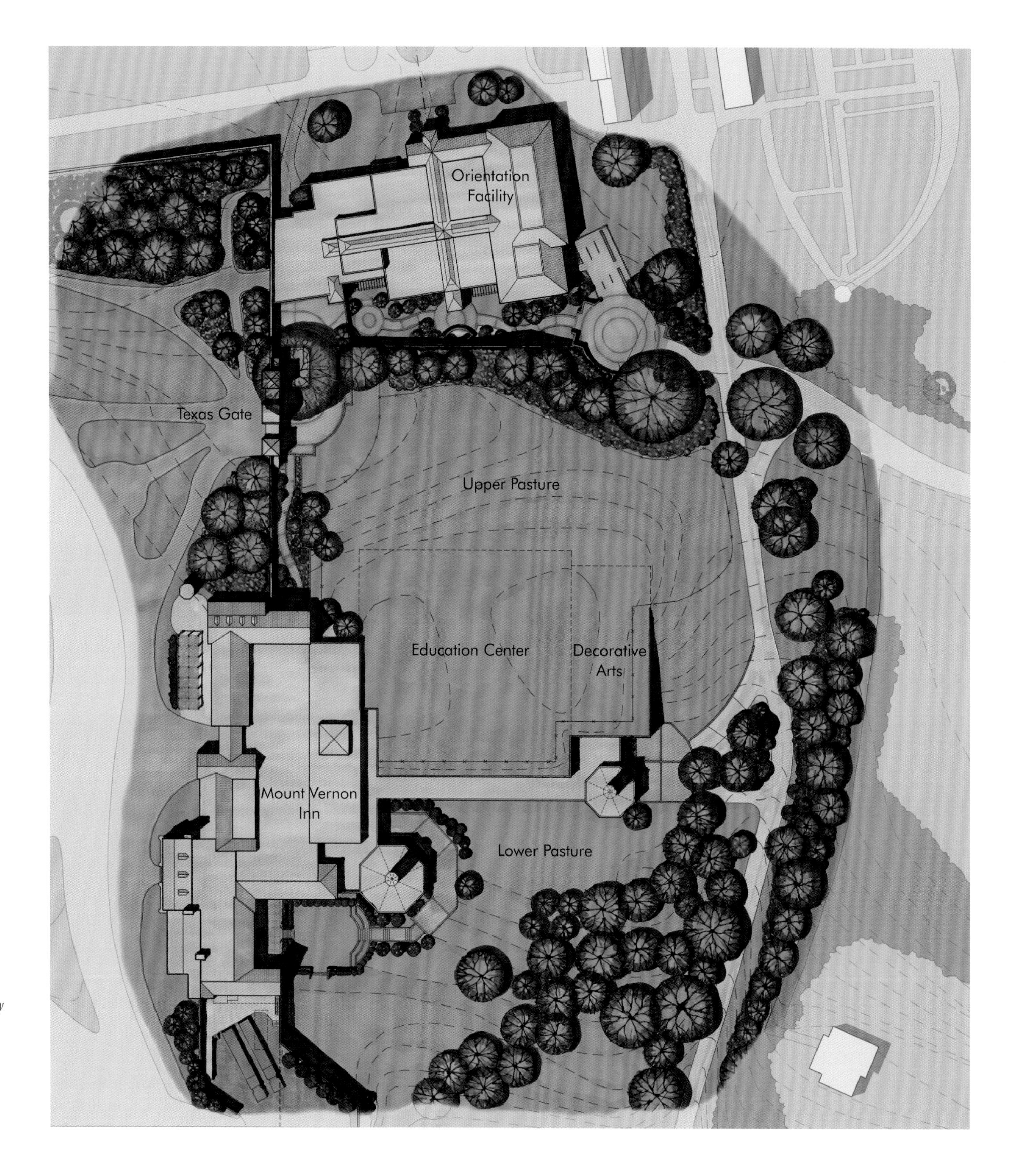

Winning master plan entry by GWWO, Inc./Architects.

Called North. Image not to scale.

GWWO that its team had been selected for the prestigious commission. Rees indicated that of all the presentations, the one conceived by GWWO demonstrated the greatest commitment to work with the Mount Vernon board and staff to create a design that respected the importance of the estate and the Association's mission.

Key to the team's success was its focus on the landscape as well as on the architecture. The building committee appreciated how the proposed design emphasized the agrarian nature of the estate. Immediately upon entering the Texas Gate, visitors would see a pasture instead of the paved entry lane that went directly from the gate to the crossroads. The committee also was intrigued by the design team's suggestion that the project be divided among multiple buildings, an approach that would spare visitors the need to descend into an underground orientation center and then ascend to reach the gardens and Mansion. Establishing the education center as the last stop for most visitors, who would exit directly into the estate-side lobby of the restaurant and gift shop area, marked an obvious division between ticketed and free zones.

On September 13, Regent Mrs. H. Taylor Morrissette offered a celebratory champagne toast to the start of an exciting and challenging design process, and then she ceremoniously signed a contract between the Mount Vernon Ladies' Association and GWWO. That evening, on the Mansion piazza, the design team joined 24 guests, including former Virginia governor A. Linwood Holton and broadcast journalist Roger Mudd, as Mount Vernon hosted the first of many capital campaign dinners.

During the next six months, the GWWO team developed the site master plan and created conceptual designs for the three building components: a new orientation center, a new education center, and an expansion and renovation of the existing restaurant and gift shop operations at the Mount Vernon Inn complex. Over time, several elements of the original proposal were altered and expanded. For example, education center space was increased by 20 percent to incorporate a 5,000-square-foot museum, which would feature exhibits of the many decorative arts objects in the estate's collection. Several additions were also planned for the Mount Vernon Inn, with new construction increasing to more than 22,000 square feet and renovation to more than 28,000 square feet. In April 1996 the design team marked the conclusion of phase one by presenting to the Mount Vernon Ladies' Association building committee its master plan, including its concept of the layout and arrangement of various structures.

The project paused at this point while the Association continued its capital campaign and assessed the income-producing capabilities of the proposed expansions and upgrades to the food, gift, and fine-dining facilities. Through the winter, spring, and summer of 1997, the architects assisted in the fundraising effort by providing conceptual views of the planned new construction.

Building Design Begins

At its fall 1997 semiannual council meeting, the Mount Vernon Ladies' Association voted to proceed with design and construction of the first segment of the master plan,

Details of new Mount Vernon Inn addition by GWWO, Inc./ Architects.
Left: *200-seat Robert H. and Clarice Smith Auditorium.*
Right: *Exterior of the dining pavilion.*

the addition to and renovation of the Mount Vernon Inn. In early 1998 GWWO began this work, which was critical because it finalized the location where visitors would exit the estate. The building had to be organized so as to provide a free (i.e., non-ticketed) zone for the estate. In addition, it would house mechanical infrastructure for the future museum and education center. Income generated by the Inn's expanded fine-dining facilities, casual restaurant, shopping, and visitor services would support the activities of the entire estate. This phase of the project also would provide amenities never before available, such as the 200-seat auditorium with multimedia projection capabilities, a new enclosed terrace for fine dining, and a private outdoor terrace available for special events. Although creating these facilities entailed the addition of a significant amount of new square footage, the team strove to maintain the character and massing of the existing National Register-listed building.

Construction on the Inn project began in the winter of 1999–2000, but even before it was finished, both the Association's building committee and the design team were drawing up a list of master plan design issues that required further investigation. Although the intended location and arrangement of the buildings were valid, certain details had to be fine-tuned. For example, it was vital that the entrance to the education center be designed, either through architecture or landscaping or both, so that visitors would be naturally drawn toward it. The congestion of arriving and departing visitors at the crossroads was another matter that needed more attention. Finally, how much if any of the new construction would be visible from the crossroads was a question of ongoing concern. In the conceptual design for the orientation center, the theaters were placed close to the historic north lane, potentially intruding upon the visitor's experience; it was necessary to study their location and massing in order to reduce their impact on the site. Before they could design the remaining new facilities, the GWWO team had to address these important issues—and a host of others.

New entrance to the Mount Vernon Inn complex from the circle drive at the terminus of the George Washington Memorial Parkway.

In the Spirit of George Washington

The challenge of designing more than 71,000 square feet of new structures on George Washington's estate required respect and reverence for this vital historic site. Since Washington's death, in 1799, many features had been added to the estate, some of which were more in keeping with its 18th-century character than others. From the beginning, the GWWO team recognized the importance of planning and situating these new buildings in ways that would not overwhelm the Mansion or the estate grounds. They also saw the need to focus on a single, unifying concept that would link the master plan with the history of the Mount Vernon Estate.

George Washington's Mount Vernon Estate and Gardens with the new Ford Orientation Center and Donald W. Reynolds Museum and Education Center. Rendering by W. G. Hook, 2006.

First, however, the team had to identify the following constraints and technical goals for the building program:

- The scale and mass of the buildings must not impose a modern-day presence on the 18th-century estate.

- The National Park Service maintains the George Washington Parkway leading to Mount Vernon. Because the vista from the terminus circle of the parkway toward the Texas Gate and the boundary walls joining it is considered historic, the Park Service required that this view not be altered by the new orientation center.

- The museum and education center experience must be engaging and nonthreatening, despite the building being sited underground.

- Although the new structures would be near the Mansion and its surrounding dependencies, they must clearly communicate to visitors they are not part of the historic ensemble.

The design team also explored ways of handling the flow of visitors to Mount Vernon. At peak times, up to 4,000 people could arrive within a few hours, and they had to be efficiently admitted. During such times, the spaces should feel comfortable and easy to navigate; during off-peak seasons, visitors should appreciate a more intimate experience. Ideally, the design of the facilities and exhibits

Thomas Oldham Barlow, The Home of Washington, *engraved after a painting by Thomas Pritchard Rossiter and Louis Remy Mignot, 1860.*

Unless some one pops in, unexpectedly, Mrs. Washington and myself will do what I believe has not been [done] within the last twenty years by us, that is to set down to dinner by ourselves.

— *George Washington, 1797*

Time-coded diagrams by GWWO representing the number of people in the orientation center throughout a day in an ***AVERAGE*** *visitation scenario.*

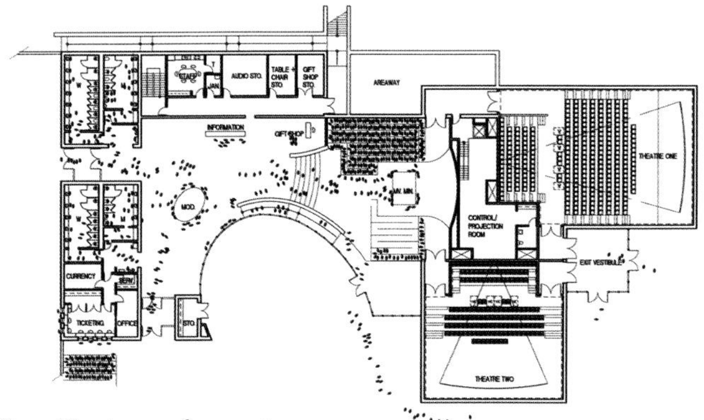

Time: 30 minutes after opening
Population milling: 336 people
Population queuing: 300 people

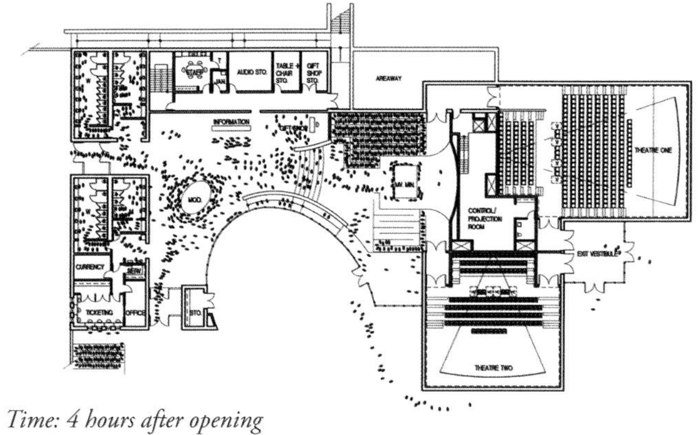

Time: 4 hours after opening
Population milling: 662 people
Population queuing: 300 people

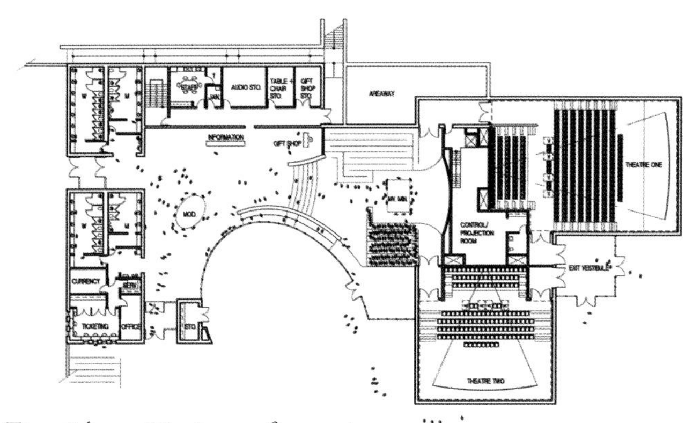

Time: 5 hours, 15 minutes after opening
Population milling: 129 people
Population queuing: 150 people

Time-coded diagrams by GWWO representing the number of people in the orientation center throughout a day in a ***PEAK*** *visitation scenario.*

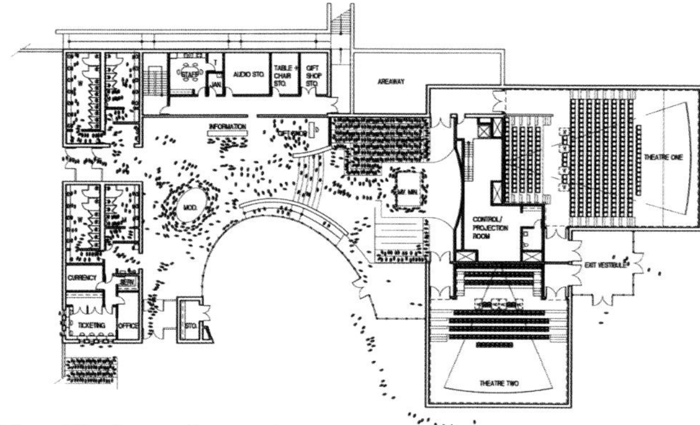

Time: 30 minutes after opening
Population milling: 638 people
Population queuing: 300 people

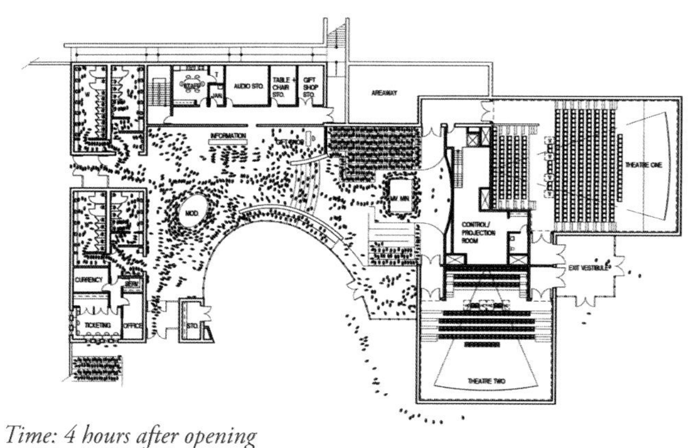

Time: 4 hours after opening
Population milling: 1,175 people
Population queuing: 300 people

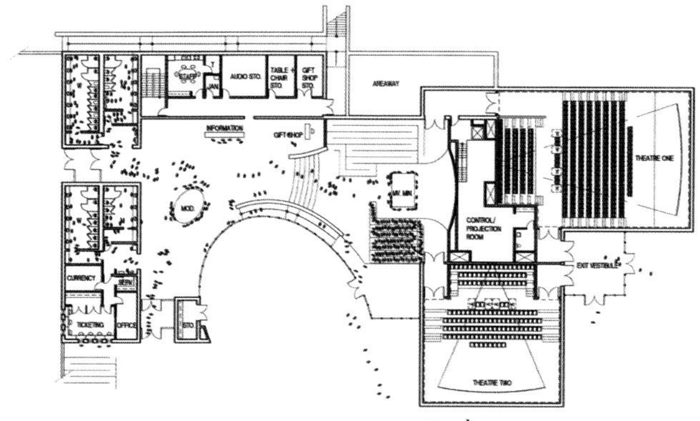

Time: 5 hours, 15 minutes after opening
Population milling: 305 people
Population queuing: 150 people

About one million visitors tour the Mansion each year. Those in line here are in the circle outside the west side of the Mansion.

would encourage visitors to proceed intuitively toward the Mansion and the historic core of the estate.

For guidance regarding these matters, the team turned to George Washington himself. An indefatigable host, he wrote in July 1797, "Unless some one pops in, unexpectedly, Mrs. Washington and myself will do what I believe has not been [done] within the last twenty years by us, that is to set down to dinner by ourselves." Obviously, he and his wife, Martha, received hundreds of visitors at the Mansion. How, the team members wondered, had Washington handled this influx of people?

Building upon the 1995 master plan, with additional research into the estate's history, the design team discovered that Washington had developed a sequence of views and paths that not only led visitors to the Mansion but also heightened their anticipation prior to arriving at the home of the nation's most prominent citizen. By carefully directing his visitors' approach along a route that was both formally designed and informally rustic, he drew attention to his considerable talents as master of a beautifully planned and maintained property.

The architects correctly concluded that several aspects of Washington's way of welcoming guests could inform the experience of modern-day visitors. Their anticipation and excitement should build as they proceed, so that the walk ultimately would enhance their experience of reaching the destination—that is, the Mansion.

1
2
3
4
5

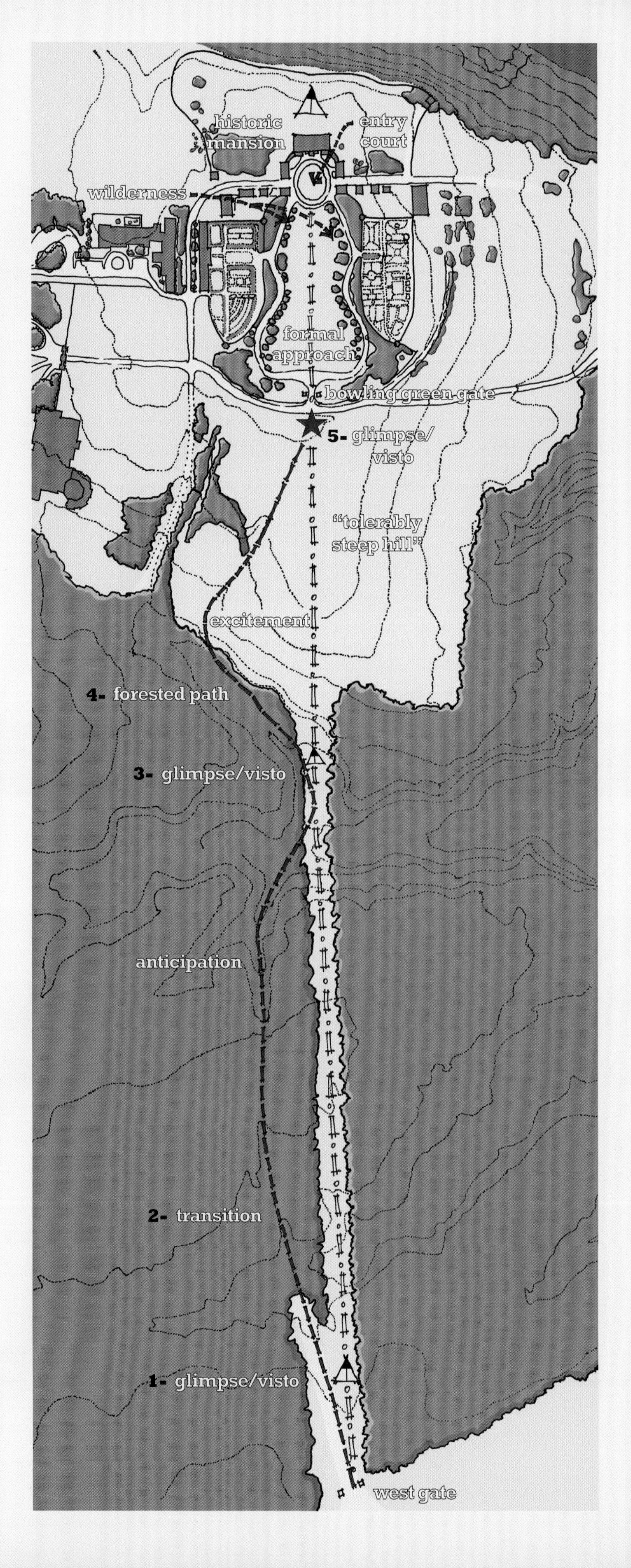

historic mansion
entry court
wilderness
formal approach
bowling green gate
5- glimpse/visto
"tolerably steep hill"
excitement
4- forested path
3- glimpse/visto
anticipation
2- transition
1- glimpse/visto
west gate

18th-Century Approach to Mount Vernon

Since the wharf on the Potomac River was reserved almost exclusively for deliveries, most of George Washington's visitors arrived overland, on roads and paths that meandered past the fields and pastures surrounding the Mansion House Farm. Washington placed a premium on first impressions. Visitors first sighted the Mansion from what is now known as the west gate. From there, they were afforded a "visto," or view, of the west face of the Mansion, in front of which were cleared land and rolling hills for about seven-tenths of a mile.

The path to the Mansion veered northward, winding through well-maintained woods and into several naturally formed deep ravines. According to visitors' accounts, letters Washington wrote to his farm manager Anthony Whiting, and historical maps, he created one or two subsequent vistos of the Mansion along the wooded path, providing visitors with glimpses of their destination. At the end of the path, they climbed a "tolerably steep hill," according to Edward Hooker, who visited in 1808, and saw the west face of the Mansion once more. Finally, visitors moved off the central axis to the left or right to travel along the north and south lanes to the Mansion, where they were received at the west front door.

Approaching Mount Vernon

Top: *Before the new buildings were constructed, visitors passed through the Texas Gate and proceeded on a straight, brick-paved, tree-lined pathway that led to the estate crossroads.* Bottom: *Through the Texas Gate, visitors now view the pasture atop the Donald W. Reynolds Museum and Education Center.*

Upon reaching the traffic circle at the end of the George Washington Parkway, a thoroughfare that affords views of wildlife, river, and woods, visitors first see a white painted-brick structure known as the Texas Gate. Since 1899, this has served as the main entrance to the estate, replacing the original approach via the west gate. Until the new buildings were opened in 2006, visitors passed through the black iron gates along a straight, brick-paved, tree-lined pathway that led directly to the north-lane crossroads. Recognizing that this direct approach was very different from the meandering path from the west gate to the Mansion that George Washington's guests took, the architects resolved to bring modern-day visitors to the historic center of the estate in a way that evoked the first president's own hospitable manner.

The designers agreed that in the new arrival sequence, visitors should be pleasantly aware of a transformation of their environment as they pass through the Texas Gate—from today's fast pace outside the historic estate to the pastoral calm within. They also should immediately experience a sense of the estate's vastness and agrarian character. The site and its natural setting, free of any modern buildings, should make an indelible first impression. Early in the design process, the architects imagined a setting that would create this desired feeling: a serene, grassy pasture gently sloping up to the southwest, grazed by sheep and bordered by thick forest. Visitor attention would focus on field, forest, and farm animals, thus suggesting a connection with the past as well as the importance of farming in Washington's life.

From the entry plaza, visitors can view the pasture atop the Donald W. Reynolds Museum and Education Center.

Perspective at the entry of the Ford Orientation Center. Watercolor by Vladislav Yeliseyev.

As a surveyor, pioneer farmer, and fisherman, George Washington earned his livelihood from the natural environment. His interest in and talent for landscape design are evident on the estate and well documented in his own writings and in books by others found in his library. To inform the team about Washington's landscape-design principles and techniques, the Mount Vernon Ladies' Association commissioned a cultural landscape study of the estate. It not only documented changes within the landscape over time but also tied them to the social and cultural context that paralleled the estate's development. The results would enable the designers to emphasize the natural world as they created the visitor experience. Based on their new knowledge, they devised imaginative ways to bring the outdoors inside—that is, to make the site design and building designs integral and the transitions between them seamless. Vistas, inspired by those Washington had cultivated as part of the approach to the Mansion, became

Entry courtyard to the Ford Orientation Center and Mount Vernon Estate.

central landscape features of the new master plan. As Washington had intended, these views reveal the vastness of the estate, offer glimpses of where one is headed, and kindle an appreciation of nature's peace and beauty.

In the plan, the centerpiece of the overall design, as well as the central element of the orientation center, would be the pasture that welcomes visitors entering the grounds. Other landscape features that Washington favored would further contribute to this first view of the estate and set the stage for visitors' educational experiences. An opening through the tree line on the far side of the pasture affords a view into the property and toward the open spaces of the great meadow in front of the Mansion grounds. A simple rail fence similar to ones used in Washington's time contains the sheep in the pasture—a green roof over the museum and education center.

Top left: *Existing column on the Mansion piazza.*
Top right: *Column in the Ford Orientation Center entry plaza.*
Bottom left: *Historic shutters on the west front door of the Mansion.*
Bottom right: *Gate designed to give visitors a glimpse from the entry plaza to the elliptical garden at the Ford Orientation Center.*

In the Spirit of George Washington

One of the most fiercely debated subjects in the field of architecture today is the so-called language of new buildings built in historic settings. Given the mission of the Mount Vernon Ladies' Association to restore the estate to its 1799 condition, the GWWO team never wavered from its position that visitors should realize the new facilities are not of George Washington's time. All team members appreciated the importance of creating a design that met three critical criteria: It must be respectful of the estate's historic significance, support the mission of the estate, and assure that visitors to the estate have a meaningful learning experience.

As the project proceeded, the team pondered how George Washington himself might have architecturally expressed the mission of the new buildings. The designers had no intention of copying the Mansion's Georgian/Federal style; such mimicry of the existing historic structures might confuse visitors about the origin of the new structures and also might dim their appreciation of Washington's own sense of style.

Top left: *Historic windows at the Mansion.*
Top right: *Window frames at the Ford Orientation Center.*
Bottom left: *Espalier on a garden wall of the estate.*
Bottom right: *Modern metal trellises, modeled on Washington's historic espaliers, attached to the Ford Orientation Center.*

Rather than copying the historic style, the architects strove to design new buildings that would generate a positive tension between the historic precedent and a modern-day expression. Toward that end, the team coined the phrase "in the spirit of George Washington," a succinct reminder always to acknowledge the influence of Washington, his landscape design, and the central role nature played in his life. In making decisions about design details, choice of materials, and spatial volumes within the new buildings and their surroundings, the team deferred to Washington's personality, his self-expression, and his progressive ideas as reflected throughout Mount Vernon.

The extended design team—architects, consultants, and the client committee—embraced the lofty ideal of acting in the spirit of George Washington. They all agreed that Washington would have explored ways to incorporate innovative technology within the new buildings. And they recognized, therefore, that the new structures should comprise a gracious, albeit contemporary, backdrop for the Mount Vernon story.

Landscape Techniques Used by George Washington

Ha-Ha Wall

A ha-ha wall, also known as a sunken fence, was an effective 18th-century device for containing livestock without obstructing views. East of George Washington's Mansion, deer often grazed in a lower pasture separated by brick retaining walls from the lawn on the higher elevation. The view from the Mansion of landscape and livestock extended over the tops of these walls, which were concealed by the grade of the lawn. Today's estate-entry view incorporates design techniques inspired by the ha-ha wall. A wall near the entry plaza but beyond the fence protects and shields a low trench drain, which carries excess rainwater away from the area. A second ha-ha wall serves as the west boundary for the pasture, which covers the roof of the museum and education center. The lower portion of the ha-ha wall becomes the serpentine glass exit-corridor wall.

Visto/Vista

Washington delighted in framing exterior views, which he called vistos, by manipulating natural features of the landscape as well as introducing elements he designed and placed. Most notably, he framed views of the Mansion at points along the entry path to the estate, including the long view from the west gate and the one from the bowling green gate over

the lawn and framed by the tree-lined serpentine paths on either side. He also carefully implemented design features, such as the ha-ha wall and the hanging wood planted on the steep slope between the east lawn and the river, to accentuate the view from the Mansion over land to the water beyond.

Espaliers

Simple wood lattices built a few inches from brick garden walls serve as a framework for fruit trees, which thrive in sunny, warm environments. With espaliers, fruit trees prosper in shallow beds located between the garden path and enclosure walls, and they provide useful ornamentation for the walls themselves. On the west side of the orientation center and the north side of the education center, modern espaliers interpret this historic gardening technique. The plants help mitigate the large scale of the brick walls and provide seasonal changes.

Brick Garden Walls

Washington enclosed his upper garden with brick boundary walls, topped with a sloping cap that sheds water and provides a pleasingly proportioned finishing crown detail. He also fashioned a tasteful curvilinear transition at the top of his lower garden wall to compensate for the drop in topography. Pilasters—a thickening of the wall expressed as an

applied column—occur at intervals along the wall face, lending a rhythm to its length and a sense of scale to its massing. The exterior brick walls of the new buildings, designed to be compatible in scale and appearance with those of the historic estate, employ similar details.

Walkways and Paving

Walkways leading to and from the new buildings are made of materials similar to those Washington used. Brick pavers and fine gravel provide comfortable walking surfaces and withstand foot traffic well, just as they did in his day. Adjacent to these paths, cobblestone gutters crafted from a historic use of granite material prevent erosion caused by storm water. The plants and trees lining the meandering woodland walk from the orientation center to the bowling green gate echo the sense of security and tranquility that characterizes the road Washington created leading through the woods to the Mansion.

Hanging Wood

The term *hanging wood* refers to the placement of trees on the lower areas of steep slopes. Washington used this 18th-century technique to underline the view from the east lawn with greenery that would not obstruct the view

of the water—in effect carpeting the base of the broad vista with treetops. At the museum and education center, the view from the entry through the two-story window and out toward the woodlands beyond extends over the tops of trees planted at the Mount Vernon Inn complex.

Shaping the Land

Washington masterfully reshaped the topography on his estate to present views of the Mansion and surrounding landscape. The vista over the bowling green to the distant west gate and beyond is an example of his efforts to create a visual effect. From the bowling green gate, a short path facing the Mansion rises steeply to the flat lawn of the bowling green; the Mansion is prominently situated on this elevated platform. The landscaping around the orientation center and the museum and education center respects the historic spirit of George Washington's designs by directing visitors' attention toward pastoral views. Similarly, the slope and form of the land adjacent to the orientation center and the pasture atop the museum and education center create views through and over the surrounding landscape. The slope of the pasture also accommodates the volume of the building below.

Groves, Wildernesses, and Shrubberies

Washington created various natural environments around the Mansion, drawing upon the instruction and terminology he gleaned from Batty Langley's book *New Principles of Gardening* (1728). He favored naturalistic, fairly densely planted arrangements that were popular in the 18th century. Mac Griswold, in her book *Washington's Gardens at Mount Vernon: Landscape of the Inner Man* (1999), discusses his interest in experimenting with plant selections and arrangements. According to Griswold, "[H]is massed and banked thickets and plantations of trees both large and small, both deciduous and evergreen, were to be understoried with shrubs frothing with bloom and fragrance." Today, the landscaping around the orientation center and the museum and education center reflects Washington's stated penchant for closely planted groves, "wildernesses," and "shrubberies."

Master Plan, 2002

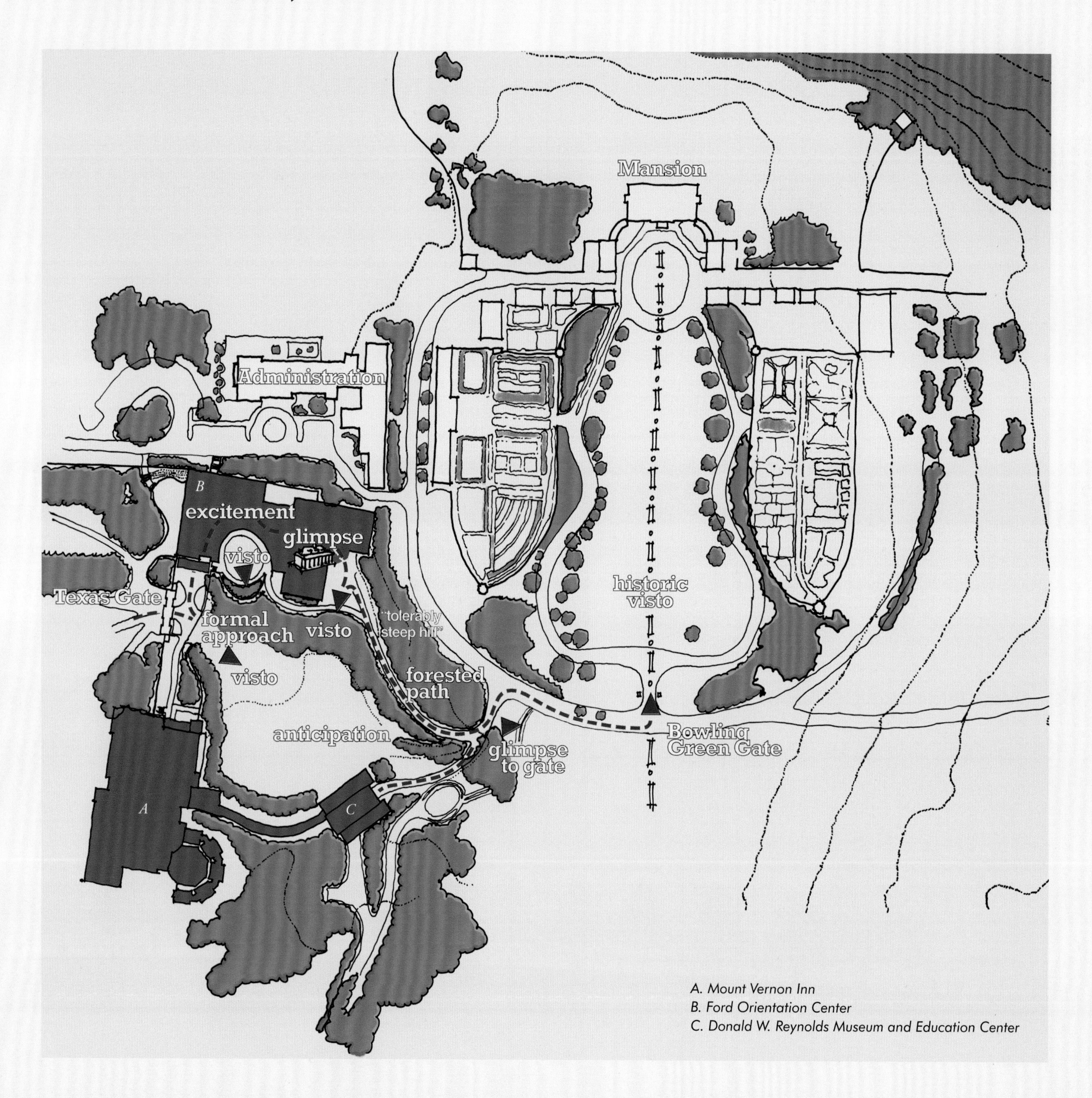

Views along a Visitor's Journey

1

As visitors enter the Ford Orientation Center, their journey unfolds clearly before them (fig. 1). Following the elliptical glass wall overlooking the clearing, they reach a vista of the pasture, orienting them along their journey (fig. 2). Glimpses of their destinations—*Mount Vernon in Miniature* and five stained-glass windows depicting "moments of truth" in Washington's life—draw visitors toward the theater lobby (fig. 3). After watching an informative 20-minute film, visitors exit to the gently sloping path leading to the historic estate (fig. 4). At the top of the path, they see their next destination, the historic bowling green gate vista of the Mansion (fig. 5).

2

4

3

5

Overleafs: *Ford Orientation Center elliptical clearing.*

Ford Motor Company Fund

The Ford Motor Company's extensive relationship with Mount Vernon has spanned the course of more than 80 years. From Henry Ford's donation of Mount Vernon's first fire truck in 1923 to its recent leadership-support of the new Ford Orientation Center, the Ford Motor Company has been a committed partner with the Mount Vernon Ladies' Association. The Ford Motor Company Fund continues its longstanding support of Mount Vernon through additional funding for the orientation center's programs and exhibits.

Right: *Keenly interested in preserving Mount Vernon, the inventor and industrialist Henry Ford donated a new fire truck to the Mount Vernon Ladies' Association in 1923 and encouraged the installation of fire hydrants and alarms.*

Henry Ford developed an early appreciation for George Washington's home at Mount Vernon. During a visit to the historic landmark in the early 1920s, he was greatly impressed by this tribute to the nation's first president. In the spirit of preserving an important icon of our nation's past, Mr. Ford commissioned a special chemical fire truck for Mount Vernon. The built-to-order American-LaFrance Combination Chemical and Hose Car arrived at the estate on September 21, 1923. This revolutionary vehicle was an essential tool in protecting Mount Vernon against its greatest threat. A blend of modern fire-fighting weapons and techniques combined with the dependability of a Ford Motor Company vehicle resulted in the ultimate in fire protection.

The arrival of this fire engine was only the beginning of the Ford Motor Company's commitment to protecting and preserving Mount Vernon as well as other aspects of American heritage. Inspired by the efforts of the Mount Vernon Ladies' Association, Henry Ford helped establish the Henry Ford Museum, now the nation's largest indoor/outdoor history museum, encompassing the Henry Ford Museum and Greenfield Village. The Henry Ford Museum strives to meet Mr. Ford's goal of documenting America's traditions of resourcefulness and innovation through its collection of American material culture.

The Ford Motor Company's dedication to the preservation of Mount Vernon continues today as part of its commitment to educating audiences about America's remarkable history. In the past decade alone, the Ford Motor Company Fund has supported a wide range of institutional educational initiatives and exhibitions, including student internship programs, the George Washington Biography Lesson curriculum guide for the State of Michigan, and major exhibitions such as *Saving Mount Vernon* and *Treasures from Mount Vernon*.

6 Ford Orientation Center: A Visitor's Introduction to the Estate

At the Ford Orientation Center, visitors to Mount Vernon can purchase tickets, gather information about all the other venues on the estate, use the restrooms, and watch a brief film about George Washington. The center is strategically located to provide a transition from the 21st century back to the 18th century. After entering the Texas Gate and viewing the pasture, which is the central organizing element of the master plan, visitors can see to the east the entry to the orientation center; landscape elements and a new garden wall conceal the rest of the building.

Ford Orientation Center entry plaza, featuring a bronze bas relief profile of Washington by Raymond J. Kaskey.

FORD
ORIENTATION
CENTER

Building Organization and Visitor Flow

Even before discussing a specific design vocabulary, the GWWO team had to refine the overall organization of the building, including its floor plan, massing, and volume. The building had to flow along in a natural way, and it had to deliver information and services efficiently.

Ticketing operations are the natural first stop on a tour of the estate. With so many exhibits and experiences available, visitors have a wealth of options from which to choose. Before leaving the orientation center, they are encouraged to watch a fast-paced, informative film about George Washington that places him in the context of 18th-century Mount Vernon.

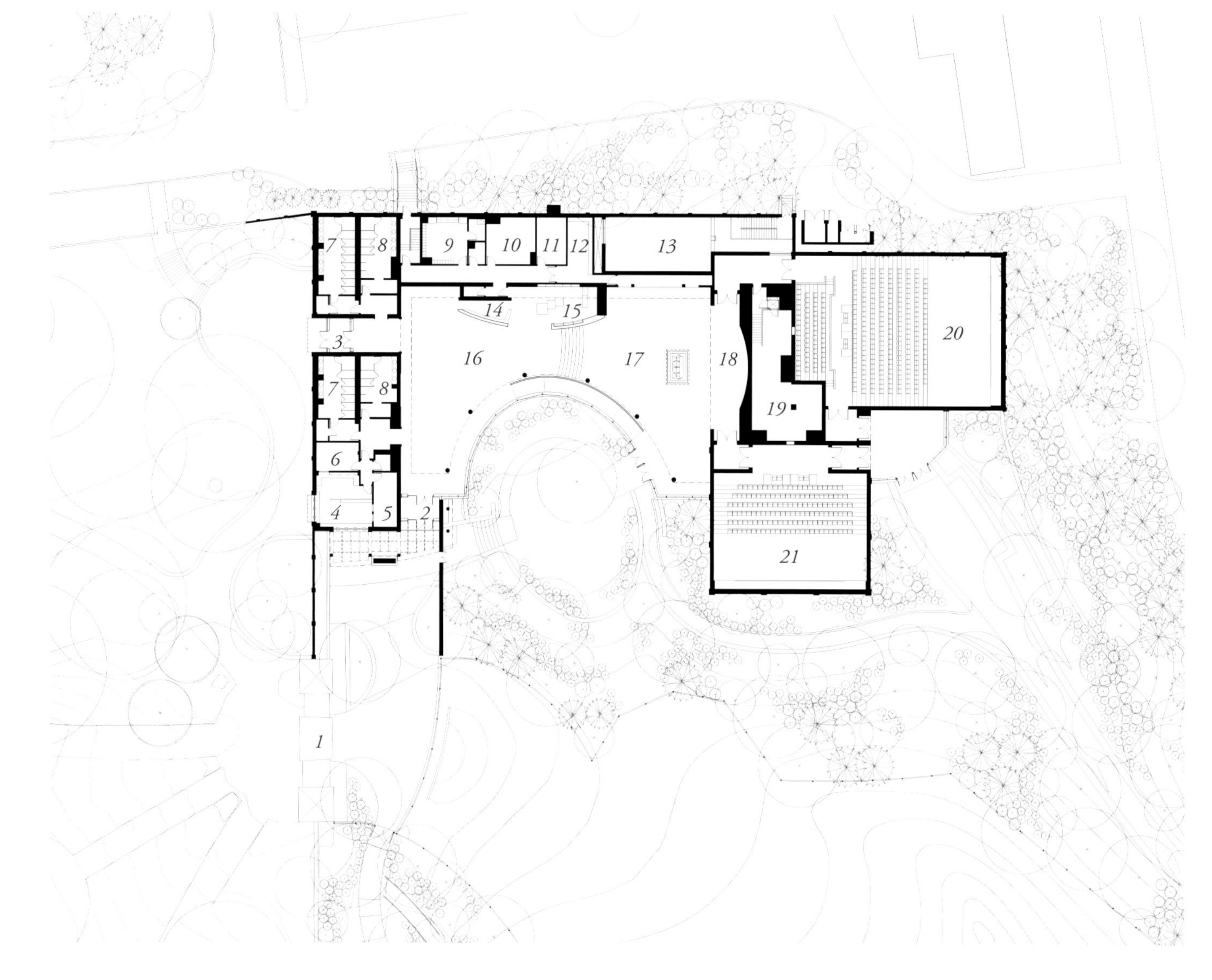

Ford Orientation Center site and building plan:
1. *Texas Gate*
2. *Individual entry*
3. *Group entry*
4. *Ticket office*
5. *Office*
6. *Office*
7. *Women's restroom*
8. *Men's restroom*
9. *Staff*
10. *Audio storage*
11. *General storage*
12. *Retail storage*
13. *Areaway*
14. *Visitor services*
15. *Retail*
16. *Upper lobby*
17. *Lower lobby*
18. *Queuing lobby*
19. *Projection room*
20. *Robert H. and Clarice Smith Theater*
21. *Edward P. Eagles Theater*

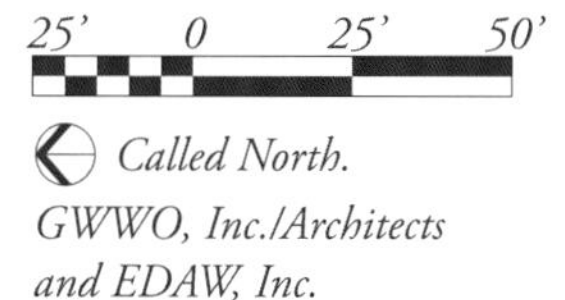

Called North.
GWWO, Inc./Architects and EDAW, Inc.

In times of peak traffic, it is necessary to prevent congestion in the building's entry lobby. The team explored ways that the design itself could encourage visitors to move—naturally and intuitively—toward the theater entrances. The solution was to position a focal point in the lower lobby at the base of the grand stairway, and an apt choice was *Mount Vernon in Miniature*, a meticulously handcrafted model of the Mansion and all its decorative contents. After being exhibited at a dozen sites across the United States, this popular replica has found an ideal permanent home.

Another feature integral to the lower-lobby décor is a set of stained-glass windows made in the mid-20th century by the studio of Karl J. Mueller and purchased for Mount Vernon in 1997 from the estate of U.S. Air Force Brigadier General Woodrow A. Abbott. The windows depict significant moments in George Washington's life and, therefore, serve both an educational and an aesthetic purpose.

Building Massing

The master plan dictated that the visual impact of new construction on the historic estate would be minimal. The theaters, located at the end of the visitors' passage through the orientation center, functionally required the largest volume and thus were the most difficult to conceal from view on the historic north lane. To address this concern, the architects determined that walls visible from the historic area would be in scale with garden walls on the estate. Because the topography rose to the south,

Educational exhibits in the theater lobby.
Top: *Lower lobby, featuring* Mount Vernon in Miniature *and stained-glass panels depicting scenes from Washington's life.*
Bottom: Mount Vernon in Miniature, *a one-twelfth-scale model of the Mansion.*

Building Concepts: Theater Volumes

In August 2002, the GWWO architects presented the Mount Vernon Ladies' Association building committee with three substantially different design schemes, all focused on the size and placement of the two theaters in the orientation center. The most extreme proposal, and one that tested the precepts of the master plan, was to construct most of the building housing the theaters underground. The ticketing operations and comfort/information facilities would surround a lobby on the same level as the Texas Gate entry. From there, visitors would descend a grand stairway to a lower lobby 10 feet below. A glass-enclosed garden courtyard in the lower lobby would be open to the sky and thus would preserve a connection to the outdoors. This lower-level space surrounding the courtyard would serve as the theater lobby. After viewing the film, visitors would exit into an underground lobby that transitioned to a well-lit, ramped exhibit space that tunneled under the service road above and emerge near the bowling green gate, ready to visit the Mansion.

Scheme one called for most of the building housing the theaters to be underground. A glass-enclosed garden, open to the sky, would bring light into the building and connect the spaces to nature.

The advantage of this scheme was that it would minimize the impact of the building size on the dimensions of the historic estate. An additional benefit of this plan was that it would provide a dramatic separation between the entrance and the lobby area.

Scheme two placed most of the building outside the existing estate-boundary walls. The Texas Gate would no longer be used as the main entry to the estate.

Another proposed plan likewise challenged project precepts but solved the issue of the theater volumes' proximity to the historic north lane. In it, the entire building would be shifted dramatically northward, thus distancing the theater volumes from the view of the historic area. Rather than passing through the Texas Gate and the boundary wall, visitors would enter the building from the north side of the wall, outside the limits of the estate. The space between the south end of the new building and the historic north lane would be heavily landscaped, disguising any views of the new building. The length of the path from the theater exit to the bowling green gate provided plenty of distance in which to slope gently up to higher ground, successfully resolving the vertical transition between the two.

Scheme three combined benefits of schemes one and two. It set the building housing the theaters into the land at the south end of the site and included vistas of the entry pasture and across the estate.

There were several downsides to this scheme: it altered the existing view from the George Washington Parkway terminus, and it relegated the Texas

Gate to a purely ceremonial role. The team also worried that the length of the path from the theater exit would be too great, possibly dampening the enthusiasm sparked in visitors at the orientation center.

After considerable deliberation about the aesthetic and technological challenges posed by the proposed size, scale, and placement of the theaters, the designers decided upon a third, and ultimately successful scheme. The theaters would be set into a grade, with a mid-level entry and exit. In contrast to the underground scheme, this one included a courtyard at grade and also provided vistas of the entry pasture and across the estate. These views of nature would help visitors orient themselves. Because the Mount Vernon Ladies' Association building committee appreciated the division between an upper and a lower lobby, as proposed in the underground scheme, the team incorporated aspects of this feature into the selected scheme.

After choosing scheme three, GWWO built a large-scale model of the orientation center to show the building spaces and forms.

they conceived the orientation center as a building set into the land rather than superimposed upon it. As in most any theater, the floor level naturally descends to provide unobstructed sightlines; this feature reduces the building volume relative to its surrounding landscape.

After lengthy consideration, the GWWO team ultimately developed an ingenious design in which the bulk of the theater is essentially "pressed down" into the hillside but is not wholly underground. A three-foot drop in the orientation center floor level creates distinct upper- and lower-lobby spaces while also diminishing the volumes of the theater. This configuration allowed the architects to limit the wall height on the south side adjacent to the north lane to a mere eight feet above exterior grade; because it is now in scale with other garden walls on the estate, this wall fits seamlessly into its historic surroundings. When visitors leave the center, they walk up a gently sloped woodland path. The space between the historic north lane and the building's south side is heavily landscaped, so people walking on the path see the wall through lush greenery.

The design team felt strongly that visitors should be aware of the natural environment, even while they are indoors. The lobby encircles the outdoor space on the west side of the building, thus making a visual connection between interior space and the pasture beyond. Working with the

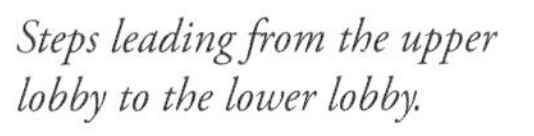

Steps leading from the upper lobby to the lower lobby.

landscape architecture firm EDAW, Inc., of Alexandria, Virginia, the GWWO designers envisioned this exterior space, which they called the clearing, as being a formally landscaped transition to the more agrarian pasture that covers the roof of the museum and education center.

Having originally envisioned the walls containing the clearing in various configurations—rectilinear, angled, or flattened arch—the design team concluded that a more organic shape, an ellipse, would create a graceful transition from the pasture to the refined interior space. The curving shape of this elliptical window wall directs the eye outward toward the pasture while lending a smooth flow to the area leading from the entrance lobby to the theater lobby.

Theater-bypass doors to the clearing and path to the theater-exit plaza.

The elliptical clearing, seen from the Ford Orientation Center lobby.

Raymond J. Kaskey's bronze bas relief profile of George Washington at the entrance to the Ford Orientation Center. Engraved on the stone panel on the brick wall behind the relief are these words of George Washington: "No estate in United America is more pleasantly situated than this . . ."

Combined with the *Mount Vernon in Miniature* model and the stained-glass exhibit, the window wall expedites the forward motion of visitors and helps prevent congestion at peak visitation times.

To enhance the sense of openness along the periphery of the orientation center, the designers installed clerestory windows. This feature, which continues from the highest bay of the window wall facing the clearing, offers a seamless view to nature from every point in the lobbies.

The final stop on the passage through the Ford Orientation Center is at one of the two theaters, where visitors can watch a film about the character and leadership contributions of George Washington. These spaces are subtly elegant and purposely understated, leading visitors to focus on the film rather than on their surroundings. Exiting the theaters, they emerge onto a woodland walk, a path that slopes gently upward toward the historic grounds and Mansion.

Character of the Building

The designers were keenly aware of visitors' initial response to the orientation center in general and to its entry space in particular. Ultimately, they decided that a low-scale entry would best convey a message of restrained elegance. To translate that idea architecturally, they designed an entry canopy that embraces a vertical stone pier displaying a bronze bas relief of Washington's profile by Raymond J. Kaskey. Carved into a stone panel on a brick wall at the building entrance are the first president's own words: "No estate in United America is more pleasantly situated than this . . ." The canopy's gentle curve leads visitors to the entry doors and then continues over the boundary/garden wall to integrate with the clearing beyond.

Materials

The resolve to proceed "in the spirit of George Washington" prompted the designers to investigate the use of natural materials. Working closely with interior designers from MFM Design of Washington, D.C., the GWWO architects chose many building materials that visitors would later encounter throughout the estate and then integrated these into the overall design of the orientation center.

Stone

For the walking surface of the Mansion piazza, George Washington used locally quarried Aquia Creek sandstone. Unfortunately, this material had not been available for many years and, furthermore, was far too porous for the anticipated exposure at the entry. Faced with these limitations, the designers sought an available and durable material that would pay homage to Washington's choice. Indiana buff-colored limestone met these criteria; its tonality is similar to the Aquia Creek stone, and it blends beautifully with the adjacent natural materials used nearby—brick and copper.

Special stone materials, hand picked by MFM designers and shipped from quarries in Spain, were used to embellish the interior floors and walls and impart an elegance evocative of Washington's style.

Copper

The canopy fascia, beams, and columns are clad in copper. Crafted with a level of detail also found in the painted wood columns of the Mansion piazza, these well-proportioned modern columns boast clean lines and crisp articulation. As the material ages, it will continually change in color and tone, minimizing maintenance and providing a subtle reminder of the links between present and past.

Copper louvered shutters on the gate that provides outside access to the clearing from the entry plaza echo the shutters on the Mansion's front doors. Even when they are closed, one can glimpse through their angled louvers to the clearing's landscape. For special occasions, the gate opens to allow direct access to the clearing and to provide a route around the orientation center, should it be closed or otherwise in use.

Details inspired by the craft and care that went into the construction of buildings in Washington's time.

Above right: *Mount Vernon historic building window-proportion study by GWWO.*
Above top: *Assembled window-frame model at GWWO's office in Baltimore.*
Above bottom: *Full-scale window-frame mock-up assembled at Mount Vernon for review.*

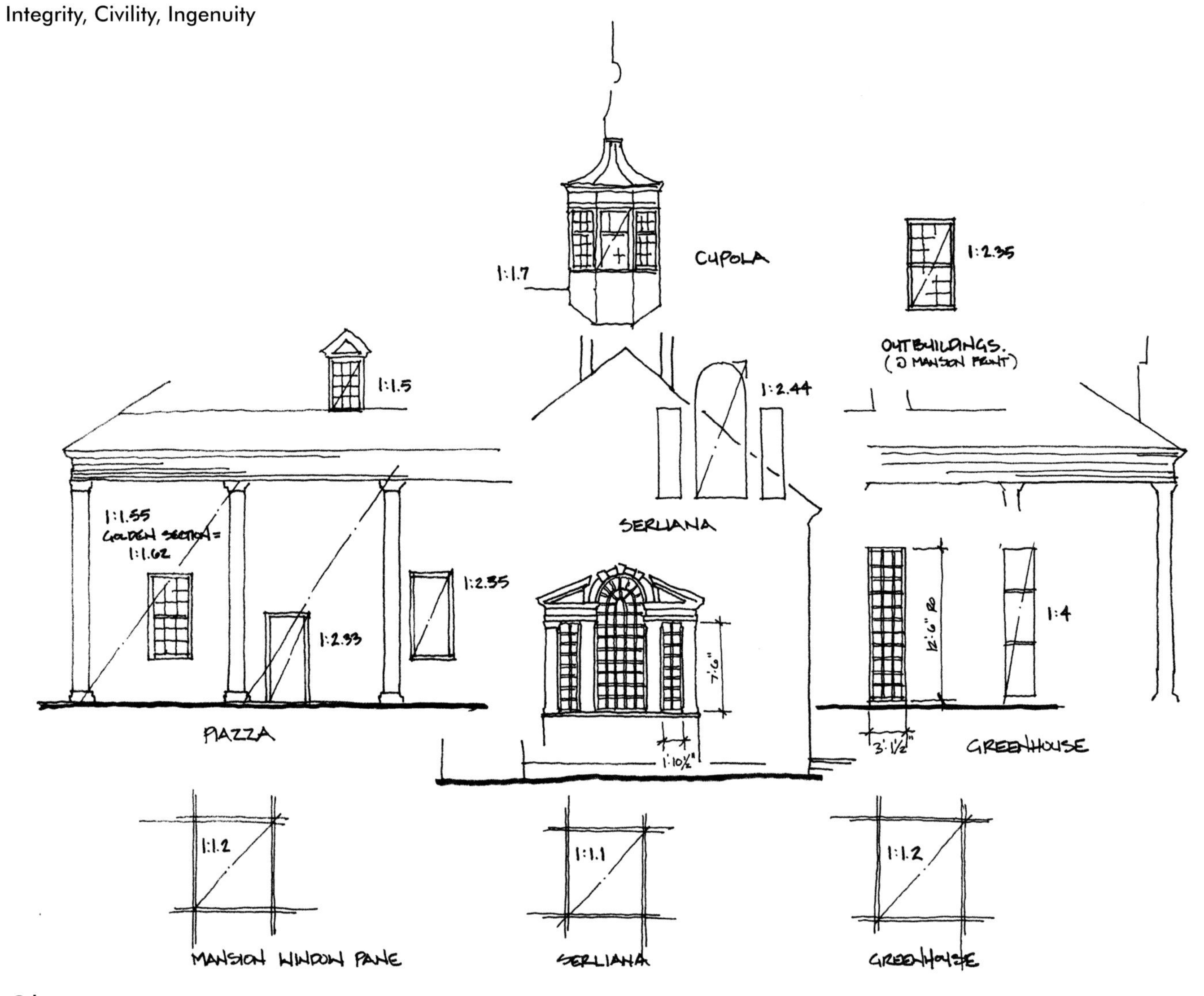

Glass

Because the indoor-outdoor connection was critical to the design concept, the window system had to provide an appropriate transition, one that respected historic precedent while affording sufficient transparency. Typical of 18th-century windows, those in the Mansion consisted of small glass panes. Obviously, this style would not be appropriate for the new transparent-glass wall because the plethora of mullions and frames would interfere with the view. Furthermore, the only window-frame material available in the 18th century was wood, which was an impractical choice for modern windows requiring very large mullion and frame cross-sections. The historic windows are notable for their delicate details and pleasing proportions. These qualities would be difficult to achieve with readily available premanufactured systems.

Ultimately, the GWWO team designed a custom aluminum system to achieve its dual goals—to honor history and maximize visibility. The overall frame proportions and the intermediate mullion patterns were based on the proportions of those of the Mansion windows but on a scale commensurate with the desired transparency. Larger mullions incorporate fluting—a light-diffusing device that reduces their visual impact, while finned smaller mullions recall the detailed historic windows.

The elliptical window wall facing the clearing is based on the proportions of the windows in the historic estate buildings.

The orientation center entrance is made spacious and welcoming by frameless glass doors and a light, airy copper-and-glass canopy.

Curly maple guest-services desk.

The designers wanted the orientation center's entry doors to convey a sense of Washington's generous hospitality. To achieve this, they selected a frameless-glass-door system. Even before visitors enter the building, they can see entirely through it, to the land and sky on the other side. They also can see the life-sized bronze statues of George and Martha Washington and their two grandchildren, who appear to be welcoming them to Mount Vernon.

Because the architects wanted the canopy over the entry to seem light and airy, they chose solar-protective glass instead of a more traditional opaque material. The detailing accentuates the trellislike nature of the copper structure and reduces the glazing support system to a minimum.

Wood

For accent walls, casework, and doors, the interior designers from MFM Design turned to the warmth and universal appeal of stained natural maple and cherry.

Brick

The material for the orientation center's exterior walls had to blend seamlessly into the historic environment. In many places, brick walls define discrete areas of the estate, a containment method Washington used in his upper and lower gardens. The design team was inspired by these 18th-century landscape features, with their sloping brick caps, rolling height transitions, well-proportioned pilasters, and subtle color variations. On the north side of the orientation center, the circa 1905 brick boundary wall

Left: *Brick garden wall with a new bonding pattern combining brick patterns found on the historic estate.*
Right: *Reconstructed brick boundary wall forming the north side of the Ford Orientation Center, based on the adjacent circa 1905 estate-boundary walls.*

was rebuilt with masonry that closely matches the original and with up-to-date insulation and detailing required for enclosing an interior space. For the east and south sides, the designers specified the walls that are compatible with those surrounding the estate gardens.

The GWWO team identified the historic stable on the south lane, with its light, dark, and even black bricks, as a model for the majority of the brickwork on the project. The historic bricks were made by putting the raw, wet masonry mix into wood molds, usually coated with sand to act as a releasing agent. The result was bricks with slightly uneven edges and a somewhat rough texture. Firing them in a hand-stoked, wood-burning kiln produced the color variations. The designers wanted the new brick walls to replicate this look—with a similar color range, surface texture, and shape—as closely as possible. They knew, of course, that the brick material and production techniques would be different from those of the 18th century, but they nonetheless wanted the ultimate assembly of bricks to be as aesthetically appealing as the original. The one exception to this approach is evident on the north wall facing the terminus of the George Washington Parkway, where a material more compatible with the existing adjoining walls is used; these bricks have a smoother texture and a more consistent color, blending with the estate's early-20th-century boundary wall. For the theater walls, the team devised a modern interpretation of Washington's espalier—a system of pins and wires supporting a network of vines—in order to soften the impact of the tall brick walls.

The historic stable wall facing the south lane was the model for the texture, craftsmanship, and color variety used in designing the new brick walls.

Orientation Center Design

It is important that visitors arriving at Mount Vernon clearly understand how to reach their destination—be it the Mansion or any of the other points of interest. If they have already seen the introductory and other educational presentations, they should be able to bypass them easily and go directly to the historic house or other estate attractions.

Mount Vernon visitors arrive either by car or by bus. In the busy seasons, spring and fall, group tours are popular, and the number of people arriving by bus during the morning can exceed 4,000. In the early stages of planning the project, the GWWO team debated whether to separate group ticketing and entry from ticketing for individuals and families or to have everyone use the same facilities. Ultimately, the sheer quantity of groups arriving during the busy seasons dictated that not only should the ticketing and entry facilities be separate (fig. 1) but so should the restrooms, with one set serving groups and another serving individuals and families. This arrangement helps minimize confusion and the possibility that small groups of visitors might feel overwhelmed by bus tours. Once they buy their tickets, enter the orientation center, and perhaps use the restrooms, everyone merges together in the upper lobby.

To help visitors plan their time at Mount Vernon, the center provides a wall map, pictorial displays (fig. 2), and brochures; paid employees and volunteers alike staff the information desk (fig. 3). The central area of the upper lobby is free of any displays or obstacles, enabling visitors to circulate freely even during the busiest periods. The one exception is an important and popular feature among visitors—four life-sized bronze statues of George and Martha Washington and their grandchildren, posed in a relaxed and welcoming fashion, as if walking toward entering guests. The idea for the statues came from Richard Molinaroli of MFM Design, Washington, D.C., the firm responsible for the interior finishes. They were executed by StudioEIS in Brooklyn, New York, based on forensic research conducted by the studio. Few modern-day families can resist having a picture taken with the Washington family. Still, visitors do not linger long in the upper lobby, because they are naturally drawn to the theater lobby, which is located down a short flight of steps (fig. 4).

1

2

3

4

5

6

7

8

As they descend to the lower lobby, visitors can look through the elliptical glass wall to see the clearing and the pasture beyond. Most will pause to enjoy *Mount Vernon in Miniature* (fig. 5) and a series of stained-glass panels depicting crucial moments in Washington's life before entering one of two theaters to watch a 20-minute orientation film. Those who have visited Mount Vernon often can decide to bypass these experiences by exiting the lower lobby through glass doors to the outdoor clearing. They then follow a path around the buildings (fig. 6) and ultimately reach the same plaza where visitors exit from the theater (fig. 7). This plaza affords a vista back toward the sheep pasture they saw when entering the Texas Gate and then saw again through the clearing from inside the building. Visitors then follow a woodland walk that eventually leads to the historic bowling green gate, where they catch their first glimpse of Washington's home.

After touring the historic area, which usually takes two hours or more, visitors follow a brick-paved path leading directly to the entrance of the museum and education center (fig. 8), where they will step down from the upper lobby to the lower lobby on a grand stairway. As they descend, they can clearly see their options: entries to the museum exhibits, the education center exhibits, and the glass-enclosed serpentine exit corridor leading to the Mount Vernon Inn, restrooms, and parking area.

The Washington Family

In the late 18th century, George and Martha Washington's hospitality was renowned throughout the land. The master of Mount Vernon once referred to his busy home as a "well-resorted tavern," and indeed in 1798 alone 677 people are known to have been guests there for at least one night. This welcoming attitude is communicated as soon as one enters the Ford Orientation Center. Four life-sized bronze statues of George Washington, Martha Washington, and their two grandchildren, Nelly and Washy, are posed in relaxed fashion, as if walking forward to greet visitors to the estate. Created by StudioEIS in Brooklyn, New York, the statues were cast at a foundry in upstate New York. The sculpture project was funded by Robert H. and Clarice Smith.

Moments of Truth in Stained Glass

The lower lobby of the Ford Orientation Center features one of the most unusual tributes to George Washington ever created, a set of five stained-glass windows depicting significant experiences in the life of the Father of Our Country. Those moments include: "I Can Not Tell a Lie," 1738; Reading of the Declaration of Independence to Washington, 1776; Washington Crossing the Delaware, 1776; Washington and Alexander Hamilton, 1777; and Washington Taking His Inaugural Oath, 1789. Created in the mid-20th century by the famed studio of Karl J. Mueller for his exhibition hall in Zephyrhills, Florida, the windows were purchased from the estate of U.S. Air Force Brigadier General Woodrow A. Abbott in 1997. Eleanor Seaman, who served as Vice Regent for Wisconsin from 1989 to 2001, and Mount Vernon Executive Director James C. Rees first viewed the dusty panels in a warehouse near Milwaukee where they had been stored for some time. Their acquisition was supported by generous gifts from the Neighborhood Friends of Mount Vernon.

Mount Vernon in Miniature

Mount Vernon in Miniature is a featured attraction of the Ford Orientation Center. This popular model of Washington's Mansion, handcrafted with a remarkable degree of authenticity and detail, was a gift from the people of the State of Washington to Mount Vernon in 1998. An international group of more than 50 craftsmen, led by Stan Ohman, Pat Olson, and the late Jean Sprague, were involved in its design and construction over a five-year period. The model was featured in the special exhibition *Treasures from Mount Vernon*, which traveled nationally from 1998 to 2000 and then continued touring for several more years as a stand-alone display. Margaret Allison, Vice Regent for Washington, organized both the original campaign to support the exhibition and a second effort to update the Miniature's furnishings based on Mansion changes after 1999.

We Fight to Be Free

The centerpiece of the Ford Orientation Center is the 20-minute dramatic film *We Fight to Be Free*, shown continuously in a pair of large-screen, Surround Sound theaters. It opens when the 23-year-old George Washington, serving as a colonial militia officer, takes charge after his British comrades have stumbled in a critical battle of the French and Indian War. Three years later, he meets a widow, Martha Custis, whose charm softens his formal military demeanor, and weds her. The closing scenes dramatize the vital moment late in 1776 when Washington led his downtrodden but determined forces across the icy Delaware River to help turn the tide of the Revolutionary War. Produced by Craig Haffner and directed by Kees Van Oostrum for the Los Angeles–based Greystone Productions, *We Fight to Be Free* features Sebastian Roché as George Washington and Caroline Goodall as Martha Washington. The production was supported by generous grants from the Ford Motor Company Fund and Donald and Nancy de Laski.

Orientation Center Construction

Overleaf: *View from the Ford Orientation Center lobby along the elliptical curtain wall.*

PLAN YOUR VISIT

Donald W. Reynolds

The largest structure ever built on the Mount Vernon estate is named for an innovative entrepreneur who established one of the world's largest and most successful media companies.

Donald W. Reynolds

Donald W. Reynolds had a modest upbringing in Oklahoma. He worked his way through the University of Missouri with jobs in a meatpacking plant and in journalism. Before long, he purchased partial ownership of a photoengraving plant and then his first newspaper. Reynolds was 34 years old when he founded the Donrey Media Group, aiming to build a big business centered on small-town newspapers.

Reynolds put his business ventures on hold to join the U.S. Army during World War II. Recognizing his talents, senior officers assigned him to establish a newspaper for enlisted soldiers. By war's end, he was a major in military intelligence and had been awarded a Legion of Merit, Bronze Star, and Purple Heart—the medal inspired by the military decoration George Washington created during the Revolutionary War. In the postwar years, Reynolds continued purchasing newspapers but also acquired radio stations and then television channels. He subsequently added cable television and outdoor advertising to the Donrey Media Group's holdings, bringing its total number of businesses to more than a hundred. Over time, he delegated an increasing degree of management responsibility to his trusted longtime colleague Fred W. Smith, who eventually became president and CEO of Donrey and chairman of the Donald W. Reynolds Foundation.

In 1954 Reynolds created the philanthropic entity that bears his name, bequeathing the majority of his estate to it upon his death in 1993. Under Smith's guidance, the Donald W. Reynolds Foundation has distinguished itself both regionally and nationally. It has formed partnerships with the country's premier cardiovascular medical research centers in an effort to find a cure for atherosclerosis. In addition, its grants to 37 of the nation's medical schools have focused on the training of physicians and health care professionals to recognize and treat problems of the aging baby boomer generation. Other landmark gifts have reflected Reynolds' personal interest in journalism and human services. Geographically, the foundation has paid special attention to charities in the three states where Donrey operations were concentrated—Oklahoma, Arkansas, and Nevada.

In 2001 Smith learned that Gilbert Stuart's iconic 1796 Lansdowne portrait of George Washington, which had long hung on loan at the National Portrait Gallery, was to be auctioned to the highest bidder and possibly removed from public view. Under his leadership the Reynolds Foundation's board acted immediately to save the painting for all Americans. But the foundation took one significant step further. As the National Portrait Gallery was then closed for several years of renovation work, the foundation underwrote a nationwide tour for the portrait, one that included the development of educational programs that would engage thousands of schoolchildren at every city where the painting stopped.

In 2002 Smith visited Mount Vernon with Reynolds Foundation President Steve Anderson. As the two men moved slowly through Washington's home, in the midst of hundreds of families and schoolchildren, it became clear that the Mount Vernon Ladies' Association's ambitious plans for a new museum and education center at the estate had a large built-in audience. The potential to "reach and teach" Americans about George Washington's life and character was tremendous—and so was the need for private support, because Mount Vernon does not receive a single dollar of government funding.

Two major grants from the Donald W. Reynolds Foundation, totaling $24 million, were critical in moving the project forward. This commitment to the estate's future was by far the most significant in Mount Vernon's 150 years of raising funds from the American people. In gratitude, this engaging new facility was named in honor of an entrepreneur and philanthropist whose life reflected George Washington's ideas about what makes America strong.

Top: *Donald W. Reynolds Foundation President Fred W. Smith, speaking at a fundraising event.*
Bottom: *Fred W. Smith talking with Gay Hart Gaines, Regent of the Mount Vernon Ladies' Association.*

Donald W. Reynolds Museum and Education Center

After leaving the historic grounds through the bowling green gate, visitors follow a path that leads downhill and northwest, toward the Mount Vernon Inn and the estate exit. The GWWO design team recognized the importance of having visitors leave the grounds via the Inn, but topography made this somewhat difficult. From an elevated part of the estate, visitors had to descend about 25 feet to reach the level of the Inn. Further complicating the situation was the need to place the entrance to the museum and education center at some point along this long, sloping path. The team felt it was critical that this route effectively—and attractively—resolve the height differential and that the building's entry be both obvious and inviting.

Right: *View through the lobby to the landscape beyond.*
Below: *Donald W. Reynolds Museum and Education Center entrance, approached from the historic estate.*

Clerestory windows and skylights visually lift the lobby ceiling.

Visitors get a close-up view of names of donors engraved in the stone wall as they descend the grand stairway to the exhibits, The glass railing of the stairway facilitates unobstructed views.

The designers' mission was to situate the larger of the two new structures in a way that did not overpower any of the existing buildings, least of all the Mansion. In keeping with the master plan they located it underground, in order to preserve visitors' pastoral views upon entering the estate. Although the building's underground location resolved the problem of concealing its mass, it was critical that visitors did not feel as though they were descending into a basement or were underground when they were inside the building. Finally, the team faced two practical concerns. First, visitors should have the option of bypassing the exhibits in favor of proceeding directly to the Mount Vernon Inn and estate exit. And second, a loading dock was necessary for the delivery of artifacts.

During the design process, the team realized that a two-story space would resolve the elevation change as well as achieve the separation between public areas and service areas. With access to only one entry lobby for both the exhibits and the Mount Vernon Inn, visitors, regardless of their final destination, would follow the same path. The designers chose to locate that entry at the halfway point on the hill between the historic area and the Mount Vernon Inn; by placing the doors closer to the historic area, they thus shortened the exterior path. Visitors enter the upper level of the two-story lobby and then descend a grand stairway to the lower level, which contains the museum and education center exhibits as well as the connection to the Mount Vernon Inn. Visually and physically separated from the visitors' path, building services are tucked under the entry level, so deliveries of artifacts can be made closer to the exhibits.

The museum and education center lobby features the wall of honor, a tall window wall with views to the landscape beyond, and the entry to the serpentine corridor leading to the Mount Vernon Inn.

From the grand stairway, the museum and education center entrances welcome visitors with a life-sized statue of Washington positioned between the two doorways.

Taking full advantage of natural light, beautiful views, and historic landscape features was critical to the team designing the museum and education center. To achieve this goal, they specified the installation of detailed clerestory windows, all-glass entrance doors, and tall window walls. On approach, the building virtually disappears, encouraging visitors to view nature on the opposite side. Upon entering the upper level, they look out over a carpet of treetops reminiscent of the hanging wood Washington favored for the Mansion. The clerestory windows afford a view of the sky and admit natural light into the space. Two grand skylights provide filtered light and accentuate two key points along the path—the bust of Donald W. Reynolds and the Wall of Honor, which lists the most generous donors to Mount Vernon's capital campaign.

Once visitors turn toward the lower lobby at the landing of the grand stairway, the entrances to both the museum and education center are visible, flanking a bronze copy of the life-sized marble statue of Washington that the French neoclassical sculptor Jean-Antoine Houdon completed for the Virginia State Capitol in 1796. Just inside the education center is an eight-foot-tall mask of Washington, lit to create a three-dimensional effect, with eyes that seem to follow you. Within the museum a celebrated terra-cotta bust of Washington, created on the estate in 1785 by Houdon as a study for the Richmond statue, beckons visitors to view a collection of objects from Washington's time.

The portal from the lower lobby to the Mount Vernon Inn and estate exit opens onto a glass-enclosed serpentine

Top: *The entrance to the education center features a reverse-lit bust of Washington.*
Bottom: *The museum entrance, featuring the celebrated Houdon bust in the first gallery.*

corridor, designed to echo the curved woodland walk visitors take when they leave the orientation center theaters. The glass west wall of the corridor overlooks a picturesque landscape between the Inn and the museum and education center. On the solid inner wall of the curving corridor is an exhibit devoted to the Mount Vernon Ladies' Association, explaining the organization's genesis, its commitment to saving the estate, and its leadership role in the American preservation movement.

The corridor ends at a modest lobby, from which visitors can either continue through to the Mount Vernon Inn and

The serpentine corridor facing the estate's beautiful landscape. Here visitors are introduced to the history of the Mount Vernon Ladies' Association.

The serpentine corridor features views of the planted landscape between the museum and education center and the Mount Vernon Inn.

Looking toward the Phoebe Apperson Hearst Learning Center entry and Mount Vernon Inn complex from inside the Texas Gate. The north wall of the education center will display a network of vines growing on a modern metal trellis, modeled after Washington's historic espaliers.

the estate exit or can enter the hands-on history exhibit designed for children ages three to eight. This area of the building also houses a distance-learning classroom and the Phoebe Apperson Hearst Learning Center, which serves as a resource-and-research room for educators and historians. In the distance-learning classroom, educators can broadcast lessons from Mount Vernon to classrooms nationwide.

Both researchers and schoolchildren reach these spaces without passing through the exhibits; their entrance is a vestibule on the northeast side of the building, opposite the orientation center.

When exiting the museum and education center, visitors have another opportunity to obtain information before entering the Mount Vernon Inn complex.

Donald W. Reynonlds Museum and Education Center site and building plan by GWWO, Inc./Architects and EDAW, Inc.

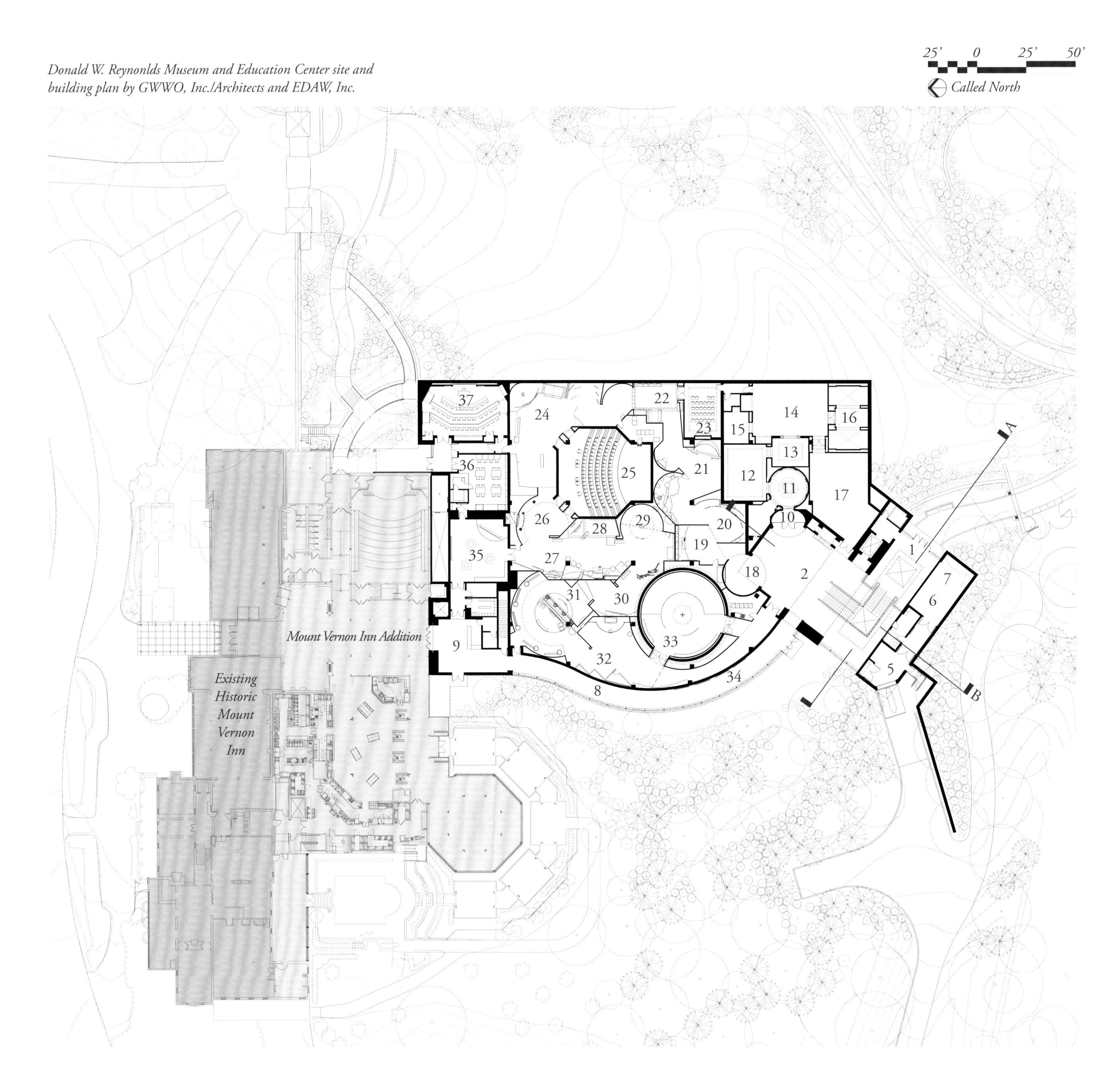

Donald W. Reynolds Museum and Education Center

1. *Upper lobby*
2. *Lower lobby*
3. *Women's restroom (below upper lobby)*
4. *Men's restroom (below upper lobby)*
5. *Loading dock*
6. *Exhibit preparation*
7. *Exhibit storage*
8. *Serpentine corridor*
9. *Information lobby*
10. *Museum entrance*
11. *The Founders, Washington Committee for Historic Mount Vernon Gallery*
12. *The A. Alfred Taubman Gallery*
13. *The John and Adrienne Mars Gallery*
14. *The Elizabeth and Stanley DeForest Scott Gallery*
15. *The L. J. Skaggs and Mary C. Skaggs Foundation Gallery*
16. *The Gilder Lehrman Gallery*
17. *The F. M. Kirby Foundation Gallery*
18. *Education Center entrance*
19. *The Mars Family and Foundation Gallery*
20. *The Mary Morton Parsons Foundation Gallery*
21. *The Richard King Mellon Foundation Gallery*
22. *The Mary Hillman Jennings Foundation Gallery*
23. *The Melody Sawyer Richardson Theater*
24. *The Gordon V. and Helen C. Smith Foundation Gallery*
25. *The Elizabeth and David Bruce Smith Theater*
26. *The Douglas and Eleanor Seaman Gallery*
27. *The Distilled Spirits Council of the United States Gallery*
28. *The J. Hap and Geren Fauth Gallery*
29. *The Donald and Nancy de Laski Gallery*
30. *The Charlotte and Walter Kohler Charitable Trust Gallery*
31. *The Richard and Helen DeVos Foundation Gallery*
32. *The Richard and Adelia Simplot Gallery*
33. *The Gay Hart Gaines Legacy Theater*
34. *The Johnston-Lemon Group Gallery*
35. *The Anne and James F. Crumpacker Family Gallery*
36. *Phoebe Apperson Hearst Learning Center*
37. *The Annenberg Foundation Classroom*

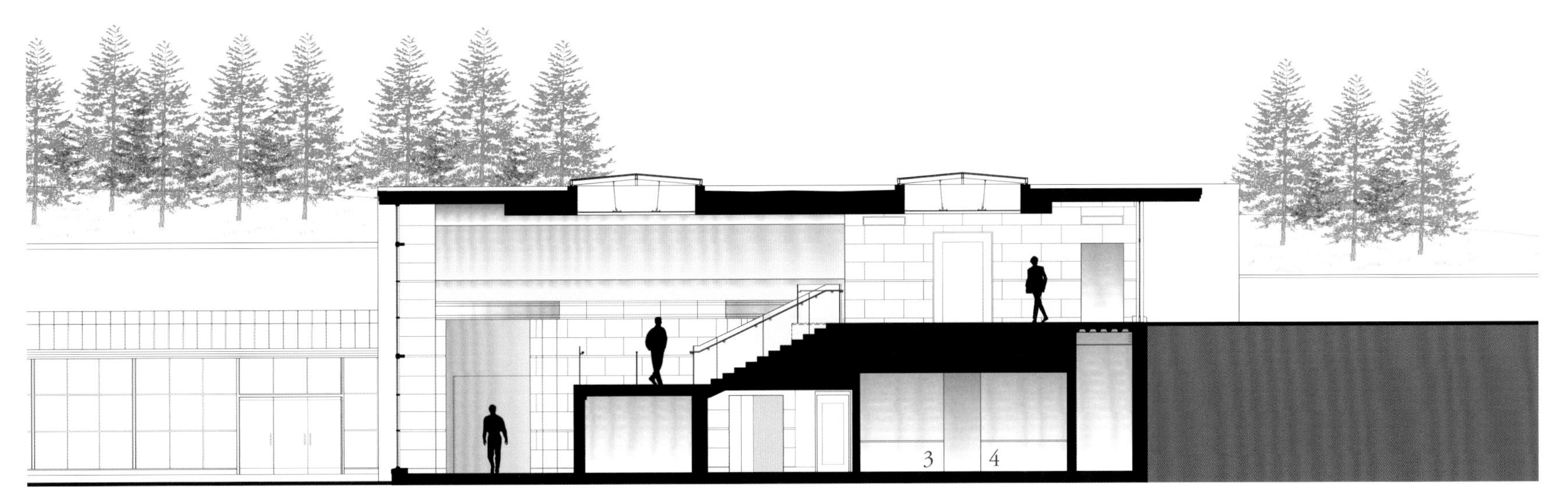

Building Section A

Building Section B

Museum and Education Center Design Approach

GWWO developed two approaches to the museum and education center design. In the first scheme, the entire new building would be level with the Mount Vernon Inn, facilitating the connection between the two structures. Locating the entrance to the building at the southwest corner would lengthen the path from the historic core of the estate but would enable one to make the transition without using steps. However, situating the entrance there would require placing either an architectural or a landscape element to help guide visitors to their next destination. The team explored various ways to address this need, including the construction of covered outdoor pavilions and the creation of a stream leading to a terminus fountain near the entry. The volume of the building, by necessity fairly tall in order to house the exhibits inside, rose from the ground as the visitor moved along its length to the point, next to the entrance, where the full height of the wall was revealed. The museum and education center building would, therefore, have a prominent presence next to the path. From the corner entrance, visitors could either enter the lobby to see the exhibits or bypass the building by walking down an outdoor colonnade leading to the Mount Vernon Inn lobby and the estate exit.

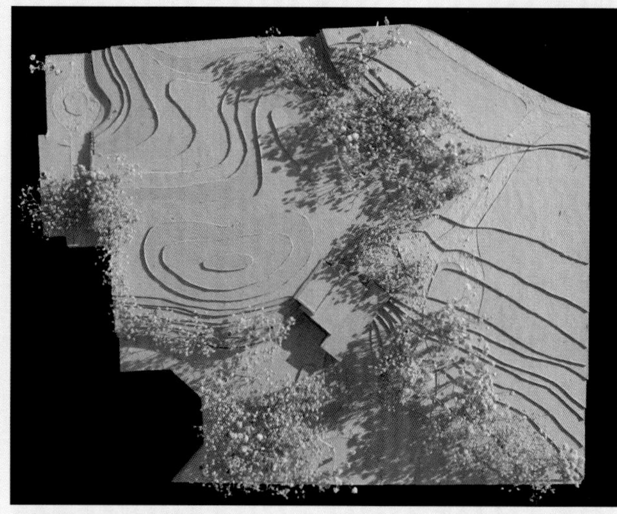

Scheme two, option one

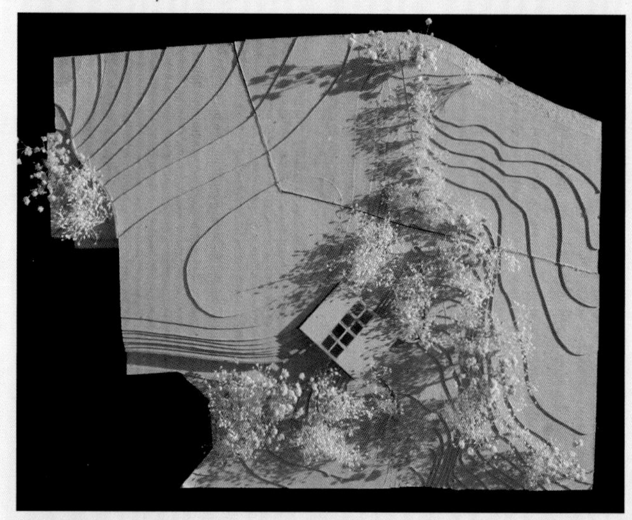

Scheme two, option two

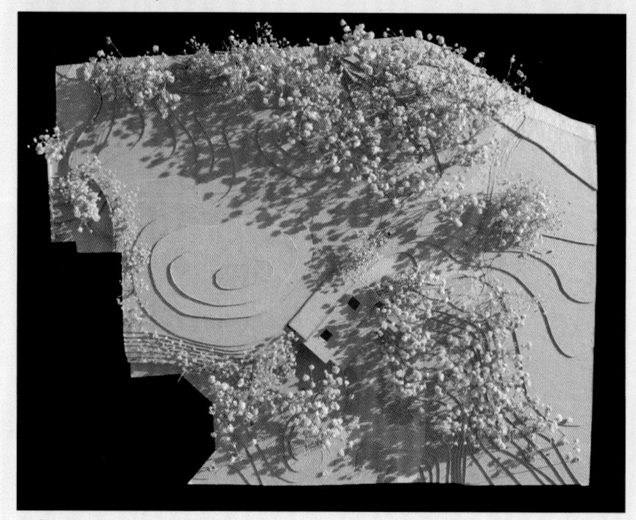

Scheme two, option three

The team had several concerns with this scheme. For example, the entrance would not be as readily apparent as one that visitors could approach directly. The height of the building adjacent to the visitor path competed with the natural setting. The grade differential made for a longer outside path. And finally, the designers feared that the covered outdoor bypass might cause some visitors unintentionally to miss the museum and education center altogether.

To deal with these concerns, the architects developed a second scheme. By locating the entrance on an upper mezzanine level, halfway up the hill, they could conceal underground half of the tall volume needed to house the exhibits. The building wall along the path would now be perceived as a brick boundary-site wall rather than as a building wall. This revised layout also oriented the entrance doors perpendicular to the direction of the building path, thus shortening the distance from the historic area to the building. Once inside, visitors could choose either to enter the museum and education center or to follow an indoor bypass corridor to the Mount Vernon Inn. By designing the visitor path so that it clearly led to the two venues, the team ensured that visitors would be aware of both educational resources.

Complete site model with the orientation center and museum and education center final schemes in place.

Museum and Education Center Construction

HITACHI
ISUZU

DEERE
160LC

Overleaf (pp. 126–27): *West facade of the Donald W. Reynolds Museum and Education Center.*
Overleaf (pp. 128–29): *View to landscape from museum and education center upper lobby.*

To Keep Him First
FORD MOTOR COMPANY FUND • DONALD W. REYNOLDS FOUNDATION
THE ANNENBERG FOUNDATION • DONALD AND NANCY DE LASKI • THE RICHARD AND HELEN DEVOS
THE GILDER FOUNDATION, INC. • WILLIAM RANDOLPH HEARST FOUNDATION
THE F. M. KIRBY FOUNDATION • JOHN AND ADRIENNE MARS • FRANK PETERS MOORE
AND ELEANOR SEAMAN • ALBERT AND SHIRLEY SMALL • THE GORDON V. AND HELEN C. SMITH
ANNE AND JAMES F. CRUMPACKER FAMILY • THE JOHNSTON-LEMON GROUP
GERRY AND MARGUERITE LENFEST • CHRISTOPHER AND BETSY LITTLE • MARS FOUNDATION
THE RICHARD S. REYNOLDS FOUNDATION • MELODY SAWYER RICHARDSON • RICHARD M. SCAIFE
RICHARD AND ADELIA SIMPLOT

AND CLARICE SMITH
MARY HILLMAN JENNINGS FOUNDATION
ELIZABETH AND STANLEY
ALFRED TAUBMAN

Galleries in the Donald W. Reynolds Museum and Education Center

Except for the initial rescue of the Mansion in the 1850s, the most significant undertaking of the Mount Vernon Ladies' Association has been the expansion of its mission to address and interpret the career and character of George Washington. For the first time, with the opening of the Ford Orientation Center and Donald W. Reynolds Museum and Education Center, not only is Washington's life on the estate interpreted—his activities within agrarian, milling, and distilling operations—but so are the challenging and demanding endeavors he faced within military, political, and diplomatic realms. Finally, the statesman's integrity, civility, and ingenuity are acknowledged and heralded so that younger generations of Americans can comprehend Washington's character and enduring significance.

The new approach to this interpretation begins in the Ford Orientation Center, where the exquisite model *Mount Vernon in Miniature* acquaints visitors with the design and scale of the Mansion, a stained-glass display highlights five key moments in his life, and an action-packed short film provides an effective introduction to the real George Washington. The Donald W. Reynolds Museum and Education Center consists of several theaters and a range of galleries, intended to satisfy visitors' various interests. The museum, for instance, houses galleries with both permanent and temporary exhibits and fine and decorative arts, books, and manuscripts linked directly to Washington's life and times. Quenroe Associates of Boulder, Colorado, and Charles Mack Designs of Baltimore provided the interior and exhibit design for these galleries, working closely with the Mount Vernon collections staff. In the education center are 15 galleries that together offer a dynamic exploration of Washington's challenges and achievements. Christopher Chadbourne and Associates of Boston, with Museum Design Associates of Cambridge, Massachusetts, designed these galleries, relying on input from Mount Vernon's permanent educational staff. Dennis Earl Moore Productions of Brooklyn Heights, New York, designed and executed the audiovisual exhibits, while producers at The History Channel created a variety of film presentations for individual galleries. MFM Design of Washington, D.C., designed the concluding interpretive exhibit, located in the glass-lined corridor, which recounts the history of the Mount Vernon Ladies' Association.

The Vira I. Heinz Endowment made major commitments to Mount Vernon's expansion of its educational programming, both within and beyond the new facilities. By supporting the creation of publication kits for classrooms, teacher-training materials, and other essential communication tools, the Pittsburgh–based foundation has enabled Mount Vernon to take its outreach efforts in this area in a variety of fresh directions. Previously forced to target select audiences because of limited resources, its educators are now able to offer many programs to the diverse groups visiting the estate.

Advanced media technology and electronics have been put to exceptionally creative uses in the galleries and other spaces of the new facilities, both enhancing visitors' enjoyment of Mount Vernon and their appreciation of the life and legacy of George Washington. The large investment that was required to develop and integrate

these up-to-date features was feasible only because of a generous grant from the Richard S. Reynolds Foundation, headquartered in Richmond, Virginia.

Finally, one of the most engaging qualities of the Donald W. Reynolds Museum and Education Center is the sense of movement and action provided by the various film programs that are on continuous view. Two years of design and production work, sponsored entirely by The History Channel and led by Dr. Libby Haight O'Connell, the channel's senior vice president, enabled Mount Vernon to include 11 films within the gallery spaces, each one different and possessed of a distinct personality.

Museum Galleries

The Founders, Washington Committee for Historic Mount Vernon Gallery

Featuring the Houdon Bust

Jean-Antoine Houdon's famous 1785 terra-cotta bust of George Washington is displayed alone with dramatic lighting in a circular, classically styled, domed gallery designed to showcase one of Mount Vernon's most prized possessions. Created at the estate by the great French neoclassical sculptor, the bust is installed precisely at Washington's height to give visitors an indication of how he towered over most of his contemporaries. Adjacent to the gallery is an area where visitors can watch a short film produced by The History Channel that tells, from the perspective of Washington's step-granddaughter, Nelly Custis, how the terra-cotta bust and life mask were created.

The A. Alfred Taubman Gallery

The World of Washington

The first gallery visitors encounter after viewing the Houdon bust features artifacts from England, Europe, China, and the United States that reflect Washington's love of beauty, refined taste, status within his world, and purchasing power. Mount Vernon's holdings include a wealth of fine and decorative arts produced by superb artists and craftspeople of the Colonial and Federal periods.

The highlight of the gallery is an elegant presidential dining table, arranged to evoke the suppers President Washington held every Thursday afternoon at four o'clock for congressmen and other government officials. This re-creation of the First Couple's table setting includes their French porcelain dinner service and table ornaments, silver candlesticks, and finger bowls. A mirror at the far end of the scene gives the appearance of a much larger gathering and reflects the images of gallery visitors as well.

Several display cases contain select examples of the Washingtons' silver, ceramics, and glass, all of which recall their tasteful hospitality. Also exhibited here is a five-piece garniture set that once adorned the mantel in the Mansion; owned by the Washingtons, these pieces have not been shown together in more than 100 years. Furniture reflecting Washington's appreciation of fine craftsmanship and his preference for understated elegance is also on display, as is Gilbert Stuart's iconic portrait of the Father of Our Country.

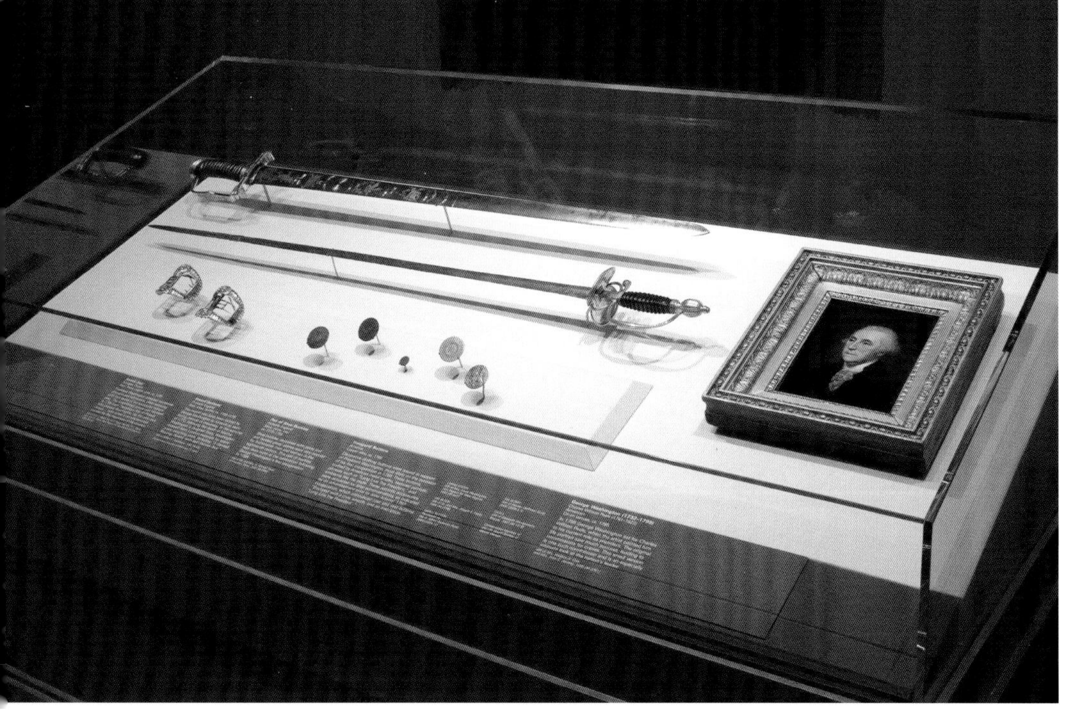

The John and Adrienne Mars Gallery

From Soldier to Statesman

This elegant gallery focuses on the public persona of George Washington—both as military leader and commander in chief. Objects pertaining to his renowned military career include his sword, silver camp cups, and a pair of silver spurs he gave to a soldier at Valley Forge; the soldier used them when he rode to Boston to fetch desperately needed supplies. There also are several pieces of Chinese export porcelain decorated with the insignia of the Society of the Cincinnati; Washington was the first president general of this hereditary order, which was composed of French and American officers who served during the Revolutionary War.

There are two paintings by Charles Willson Peale in this gallery, one of which, from 1780, portrays Washington at the Battle of Princeton. The other, a formal portrait of the first president painted after 1795, hangs near a stylish side chair that Washington purchased from a departing French ambassador and an armchair from the first Congress over which he presided. Also featured are smaller-scale pieces, such as a pair of silver-and-paste shoe buckles worn by Washington to his first inauguration, buttons made to commemorate that event, and a cabinet-sized portrait of President Washington from about 1795. This elegant room is further complemented by an exquisite parquet floor that incorporates an outstanding rendering of the U.S. presidential seal.

The Elizabeth and Stanley DeForest Scott Gallery

At Home with the Washingtons

In this largest of the permanent-collection galleries, visitors get a sense of the Washingtons' daily lives. Among the first artworks they encounter is a 12-foot-wide painting created in 1859 by Louis Remy Mignot and Thomas Pritchard Rossiter depicting George Washington, his family, and the Marquis de Lafayette on the Mansion piazza. Complementing this image, which is on loan from the Metropolitan Museum of Art in New York City, is a piazza setting composed of the Washingtons' breakfast table, a Sevres porcelain tea service, a silver hot-water urn, and two Windsor chairs. Two paintings of circa 1787–92 attributed to Edward Savage record how Mount Vernon looked in Washington's day. A weather vane in the form of a dove of peace, which once topped the Mansion, and a sundial are among the original artifacts on view.

A domestic vignette exploring women's activities at Mount Vernon includes one of the shell-patterned needlepoint cushions that Martha Washington fashioned, along with her sewing basket and worktable. Another area of the gallery, highlighting outdoor activities, features Washington's duck-hunting gun, hunting horn, riding crop, and fishing-tackle box. The folding camp bed, trunk, razor, and telescope he used during the Revolutionary War are displayed in an inset wall case to give viewers an idea of how their owner lived when he was away from home.

The L. J. Skaggs and Mary C. Skaggs Foundation Gallery

The Washington Style

Cases in this gallery contain personal objects that George and Martha Washington, their children, and grandchildren used and wore daily and on special occasions. Together, they reveal an emphasis on physical appearance and on the significance of particular imports in colonial America and, eventually, the new republic.

Items on view range from George Washington's shoe and knee buckles to Martha Washington's earrings and necklaces. One case is devoted to rare surviving textiles worn by the family. These light-sensitive fabrics, including the brown broadcloth American-made suit Washington wore to his first inauguration, will be rotated regularly in order to combat deterioration. Jewel-like miniature portraits of Washington family members, by artists including Charles Willson Peale, are shown in another case.

The elegant, paneled-wood cases housing the objects enhance the opulent, treasury-like feel of this small and intimate gallery, which is anchored by a pair of mid-nineteenth-century Rembrandt Peale "Porthole" portraits of George and Martha Washington.

The Gilder Lehrman Gallery

Books and Manuscripts

Graced by beautiful wood paneling and a groin-vaulted ceiling, this gallery focuses on Washington's seemingly insatiable curiosity and desire to increase his understanding of the world around him, as evidenced by his manuscripts, maps, prints, and books. These rare objects from two premier collections—one belonging to Mount Vernon and the other to the Gilder Lehrman Institute of American History in New York City—include a wide range of documents related to such topics as the establishment of the United States, slavery, and the Washington family. The first president's last will and testament is also on display. The gallery is enriched by loans from the Boston Athenaeum, which has the largest collection of Washington's own books. With its tranquil aura of an elegant library, the Gilder Lehrman Gallery also contains his recently restored terrestrial globe, inkstand, and one of his Argand lamps.

The F. M. Kirby Foundation Gallery

Temporary Exhibitions

This 1,100-square-foot space can accommodate one or two exhibitions per year, organized either by Mount Vernon or by other institutions. The inaugural exhibition, *A Son and His Adoptive Father*, which opened in October 2006, focused on the close relationship between George Washington and the Marquis de Lafayette. This popular display of French and American artifacts was the first major show mounted by Mount Vernon; it traveled to Lafayette College in Easton, Pennsylvania, and the New-York Historical Society. *Setting the President's Table: American Presidential China*, opening in February 2008, features porcelain objects related to presidents from Washington to Ronald Reagan. The exhibition is drawn from the Robert L. McNeil, Jr., Collection at the Philadelphia Museum of Art.

A NATIONS THANKS TO
FREEDOMS FRIEND

Education Center Galleries

As visitors enter the education center, they are greeted by an eight-foot-high bust of George Washington, whose gaze (courtesy of special effects lighting) appears to follow the visitor. The larger-than-life-size head symbolizes the intangible Washington, intriguing and alive, and represents the interactive nature of the education center experience.

The Mars Family and Foundation Gallery

Reconstructing George Washington

This unique gallery is designed to resemble a state-of-the-art forensics laboratory. Vials, facsimiles of body parts, and stainless-steel work surfaces surround a flat-panel monitor that displays images from the million-dollar forensic investigation, conducted from 2003 to 2005, which resulted in the creation of three life-sized figures of George Washington. Visitors can watch a History Channel film that describes the scientific processes involved in creating the figures based on anthropological study, historical research, and artistic interpretation. Showing Washington at ages 19, 45, and 57, these figures appear in the three education center galleries centered on those periods of his life.

The Mary Morton Parsons Foundation Gallery

Young Virginian

In this gallery, multimedia exhibits and artifacts offer insights into George Washington's childhood and examine his youthful ambitions, struggles, and hardships. An animated timeline of significant events in his early years is projected on a wall.

The first of three forensically based life-sized figures of Washington—here as a 19-year-old surveyor—is showcased, standing amid a full-scale diorama that includes trees, animals, and forest sounds. Washington's original surveying tools are exhibited in a nearby case. Other displays focus on his youthful travels and on the 110 *Rules of Civility and Decent Behavior in Company and Conversation*, which the adolescent Washington carefully copied out by hand and took permanently to heart.

The Richard King Mellon Foundation Gallery

Upstart Colonial Officer

Here the light dims to convey an impression of the forbidding wilderness in which Washington soldiered as a British colonial officer during the French and Indian War. A diorama set into the wall and containing hundreds of small figures and animals conjures up the Battle of Fort Necessity in July 1754, when Washington's tactical misjudgments, as well as rainy weather, combined to help the French defeat the English in the first major engagement of the war. A large lighted map shows the sequence of battle sites spreading across colonial North America and Europe. Washington's sword from this war is also on display.

The Mary Hillman Jennings Foundation Gallery

Gentleman Planter/Revolutionary

Leaving the French and Indian War exhibits, visitors encounter displays that evoke Washington's journey home from the war. His marriage to Martha Dandridge Custis in 1759 is represented by her jewelry and reproductions of her gold wedding dress and rhinestone-covered shoes.

Martha Washington's Bible, among other religious artifacts, is displayed here. Visitors can sit on a reproduction of the Washington family box pew from Pohick Church in nearby Lorton, Virginia, as they watch a History Channel film about the role religion played in Washington's life and his views on religious freedom.

This gallery also touches upon Washington's service in the Virginia House of Burgesses and his involvement in Freemasonry, which lasted his entire adult life.

The Melody Sawyer Richardson Theater

A 40-Year Romance

What seems at first glance to be a tastefully decorated sitting room or parlor—with a fireplace, portraits, fine draperies, and rows of Windsor chairs—is actually a 30-seat theater. Here visitors can watch a History Channel film, narrated by actress Glenn Close and shot at several historic locations, which highlights the Washingtons' courtship and marriage. The film focuses primarily on the important role played by Martha Washington, showing, for example, how she spent more than half of the Revolutionary War years at her husband's side, often performing nursing duties.

The Gordon V. and Helen C. Smith Foundation Gallery

First in War

The daunting challenges George Washington encountered as commander in chief of an untrained, under-funded ragtag citizen army are the subjects of the exhibits here. By viewing a three-dimensional mural of British forces, visitors get a vivid sense of how superior the British army and navy seemed in comparison to the Continental Army. A soldiers' hut at Valley Forge illustrates the hardships inflicted by the long, bitter winter of 1777–78. Viewable through the window of the hut is a History Channel film depicting Baron von Steuben drilling fledgling troops.

The second of the forensically based life-sized figures of Washington shows him at age 45, looking stoically determined astride his horse Blueskin. Nearby is an interactive map with buttons that, when pressed, illuminate major battles of the Revolutionary War. A large display of period weapons used by American, British, French, Hessian, Native, and Spanish forces emphasizes the international scope of the conflict; an adjacent display case contains several objects related to the Marquis de Lafayette and the French contribution to the war effort. Projected on a screen inset in the front doorway of an 18th-century tavern is a History Channel film that covers Washington's innovative use of espionage during the war.

The Elizabeth and David Bruce Smith Theater

General Washington, Commander in Chief

Presented in this 110-seat theater is a fast-paced, "4-D" multimedia production tracing Washington's important military victories at Boston, Trenton, and Yorktown. Representing the course of the Revolutionary War, it combines concise narration, changing battle maps, and visual effects to create a "strategy and tactics" show that evokes a sense of actual conditions and events. When cannons fire, for example, seats vibrate and "smoke" wafts through the room. Fog drifts through the theater at three key moments to help dramatize how the Americans used inclement weather to their advantage in several battles. When Washington and his troops cross the Delaware River, "snow" falls from above the viewers and dissipates just before landing on them.

Further enhancing the impact are stadium-style seating and ultra-high-definition video, projected on a pair of large, angled screens—one oval, the other rectangular. The veteran IMAX film team at Dennis Earl Moore Productions in Brooklyn Heights, New York, created the multimedia presentation and designed the screen and theater configurations.

The Douglas and Eleanor Seaman Gallery

Citizen Soldier

Featuring an elaborate cornice molding with heavily detailed neoclassical elements and marbleized, gilded Ionic columns, this gallery boasts a life-sized figure of "King Washington" that appears to dissolve into an image of him as a farmer and private citizen. This effect is created by means of special lighting and the use of mirrors—a long-standing theatrical technique known as Pepper's Ghost. The message conveyed is that after the Revolutionary War, Washington, unlike other historic military men pictured here, resisted the temptation to become a king or dictator. Instead, he transferred his power to the civilian government and returned to a tranquil agrarian private life at Mount Vernon.

The Distilled Spirits Council of the United States Gallery

Visionary Entrepreneur

George Washington's post–Revolutionary War life at Mount Vernon as a landowner, experimental farmer, and businessman is examined here. A lifelike mule illustrates how Washington introduced that hardworking animal to American agriculture and also pulls a Rotherham plow, one of the advanced farming tools he utilized. A model of Washington's gristmill demonstrates his use of sophisticated milling processes, and an 18th-century whiskey still reflects his entrepreneurial spirit in that burgeoning industry. Mount Vernon's lucrative fishery is also depicted, and a History Channel film explains how important the Potomac River and regional canals were to Washington's vision for the new nation.

The J. Hap and Geren Fauth Gallery

The Dilemma of Slavery

This gallery explores slavery as an important aspect of daily life at Mount Vernon in the 18th century. An audio recitation of individual slaves' names and the tasks they performed at the estate sets an appropriately somber tone. Interpretive texts and original tools introduce visitors to the personal stories of several enslaved residents, two of whom are represented by portraits.

A History Channel film features scholars and descendants of Mount Vernon slaves discussing questions that visitors can "ask" from an interactive rail: Did George Washington treat his slaves well? What was it like to be a slave at Mount Vernon? Why did he specify that his slaves were not to be freed until after his death? What were the effects of his decision to free his slaves? A timeline illustrates the evolution of Washington's views on slavery and his increasing awareness that it contradicted the ideals of the new nation.

The Donald and Nancy de Laski Gallery

A Leader's Smile

George Washington's dentures are the highly personal focal point of this gallery. Surrounding them is a timeline detailing his dental agonies—from losing two teeth during the French and Indian War to acquiring his last set of dentures, in 1798. As visitors consider this chronicle, they better understand the daily pain Washington endured for much of his life. A toothbrush and other cleaning tools that were used in vain to try to save his teeth also are exhibited. Finally, a History Channel film describes how Washington's dentures were fabricated and assures viewers that, contrary to myth, they were not made of wood.

The Charlotte and Walter Kohler Charitable Trust Gallery

Indispensable American

Visitors enter this gallery under an archway consisting of large three-dimensional cutouts of the 13 original states, which collide chaotically to underscore the fractious and confusing state of the nation in the years that immediately followed the Revolutionary War. Interactive computer monitors near a painting of Washington presiding over the Constitutional Convention of 1787 enable visitors to learn about the men who shaped the U.S. Constitution and about the issues and areas of contention that were involved in the writing of this monumental document. Visitors also hear lively debate by convention delegates on topics such as executive powers and slavery. The gallery underscores the fact that, at the convention, George Washington was once again the "indispensable man."

The Richard and Helen DeVos Foundation Gallery

The People's President

In this interactive depiction of the first U.S. presidential inauguration, visitors see the last of three forensically based life-sized figures of Washington—this one at age 57—taking the oath of office in April 1789 on a replica of the balcony of Federal Hall in New York. They can place their hands on a reproduction of the Bible he used in the ceremony. They even can recite the oath and, when finished, raise their hands to signal the roar of a cheering crowd.

Beyond the Federal Hall facade are screened two History Channel films. The first illuminates the precedents Washington set as the nation's first president, and the second includes a montage of U.S. senators reading his Farewell Address from the Senate floor, as they have done every year on his birthday since 1896. On the wall, original drawings by modern political cartoonists illustrate major issues of Washington's presidency. And visitors can learn about members of his cabinet by opening the doors of, appropriately, a wooden cabinet.

The Richard and Adelia Simplot Gallery

Private Citizen

The mood is somber and reverential in this gallery, which details Washington's final hours. There is a display of 18th-century medical instruments similar to those used to treat the former president during his brief fatal illness. An 18th-century bier and a reproduction of his coffin are surrounded by cases displaying objects commemorating Washington's life and death. Visitors can watch a video of the historic 1999 reenactment of his funeral and see a re-creation of the Mansion's third-floor garret chamber, where Martha Washington sought refuge following her husband's death and probably burned all their personal correspondence in order to safeguard their privacy. Inspiring words from famous Washington eulogies are projected on a black gallery wall.

The Gay Hart Gaines Legacy Theater

The Grand Finale

Visitors conclude their education center experience with a film finale that envelops them in images, music, and the spoken word. Created by Dennis Earl Moore Productions of Brooklyn Heights, New York, it is presented in ultra-high-definition video and 19-channel audio on a seven-foot-tall hemispherical screen employing 13 projectors. The Grammy–winning Brooklyn Youth Chorus performs "America the Beautiful" as the backdrop for a series of historic images and modern photographs that capture the breadth and essence of American cities, symbols, landmarks, and landscapes. Drawing upon the famous eulogy delivered at Washington's funeral by General Henry ("Light Horse Harry") Lee, Pulitzer Prize–winning historian David McCullough and former U.S. Secretary of State General Colin Powell offer pertinent reflections on Washington's remarkable character, enduring legacy, and vital contributions to the nation's history and culture.

The Johnston-Lemon Group Gallery

Saving Mount Vernon

All visitors to Mount Vernon pass through this gallery, which is incorporated into the serpentine corridor connecting the Donald W. Reynolds Museum and Education Center to the Mount Vernon Inn, gift shops, and estate exit. Here they can learn the inspiring story of the Mount Vernon Ladies' Association—how its members rescued the estate and set the standard for historic preservation in America. In a time when American women were rarely able to play leadership roles, the original Regent, Ann Pamela Cunningham, and her Vice Regents took the initiative and began the careful restoration of Washington's dilapidated home.

Combining contemporary and historic images with text, the exhibit also examines the Association's determination to preserve the remarkable view across the Potomac River that Washington so enjoyed. The exhibit concludes by underscoring the Association's commitment to educating the broadest possible audience about the life, leadership, and contributions of George Washington.

Phoebe Apperson Hearst Learning Center

Supported by the William Randolph Hearst Foundation

Visitors to this multifaceted educational laboratory, whether in person or online, have access to scholarly annotated editions of Washington's voluminous correspondence and other relevant materials about the man, his life, and his times. Four computers with links to these records serve the estate's historical interpreters, who also staff the Hearst Learning Center, as well as the hundreds of teachers and scholars who visit Mount Vernon annually. These resources enable the center to take an important step toward becoming the equivalent of a Washington presidential library. Website guides, bibliographies, innovative lesson plans, and other learning materials, many of which can be purchased in the Shops at Mount Vernon, are also available here.

The Annenberg Foundation Classroom

Distance-Learning Center

This area of the education center connects communities worldwide to Mount Vernon through a series of virtual programs and teacher workshops, all of which focus on George Washington's inspiring example as a leader of character. Pulitzer Prize–winning author David McCullough and award-winning journalist Cokie Roberts hosted the classroom's first distance-learning broadcast on October 26, 2006; it reached some 11 million students via international satellite feed. The state-of-the art equipment here enables Mount Vernon to reach special audiences in a highly cost-effective manner.

The Anne and James F. Crumpacker Family Gallery

Hands-on History

Created with Mount Vernon's youngest visitors (ages three to eight) in mind, this hands-on history area exposes children to the same themes and ideas presented in the galleries. But here, touching objects is both allowed and encouraged. Children can learn about Washington and his era by dressing in 18th-century clothing and putting on short impromptu plays on a stage fitted with a historic view of the Mansion. They can also read books, explore activity boxes, and "meet" the first president's farm animals—a noisy lot! Other learning materials include games of the 18th century, puppets, puzzles that use shards to teach about archaeology, and a miniature Mansion that can be fitted with "period furniture."

The Grand Opening

9

For two weeks preceding the grand opening of the Ford Orientation Center and the Donald W. Reynolds Museum and Education Center, the Mount Vernon Ladies' Association hosted several preview parties to recognize its many constituencies.

Museum colleagues and workers in the tourism industry attended the first preview, at which they heard remarks by Roger Dow, president and CEO of the Travel Industry Association of America. This was followed by an evening for educators, which began with a reception for Edward P. (Ted) Eagles, a retired teacher from St. Albans School, Washington, D.C.; one of the Ford Orientation Center theaters is named in his honor, thanks to generous support from the Robertson Foundation. There was an evening for many of the capital campaign's generous donors,

Friends, former students, and members of the Robertson Foundation were among the many well-wishers attending the dedication and reception for the Edward P. Eagles Theater in the Ford Orientation Center. From left to right: Terence W. Collins, Edward P. (Ted) Eagles, Julian H. Robertson, Jr., Josie Robertson, William R. Goodell, Sarah Collins Robertson, and Beth Collins.

Guests attending the October 22 preview included many of the capital campaign's larger donors as well as Washington and Custis family descendants.

RECONSTRUCTING

Left top: *Guests at the October 22 donor party were greeted by (left to right) Mount Vernon Executive Director James C. Rees, Regent Gay Hart Gaines, and the evening's special co-hosts, Kathleen Matthews and Chris Matthews.*
Left bottom: *Robert H. Smith, one of Mount Vernon's most generous benefactors; Regent Gay Hart Gaines; and Ann Bookout, Vice Regent for Texas.*

Below top *(left to right): Mary Ourisman, U.S. Ambassador to Barbados; J. Hap Fauth; Geren Fauth, Vice Regent for Minnesota; and Mandell J. Ourisman.*
Below bottom: *Lois Chiles; Lewis Lehrman, co-founder of the Gilder Lehrman Institute of American History; and Martha-Ann Alito, wife of U.S. Supreme Court Associate Justice Samuel Alito.*

which also included a number of Washington and Custis family descendants. There was also a premiere of the film *George Washington and the Dilemma of Slavery*, presented in partnership with Black Women United for Action, as well as a party for all craftspeople, artisans, and designers who worked on the project. A behind-the-scenes tour was organized for the Friends of Mount Vernon.

The culminating preview event was aptly called "A Night of Celebration," a triumphant gala heralding both the tremendous success of the capital campaign and the long-awaited opening of the state-of-the-art facilities. Chaired by Melody Sawyer Richardson, the Association's Vice Regent for Ohio, the event, held on October 26, 2006, attracted some 520 guests from across the nation—major campaign donors, museum lenders, friends of the Association, and individuals who applied their talents to creating the buildings, galleries, and films. Guests were greeted in the Ford Orientation Center lobby by Regent Gay Hart Gaines, Executive Director James C. Rees, and even George and Martha Washington themselves—life-sized, cast in bronze, and certain to become iconic symbols of Mount Vernon for the millions of visitors they will greet in the years to come.

On a crisp, beautiful fall evening, guests toured the new buildings and galleries and then proceeded to the Mansion for tours, entertainment by a fife-and-drum corps attired in period costume, and an impressive display of fireworks over the Potomac River. Inside a heated tent, they dined on delicacies inspired by 18th-century menus, including wild mushroom colonial "pye," duck with Mount Vernon

Above (left to right): *Television host Pat Sajak, author and historian David McCullough, and Donald W. Reynolds Foundation Chairman Fred W. Smith enjoy "A Night of Celebration."*
Left: *Boyce Ansley, Vice Regent for Georgia and chair of the New Facilities Oversight Subcommittee, with her husband, Shepard B. Ansley.*

herbs and cherry sage sauce, and chocolate soufflé cake. The elegant space, decorated by premier floral and event designer Don Vanderbrook of Cleveland, included a hedge-bordered dance floor and fruit-espaliered walls—all inspired by George Washington's gardens, just a stone's throw away. Stunning alternating centerpieces in silver Revere bowls complemented exquisite tablecloths custom made by Fabricut.

Betsy Nalty, Vice Regent Emerita for Louisiana; Eleanor Seaman, Vice Regent Emerita for Wisconsin; and Jeannie Kilroy, a member of the Friends of the Collections.

Donald de Laski, one of Mount Vernon's most generous benefactors, cutting a rug with Melody Sawyer Richardson, Vice Regent for Ohio and chair of "A Night of Celebration."

Top: *Stanley and Elizabeth Scott, benefactors of the largest museum gallery, arrive on the red carpet.*
Bottom: *Mitzi Perdue (center) arrives at the Texas Gate and welcomes Margot and Ross Perot.*

Through the use of special effects lighting, this larger-than-life concave bust of George Washington watches over the preview gala festivities.

The Ford Orientation Center officially opened to the public with a ribbon-cutting ceremony on October 27, 2006. From left to right: James C. Rees, Executive Director of Mount Vernon; Gay Hart Gaines, Regent of the Mount Vernon Ladies' Association; Sandy Ulsh, then-president of the Ford Motor Company Fund; and author David McCullough.

Guests included U.S. Supreme Court Associate Justice Samuel Alito, political commentator and news analyst Cokie Roberts, two-time U.S. presidential candidate H. Ross Perot, broadcast journalist Roger Mudd, U.S. Ambassador to Barbados Mary Ourisman, Missouri Governor Matt Blunt, and television game-show host Pat Sajak. Washington, D.C., television newscaster Kathleen Matthews served as the evening's master of ceremonies, and celebrated soprano Nicole Cabell provided a dramatic musical interlude. Following dinner, historian David McCullough delivered a stirring talk in which he emphasized George Washington's importance to the nation. "He exemplified courage and strength," McCullough reminded his listeners, "and he would not give up." He then underscored the value of those qualities in today's world, "as we try to cope with our own uncertainties."

Fred W. Smith, Chairman of the Donald W. Reynolds Foundation, cut the official ribbon, opening the new Donald W. Reynolds Museum and Education Center. He was assisted by author David McCullough, Reynolds Foundation President Steve Anderson, and foundation board member Wes Smith.

The next morning, the public gathered outside the gates, waiting for the ceremonial ribbon cutting. Nearly 1,000 invited guests also were on hand to observe the official opening of the new buildings. Held on the "green roof" of the Donald W. Reynolds Museum and Education Center, the ceremony included patriotic speeches and a rousing performance of "America the Beautiful" by the Brooklyn Youth Chorus. A ceremonial cannon blast marked the end of the ceremonies—and signaled the beginning of a new chapter in the history of Mount Vernon.

Right: *On October 27, guests gathered in the pasture that serves as the green roof of the Donald W. Reynolds Museum and Education Center to celebrate the opening of 71,000 square feet of new educational facilities. The program featured remarks by author David McCullough, a performance by soprano Nicole Cabell, and a special appearance by the Brooklyn Youth Chorus. On stage in front of the chorus are (left to right): David McCullough; Regent Gay Hart Gaines; Fred W. Smith, Chairman of the Donald W. Reynolds Foundation; Sandy Ulsh, then-president of the Ford Motor Company Fund; Melody Sawyer Richardson, Vice Regent for Ohio; The Honorable Togo D. West, Jr., new chairman of the Mount Vernon Advisory Committee; Steve Anderson, President of the Donald W. Reynolds Foundation; and James C. Rees, Executive Director of Mount Vernon.*

The Brooklyn Youth Chorus sang "America the Beautiful."

As part of the October 27 public opening, the torch is passed to David McCullough, who fires the ceremonial cannon, signaling the opening of the new facilities and the beginning of a new era in the history of Mount Vernon.

We are on hallowed ground. . . .
He exemplified courage and strength
. . . and he would not give up.

—David McCullough

Overleaf: *Ford Orientation Center lobby.*

GEORGE WASHINGTON
AGE 53
"WASHY"
AGE 4

Epilogue

James C. Rees

The opening of the Ford Orientation Center and the Donald W. Reynolds Museum and Education Center in October 2006 marked the end of a critical—and very exciting—chapter in Mount Vernon's quest to become a world center for the study of George Washington. But truth be told, it's probably a fairly early chapter in what will be a very lengthy book.

We at the Mount Vernon Ladies' Association are proud that one major challenge has been addressed in a dynamic fashion during the year the two new buildings have been up and running: the visitor experience on the Mount Vernon Estate has been enhanced dramatically. Our new galleries and theaters focus on aspects of Washington's life that previously were impossible to cover. We have been able to integrate and utilize to great effect a myriad of new technologies without compromising the authenticity of Washington's buildings and grounds. As a result, Mount Vernon is now much more "family friendly," not to mention far more accommodating during the coldest and warmest times of the year.

We promised to introduce visitors to "the real George Washington" through our new exhibits. Responses to initial visitor surveys indicate that we can claim a considerable degree of success in that regard. The average visit time has increased by about 55 percent, or 85 minutes. Most visitors tell us they leave the estate with a better idea of how Washington really looked. They also say they comprehend why he was the "indispensable man" in the founding of the nation, and understand that his true personality was not the staid one reflected on the dollar bill. The "old" Mount Vernon did little to dispel this iconic image of the sedentary and stiff senior statesman whom far too many Americans feel was "great but boring." The "new" Mount Vernon strives to reveal Washington's creativity, vitality, and inventiveness—even to the point, in several spots, of portraying the Father of Our Country as America's first action hero. Admittedly, we have pushed the envelope here and there in an effort to reach younger audiences, just as we have in places tried to inject a sense of humor—not one of the first president's strong suits. But in our eyes, the new exhibits do not trivialize Washington.

On the other hand, there are places where we could not resist the temptation to idolize George Washington. It is hard not to—he is one of those rare individuals for whom we feel more admiration and respect the more closely we examine him. His character is sterling, and the reputation he fought so mightily to protect remains remarkably steadfast.

While conducting research for our new galleries, we were surprised to discover that many questions about Washington remain unanswered, despite the countless biographies and studies that have been written over the past two centuries. We came to understand that "the real George Washington" was still being revealed one clue at a time, and that continuing study of him was an important factor in safeguarding his legacy to our nation.

In successive meetings with America's most respected Washington experts we could not help but notice a growing enthusiasm, not just for the galleries we were creating but also for the man himself. Part of this excitement was clearly engendered by a sense of place; the 500 acres that the Mount Vernon Ladies' Association is so blessed to own comprise a singular, retreat-like sanctuary. It is always inspiring to discuss Washington with these scholars in this environment, because during his 40 years of living here, he likely walked every acre, supervised the planting of hundreds of trees, and frequently watched the sunrise over the Potomac River from a chair on the piazza—just as we often do today. His presence continues to be felt. And the million-plus people who tour Mount Vernon each year only enhance our sense of inspiration, because their presence demonstrates that Washington is still revered as the founder of our nation.

We realize that the opening of the Ford Orientation Center and the Donald W. Reynolds Museum and Education Center did not represent the end of a journey but rather the beginning of a far greater one. Mount Vernon—the place and its people—has the potential to become a world-class learning center, with a focus unlike that of any other institution. Simply put, George Washington is the greatest leader of the greatest nation the world has ever known. The lessons his life can teach us would fill a hundred galleries, a thousand books, and countless websites. It is our job to take the study of Washington to a new level and to disseminate this knowledge in imaginative ways to the widest range of audiences possible. Just as George Washington was ever eager to break new ground, we must call upon the most creative minds and the most advanced technologies to make him relevant and meaningful to different people and cultures. In doing so, however, we must keep one foot firmly rooted in the past, recognizing that his achievements continue to shape the United States today.

We hope you have enjoyed this book about a new set of experiences at Mount Vernon and the people who made them possible. The determination, generosity, and patriotism of everyone involved give us the courage to move forward, knowing that Washington remains "first in the hearts of his countrymen."

James C. Rees has been Mount Vernon's Executive Director since 1994, having worked for the estate since 1983. Previously, he held positions at the National Trust for Historic Preservation and the College of William and Mary, Williamsburg, Virginia. He has served as president of the Virginia Association of Museums and the Friends of the Potomac River and is the author of *George Washington's Leadership Lessons* (2007).

Overleaf: *Ford Orientation Center at dusk.*

FOR KIDS

Project Credits

Projects of this scope and complexity require the attention and expertise of numerous individuals and organizations in many fields, from initial planning through design and construction. The most successful such projects are realized only with intense focus from the owner and owner's representatives who define the needs, from the design team that conceives the formal and functional expression, and from the contractors and builders. Here we acknowledge the people and firms who collectively made our vision a reality. We have attempted to assemble a complete listing, and we apologize if we have inadvertently omitted any contributors.

View of the Ford Orientation Center from the clearing.

OWNER:

Mount Vernon Ladies' Association of the Union
Mount Vernon, Virginia

(as of the Grand Opening, October 27, 2006)

Mrs. Stanley N. Gaines, *Regent*
Mrs. James Evan Allison, *Vice Regent for Washington*
Mrs. Shepard B. Ansley, *Vice Regent for Georgia*
Mrs. John F. Bookout III, *Vice Regent for Texas*
Mrs. William H. Borthwick, *Vice Regent for California*
Mrs. Sam Buchanan, *Vice Regent for Arkansas*
Ms. Elizabeth M. Chapin, *Vice Regent for Massachusetts*
Mrs. James F. Crumpacker, *Vice Regent for Oregon*
Mrs. Jared I. Edwards, *Vice Regent for Connecticut*
Mrs. J. Hap Fauth, *Vice Regent for Minnesota*
Mrs. Stewart Gammill III, *Vice Regent for Mississippi*
Mrs. Randolph H. Guthrie, *Vice Regent for New York*
Mrs. Jospeph W. Henderson III, *Vice Regent for the District of Columbia*
Mrs. Frank X. Henke III, *Vice Regent for Oklahoma*
Mrs. Charles G. Lane, *Vice Regent for South Carolina*
Mrs. Robert W. Lawson III, *Vice Regent for West Virginia*
Mrs. Robert E. Lee IV, *Vice Regent for Maryland*
Mrs. John F. Mars, *Vice Regent for Wyoming*
Mrs. Charles B. Mayer, *Vice Regent for Louisiana*
Mrs. P. William Moore, Jr., *Vice Regent for Virginia*
Mrs. Melody Sawyer Richardson, *Vice Regent for Ohio*

Mrs. J. Schley Rutherford, *Vice Regent for Alabama*
Mrs. Lloyd A. Semple, *Vice Regent for Michigan*
Mrs. Everette C. Sherrill, *Vice Regent for North Carolina*
Mrs. Richard Simplot, *Vice Regent for Idaho*
Mrs. P. Coleman Townsend, Jr., *Vice Regent for Delaware*
Mrs. James M. Walton, *Vice Regent for Pennsylvania*

James C. Rees, *Executive Director*
Stephen A. McLeod, *Assistant to the Director*
Anne M. Johnson, *Desktop Publishing Specialist*
Sandra N. Newton, *Board Secretary*

Charles A. Menatti, *Chief Operating Officer*
William Robertson, *General Manager, Mount Vernon Inn*
Julia Mosley, *Director of Retail*
Beverly Addington, *Director of Licensing*
Megan Dunn, *Director of Human Resources*
James C. Simms, *Director, Special Projects*
J. Dean Norton, *Director, Horticulture*
Joseph Sliger, Jr., *Director, Operations and Maintenance*

Barton Groh, *Chief Financial Officer*
Scot A. Bryant, *Director, Information Technology*

Linda Ayres, *Associate Director for Collections*
Carol Borchert Cadou, *Robert H. Smith Senior Curator*
Barbara McMillan, *Librarian*
John Rudder, *Special Projects Manager*
Mary V. Thompson, *Research Specialist*
Gretchen Goodell, *Assistant Curator*
Christine H. Messing, *Assistant Curator*
Mary Margaret Carr, *Collections Manager*
Dawn Bonner, *Collections Administrative Specialist*

Ann Bay, *Associate Director for Education*
Nancy Hayward, *Assistant Director for Education*
Elizabeth Maurer, *Project Coordinator*
Christina Hills, *Research Assistant*

Stephanie Brown, *Associate Director for Public Affairs*
Emily Coleman Dibella, *Director of Media Relations*
Daniel E. Froggett, *Manager, Special Events*

Julie D. Carter, *Associate Director for Development*
Wendy Arminio, *Manager, Research and Records*
Pam Townsend, *Manager of Membership Programs*

Dennis J. Pogue, *Associate Director for Preservation*
Esther White, *Director, Archaeology*

Two former staff members, Thomas A. Lillis, *Director of Design and Construction*, and Richard Dressner, *Associate Director for Development*, also played major roles in the success of the project.

Mount Vernon's Consultants for the Ford Orientation Center and Donald W. Reynolds Museum and Education Center

Regan Associates — Herndon, Virginia
Owner's Representative
Sean Regan

John Hill, FAIA — Baltimore, Maryland
Owner's Design Advisor

Oculus — Charlottesville, Virginia
Cultural Landscape Study
Liz Sargent, ASLA, *Principal*

Project Cost, Inc. — Alexandria, Virginia
Cost Consulting
John Billings

Turner Construction Company — Arlington, Virginia
Cost Consulting
Tommy Thomas, *Lead Estimator*
Don Denman, *Preconstruction Planning Executive*
Susan B.W. Boggs, *Manager, Business Development*

James G. Davis Construction Corporation — Rockville, Maryland
Cost Consulting
Barry D. Perkins, *Project Executive*

Kjellstrom and Lee, Inc. — Richmond, Virginia
Cost Consulting
Peter S. Alcorn, *President*

Steven R. Keller & Associates, Inc. — Ormond Beach, Florida
Security Design
Steve Keller, *Project Manager*
Steven Swen, *Project Engineer*

Garrison/Lull, Inc. — Allentown, New Jersey
Environmental Systems
William P. Lull, *Senior Conservation Environment Consultant*

PPI Consulting Presentation Planning, Inc. — Washington, D.C.
Audio Visual Consulting
Bob Haroutunian

Mount Vernon's Consultants for the Master Plan and Mount Vernon Inn Renovations and Additions

Herbert Sprouse Consulting — Grantham, New Hampshire
Preliminary Programming

Operations Research Consulting Associates — Ocoee, Florida
Circulation and Pedestrian Flow

Kjellstrom and Lee, Inc. — Richmond, Virginia
Design and Construction Advisor

Arch Et Al — Chevy Chase, Maryland
Retail Design

Hugh Frost — Potomac, Maryland
Signage

Barton Malow Company — Baltimore, Maryland
Cost Estimating

BUILDING AND SITE DESIGN:

GWWO, Inc./Architects — Baltimore, Maryland
Architect and Project Coordinator

Architectural Design Team for the Master Plan and Mount Vernon Inn Renovations and Additions

GWWO, Inc./Architects — Baltimore, Maryland
James R. Grieves, FAIA, *Principal for Design*
David G. Wright, AIA, *Principal-in-Charge*
Terry A. Squyres, AIA, *Project Manager*

Architectural Design Team for the Ford Orientation Center and Donald W. Reynolds Museum and Education Center

GWWO, Inc./Architects — Baltimore, Maryland
Alan E Reed, AIA, *Principal for Design*
Phillip W. Worrall, AIA, *Principal-in-Charge*
Terry A. Squyres, AIA, *Project Manager*
Lisa A. J. Andrews, AIA, *Project Architect*

Design Consultants for the Master Plan and Mount Vernon Inn Renovations and Additions

LDR International, Inc. — Columbia, Maryland
Site and Landscape Design

Walter L. Phillips, Inc. — Falls Church, Virginia
Site Engineering

Henry Adams, Inc. — Towson, Maryland
Mechanical/Electrical Engineering

Robert Silman Associates, PC — New York, New York and Washington, D.C.
Structural Engineering

Woodburn & Associates, Inc. — Annandale, Virginia
Foodservice Design

Bill Wright, AHC — Baltimore, Maryland
Door Hardware

Design Consultants for the Ford Orientation Center and Donald W. Reynolds Museum and Education Center

EDAW, Inc. — Alexandria, Virginia
Original Site and Landscape Design
Roger J. Courtenay, ASLA, *Principal-in-Charge*
Jon Pearson, *Project Manager*
Kirsten Melberg, *Project Landscape Architect*
Ann Anderson, *Plant Design*
Aiman Duckworth, *Construction Administrator*

Donovan, Feola, Balderson & Associates, Inc. — Montgomery Village, Maryland
Additional Landscape Design
Andrew Balderson, *President*

James Posey Associates, Inc. — Baltimore, Maryland
Mechanical/Electrical Engineering
Steven J. Hudson, PE, CIPE/CPD, *President, Principal-in-Charge*
Craig N. Rasmussen, PE, *Senior Project Manager, Mechanical Engineering*
Richard E. Lang, PE, *Principal, Electrical Engineering*
David M. White, *Project Manager, Electrical Engineering*

Thornton Tomasetti, Inc. — Washington, D.C.
Structural Engineering
Mark Tamaro, PE, *Vice President*

Urban Engineering & Associates, Inc. —
Annandale, Virginia
Site Engineering
David T. McElhaney, PE, *Principal*

MFM Design — Washington, D.C.
Interiors
Richard Molinaroli, *Principal-in-Charge, Lead Designer*
Harry Raab, *Project Detailer*

Auerbach Pollock Friedlander — New York, New York
Theater Design
Paul Garrity, *Principal-in-Charge*
Daniel Mei, *Project Manager and Audio Visual Systems Designer*
Don Guyton, *Theater Systems Designer*

George Sexton Associates — Washington, D.C.
Lighting Design

Acoustical Design Collaborative, Ltd. —
Ruxton, Maryland
Acoustical Design
Neil Thompson Shade, *Principal*

Simpson Gumpertz & Heger, Inc. —
Waltham, Massachusetts
Waterproofing Consulting
Kevin Cash, PE, *Principal*

Chesapeake Door Security Solutions —
Harpers Ferry, West Virginia
Door Hardware
Bryan Myers, *Senior Specification Consultant*

EXHIBIT DESIGN:

Donald W. Reynolds Museum Exhibits

Quenroe Associates and Charles Mack Designs —
Boulder, Colorado and Baltimore, Maryland
Exhibit Design
ElRoy Quenroe and Charles Mack

Steven Hefferan — Boulder, Colorado
Exhibit Lighting Design

Donald W. Reynolds Education Center Exhibits

Christopher Chadbourne & Associates, Inc. and Museum Design Associates, Inc. —
Boston, Massachusetts and Cambridge, Massachusetts
Exhibit Design
Christopher Chadbourne (CCA)
David Seibert, *President* (MDA)
Peter Barton (CCA)
Eric Getz (CCA)

Dennis Earl Moore Productions —
Brooklyn Heights, New York
Audiovisual Media Design
Dennis Earl Moore
Steven Meyer

Available Light — Salem, Massachusetts
Exhibit Lighting Design
Steven Rosen, IALD, *Principal*
Mary Barnwell, *Senior Designer*
Derek Barnwell, *Project Manager*

StudioEIS — Brooklyn, New York
Forensic Models
Elliot Schwartz and Ivan Schwartz

Ford Orientation Center Exhibits and Building Signage

MFM Design — Washington, D.C.
Richard Molinaroli, *Principal-in-Charge, Lead Designer*
David Fridberg, *Principal, Co-Designer*
Elizabeth Miles, *Interpretive Planner*
Harry Raab, *Project Detailer*

StudioEIS — Brooklyn, New York
Bronze Sculptures
Elliot Schwartz and Ivan Schwartz

Greystone Productions — West Hollywood, California
We Fight to Be Free
Craig Haffner, *Producer*
Kees Van Oostrum, *Director*
Sebastian Roché, *George Washington*
Caroline Goodall, *Martha Washington*
Stephen Lang, *Dr. James Craik*

Mount Vernon Ladies' Association Exhibit

MFM Design — Washington, D.C.
David Fridberg, *Principal-in-Charge, Lead Designer*
Richard Molinaroli, *Principal, Co-Designer*
Elizabeth Miles, *Interpretive Planner*
Harry Raab, *Project Detailer*

BUILDING AND EXHIBIT CONSTRUCTION:

Mount Vernon Inn Renovations and Additions

The Whiting-Turner Contracting Company — Chantilly, Virginia
Construction Manager
K. C. Haile, *Vice President*
Jill Bushkoff, *Project Manager*
Eric Feining, *Project Engineer*

Ford Orientation Center and Donald W. Reynolds Museum and Education Center

Turner Construction Company — Arlington, Virginia
General Contractor
Ben Shor, *Vice President/Operations Manager*
Jack Rogers, *Project Executive*
Daniel A. Lavanga, *Project Manager*
John McManus, *Senior Project Manager*
Gene Brown, *Project Superintendent*
Travis Sumlar, *Superintendent*
Robert House, *Superintendent*

Donald W. Reynolds Museum Exhibit Construction

United Services Associates, Inc. — Van Nuys, California

In-Depth, Inc. — Philadelphia, Pennsylvania

Lexington Acquisitions, Inc. — Pacoima, California

Barbizon Capital, Inc. — Alexandria, Virginia

Helmut Guenschel, Inc. — Baltimore, Maryland

Donald W. Reynolds Education Center Exhibit Construction

Art Guild, Inc. — Thorofare, New Jersey
John Ochipinti, *Project Manager*
Gifford Eldridge, *Project Manager*

To Keep Him First Capital Campaign

Right: *Board members of the Donald W. Reynolds Foundation gathered at Mount Vernon.*

The following list includes commitments of $25,000 or more to the *To Keep Him First* capital campaign as of May 15, 2007.

Benefactors ($5,000,000 +)

Ford Motor Company Fund
Donald W. Reynolds Foundation
Robert H. and Clarice Smith

Patrons ($1,000,000–$4,999,999)

The Annenberg Foundation
Donald and Nancy de Laski
The Richard and Helen DeVos Foundation
J. Hap and Geren Fauth
Gay Hart Gaines
The Gilder Foundation, Inc.
William Randolph Hearst Foundation
Vira I. Heinz Endowment
The Mary Hillman Jennings Foundation
The F. M. Kirby Foundation
John and Adrienne Mars
Frank Peters Moore
Robertson Foundation
Douglas and Eleanor Seaman
Elizabeth and Stanley DeForest Scott
Albert and Shirley Small
The Gordon V. and Helen C. Smith Foundation
A. Alfred Taubman

Sponsors ($500,000–$999,999)

Anne and James F. Crumpacker Family
The Johnston-Lemon Group
William R. Kenan, Jr. Charitable Trust
The Charlotte and Walter Kohler Charitable Trust
Gerry and Marguerite Lenfest
Christopher and Betsy Little
Mars Foundation
Richard King Mellon Foundation
The Mary Morton Parsons Foundation
The Richard S. Reynolds Foundation
Melody Sawyer Richardson
Richard M. Scaife
Sarah Scaife Foundation Incorporated
Dr. and Mrs. Robert C. Seamans, Jr.
Richard and Adelia Simplot
The L. J. Skaggs and Mary C. Skaggs Foundation

Sustainers ($100,000–$499,999)

Allegheny Foundation
Margaret and James Allison
Bruce and Joy Ammerman
Argyros Foundation
Jean van Horne Baber
Bacardi USA, Inc.
Royce and Kathryn McCormick Baker
Carol and Les Ballard
John and Ann Bookout
Maribeth and Wm. Harold Borthwick
Brown-Forman Corporation
Alison W. Burdick

Castle Rock Foundation
The Chisholm Foundation
Christie's
Theodore and St. Clair Craver
Diageo
Distilled Spirits Council of the United States
Dominion
Robert and Lori Duesenberg
Bruce and Katherine Eberle
Hart and Nancy Fessenden
The Founders, Washington Committee for Historic Mount Vernon
Future Brands LLC
Lynn and Stewart Gammill
Douglas and Carolyn Gomez
The Florence Gould Foundation
Taffy Gould
Beatrice H. and Randolph H. Guthrie
The Phil Hardin Foundation
Joseph and Lucia Henderson
Bonnie and Frank X. Henke III
Hickory Foundation
The History Channel
Michael and Lindy Keiser
The Kiplinger Foundation
Jay and Jean Kislak
David H. Koch
Charles Glover and Virginia Dawson Lane
Robert and Carew Lee
Mr. and Mrs. Lewis E. Lehrman
Frayda and George L. Lindemann
A. Michael and Ruth C. Lipper
Mr. and Mrs. Benjamin F. Lucas II
The Lynch Foundation
Dr. Leonard and Geana Madison
Jacqueline B. Mars
Cameron and Charles Mayer
Mellon Financial Corporation
Mr. and Mrs. A. Malachi Mixon III
Mr. and Mrs. P. William Moore, Jr.
Mrs. H. Taylor Morrissette
Eleanor and Phillip Newman
George and Carol Overend
Owens-Illinois, Inc.
Pernod Ricard USA
Perot Foundation
The Pohanka Family
Dr. and Mrs. James S. Reibel
Mrs. George Revitz
Everette and Jean Sherrill
Mrs. Nelson T. Shields III
T. Eugene and Joan H. Smith
Mr. and Mrs. Franz T. Stone
Jack and Susan Taylor
Byron and Sissy Thomas
Thomas and Iris Vail
Robert A. and V'Etta Virtue
Letitia Rees Wallace
Ellen C. and James M. Walton
Mr. and Mrs. W. Temple Webber, Jr.
The Joe Weider Foundation
Chuck and Lucy Weller
Ann and Stephen West
Constance and Frederick West
Wine & Spirits Wholesalers of America

Fellows ($25,000–$99,999)

The Achelis Foundation
Hugh Trumbull Adams
Warren and Sonia Adler
John and Stavroula Alachnowicz
Albert A. Alley Family Foundation
American Village Citizenship Trust
Richard and Susan Ammerman
Stephen Cary Ammerman
Helen Sharp and Thomas Dunaway Anderson
Jack R. and Rose-Marie Anderson

Cyrus and Janet Ansary
Shepard and Boyce Ansley
C. Joseph Arbogast
Mr. and Mrs. Andrew W. Armour IV
John and Jane Aurell
Mr. and Mrs. Judson C. Ball
Barton Inc.
Thomas and Charlotte Battle
Joe and Lois Rae Beall
Ted and Page Lee Bell
Robert Bernasconi
Robert and Kathleen Best
Mr. and Mrs. Clarence M. Bishop, Jr.
Carl Wood Brown
David R. and Ann N. Brown
Brunschwig & Fils
Lissy and Stewart Bryan
Frances and Sam Buchanan
Ruth H. Buchanan
Karen Buchwald and Tom Rastin
Burke & Herbert Bank & Trust Company
Tyler and Talbot Cain
Philip and Betsey C. Caldwell
Richard and Nancy Banning Call
Elizabeth and Craig Campbell
Mr. and Mrs. Edmund N. Carpenter II
Carrier Corporation
John K. Castle
Richard B. and Lynne V. Cheney
Chevy Chase Bank
David R. and Margaret C. Clare Foundation
Alfred C. Clark
Dr. and Mrs. Thomas F. Cleary
Stephen and Ellen Clouse
Mr. and Mrs. Benjamin Coates
The Coca-Cola Company
Robert M. and Annetta J. Coffelt
Robert M. Coffelt, Jr.
Judith Stanley Coleman
Mrs. LeRoy Collins
Clement and Lianne Conger
Stephania and Donald Glover Conrad
Lovick P. and Elizabeth T. Corn
Dr. and Mrs. Delos M. Cosgrove
Rear Admiral and Mrs. Peter Cressy
Richard A. and Janice M. Crosby
Cruzan International, Inc.
Mr. and Mrs. Lipscomb Davis, Jr.
The Shelby Cullom Davis Foundation
Ruth and Bruce Dayton
Mr. and Mrs. W. Alan Dayton
Lawrence and Florence De George
Mr. and Mrs. Wesley M. Dixon
Mrs. John R. Donnell
The William H. Donner Foundation, Inc.
Elizabeth and W. John Driscoll
The Dwoskin Family Foundation
Eastman Kodak Company
Jared I. and Clare C. Edwards
Donald G. and Sandra Brown Everist
E-Z-GO
Buford L. Farrington
Mr. and Mrs. Walter L. Farrington, Jr.
Lorraine H. Finch
The Fisher Fund of The Pittsburgh Foundation
Mr. and Mrs. John W. Fisher
Jo Ellen and Joe Ford
Mr. and Mrs. C. Thomas Fuller
Ralph Hart Gaines
Stanley Noyes Gaines, Jr.
The Honorable and Mrs. Joseph V. Gartlan, Jr.
Brian Gazeley and Joseph Landry
Robert B. and Anne W. Gibby
Robert B. Gibby, Jr.
Stewart and Gene Gilchrist
Newt and Callista Gingrich
John M. and Julia Cobey Gluck
Audrey and Martin Gruss Foundation

Mrs. John H. Guy, Jr.
Mrs. Randall H. Hagner, Jr.
Benjamin and Derrill Hagood
Cheryl and Fred Halpern Family
Colonel G. F. Robert Hanke USMC
Stephen and Norma Hartwell
Albert and Kelley Hawk
Ray and Reba Hawn
Hugh and Barbara Heishman
Dorothy Compton Marks and W. Gibbs Herbruck
The Heritage Foundation
Henry L. Hillman Foundation
Margaret Mellon Hitchcock Foundation

Charles Hoeflich
R. Faye Holland
Arthur Hollins III
David and Mary Davis Holt
John and Ruth Huber
Page Hufty
Elizabeth and George Wesley Huguely III
Mr. and Mrs. Benjamin R. Huske III
Patricia and Benjamin Jaffray
Betty Wold Johnson
Colonel Erik G. Johnson USA
Gerald and Judith Kahler
Governor and Mrs. Frank Keating
Jorie Butler Kent
Walter S. Kiebach
William S. and Lora Jean Kilroy Foundation
Mr. and Mrs. Norman V. Kinsey
James R. and Betsy S. Kleeblatt
Kohler Co.
Mills Bee Lane Memorial Foundation
Anna L. and Robert M. Lawson, Jr.
Robert W. and Priscilla P. Lawson
Sherri Parker Lee
Martin Lewis and Diane Brandt
Colonel Ruby W. Linn USA
James Lintott and J. May Liang
Harry and Erika Lister
Luxco, Inc.
Mrs. William G. Marr
Nancy Peery Marriott Foundation, Inc.
Mr. and Mrs. Forrest E. Mars, Jr.
Linda A. Mars
Lowell B. Mason, Jr., and Elizabeth Upjohn Mason
William W. Massey
Christopher and Kathleen Matthews
Mr. and Mrs. William M. Matthews
Mr. and Mrs. Ernest N. May, Jr.
George and Sally Mayer
Harold and Babs McClendon

Left: *Living-history interpreter Katie Pohlman brings the 18th century to life for visitors strolling along the serpentine path.*

McCormick Distilling Co.
Mr. and Mrs. John J. McDonnell, Jr.
Jennifer and Kennard McKay
Heath and Judy McLendon
Denman and Babs Gaillard McNear
Mr. and Mrs. James C. Meade
Charles and Karen Menatti
Ira Lebeck Mendell
Mercantile Potomac Bank
Dr. and Mrs. Keith Merrill, Jr.
The Middendorf Foundation, Inc.
Michael and Edith Miller
Gretchen and Marshall Milligan
Moët Hennessy USA
Jacqueline G. Moore
Mr. and Mrs. David B. Morgan
Roger and E. J. Mudd
Mr. and Mrs. Donald J. Nalty
James and Georgianne Neal
The Samuel Roberts Noble Foundation
The Oakmead Foundation
The John M. Olin Foundation, Inc.
Dr. Ronald Neil Ollstein
The Payne Fund
Robert and Virginia Payne
Nona Rust Peebles
Pamela J. and James D. Penny
Mrs. Frank Perdue
Robert S. Perkin
David and Marianne Pfaelzer
Jim and Jo Carol Porter
James T. and Lois J. Pott
Mr. and Mrs. A. Russell Quilhot
Robert and Christine Ralphs
Lawrence and Lee Ramer
Gayle and Randy Randol
Henrietta K. and James G. Randolph
James C. Rees IV
Remy Cointreau USA, Inc.
Leonard and Louise Riggio
Ruby Stewart Rinker
Barbara and Doyle Rogers
Michele Metrinko Rollins
Elaine Decker Rosensweig
Dr. Tamzin A. Rosenwasser
Dorothy B. Russ and Robert J. Conley
Laura and J. Schley Rutherford
Mr. and Mrs. B. Francis Saul II
Dorothy McIlvain Scott
Laura Gaines Semler
Lloyd and Cynthia Semple
Dan and Mary Jane Sheppard
Sidney Frank Importing Co. Inc.
Maida Pearson Smith
Stamm-Woodruff Family
Ted and Vada Stanley
The Starr Foundation
Ken and Alice Starr
Reverend and Mrs. Richard C. Stazesky
Jeffrey and Barbara Steele
Ben S. Stefanski Family
Mrs. James G. Stevens
Jacqueline Gaines Stitt
Barry and Ann Sullivan
Melinda and Paul Sullivan
Mr. and Mrs. Richard J. J. Sullivan, Jr.
Suntory International Corp.
Susan M. and P. Coleman Townsend, Jr.
Vincent and Patricia Trosino
Harlow G. Unger
The I. N. and Susanna H. Van Nuys Foundation
John D. and Patricia C. Veatch
Wachovia Foundation
Rachel Mellon Walton
Jonathan Westervelt Warner
The George Washington Society, Ltd. of Delaware
Lawrence and Evelyn Washington
Martha Gaines Wehrle

The John L. and Sue Ann Weinberg Foundation
The Honorable and Mrs. Togo D. West, Jr.
Mr. and Mrs. George Y. Wheeler III
James H. and Ann R. Wiborg
Calhoun and Ann Wick
Louise and Henry K. Willard II
Brian J. Williams
Douglas and Priscilla Williams
Elizabeth Fitger and Harold B. Williams
Penelope P. Wilson
Diana Strawbridge Wister
WorldStrides
Jerry A. Yarbrough
Christine and Jaime Yordán

View of the Texas Gate and Ford Orientation Center from the pasture above the Donald W. Reynolds Museum and Education Center.

Notes to Chapter One

1. For a recent, very readable biography of George Washington, see Joseph J. Ellis, *His Excellency, George Washington* (New York: Alfred A. Knopf, 2004); for the development of the Mount Vernon plantation and building and expansion of the Mansion and associated outbuildings, see Robert F. and Lee Baldwin Dalzell, *George Washington's Mount Vernon: At Home in Revolutionary America* (Oxford: Oxford University Press, 1998); for a study of Washington's gardening and landscaping activities, see Mac Griswold, *Washington's Gardens at Mount Vernon: Landscape of the Inner Man* (Boston and New York: Houghton Mifflin, 1999); for Washington's farming activities, see Alan and Donna Jean Fusonie, *George Washington, Pioneer Farmer* (Mount Vernon, Va.: Mount Vernon Ladies' Association, 1998).

2. Julian U. Niemcewicz, "Under Their Vine and Fig Tree: Travels through America in 1797–1799 and 1805," reprinted in *Experiencing Mount Vernon: Eyewitness Accounts, 1784–1865*, edited by Jean B. Lee (Charlottesville and London: University of Virginia Press, 2006), 69–88.

3. For the history of the Mount Vernon Ladies' Association, see James C. Rees, "Preservation: The Ever-Changing Frontier," in *George Washington's Mount Vernon*, edited by Wendell Garrett (New York: Monacelli Press, 1998), 218–44.

4. For a report on the findings of the Mount Vernon paint study, see Matthew Mosca, "Paint Decoration at Mount Vernon: The Revival of Eighteenth-Century Techniques," in *Paint in America: The Colors of Historic Buildings*, edited by Roger W. Moss (New York: John Wiley and Sons, 1994), 104–27.

Acknowledgments

The Mount Vernon Ladies' Association (MVLA) and GWWO, Inc./Architects would like to recognize the following individuals, who made significant contributions to *Integrity, Civility, Ingenuity: A Reflection of George Washington*. This book would not have been possible without their time, energy, and dedication.

Gay Hart Gaines, *Regent, MVLA*
James C. Rees, *Executive Director, MVLA*

Wendy Arminio, *Manager, Development Research and Records, MVLA*
Linda Ayres, *Associate Director for Collections, MVLA*
Ann Bay, *Associate Director for Education, MVLA*
Dawn Bonner, *Collections Administrative Specialist, MVLA*
Anne M. Johnson, *Desktop Publishing Specialist, MVLA*
Stephen McLeod, *Assistant to the Executive Director, MVLA*
Dennis J. Pogue, *Associate Director for Preservation, MVLA*
Mary V. Thompson, *Collections Research Specialist, MVLA*

Alan E Reed, *President, GWWO*
Phillip W. Worrall, *Principal, GWWO*
David G. Wright, *Principal, GWWO*
Terry A. Squyres, *Senior Associate, GWWO*
Laura M. Werther, *Senior Associate, GWWO*
Jodi T. Hume, *Senior Associate, GWWO*
Kyle J. Isfalt, *Graphic Designer, GWWO*

Phil Freshman, *Manuscript Editor and Proofreader*
Susan C. Jones, *Initial Text Revision*
Arlene Prunkl, *Preliminary Editor, PenUltimate Editorial Services*

Overleaf: *Window details on the Ford Orientation Center.*

Illustration Credits

Numbers refer to page numbers.

Fabian Bachrach: 106

Keith Barraclough: 36, 37 left, 40, 107 top

Harry Connolly: 15 bottom, 16 top, 19 right, 21 right

Hal Conroy: 18 right, 49 top, 58

Robert Creamer: 1, 2–3, 4–5, 6 top, 6 bottom, 7, 9, 13, 29 left, 29 bottom right, 33, 38–39, 52 left, 52 right, 53, 67, 69, 71 bottom left, 71 bottom right, 74–75, 76, 77, 80–81, 83, 86, 87 top, 88, 89 top, 90, 91, 92, 93 top right, 93 bottom right, 95, 96, 97 top left, 97 top right, 98, 99, 108–09, 110, 111, 112, 113, 115, 116, 117, 126, 127, 128, 131–59, 170–71, 174–75, 187, 190, 191

Luke C. Dillion: 17 right, 28 top

Kent Eanes: 91 left

EDAW, Inc.: 49 bottom

Courtesy of Clare Edwards: 24

Nancy Emison: 60

Mark Finkenstaedt: 15 middle, 29 top, 34

William Geiger: 26

GWWO, Inc./Architects: 37 right, 46, 47, 48, 50, 54, 59, 66 top, 68, 70 top left, 70 bottom left, 71 top left, 72, 73, 82, 84, 85, 94, 97 bottom right, 100–01, 102, 103, 118, 119, 120, 121, 122–23, 124, 125

Maureen Hannam: 166

Carol Highsmith: 63 top

Russell Hirshon: 162 bottom left, 162 bottom right, 162 top right, 163 top, 163 bottom, 164 bottom right, 164 bottom left, 164 top right, 164 top left

Kyle Isfalt: 11, 66 bottom, 70 bottom right, 87 bottom, 114, 176

Paul Kennedy: 12, 30, 61

Robert C. Lautman: 14, 19 left, 21 left, 22–23, 28 bottom, 43, 63 bottom right, 64 top, 65 bottom

Taylor Lewis: 62 left, 62 right

Howard Marler: 17 left

Cal McWhirter: 107 bottom, 169, 168 top

Artis Mooney: 167 right

Matt Mosca: 20 bottom

Mount Vernon Ladies' Association: 15 top, 16 bottom, 20 top, 25, 27 left, 27 right, 31 top, 32, 35 left, 35 right, 42, 45, 56–57, 64 left, 64 bottom right, 79, 185

James C. Rees: 31 bottom

Kyle Samperton: 160, 160–61, 162 top, 165, 167 left, 168 bottom, 182

Ted Vaughan:18 left, 63 left, 65 top

Laura Werther: 6 middle, 70 top right, 71 top right, 89 bottom

Brady Willette: 41